FORTSCHRITTE DER CHEMIE ORGANISCHER NATURSTOFFE

PROGRESS IN THE CHEMISTRY OF ORGANIC NATURAL PRODUCTS

PROGRÈS DANS LA CHIMIE DES SUBSTANCES ORGANIQUES NATURELLES

HERAUSGEGEBEN VON EDITED BY RÉDIGÉ PAR

L. ZECHMEISTER
CALIFORNIA INSTITUTE OF TECHNOLOGY, PASADENA

SECHSTER BAND SIXTH VOLUME SIXIÈME VOLUME

VERFASSER AUTHORS AUTEURS

J. BONNER · H. J. DEUEL JR. · CH. DHÉRÉ · S. M. GREENBERG
O. HOFFMANN-OSTENHOF · E. LEDERER · L. RETI

MIT 32 ABBILDUNGEN WITH 32 ILLUSTRATIONS AVEC 32 ILLUSTRATIONS

WIEN · SPRINGER-VERLAG · 1950

ISBN-13:978-3-7091-7176-9 e-ISBN-13:978-3-7091-7175-2
DOI: 10.1007/978-3-7091-7175-2

Softcover reprint of the hardcover 1st edition 1950

Titel Nr. 8212

Inhaltsverzeichnis.
Contents. — Table des matières.

Some Biochemical and Nutritional Aspects in Fat Chemistry.

Some Biochemical and Nutritional Aspects in Fat Chemistry.

By H. J. DEUEL, JR. and S. M. GREENBERG, Los Angeles, California.

With 7 Figures.

Contents.

I. Introduction.

1. General Remarks on Fat Composition.

The composition of fats is of considerable importance in assessing their nutritional value. Not only may the types of fatty acids be related to the physiological activity, but the arrangement of these fatty acids in the triglyceride molecules appears to be involved. In addition to the triglycerides, which compose the major portion of natural fats, a number of substances are present in the dissolved state which may contribute to the dietary value of the fat. These include the phospholipids, which aid in the emulsification of fats since they are soluble in both water and lipids; and they are excellent sources of such important constituents in the animal economy as choline and phosphoric acid. In addition, there occur a number of components which may be of some nutritional importance in the non-saponifiable extract of fat. Such substances are the fat-soluble vitamins, the sterols, the hydrocarbons, and in some cases, the higher alcohols.

The fatty acids found in the common animal and vegetable fats are relatively few in number. In the group of saturated acids which have the characteristic formula, $C_nH_{2n+1} \cdot COOH$, palmitic and stearic acids are almost invariably present, while many fats contain large proportions of myristic and lauric acids. In the list of unsaturated acids, oleic acid must be regarded as an almost universal component of natural fats. Linoleic and linolenic acids likewise are frequently present in fairly large amounts in vegetable fats, although they occur usually in small proportions if at all in animal fats. The chief fatty acids which have been identified in various natural fats are listed in Tables 1 and 2.

Table 1. Summary Table Listing Saturated Fatty Acids Found in Common Fats (70).

Fatty acid	Formula	Source with highest content		Other good sources
		Source	% of total fat	
Butyric	$CH_3(CH_2)_2COOH$	Butter (cow)	4	Milk fat of various species
Caproic	$CH_3(CH_2)_4COOH$	Coconut	2	Butter, palm nut oils
Caprylic	$CH_3(CH_2)_6COOH$	Coconut	10	Butter, palm nut oils
Capric	$CH_3(CH_2)_8COOH$	Elm	50	Coconut, butter, palm nut oils
Lauric	$CH_3(CH_2)_{10}COOH$	Cawal-Kurundu	86	Laurel oil, spermaceti, babassu, palm kernel
Myristic	$CH_3(CH_2)_{12}COOH$	Nutmeg butter	77	Kombo, dika, ucuhuba
Palmitic	$CH_3(CH_2)_{14}COOH$	Japan wax	77	Many animal and vegetable fats
Stearic	$CH_3(CH_2)_{16}COOH$	*Allanblackia*	57–63	Many animal and vegetable fats
Arachidic	$CH_3(CH_2)_{18}COOH$	Rambutan tallow	35	Peanut oil
Behenic	$CH_3(CH_2)_{20}COOH$	*Xylia xylocorpa*	17.3	Niam, peanut oil, oil of Ben
Lignoceric	$CH_3(CH_2)_{22}COOH$	Coral tree	26	Peanut oil, rapeseed oil, cerebrosides
Carnaubic	$C_{24}H_{48}O_2$(?)	—	—	Carnauba wax
Cerotic	$CH_3(CH_2)_{24}COOH$	—	—	Beeswax, wool fat, opium wax

Within the last few years the methods for identification and for determination of the individual fatty acids have vastly improved so that at present they have considerable sensitivity and precision. DAUBERT (70)

Table 2. Summary Table Listing Unsaturated Fatty Acids Found in Common Fats (*70*).

Common name	Systematic name	Source with highest content		Other sources
		Source	% of total fat	
Crotonic	2-Butenoic	—	—	Croton oil
Tiglic	$C_5H_8O_5$	—	—	Croton oil
Oleic	*cis*-9-Octadecenoic	Coula	95	Many animal and vegetable fats
Elaidic	*trans*-9-Octadecenoic	—	—	Hydrogenated fats
Petroselenic	6-Octadecenoic	Parsley	76	Umbelliferae oils
Erucic	13-Docosenoic	Nasturtium	82	Cruciferae oils
Linoleic	9,12-Octadecadienoic	Safflower	78	Linseed and cottonseed oils
Tariric	6-Octadecinoic	Tariri (Picramnia Sow)	95	—
Linolenic	9,12,15-Octadecatrienoic	Perilla oil	70	Linseed oil
Elaeostearic	9,11,13-Octadecatrienoic	Tung oil	75–95	—
Arachidonic	5,8,11,14-Eicosatetraenoic	—	—	Animal fat, phosphatides
Clupanodonic	4,8,12,15,19-Docosapentaenoic	—	—	Japanese sardine oil

has summarized the most widely used methods for determination of fatty acids as follows:

a) Non-solvent Crystallization, whereby higher melting fatty acids can be separated by chilling from those having lower melting points.

b) Crystallization from Solvents. This can be brought about by separations of the barium, magnesium, or lead soaps of the solid fatty acids from those of the liquid fatty acids. This procedure is successful only to a limited extent.

A second method involves the separation of the fatty acid bromine addition products such as the solid tetrabromide of linoleic acid or the hexabromide of linolenic acid, both of which are insoluble in petroleum ether. Due to complex mutual solubilities of isomers this test is only of limited use. A third technique employing solvent crystallization which has been especially successful of late is low-temperature crystallization from acetone. The procedure has been developed in the laboratory of J. B. BROWN (*44*). By this method many naturally-occurring unsaturated fatty acids such as oleic, linoleic, linolenic, erucic, and ricinoleic, can be separated from each other as well as from the saturated acids.

c) Fractional Distillation of methyl or ethyl esters is by far the most accurate and widely-used method for the isolation and the determination of the individual fatty acids (*13, 252, 269, 355*).

d) Chromatographic Separation. Only a condensed extract of recent developments in this field can be given and the reader must be referred to special monographs (*363*). A number of procedures for applying partition chromatography have recently been described which effectively separate fatty acids from each other. SMITH (*318*) first employed silica gel partition chromatography for the resolution of formic, acetic, propionic, *n*-butyric, and *n*-valeric acids mixtures, in chloroform + 1% butanol, into the component acids. Although the butanol-chloroform-water system has been satisfactorily used by others (*106, 270*) for the separation of C_1 to C_4 acids, it is not effective for longer-chain compounds; however, on buffering the silica gel tube, MOYLE *et al.* (*246*) were able to effect a separation of the C_2 to C_8 acids and also to separate partially the isomeric forms of valeric acid. PETERSON and JOHNSON (*262*), by employing a benzene-aqueous sulfuric acid in Celite-packed tubes, could separate the formic acid and the even-chain acids from C_2 to C_{10} from each other. This technique is also applicable for the separation of unsaturated acids from each other (*273*). Thus, it offers a method for separation of oleic from linoleic acid which cannot be accomplished by fractionation. A methanol-isooctane mixture has found application in partition chromatography for saturated fatty acids of intermediate length (*271*). When furfuryl alcohol and 2-aminopyridine are used as the immobile solvent and *n*-hexane as the mobile phase on a silica gel column, C_{11} to C_{19} fatty acids can be effectively determined (*272*), *cf.* also BOLDINGH (*35 a*).

Although the elution technique has been applied to chromatography by CASSIDY (*55, 56*) and by KAUFMANN and WOLF (*201*), the separations were not complete. A modification of this procedure called "displacement analysis" appears to be much more effective in the separation of components of members of the homologous series. By the use of silica gel columns and heptane as the solvent, CLAESSON (*59*) has been able to separate saturated, unsaturated, and branched-chain fatty acids. HOLMAN and HAGDAHL (*177*) succeeded in making a fair separation of the C_{12} to C_{22} acids by displacing the fatty acids from a column of charcoal and Hyflo Super Cel with cetyl pyridinium chloride or picric acid. In a later study, the same authors (*178*) used the fatty acids themselves as the displacing agents and effected a separation of the acids from C_1 to C_{22}. Adsorption on silicic acid has likewise been found to be an effective method for the isolation of methyl linoleate and methyl linolenate in pure form (*278*).

e) Counter current Distribution. This method, which has recently come into prominence as a new analytical procedure (*62*, *63*), can likewise be applied for the separation and quantitative estimation of C_2 to C_5 acids (*297*). Using an isopropyl ether —2,2 *M*-phosphate buffer system at p_H of 5,17, the determination can be made with a 2 to 3% accuracy.

f) Ultraviolet Spectroscopy is another method which has proved of use in identification of some fatty acids. It is possible from such data to distinguish not only the fatty acids containing 2, 3, or 4 double bonds, respectively, but also to differentiate compounds containing conjugated and non-conjugated double linkages.

g) Raman Spectroscopy. The determination of the patterns of the Raman spectrograms are useful not only in distinguishing conjugated bonds but also in differentiating geometrical isomers. Thus, VAN DEN HENDE (*344*) has reported that oleic acid and oleates can readily be identified as *cis* compounds, and elaidic acid and elaidates as the corresponding *trans* forms, on the basis of the lower frequencies in the spectrogram of the former type.

h) Determination of the Triglyceride Structure. The nutritional value of a fat is determined not only from its fatty acid composition but also from the arrangement of these fatty acids in the glyceride molecules. Most natural fats contain only small quantities of "simple" triglycerides in which the three fatty acid molecules are identical. It was formerly believed that the entire fat was so constituted; this arrangement of fatty acids is termed the "mono-acid triglyceride" theory.

By far the greatest portion of any fat is composed of "mixed" triglycerides in which two or in some cases three different fatty acids are present as esters of a single glycerol molecule. The determination of the component triglycerides is much more difficult than the establishing of the composition of a fat in terms of its fatty acids. The structurally-adjacent members in the analogous series of triglycerides differ so little in physical and chemical properties from each other that their separation is virtually impossible by any direct method. Furthermore, such an excellent procedure as fractional distillation is not applicable for the resolution of triglyceride mixtures.

One of the most widely employed methods for the determination of triglyceride structure is the oxidation technique of HILDITCH and LEA (*162*) which is described in the first survey by HILDITCH in these *Fortschritte* (*161*). When the fats are oxidized by permanganate in acetone solution, the ethylenic linkages are split and two carbons are oxidized to carboxyl groups. The water-soluble azelaeo-glycerides thus obtained can be readily separated in form of their alkali salts from the stable saturated triglycerides.

In their earlier work, HILDITCH and his co-workers likewise employed complete and partial hydrogenation to study glyceride composition. However, since the exact sequence of events in the hydrogenation of a complex mixture of glycerides, having different degrees of unsaturation, is uncertain, HILDITCH now believes that this technique does not contribute much to the knowledge of the triglyceride structure. In his more recent studies, HILDITCH has employed the low-temperature crystallization method of BROWN (*44*) for investigation of the mixed glycerides in natural fats. By the use of different solvents and temperatures, some mixed triglyceride fractions can be obtained which are much less complex than those in the original fat. By the investigation of the composition of these fractions it is possible to make deductions on the probable composition of the triglyceride mixture.

The occurrence of a number of simple triglycerides in fats has been proved by actual isolation of these compounds. Such representatives as trilaurin, trimyristin, tripalmitin, triolein, trilinolin, trilinolenin, trierucin, triricinolin, trielaeostearin, as well as traces of tristearin have been proved to be present in fat. On the other hand, only two mixed triglycerides have been proved to occur with certainty. These include 2-oleyldistearin, present in kokum butter, cocoa butter, and the tropical *Allanblackia* fat; and 2-oleyldipalmitin which has been reported in *Stillingia* and piquia fat. 2-Palmityloleostearin probably occurs in lard (*70*).

Marked differences exist in the type of glyceride sfound in vegetable fats as compared with animal fats. In most cases the distribution of the fatty acids in the triglycerides of vegetable fat differs from that of a random or chance distribution, but follows more nearly a so-called "*even distribution*". The rule of even or "widest" distribution was first proposed by HILDITCH and LOVERN (*166*) in 1936 and has been discussed by HILDITCH (*159*) in the light of more recent developments in his excellent survey in these *Fortschritte*. According to his scheme, if about one-third of the total fatty acids consists of an acid A, then that particular acid will occur in practically all of the triglyceride molecules to the extent of one molecule. If between 33 and 67% of the total acids consists of acid A, then it will be present with increasing frequency in the triglyceride molecules to the extent of two molecules but will not form in any appreciable number of cases a simple triglyceride. It is only after the A-content of the total acids exceeds 67% that one begins to find A-triglycerides.

This is in line with the earlier observation that completely saturated triglycerides appear to any appreciable extent in vegetable fats only if the saturated acid content reaches or surpasses 60% of the total fatty acids. HILDITCH (*159*) admits that the rule of even distribution may not be adhered to with mathematical rigidity. In fact, HILDITCH and MADDIson (*167*) as well as Dunn, Hilditch, and Riley (*99*) have pointed out that cottonseed oil and certain *Citrus* seed fats in which only 25% of the acids are saturated, nevertheless contain small amounts of disaturated-mono-unsaturated triglycerides.

In the case of such animal fats as milk fat and depot fats which are rich in stearic acid, the principle of even distribution apparently does not hold. Hilditch and co-workers (*15, 16, 26*) first assumed that in these cases random distribution obtains. However, more recently the hypothesis has been modified with the suggestion that first an even distribution pattern of oleic and palmitic acid is laid down, and this is followed by a partial bio-hydrogenation which results in the acid distribution observed. Hilditch (*160*) has also pointed out that even though a similarity in the distribution of saturated fatty acids to a random one may be the case, there is no evidence to indicate that such a chance distribution obtains with the unsaturated fatty acids (*1c*).

On the other hand, Longenecker (*216*) has supported the idea that animal fats are built up according to the theory of probability. Norris and Mattil (*251*) have suggested that the enzyme systems concerned with the synthesis of triglycerides probably differ markedly in the vegetable and animal kingdoms. For this reason the triglycerides are constructed on different principles. In his recent contribution, Hilditch (*159*) has stated that further evidence will shortly be published from his laboratory, indicating that the animal glycerides constitute merely special cases of the general principle of even distribution.

Doerschuk and Daubert (*97*) have recently suggested that fatty acids are assembled in corn oil according to a "*partial random distribution*". This theory would appear to be somewhat of a compromise between the random and the even patterns of triglyceride structure. Making use of the method of low temperature precipitation from acetone, these workers were able to prepare 19 different fractions of corn oil. The fatty acid make-up of each triglyceride was determined and the results compared with the composition of the triglycerides which should obtain if fats were assembled according to: the mono-acid theory, the even distribution arrangement, a random pattern, or a partial random distribution scheme.

The chief differences between this new theory and the even distribution pattern of Hilditch are that, in spite of the fact that linoleic acid is present to the extent of about 60%, some triglyceride molecules may be included in the fat, which do not contain linoleic acid residues, furthermore, some trilinolein may be present. According to the even pattern, at least one linoleate molecule would occur in every molecule and two such molecules would be found in most triglycerides. Another new postulate is that oleic acid and the saturated acid residues may occur twice in a limited number of molecules, in spite of the fact that each accounts for considerably less than 33% of the total acids. The even distribution pattern would allow only a single molecule of either of these fatty acids per triglyceride.

A comparison of the triglyceride composition found with that which might be expected on the basis of the four discussed theories is given in Table 2a.

Table 2a. A Comparison of the Triglycerides Found in Corn Oil with Those Calculated on the Basis of Different Theories on Triglyceride Patterns of Fat.

Adapted from DOERSCHUK and DAUBERT (97). Ln = linolenic acid radical; Lo = linoleic acid radical; Ol = oleic acid radical; and S = saturated acid radical.

Triglyceride structure	Number of positional isomers	Triglycerides from 1 kg. corn oil calculated and found				
		Mono-acid triglyceride	Random distribution	Even distribution	Partial random pattern	Experimentally determined
		grams				
Ln, Ln, Ln	1	6,15	0,003	0	0	0
Lo, Lo, Lo	1	596	223	0	8,0	8,9
Ol, Ol, Ol	1	230	12,5	0	0	0
S, S, S	1	147	3,24	0	0	0
Ln, Ln, S	2	0	0,02	0	0	0
Ln, Ln, Ol	2	0	0,03	0	0	0
Ln, Ln, Lo	2	0	0,07	0	0	0
Lo, Lo, S	2	0	164	291	332	335
Lo, Lo, Ol	2	0	257	524	485	482
Lo, Lo, Ln	2	0	6,96	0	0,5	0
Ol, Ol, S	2	0	23,9	0	1,69	3,60
Ol, Ol, Lo	2	0	98,0	0	70,5	72,7
Ol, Ol, Ln	2	0	1,02	0	0,00	0
S, S, Ol	2	0	15,3	0	4,56	0,18
S, S, Lo	2	0	39,9	0	17,5	16,8
S, S, Ln	2	0	0,41	0	3,20	4,10
S, Ol, Lo	3	0	125	146	40,8	41,5
Ol, Lo, Ln	3	0	5,34	15,5	1,00	0
S, Ol, Ln	3	0	1,30	0	10,5	11,8
S, Lo, Ln	3	0	3,40	3,00	2,94	3,19

The experimental results reported by DOERSCHUK and DAUBERT (97) for corn oil agree excellently with those calculated on the "partial random" basis, and show very poor correlation with the conception of either even distribution or random arrangement. The data indicate that the mono-acid theory cannot be valid. It will be interesting to determine whether other vegetable fats or any animal fats can be shown to exhibit a pattern similar to that of corn oil when this newer analytical procedure is applied. This compromise proposal of partial random distribution would seem to offer an explanation for fatty acid arrangement in triglycerides which is superior to any one of the current theories.

2. Composition of Substances Associated with Triglycerides.

As is well known, phospholipids or phosphatides are present in many crude fats although they are largely removed during the process of refining. The proportion of these compounds present in animal fats is usually quite small, while fairly large amounts are found in many vegetable fats, especially in soybean and corn oils. In addition to the α- and β-lecithins, it has recently been demonstrated that several different types of cephalins occur in vegetable and animal fats (*129*), such as phosphatidyl-ethanolamine, phosphatidyl-serine, and lipositol.

$$H_2C{-}O{-}C({=}O)\cdot(CH_2)_7CH{=}CH(CH_2)_7CH_3$$
$$HC{-}O{-}C({=}O)\cdot(CH_2)_{16}CH_3$$
$$H_2C{-}O{-}P({=}O)(OH){-}O{-}CH_2CH_2\overset{+}{N}(CH_3)_3 \quad OH^-$$

α-Lecithin.

$$H_2C{-}O{-}C({=}O)\cdot(CH_2)_7CH{=}CH(CH_2)_7CH_3$$
$$HC{-}O{-}P({=}O)(OH){-}OCH_2CH_2\overset{+}{N}(CH_3)_3 \quad OH^-$$
$$H_2C{-}O{-}C({=}O)\cdot(CH_2)_{16}CH_3$$

β-Lecithin.

$$H_2C{-}O{-}C({=}O)\cdot(CH_2)_7CH{=}CH(CH_2)_7CH_3$$
$$HC{-}O{-}C({=}O)\cdot(CH_2)_{16}CH_3$$
$$H_2C{-}O{-}P({=}O)(OH){-}O{-}CH_2CH_2NH_2$$

Phosphatidyl-ethanolamine.

$$H_2C{-}O{-}C({=}O)\cdot(CH_2)_7CH{=}CH(CH_2)_7CH_3$$
$$HC{-}O{-}C({=}O)\cdot(CH_2)_{16}CH_3$$
$$H_2C{-}O{-}P({=}O)(OH){-}O{-}CH_2CH(NH_2)COOH$$

Phosphatidyl-serine.

Phosphatidyl-serine and phosphatidyl-ethanolamine have been found only in brain and other animal lipids, while the third type, lipositol, has been reported to occur not only in brains but also in soybean oil. In addition to fatty acids, glycerol and phosphoric acid, lipositol contains also the vitamin, inositol.

The fat-soluble vitamins are found in the non-saponifiable extract. These include vitamin A, $C_{20}H_{29}OH$, which is present in a limited number of animal fats including butter fat and fish liver oils; carotene, $C_{40}H_{56}$, present in butter fat and in many vegetable extracts including fats such as palm oil; vitamins D_2 and D_3 which are found in appreciable quantities in various fish liver oils, but which may also be synthesized *in vitro* by the irradiations of suitable provitamins; and, finally, the tocopherols or vitamins E occurring chiefly in vegetable fats. The tocopherols are of considerable importance since they are helpful in preventing oxidative destruction of the fat.

Sterols are also present in non-saponifiable extracts. The component sterol found in most animal fats is cholesterol, $C_{27}H_{45}OH$, while the vegetable fats contain closely related substances, β-sitosterol and stigmasterol. It is noteworthy that both vegetable and animal fats may contain hydrocarbons. For example, squalene, $C_{30}H_{50}$, a compound with isoprenic structure, is common in the liver oils of many different species of Elasmobranch fishes. TSUJIMOTO (*338*, *339*) has reported the presence of squalene in sixteen out of thirty-six species of such fishes obtained from Japanese waters. In some instances squalene may make up from 50 to 80% of the total unsaponifiable matter. A related iso-octodecane called pristane, $C_{18}H_{36}$ (*337*), likewise occurs in smaller proportions. Three glycerol ethers, namely, selachyl alcohol, chimyl alcohol, and batyl alcohol (*333*, *340*), were found in the liver fat of gray dog fish and rat fish where they amount to as much as 10% and 37%, respectively, of the total fatty material.

$$CH_3(CH_2)_{14}CH_2\cdot O\cdot CH_2 - HOCH - HOCH_2$$

Chimyl alcohol.

$$CH_3(CH_2)_{16}CH_2\cdot O\cdot CH_2 - HOCH - HOCH_2$$

Batyl alcohol.

$$CH_3(CH_2)_7CH{=}CH(CH_2)_7CH_2\cdot O\cdot CH_2 - HOCH - HOCH_2$$

Selachyl alcohol.

BAER and FISCHER (*12 a*) reported that natural batyl and chimyl alcohol belong to the *d*-series and that selachyl alcohol, because of its close relationship to batyl alcohol, can also be assigned to the same series. These investigators, in collaboration with RUBIN (*12 b*) were later able to synthesize the α-oleyl glycerol ether (selachyl alcohol) that corresponds the natural product.

The glyceryl ethers almost certainly do not occur in nature in the form of free alcohols but are esterified with fatty acids, *e. g.* according to the following formula:

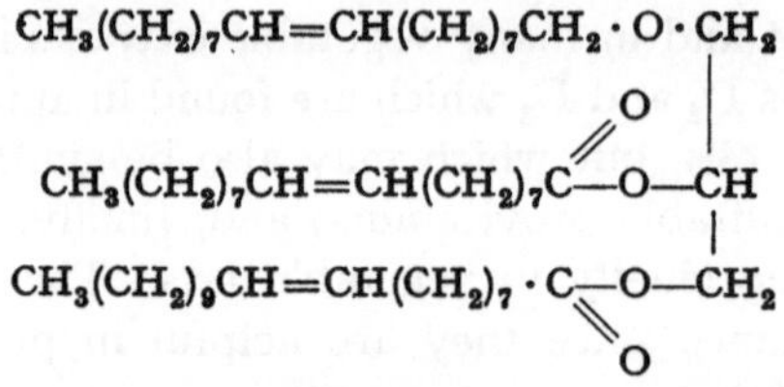

$$
\begin{array}{l}
CH_3(CH_2)_7CH{=}CH(CH_2)_7CH_2\cdot O\cdot CH_2 \\
\qquad\qquad\qquad\qquad\qquad\qquad\qquad\;\; | \\
\qquad\qquad\qquad\qquad\qquad\qquad\quad O \\
\qquad\qquad\qquad\qquad\qquad\qquad\quad /\!/ \\
CH_3(CH_2)_7CH{=}CH(CH_2)_7C{-}O{-}CH \\
\qquad\qquad\qquad\qquad\qquad\qquad\qquad\;\; | \\
CH_3(CH_2)_9CH{=}CH(CH_2)_7\cdot C{-}O{-}CH_2 \\
\qquad\qquad\qquad\qquad\qquad\qquad\quad \backslash\backslash \\
\qquad\qquad\qquad\qquad\qquad\qquad\quad\; O
\end{array}
$$

Selachyl alcohol esterified with oleic and gadoleic acids.

In addition to the alcohols mentioned, some other higher alcohols may be present to a considerable extent as esters in fats obtained from marine animals. Thus, HILDITCH and LOVERN (*163–165*) have reported that sperm head oil contains 74% of wax esters and 26% of triglycerides, while the blubber oil of the same whale contained 66% and 34% respectively of wax esters and triglycerides. The saturated alcohols present were tetradecyl, $CH_3(CH_2)_{12}CH_2OH$, hexadecyl or cetyl, $CH_3(CH_2)_{14}CH_2OH$, and octodecyl alcohol, $CH_3(CH_2)_{16}CH_2OH$. Hexadecyl alcohol is present in the greatest amount. Among the unsaturated alcohols are hexadecenyl and octadecenyl alcohol, $CH_3(CH_2)_7CH{=}CH(CH_2)_7CH_2OH$, and also a C_{20}-unsaturated alcohol. It is not surprising that alcohols corresponding with most of the known natural fatty acids have been reported, since in all probability these originate by metabolic changes from the fatty acids.

The bacterial lipids contain branched-chain fatty acids, ketones, and alcohols not ordinarily found in lipids from other sources (*3*).

3. Factors Influencing the Composition of Fats.

Slight variations in the composition of *vegetable fats* obtain in the same species grown under different conditions as well as in related plant species raised in the same environment. Although such variations are minor in most cases, they may be of sufficient magnitude to be of commercial importance. This is particularly true with linseed oils in which wide variations in the linoleic and linolenic acid content have been reported by a number of investigators. Thus, variations in iodine number from 155 to 196 have been noted in the Bison variety of linseed raised in different localities. The figures for the linolenic acid content lay between 35 to 59% for the samples having the lowest and highest iodine number respectively (*256–258*). The extreme variations found by PAINTER and NESBITT (*256–258*) with linseed oils of different varieties range from samples having iodine numbers of 128 to those showing values of 203. The authors indicate that climatic conditions, high temperature, and insufficient moisture while the seed is ripening may produce linseed oils which have very low iodine values. The variations found in different samples of this oil are certainly far greater than those which occur in most vegetable fats.

On the other hand, it has been known for many years that wide variations in the composition of ***animal fats*** can be produced in certain conditions. Of the factors which alter the type of fat, diet seems to be the most important one. LEBEDEFF (*212*) many years ago, demonstrated that the fat of a fasted dog was profoundly affected by feeding the animal on mutton tallow or linseed oil. In the first case, a very hard fat was produced and deposited, while in the second instance, the quantity of unsaturated fatty acids was much higher than for a normal animal; thus, the body fat was practically liquid. Erucic acid, an acid ordinarily not present in the dog fat, has been shown to occur in such fat after feeding a previously fasted dog large amounts of rapeseed oil which is particularly rich in erucic acid (*249*). Although poly-unsaturated acids are present in tissue fats of animals on a fat-free diet, increased levels of tetraenoic, pentaenoic, and hexaenoic fatty acids are deposited in the phosphatides to a very considerable extent after feeding an adequate source of them such as cod liver oil (*277*).

Neither butyric, caproic (*100, 101*) or caprylic acid (*264*) can be deposited in the tissues of rats, even when fed in large doses. However, capric and lauric acids (*264, 265*) as well as myristic acid (*100*) can be built up into the fat depots in rather large proportions when these acids comprise an important part of the diet. VISSCHER (*347*) has recently demonstrated in a convincing manner that an odd-carbon fatty acid can be laid down in the triglycerides when fed in a sufficient amount and over a prolonged period. As much as 24% of the depot fatty acids of rats were found to consist of undecylic acid after the animals had been placed for six weeks on a diet containing the triglyceride of this acid.

The administration of diets high in carbohydrates or proteins has long been believed to cause the formation of hard body fat. This was beautifully demonstrated by ANDERSON and MENDEL (*5*) who showed that when corn starch was substituted equi-calorically for fat in the diet, a progressive hardening of the body fat ensued. Variations in the diet of farm animals may produce changes in the composition in their body fat which may be of great commercial importance. Thus, the so-called "soft pork" was a product which was unacceptable to the consumer. As a result of the studies in the United States Department of Agriculture (*151, 152*) it was shown that the low-melting pork fat resulted from the consumption of a large proportion of peanuts. ELLIS and ISBELL (*104*) found later that when peanuts or soybeans were fed in large proportion to hogs, fats having a high iodine number and low melting point were produced. On the other hand, the fat deposited in hogs, receiving diets of corn and skimmed milk or brewer's rice and tankage, had a low iodine number and a relatively hard consistency. The use of corn for hardening the fat of cattle is regularly employed by stock men.

II. Nutritional Evaluation of Fats in General.

1. Functions of Fat in the Animal.

There are a number of functions which fat may fulfill in animal nutrition. In the first place, it possesses the highest caloric density of any of the known foodstuffs. In the dry state, it yields 9,3 kilocalories/gram compared with 4,1 Kcal/g. which is given by carbohydrate and protein under similar conditions. Since fat is hydrophobic, while carbohydrates and proteins are usually hydrophilic, the comparative values show a far greater variation than the 2,3 : 1 caloric ratio which exists between purified fats on the one hand, and purified carbohydrates or proteins on the other. Fat is therefore the foodstuff to be employed in greatest proportion when the bulk of a diet must be restricted.

Fats also serve as a carrier of fat-soluble vitamins. Even when they do not contain such vitamins, they are of importance in improving the absorption of these compounds from the gastrointestinal tract. In addition, the fats also may be a source of choline, provided that an appreciable proportion of phospholipids is present. EVANS and LEPKOVSKY (*117*) showed a number of years ago that high-fat diets likewise exert a marked sparing action on the thiamin requirement.

Fats are the only natural source of certain unsaturated (so-called "essential") fatty acids required for the structural development of the animal tissues.

Fats are also important in the diet in order to improve palatability. Low-fat diets prove extremely unappetizing and their use is often associated with a voluntary reduction of the caloric intake. For example, it was stated by STARLING (*321*) that during World War I, the civilian population of Great Britain lost weight partly as a result of their failure to consume as high a number of calories on the high-carbohydrate, low-fat diet as were needed to maintain caloric equilibrium.

Finally, fats are most important adjuncts in the preparation of many baked foods.

2. The Digestibility of Fats.

LUSK (*220*) has defined a foodstuff as a substance which is capable of being added to the body's substance, or, when absorbed into the blood stream, can prevent or reduce the wasting of some necessary constituents of the organism. It is thus obvious that for a substance to be placed in this category, it must be capable of absorption from the gastrointestinal tract. The nutritional value of a fat must therefore be in proportion to its digestibility. The extent of digestibility is usually expressed as the "coefficient of digestibility". This term refers to the percentage of foodstuff given which is absorbed. Corrections are ordinarily made for metabolic fat based on control experiments with a fat-free diet.

Table 3. Average Coefficients of Digestibility of Vegetable Fats and Oils (Melting Point Below 50° C.) in Human Subjects.

Fat tested	Reference	Number of tests	Average daily fat intake (g.)	Coefficient of digestibility (%)
Almond	(*181*)	4	70	97
Apricot kernel	(*184*)	4	68	98
Avocado	(*185*)	3	100	88
Black walnut	(*181*)	4	56	98
Brazil nut	(*181*)	3	81	96
Butternut	(*181*)	3	43	95
Charlock	(*182*)	4	60	99
Cherry kernel	(*185*)	4	57	98
Cocoa butter	(*207*)	11	51	95
Coconut	(*207*)	12	64,6	98
	(*187*)	3	28	89
Cohune	(*185*)	4	52	99
Corn	(*184*)	7	80	97
	(*187*)	2	29,8	97
Cottonseed	(*207*)	12	86	98
	(*88*)	9	60	97
Cupuassu	(*185*)	4	41	94
English walnut	(*181*)	3	78	98
Hemp seed	(*185*)	3	57	98
Hickory nut	(*181*)	4	95	99
Japanese mustard seed	(*185*)	3	79	96
Java almond	(*87*)	2	60	97
Melon seed	(*183*)	3	40	98
Olive	(*207*)	10	73	98
Palm kernel	(*207*)	4	100	98
Peach kernel	(*184*)	3	60	97
Peanut	(*207*)	5	98	98
Pecan	(*181*)	4	104	97
Poppy seed	(*185*)	7	50	96
Pumpkin seed	(*183*)	2	75	98
Rapeseed	(*182*)	4	82	99
	(*88*)	8	60	99
Sesame	(*207*)	5	90	98
Soybean	(*182*)	7	81,6	97,5
	(*187*)	2	21	94
Sunflower seed	(*182*)	4	90	96
Tea seed	(*87*)	1	49	91
Tomato seed	(*183*)	3	57	96
Watermelon seed	(*87*)	3	30	95

a) The Digestibility in Man of Vegetable and Animal Fats with Melting Points Below 50° C. Practically all fats belonging to this category have been found to be largely digested by the normal individual. Although no allowance was made for the fat which is excreted as soap in the long series of experiments conducted by the Office of Home Economics of the United States Department of Agriculture, it is not thought that this

Table 4. Average Coefficients of Digestibility of Animal Fats and Oils (Melting Point Below 50° C.) in Human Subjects.

Fat tested	Reference	Number of tests	Average daily fat intake (g.)	Coefficient of digestibility (%)
Beef	(*206*)	10	100	93
Brisket	(*208*)	7	80	97
Butter	(*206*)	8	100	97
	(*187*)	35	26	89
	(*72*)	3	49	97
Chicken	(*208*)	8	95	97
Cod-liver	(*87*)	4	47	98
Egg yolk	(*208*)	6	83	94
Fish	(*208*)	3	60	95
Goat's butter	(*180*)	4	43	98
Goose	(*208*)	7	95	95
Hardpalate	(*180*)	3	88	94
Horse	(*180*)	3	63	94
Kid	(*180*)	3	60	95
Lard	(*206*)	9	90	97
Milk cream	(*208*)	7	78	97
Oleo	(*180*)	8	57	97
Ox marrow	(*180*)	4	87	94
Ox tail	(*180*)	3	74	97
Turtle	(*180*)	4	46	99

correction would cause any marked alterations in the previously calculated coefficients of digestibility. Recently it has been shown by DEUEL *et al.* (*88*) that when the fatty acids excreted as soaps in the feces are also considered, the digestibility of cottonseed or rapeseed oils is practically identical with that determined in the earlier tests where the mentioned factor was omitted from consideration. Such data are summarized in Tables 3 and 4.

It would thus appear that most animal and vegetable fats are well digested by man when taken in amounts of 50 to 100 grams daily. With the exception of avocado fat and tea seed oil, which were digested to the extent of only 88% and 91%, respectively, the average coefficient of digestibility of the thirty-four vegetable fats studied varied between 94 and 99%. In the case of the eighteen animal fats investigated, the coefficient lay between 93% and 99%. Most of these data have been summarized by LANGWORTHY (*205*).

b) The Digestibility in Man of Vegetable and Animal Fats with Melting Points Above 50° C. In contrast to the almost complete digestibility of vegetable and animal fats with melting points under 50° C., there are numerous reports which demonstrate that fats having higher melting points are less completely utilized in man. LANGWORTHY and HOLMES (*206*) stated that "of the fats tested, the fats of low melting point are

Table 5. Average Coefficients of Digestibility in Man of Some High-melting Animal Fats and Some Hydrogenated Vegetable Fats.

Fat tested	Melting point °C.	Reference	Number of tests	Average daily fat intake	Coefficient of digestibility (%)
Deer	51,4	(*87*)	3	46	82
Mutton	50	(*206*)	7	53	88
Oleostearin	50	(*180*)	3	66	80
Hydrogenated fats*	—	(*183*)	—	—	—
Cottonseed	35	—	5	—	97
	46	—	3	—	95
Peanut	37	—	5	—	98
	39	—	3	—	96
	43	—	5	—	96
	50	—	4	—	92
	52	—	3	—	79
Corn	33	—	5	—	95
	43	—	5	—	95
	50	—	5	—	88
Blended hydrogenated fats**	—	(*87*)	—	—	—
Cottonseed:					
(13 : 88)***	45,8	—	2	53	96
(19 : 81)	47,8	—	4	76	94
(24 : 77)	48,1	—	2	49	94
(22 : 78)	50,0	—	3	57	87
Peanut:					
(6 : 94)***	43,0	—	5	74	97
(9 : 91)	43,2	—	4	80	97
(33 : 67)	51,1	—	4	90	93
Corn:					
(9 : 91)***	39	—	4	103	95
(25 : 75)	49	—	3	105	93
(31 : 69)	54	—	3	92	92

* In these tests the total oil was subjected to partial hydrogenation to produce fats having varying melting points.

** A portion of the fat which was completely hydrogenated was blended with sufficient untreated oil to give mixtures of the melting points indicated.

*** The values in parentheses indicate the following ratio: completely saturated fat: untreated oil (used in the fat blend).

capable of more complete assimilation than those of high melting point". HOLMES and DEUEL (*184*), also suggest that an inverse relationship exists between the extent of absorbability and the melting point of fats.

Although the average coefficients of digestibility of deer fat, mutton fat, oleostearin, and several of the higher melting hydrogenated fats are distinctly lower than those of the vegetable and animal fats (melting below 50° C.), these values are to be regarded as the maximum rather than the minimum for human subjects. It has been pointed out earlier

Table 6. Average Coefficients of Digestibility for Some Oleomargarines in Man.

Oleomargarine sample tested and its composition	Reference	Number of tests	Average daily fat intake (g.)	Coefficient of digestibility (%)
No. 1. 59% oleo oil, 7% lard, 22% vegetable oils (cottonseed and peanut), 12% milk fat	(179)	9	—	97
No. 2. 41% oleo oil, 32% lard, 24% peanut oil, 3% milk fat	(179)	7	80	93
No. 3. 67% oleo oil, 33% cottonseed oil, 0,1% milk fat	(179)	4	—	97
Hydrogenated vegetable oil margarine (melting point 94–95° F.; Wiley)	(72)	7	86	97

that no allowance has been made in these tests for the fatty acids excreted as soaps in the feces. The fact that markedly lower values for digestibility of high melting fats are recorded in experiments on rats (Table 7, p. 20) than for man is largely to be attributed to the circumstances that the fecal soap fraction was included in the calculations of the tests on rats.

It seems probable that considerable amounts of the original fat would be lost in this fraction also in man. Further work would be needed to clarify this point using fats with relatively high melting points.

c) Digestibility of Oleomargarines. Margarines have been found to be well digested in man. Some of the data on human subjects are recorded in Table 6, and tests on animals in Table 7. A high digestibility was found in the margarine samples 1, 2, and 3, which had been commercially produced more than thirty years ago, while a modern preparation manufactured from hydrogenated cottonseed oil was shown to have a digestibility of 97%. Similar values are valid for the fourth margarine sample in rats in which the soap fraction has also been taken into account.

d) Digestibility of Polymerized Oils. Although edible polymerized oils have had a limited use in the food industry for a number of years, it is only recently that attention has focussed on them. Polymerization is doubtlessly associated with profound changes not only in physical properties but also in the chemical structure. By a suitable polymerization process the characteristic odor and taste of unsaturated glycerides can be removed. On extensive treatment, a considerable viscosity develops in the oil; that this is caused by a chemical alteration is indicated by the fact that the degree of unsaturation, as determined from the iodine number, decreases markedly during the change.

Bradley (*41*) has suggested that the main chemical alterations during polymerization of unsaturated fats involves the double bonds. According to Brocklesby (*43*), the most probable reaction is concerned with a preliminary formation of an unstable 4-carbon ring which opens and then rearranges as follows:

$$\begin{array}{ccccc} R\cdot CH{=}CH\cdot R' & & R\cdot CH{-}CH\cdot R' & & R\cdot CH_2{-}CH\cdot R' \\ + & \rightarrow & |\quad\quad | & \rightarrow & | \\ R\cdot CH{=}CH\cdot R' & & R\cdot CH{-}CH\cdot R' & & R\cdot C{=}CH\cdot R' \end{array}$$

Such a reaction may take place between two unsaturated acids which are either located in the same glyceride molecule or in different molecules. Lassen *et al.* (*210*) believe that the second possibility is the more likely one.

Several interesting studies have recently appeared which suggest that polymerization may bring about an alteration in the physiological behavior, as well as in physical and chemical properties; for example, Lassen, Bacon, and Dunn (*210*) have found that the fat may become less digestible.

When sardine oil had been polymerized by heating in a glass vessel at 250° C. in a nitrogen atmosphere, only a slight decrease in digestibility was observed in rats in a sample having an iodine number of 155,5 (iodine number of the control oil, 177,7). The coefficients of digestibility were 98,3 and 94,9, respectively, for the unheated and the first fraction of polymerized oil. On further heating the sardine oil digestibility was considerably reduced. The second fraction (iodine number, 138,1) was digested only to the extent of 89,5%, while in the third fraction (iodine number, 124,1), the coefficient of digestibility had fallen to 84,8.

Crampton and his associates (*64*) reported that high mortality results in rats fed on linseed oil which had been previously subjected to polymerization. The death of the rats occurred within a very short time, and was associated with a lowered digestibility. On the other hand, Deuel, Greenberg, and Savage (*75*) found no decreased growth rate in rats fed on diets containing 40% of cottonseed oil which had been kept at 205–210° C. for a long period, during which a number of batches of French fried potatoes were prepared.

e) Digestibility of Fats in Animals Other than Man. According to McCay and Paul (*224*), the extent of digestibility of fats in the guinea pig is usually somewhat less than in man. The inverse relationship between digestibility and melting point of a series of hydrogenated cottonseed oils and hydrogenated lards was convincingly demonstrated in rats. With both of these types of fat, there appears a progressive decrease in the digestibility, coincident with the rise in the melting point.

Hydrogenated cottonseed oil melting at 65° C. was digested to the extent of only 24%, while hydrogenated lard melting at 61° C. had a coefficient of digestibility of 21%. The fat which was digested to the lowest extent, was hydrogenated perilla oil melting at 67,5° C., in which

Table 7. Average Coefficients of Digestibility of Some Fats in Several Species of Animals.
(Figures in parentheses give the references to the authors reporting these values.)*

Fat fed	Melting point °C.	Coefficient of digestibility (%)				
		Guinea pig (224)	Rat	Rabbit	Sheep	Dog
Beef tallow	—	72,0	—	—	—	—
Butter	—	91,0	88,3 (172)	—	—	—
Castor	—	96,2	98 (323)	92,1 (259)	99 (259)	—
Cocoa butter	—	—	63,3 (172)	—	—	—
Coconut	—	94,0	98,9 (172)	—	—	—
Cod-liver	—	93,8	—	—	—	—
Corn	—	86,5	97,5 (172)	—	—	99 (285)
Cottonseed	—	87,4	—	—	—	99 (285)
hydrogenated	38	—	91 (121)	—	—	—
,,	46	—	83,8 (11)	—	—	—
,,	54	—	68,7 (11)	—	—	—
,,	62	—	38 (121)	—	—	—
,,	65	—	24,0 (11)	—	—	—
Crisco	—	73,8	97,3 (65)	—	—	—
Lard	—	75,2	96,6 (65)	—	—	—
bland	48	—	94,3 (65)	—	—	—
hydrogenated	55	—	63,2 (65)	—	—	—
,,	61	—	21,0 (65)	—	—	—
Margarine	34	—	97,0 (65)	—	—	—
,,	—	—	97 (73)	—	—	—
Mutton tallow	—	79,8	74,6 (172)	—	—	—
Neats foot	—	93,5	—	—	—	—
Oleo stock	—	—	74 (172)	—	—	—
Olive	—	94,5	—	—	—	—
Peanut	—	91,8	—	—	—	—
Perilla (hydrogenated)	67,5	—	6 (121)	—	—	—
Rapeseed	—	—	82 (73)	—	—	—
Salmon	—	94,0	—	—	—	—
Soybean	—	94,5	98,5 (172)	—	—	—

* Soaps were included in the determination in all reports except in reference numbers (121), (224), (259), and (285).

case only 6% was absorbed from the gastrointestinal tract. The behavior of castor oil differs in most of the lower animals from that in man. In the case of the guinea pig, rat, rabbit, and sheep, castor oil is practically completely digested, and shows none of the usual purging action it exerts in man.

The result of tests on a number of fats in several species of animals are recorded in Table 7.

A number of pertinent studies have also been made with some simple triglycerides and the corresponding fatty acids. Trilaurin has been found to be practically completely digested in the rat, while trimyristin is digested only to the extent of about 75%. The data on tripalmitin are conflicting. While HOAGLAND and SNIDER (*173*) found digestibility of 83% for tripalmitin in the rat, CHENG *et al.* (*57*) reported that it was absorbed to the extent of only 28%. Tristearin has been found to be only slightly absorbed in the rat and in the dog (*10*). Free palmitic and stearic acids are utilized to only about the same extent as their triglycerides, whereas oleic and elaidic acids are both completely digested in the rat (*259*). Elaidic acid has a low coefficient of digestibility in the guinea pig, as compared with the high value obtained with oleic acid in the same species.

Table 8. Average Coefficients of Digestibility of Some Purified Simple Triglycerides and Fatty Acids in Several Species of Animals.
(The figures in parentheses give the bibliographic references.)

Triglyceride or acid fed	Melting point ° C.	Guinea pig	Rat	Dog.
Trilaurin	49	—	97,3 (*57*)	—
Trimyristin	56	—	76,6 (*57*)	—
Tripalmitin	66,5	—	83 (*173*) 27,9 (*57*)	95 (*222*)
Palmitic acid	63	—	35,6 (*57*)	82 (*222*)
Tristearin	70	—	7 (*173*) 18,9 (*57*)	10 (*10*)
Stearic acid	69	—	15,8 (*57*)	—
Oleic acid	—	95,0 (*259*)	95,4 (*259*)	—
Elaidic acid	—	55,6 (*259*)	95,6 (*259*)	—

f) Factors Influencing the Digestibility of Fats. There are a number of factors other than melting point, species of animal, and polymerization, which may influence the extent to which a fat is absorbed. One of them is the concentration of fat in the diet. Thus, HOAGLAND and SNIDER (*172*) reported increased digestibility coefficients for the least digestible fats when the total fat intake was increased.

Mixed triglycerides have been shown to be more digestible in the rat than a corresponding mixture of simple triglycerides. MATTIL and HIGGINS (*234*) report that when fat is fed at a 15% level in the diet, larger quantities are absorbed from mixed triglycerides than from equivalent mixtures of tristearin and triolein.

Another factor which may markedly alter the absorption of fats is the proportion of calcium present in the diet. Although CHENG and her associates (*57*) found that no appreciable effect on the high digestibility of bland lard resulted from the presence of calcium salts, however, in the case of high-melting fats, the extent of absorption was markedly reduced when considerable amounts of calcium were present.

Table 9. Digestibility of Stearic and Oleic Acids fed as Simple and Mixed Triglycerides [MATTIL and HIGGINS (*234*)].

Fat in diet	Fat In food (g.)	Fat In feces (g.)	Coefficient of digestibility (%)
Tristearin, 10%; triolein, 5%	51,0	29,5	42,2
	79,2	49,4	37,6
Distearo-monoolein, 15%	94,2	39,7	57,9
	24,5	9,0	63,3
Tristearin, 5%; triolein, 10%	87,5	27,3	68,7
	35,9	11,3	68,5
Monostearo-diolein, 15%	89,4	24,3	72,8
	14,6	3,9	73,3

Thus, a hydrogenated lard melting at 55° C. was absorbed to the extent of only 58% in the presence of calcium and 78% in its absence. Trimyristin was ingested to the extent of 38% when calcium was present and 77% in its absence. Moreover, the depressing effect of calcium salts on fat digestibility was found to be proportional to the amount included in the diet. Thus, trilaurin was digested to the extent of 70,5, 87,2, 89,5, and 97,3%, respectively, when the diets contained 6,1, 2,5, 1,17, and 0 mg. of calcium per gram of food (*57*).

It is possible that components in the diet other than calcium may also influence the extent to which the fat is digested. BARNES, PRIMROSE, and BURR (*19*) have reported a slightly higher digestibility of butter or hydrogenated lard when a high-protein diet was employed than when a low-protein regime was administered, although the digestibility of lard itself was unchanged on the two diets. More information is needed on this point, especially since no tests on man are available which would bear out these results.

The presence of emulsifying agents has likewise been shown to improve the extent of digestibility of fat. Crude lecithin increases the extent of digestibility of otherwise poorly utilized high-melting fats in the rat, and it also increases the tolerance for cottonseed oil (*11*). Although no improvement in fat absorption is to be expected or needed in normal men where the fecal fat comprises only about 4% of the ingested fat (*359*, *360*), improvement might be expected in sprue, in certain disorders of the pancreas, and in infections of the gastrointestinal tract where as much as 40 to 50% of ingested lipids may be lost in the feces (*359*). JONES and his collaborators (*196*) were able to demonstrate that the poor fat digestibility in a number of patients markedly improved when "Tween 80" (PSM or polyoxyethylene-sorbitan monooleate) was added to the diet. Further evidence of the beneficial effects of emulsifying agents on utilization of fats is afforded by the demonstration that the rate of absorption is increased by such means.

It seems to be rather generally believed that the digestibility of fats is identical in man and the rat. However, one discrepancy in addition to the variation in purging effect of castor oil between the two subjects has been recently pointed out. Whereas, rapeseed oil was shown to be digested to the extent of 99%, in two independent series of tests carried out on man (*88*, *182*), crude rapeseed oil was found to be utilized to the extent of only 77% in the rat, while a highly refined product gave a digestibility of 82%. Whether or not species variations can offer an explanation for the discrepancies between rat and man in the digestibility of high-melting hydrogenated fats, should be decided in further experiments on man where account is taken of the soaps excreted in the feces.

g) Utilization of Fats Introduced Parenterally. Although DUNHAM and BRUNSCHWIG (*98*) were unable to demonstrate any change in depot fat or any nitrogen-sparing action when fats were introduced by vein, McKIBBEN and co-workers (*229*) showed that corn oil emulsions, stabilized with a variety of reagents, were utilized for energy. In a later study (*228*) it was found that the injection of an emulsion of coconut oil resulted in dogs in a change of a negative nitrogen balance to a positive one. Soybean phosphatides have been used as emulsifying agents (*139*) as well as a mixture (*242*) of "Span 20" (0,5%), Asolectin, a soya-phosphatide preparation (0,4%), and sodium cholate (0,1%). Fat emulsions containing as much as 30% fat have been successfully employed (*60*). For a review, see (*320*).

That such parenterally introduced fat preparations are utilized is also indicated by a recent report of GEYER, CHIPMAN, and STARE (*137*). When trilaurin containing radioactive carbon is injected intravenously into rats, the quick and extensive excretion of labeled carbon dioxide in the respiratory gases indicates conclusively that the fat administered is utilized. Confirmatory results on the rapid oxidation of parenterally introduced fats have recently been reported from the CHAIKOFF group (*214*) who administered tripalmitin containing radioactive carbon.

3. The Rate of Absorption of Fats from the Gastro-intestinal Tract.

Not only is the total quantity of fat which can be absorbed from the gastrointestinal tract an important index of the nutritional value, but also the rate at which such absorption takes place can be used for the evaluation of the nutritive effect. Although no differences in total digestibility of two fats which are absorbed at widely different rates might be noted when they are fed in small doses, a completely different picture might appear if the same fats were taken in large quantities. Thus, the maximum amount of fats which could be utilized without causing a digestive disturbance would vary widely. In the case of slowly absorbed fats, diarrhea would be expected to occur on lower doses than with such fats which are rapidly removed from the gastrointestinal tract. The total utilization under such conditions would be markedly reduced.

There is a paucity of data on the relative rates of fat absorption. Unfortunately, the units in which this factor is expressed are not uniform and, thus, one series of tests cannot always be compared with another in which a different method of calculation is involved. The intestinal loop method is not very satisfactory, since it is far from a physiological condition, especially if pancreatic juice and bile are excluded from the loop. Moreover, the indirect method, which involves a determination of the rate of increase of blood lipids and of the fat in thoracic lymph, is not sufficiently quantitative to be of much value, although it does give considerable evidence about the nature of the fat absorbed. Probably, the most satisfactory procedures are the direct ones which involve an estimation of the amount of fat remaining in the intestine at several periods of time after feeding of known amounts. Such a method has been used by IRWIN, STEENBOCK, and TEMPLIN (*191*) as well as by DEUEL, HALLMAN, and QUON (*83*). The former workers observed that fat absorption in rats is not affected by age (in a range of four to seven months), by sex, by pregnancy, or by the season of the year.

Most uniform results on absorption of fat are obtained if one expresses the value in terms of the fat absorbed per square centimeter of body surface per hour. This procedure of calculation was shown to give essentially constant results when employed on rats of widely varying sizes and with different doses of fat (*80*). On the other hand, STEENBOCK, IRWIN, and WEBER (*322*) prefer to use the percentage of ingested fat which is absorbed in a four-hour period as the index of absorption rate. They conclude that the rate of fat absorption in the rat, expressed in percentage of ingested fat which disappears from the gastrointestinal tract, occurs in the following order (%):

Butter (oil), 71,0; halibut liver oil, 70,2; cod liver oil, 67,7; raw linseed oil, 67,0; olive oil, 63,4; lard, 57,0; rancid lard, 53,8; commercial shortening A, 53,8; cottonseed oil, 53,7; commercial shortening B, 52,8; cocoa butter, 47,9; and coconut oil, 47,4.

When the calculation of absorption is based on the milligrams absorbed per 100 square centimeters of body surface per hour, little differences obtain between a number of common fats.

The results of several groups of investigators follow: butter fat, 42,1 (*80*); coconut oil, 53,2 (*80*); cottonseed oil, 39,8 (*80*); margarine fat, 46,1 (*80*) and 34,7 (*65*); prime steam lard, 38,3 (*65*); and rapeseed oil, 30,0 (*80*). Although such hydrogenated fats as "Crisco" and bland lard, which are absorbed at the rate of 34,6 (*65*) and 31,4 (*11*) milligrams per 100 cm²/hour, respectively, fall in the lower range of the common fats, the results on a series of high-melting, hydrogenated fats are much lower. Thus, in three-hour tests on hydrogenated lard (melting point 55° C.), a value of 20,7 (*65*) was obtained, while with hydrogenated cottonseed oil (melting point 46° C.), the value was 24,7 (*11*). Another sample of hydrogenated cottonseed oil melting at 54° C. had an absorption rate of only 8,5 milligrams per 100 cm^2/hour (*11*).

The slow rate of absorption of rapeseed oil is in line with its lower digestibility (*73*), while the decreased absorption rates of some high-melting hydrogenated fats also fall in line with markedly lower coefficients of digestibility.

As might be expected, the rate of absorption of the shortchain triglycerides is much more rapid than that of the natural fats. This is probably partly to be ascribed to the fact that fatty acids so constituted are water-soluble and, therefore, are able to penetrate the intestinal mucosa more readily than do fat-soluble particles. In the study of a number of synthetic triglycerides, DEUEL and HALLMAN (*79*) found that triacetin was most rapidly absorbed. The absorption rate became progressively slower with the lengthening of the chain.

The values expressed in milligrams per 100 cm^2/hour are as follows: triacetin, 68,1; tributyrin, 65,0; triisovalerin, 45,7; tricaproin, 54,5; tricaprylin, 45,9; and trilaurin, 21,9.

Another interesting variation was shown to result when the absorption rates of triglycerides of odd- and even-chained fatty acids were compared. Thus, the figures reported for tripropionin (C_3) of 31,4, for trivalerin (C_5) 32,9, and triheptylin (C_7) 28,0, are approximately 50% of those of the triglycerides composed of the even-chained fatty acids of comparable length. Such variations in the absorption rate of triglycerides apparently are to be ascribed to the differences in the rate of absorption of the component fatty acids rather than to alterations in the speed of hydrolysis of their glycerides (*84*).

Changes in the rate of absorption under *pathological* conditions are evidently interrelated with chemical structure. VERZÁR and LASZT (*345*, *346*) were the first to demonstrate that the speed of fat absorption in rats is markedly lowered following adrenalectomy. This result was later confirmed by BAVETTA *et al.* (*24*) who also demonstrated that the delayed rate of absorption is associated with a slower speed of removal of the fatty acids from the gastrointestinal tract, since the rate of lipolysis was not decreased. Moreover, the absorption returned to normal if the animals were pretreated with cortical hormones.

Although BARNES and his co-workers (*18*) failed to observe an inhibitory effect of adrenalectomy on the absorption of corn oil, they subsequently reported a lowering in the absorption rate of emulsified hydrogenated cottonseed oil (*20*). That variations in absorption after adrenalectomy are related to the chemical nature of the fatty acid was later observed by BAVETTA and DEUEL (*23*). These workers reported no alteration from normal in the absorption rate of sodium butyrate or of its triglyceride in adrenalectomized rats. It was subsequently shown that the speed of absorption of two other triglycerides in which the fatty acids are water-soluble, namely, tricaproin and tricaprylin, was not affected by removal of the adrenals. In the case of the next higher homolog, tricaprin, a reduction in absorption rate was obtained after adrenalectomy (*22*). Thus, it would appear that the cortical hormones are necessary to insure the prompter absorption of those fatty acids which are not soluble in water, but they are not concerned with the removal of the water-soluble acids from the intestine.

The speed at which fats are absorbed in normal animals can be improved when an emulsifying agent such as lecithin is given (*11*).

In the case of cottonseed oil, the rate of absorption was increased from 47,8 to 61,8 milligrams per 100 cm^2/hour by lecithin. The comparative figures for hydrogenated cottonseed oil samples melting at 46° C. and 54° C. were 26,5 and 18,0, respectively, in the absence of lecithin, compared with 48,9 and 30,1 when lecithin was given with the fat.

A similar improvement in the speed of removal of fat (*2, 196*) and vitamin A (*2*) from the intestine has been noted in man when lecithin was added to the diet; JONES *et al.* (*196*) were likewise able to demonstrate a considerable improvement in vitamin A absorption in patients suffering from sprue, pancreatic fibrosis, or regional enteritis. In all tests on man, the estimation of the speed of absorption was based on the changes in the vitamin A or fat tolerance curves rather than the direct method used in rats.

4. Fats as Sources of "Essential" Fatty Acids.

The first definite proof of a fat-deficiency disease resulting from complete exclusion of this foodstuff from the diet was demonstrated in the rat by EVANS and BURR (*116*). Although MCAMIS and co-authors (*223*) did not report a deficiency syndrome in rats on a fat-free diet, they did note that their animals receiving fat showed a superior growth. Shortly thereafter, BURR and BURR (*51*) described the characteristic pathological symptoms of rats from whose diet fat had been rigidly excluded for some time. The diet employed consisted of alcohol- and ether-extracted casein, fat-free yeast, sucrose, salts, vitamins A und D.

When young rats were subjected to such a regime, a preliminary period ensued during which they gained weight. Their body weight then became constant prematurely and, finally, as the deficiency was continued, the animals lost weight rapidly and died. BURR (*49*) lists the typical symptoms of this deficiency in the rat as follows: 1. marked retardation of growth; 2. development of scaly skin and caudal necrosis; 3. kidney lesions and hematuria; and 4. death. Kidney lesions were found in 100% of the cases at autopsy while the incidence of caudal necrosis was variable.

Other less specific lesions may include histological changes in the ovaries, uterus, and other tissues; poor ovulation, reproduction, and lactation in the female, and sterility in the male. An excess water consumption, high respiratory quotient, and a high metabolic rate were ordinarily noted.

When a relatively small proportion of lard, cod liver oil, or liver was fed to rats showing the fat-deficiency symptoms mentioned, an almost dramatic cure followed. That the active curative agent was linoleic, linolenic, or arachidonic acid, was postulated by BURR and BURR (*52*) who pointed out that the varying potencies of the fat in curing the fat-deficiency syndrome were related to the proportions of the component "essential" fatty acids present.

The saturated fatty acids, as well as oleic acid, were found to be without curative effect. Linoleic (*51, 53*) and arachidonic acid (*50, 190, 317, 341*) have been shown to have the highest therapeutic activity while linolenic acid (*53*) has been reported to have a somewhat lower potency. BURR, in his excellent review of the essential fatty acids (*49*), lists all the compounds giving positive curative effects and the many which have been tested and proved inactive. These acids are listed in Table 10 (p. 28—29), along with some additional ones which were investigated by KARRER and KOENIG (*198*).

There is excellent evidence from an examination of the essential fatty acid content of rat tissue as well as from bioassays that the rat is unable to synthesize the "essential" fatty acids. In animals maintained on a fat-free diet, a depression of the acids has been observed in body lipids (*123*). A number of workers (*17, 102, 217*) have shown that the amount of the essential fatty acids in animal tissue fats is directly dependent on their content in the diet and that no synthesis of them can occur *in vivo*. The most striking proof of this fact was given by SCHOENHEIMER and RITTENBERG (*306*) using deuterium as a tracer. When deuterium oxide was introduced into rats, no exchange of hydrogen and deuterium took place with the essential fatty acids but it did occur with the other fatty acids.

Although BURR and his co-workers (*50*) first suggested that linoleic, linolenic, and arachidonic acids and cod liver oil fatty acids exhibit both qualitative and quantitive differences in curative action, he later stated that inloleic acid alone is essential (*49*). He indicated that one should speak of the essential acids in the singular since any one of the three compounds by itself may prevent the deficiency. However, it is suggested that there may be some qualitative differences in the action of the acids mentioned. Not all compounds which will produce growth will clear the skin disturbances. On the other hand, all compounds having a curative action on the skin will induce growth. Linolenic acid or the cod liver oil acids bring about growth without healing the dermatitis or

Table 10. Purified Fatty Acids Tested for Activity in Curing Fat Deficiency Symptoms.

Common name	Systematic name	Formula	Reference
		Positive Response:	
Linoleic acid	9,12-Octadecadienoic acid	$CH_3(CH_2)_4CH : CHCH_2CH : CH(CH_2)_7COOH$	(*49*)
(Linoleyl alcohol)	9,12-Octadecadienol	$CH_3(CH_2)_4CH : CHCH_2CH : CH(CH_2)_7CH_2OH$	(*341, 342*)
Linolenic acid	9,12,15-Octadecatrienoic acid	$CH_3CH_2CH : CHCH_2CH : CHCH_2CH : CH(CH_2)_7COOH$	(*53, 233, 330, 331*)
Arachidonic acid	5,8,11,14-Eicosatetraenoic acid	$CH_3(CH_2)_4CH : CHCH_2CH : CHCH_2CH : CHCH_2CH : : CH(CH_2)_3COOH$	(*50, 190, 317, 341*)
	Docosahexaenoic acid	Formula in doubt	(*189*)
Hexahydroxystearic acids (Linusic, m. p. 203° C.) (Isolinusic, m. p. 175° C.)	9,10,12,13,15,16-Hexahydroxy-octadecanoic acid	$CH_3CH_2(CHOH)_2CH_2(CHOH)_2CH_2(CHOH)_2(CH_2)_7COOH$	(*189*)
		Negative Response:	
Oleic acid	*cis*-9-Octadecenoic acid	$CH_3(CH_2)_7CH : CH(CH_2)_7COOH$	(*53, 119, 330*)
Elaidic acid	*trans*-9-Octadecenoic acid	$CH_3(CH_2)_7CH : CH(CH_2)_7COOH$	(*315, 331*)
12-Oleic acid	12-Octadecenoic acid	$CH_3(CH_2)_4CH : CH(CH_2)_{10}COOH$	(*341*)
Erucic acid	13-Docosenoic acid	$CH_3(CH_2)_7CH : CH(CH_2)_{11}COOH$	(*341*)
Ricinoleic acid	12-Hydroxy-9-octadecenoic acid	$CH_3(CH_2)_5CHOHCH_2CH : CH(CH_2)_7COOH$	(*341*)
Linolelaidic acid	*trans*-9-*trans*-12-Octadecadienoic acid	$CH_3(CH_2)_4CH : CHCH_2CH : CH(CH_2)_7COOH$	(*48*)
9,11-Linoleic acid	9,11-Octadecadienoic acid	$CH_3(CH_2)_5CH : CHCH : CH(CH_2)_7COOH$	(*48*)
α-Eleostearic acid	9,11,13-Octadecatrienoic acid	$CH_3(CH_2)_3CH : CHCH : CHCH : CH(CH_2)_7COOH$	(*53*)
Dioxidostearic acid	9,12-Dioxidooctadecanoic acid	$CH_3(CH_2)_4CHCHCH_2CHCH(CH_2)_7COOH$ (with O bridges: \/ O under each CHCH)	(*189*)

Trihydroxystearic acid	9,10,12-Trihydroxyoctadecanoic acid	$CH_3(CH_2)_5CHOHCH_2CHOHCHOH(CH_2)_7COOH$	(*189*)
Tetrahydroxystearic acid	9,10,12,13-α- and β-Tetrahydroxy-octadecanoic acid; 9,10,12,13-γ- and δ-Tetrahydroxy-octadecanoic acid	$CH_3(CH_2)_4CHOHCHOHCH_2CHOHCHOH(CH_2)_7COOH$	(*189*)
Chaulmoogric acid	13-(2-Cyclopentenyl)-*n*-tridecanoic acid	$CH_2 \cdot CH_2 \cdot CH(CH_2)_{12}COOH$ (ring: CH_2–CH=CH–CH)	(*189, 341*)
Clupanodonic acid	4,8,12,15,19-Docosapentaenoic acid	$CH_3CH_2CH : CH(CH_2)_2CH : CHCH_2CH : CH(CH_2)_2CH : CH(CH_2)_2CH : CH(CH_2)_2COOH$	(*330*)
—	2-Phytenic acid	$(CH_3)_2CH(CH_2)_3CH(CH_3)(CH_2)_3CH(CH_3)(CH_2)_3C(CH_3) : CHCOOH$	(*198*)
—	2,6-Phytadienic acid	$(CH_3)_2CH(CH_2)_3CH(CH_3)(CH_2)_3C(CH_3) : CH(CH_2)_2C(CH_3) : CHCOOH$	(*198*)
—	10,13-Nonadecadienoic acid	$CH_3(CH_2)_4CH : CHCH_2CH : CH(CH_2)_8COOH$	(*198*)
—	11,14-Eicosadienoic acid	$CH_3(CH_2)_4CH : CHCH_2CH : CH(CH_2)_9COOH$	(*198*)

lowering the abnormally high water consumption. Linoleic acid, and arachidonic acid to a lesser extent, do promote growth, cure the dermatitis, and lower the water consumption. Inasmuch as these studies were made before the interrelations of the essential fatty acids, biotin and pyridoxine, in clearing the dermatitis, were recognized, it would seem that this problem should be re-investigated in light of our more recent knowledge.

It is believed by most workers that the key acid is arachidonic. TURPEINEN (*341*) suggested this some years ago, and many recent experiments have supported his contention. Although it is suggested by BURR (*49*), that linoleic acid is active *per se*, it has been shown that it is used in the body for the synthesis of arachidonic and possibly clupanodonic acids (*189, 316*). WIDMER and HOLMAN (*354*) have come to the same conclusion. When fat-deficient rats were given linoleate supplements for two months, an increase in arachidonate content appeared in the heart and liver. No change over the negative controls was noted in rats receiving stearate and oleate, although linolenate appeared to cause an increase in the poly-unsaturated acids in the organs just mentioned.

No absolute values can be quoted for the optimum dosage of essential acids for the rat or other animals. A maximum growth rate in curative experiments has been reported for dosages as low as 20 mg. (*233*) and as high as 100 mg. per day (*341*). Some data from our laboratory indicate that the requirement of linoleic acid for the growing male rat exceeds 50 mg. per day and that for the female is approximately 20 mg.

The *requirement for the essential acids* apparently varies not only with sex, but also with the foods being consumed concomitantly. EVANS and LEPKOVSKY (*120*) reported that the requirement was higher when other fatty acids and fats were present in the diet. Under such conditions, rats may exhibit poorer growth than is found with animals receiving a diet devoid of fat. The same phenomenon has been noted when rats were fed a regime containing a considerable proportion of elaidin (*315*) or hydrogenated coconut oil (*49*).

Although most of the experimental work on essential fatty acid-deficiency has been carried out on rats, a number of other species are also susceptible to the lack of dietary fat. WHITE *et al.* (*352*) reported a syndrome in mice deprived of the essential fatty acids, similar to that in rats. Although RUSSELL and his collaborators (*290*) could not demonstrate the deficiency in chickens, this failure may have been due to the presence of corn starch in the diet. This polysaccharide contains enough unextractable linoleic acid to prevent severe deficiency symptoms in the rat (*51, 119*).

While the hog is sufficiently resistant to fat deprivation so that the deficiency syndrome does not occur, ELLIS *et al.* (*102-105*) did find that the linoleic acid content of the lard dropped from a normal value of

Table 11. Arachidonate and Linoleate in Blood Serum of Normal Subjects and of Patients with Eczema (*45*).

Condition of patient	Number of samples (pooled)	% of total fatty acids	
		Arachidonate	Linoleate
Young patients with eczema	18	1,34	3,20
Adult patients with eczema	8	1,60	4,20
Control subjects......................	35	2,83	4,80
,, ,, ...	12	2,90	5,20

7–15% to 1,2% when the diet contained as little as 0,8% of extractable fat. No clear-cut evidence of the need of essential fatty acids has been adduced for the calf (*148*) or the cow (*141*, *235*), although the iodine numbers of the blood lipids were shown to fall on fat-free diets.

Dogs appear to be readily susceptible to lack of fat in their diet (*156*). The syndrome is quite similar to that in the rat. The dog is an excellent animal for the study of the distribution of the fatty acids in the lipid fractions since these respond markedly to the fat-free regime.

Although there is no convincing evidence for the requirement of the essential fatty acids in adult man, several workers have noted the need for these components in the baby (*146*, *153*). Infants kept on a low-fat diet developed an eczema which was readily cured by fat feeding.

Some workers have failed to show a relationship between the essential fatty acids and eczema (*143*, *332*) but the eczema studied by them may have been of a different type. Hansen and Burr (*155*) have reported that on a fat-deficient diet the iodine number of the blood serum lipids of rats falls concomitantly with the development of the scaly dermatitis of the skin. In human subjects, the eczematous condition is also usually accompanied by a lowering in serum linoleate and arachidonate (*45*, *61*, *127*, *128*, *154*, *156*) as well as by a drop of about 25% in the iodine number of the blood serum lipids (*61*). Finnerud *et al.* (*128*) report that, concomitantly with the cure of the eczema by the use of unsaturated fats, there is a rise in the linoleate but not in the arachidonate in the blood serum lipids. Table 11 shows the results of determinations of the polyunsaturated acids, by the bromine precipitation method, on pooled samples of lipid extracts of blood serum from eczematous and control patients.

Although it would seem quite probable that a relationship may exist between essential fatty acid deficiency and certain types of eczematous conditions in the baby and even in adult man, there is little other evidence of the necessity of the essential acids in the human subject. Brown *et al.* (*46*) report an experiment on one adult human subject who maintained himself for six months on a low-fat diet without any apparent harmful effects. Some change in respiratory metabolism similar to that found in the rat was noted and a 50% reduction in blood arachidonate and linoleate resulted. This was out of proportion to the drop in total blood lipids. Although one can hardly question the necessity of the essential acids in man, it is probable that the high reserves of these compounds in the body fat stores preclude the ready demonstration of a deficiency. Possibly, a wider study of this problem might reveal some individuals with a lower reserve of these acids who might respond with clear deficiency symptoms when fat is absent from their diet.

Other than the necessity for maintaining the normal nutrition of the skin, it is difficult at the present time to assign a function to the essential unsaturated acids. ENGEL (*108*) has suggested that the lipotropic action of choline is evident only when linoleic acid is present. The problem of the requirements of the essential fatty acids and the mechanism of their action in the animal organism is thus as yet largely unsolved.

5. Fat-Vitamin Interrelationships.

a) Thiamin and Fat. The interrelation of fat to vitamin requirements and to vitamin utilization in the animal organism has become an increasingly fruitful field of investigation for the biochemist. One of the early observations of such an interrelation was made by EVANS and LEPKOVSKY (*117, 125*) who found that, on diets adequate except in vitamin B_1, the onset of polyneuritic symptoms was markedly retarded on high-fat diets. Subsequently, the same investigators (*118*) pointed out that, when on a high-carbohydrate diet containing, presumably, an adequate amount of yeast as well as being a source of the unsaturated fatty acids, animals exhibited a favorable response to the feeding of additional fat. Many vegetable and animal fats were studied, including lard, coconut oil, synthetic coconut oil, partially hydrogenated cottonseed oil (Crisco), cottonseed oil, hydrogenated coconut oil, and perilla oil. All except certain high-melting fats (hydrogenated perilla oil, m.p. 67,5° C.) whose absorption was poor (*121*), possessed this thiamin-sparing action. The most effective fat in sparing vitamin B_1 has been reported to be coconut oil.

EVANS and LEPKOVSKY also observed that saturated fats, with iodine numbers of only 8, spare thiamin as effectively as a highly unsaturated oil having an iodine number of 187. Certain of the purified triglycerides, such as trimyristin and tricaprylin, were very effective, while tristearin was inactive (*122*). Neither the precise melting point nor the degree of saturation of the several natural fats seemed to influence this activity. Synthetic esters of the fatty acids were reported by SALMON and GOODMAN (*294*) as active, the maximum vitamin-sparing action being found with the C_8-fatty acid.

In 1940, McDONALD and McHENRY (*230*) pointed out that the presence of fat in the diet delayed the onset of bradycardia characteristic of thiamin deficiency in the rat. This finding was confirmed by BANERJI (*14*) who also observed that inclusion of various proportions of fat in thiamin-deficient diets counteracts the augmented excretion of bisulfite-binding substances which is typical of the polyneuritic syndrome. The sparing effect does not appear to be caused by any interaction between the fat and thiamin in the intestine, since it was also observed when the vitamin was administered parenterally (*124*). Furthermore, MELNICK and FIELD (*239*) showed by means of chemical analyses that thiamin was absent from fats which displayed a sparing action.

WHIPPLE and CHURCH (*350*) demonstrated that the fat content of the ration can influence bacterial synthesis of thiamin in the intestinal tract of the rat. When lard was incorporated into a thiamin-deficient diet containing sucrose as the carbohydrate, the feces of the rats showed the presence of thiamin; however, when the lard was omitted from the diet, the thiamin activity of the stools was lost. Carbohydrate intake in human subjects has been found to cause a diminution of the urinary output of thiamin which probably indicates that there is a greater need for this vitamin under such conditions. Despite the fact that in experimental animals a high-fat diet reduces the thiamin requirement (*9*, *117*, *324*), marked increases in fat consumption fail to augment the urinary excretion values (*54*, *275*) of this vitamin. SCHOENHEIMER and RITTENBERG (*305*) demonstrated, by means of tracer methods, in mice the dynamic condition not only of the phosphorus-containing lipids of kidney and liver but also of the deposited fat. The rapid turnover of fatty acids shows that there is a continuous conversion of carbohydrates into fatty acids under normal dietary conditions.

If one were able to exclude the evidence of WHIPPLE and CHURCH (*351*) indicating that thiamin has a function in fatty acid synthesis, the explanation by SCHOENHEIMER and RITTENBERG (*305*) would aid in the clarification of the relation of fat to its thiamin-sparing activity. The latter workers point out that, since in the case of most animals food is taken in only at intervals, the absorbed constituents of the diet are not oxidized immediately but must be deposited for a short period. The quantity of glycogen which may be stored in the organs is relatively small; most of the absorbed carbohydrate is therefore transformed immediately into fatty acids, which are laid down in the fat depots to be utilized for combustion during the post-absorptive periods. McHENRY and CORNETT (*225*) assume that fat is the fuel for basal metabolism, while carbohydrate may be the more ready source of energy for muscular work. If such were the case, the animal would require fat which could be supplied in the diet, or which could be biosynthesized from carbohydrates under the influence of thiamin and perhaps other members of the vitamin B group. With reduced fat intake the amount of thiamin required to bring about the synthesis of fat would be increased, and, conversely, the need for thiamin would be lessened by the presence of fat in the food.

b) Riboflavin and Fat. The interrelated effects of fats and riboflavin (vitamin B_2) are not clearcut, nor has as much work been done in this field as in the case of thiamin. In 1934, EVANS *et al.* (*126*) reported that fat has no sparing action on the riboflavin requirements of the rat. MANNERING and co-workers (*231, 232*) postulated that the riboflavin needed by rats is increased on high-fat diets.

ELVEHJEM (*107*) has suggested that the intestinal synthesis of riboflavin is inhibited when fat is substituted for dextrin in the diet. On the other hand, CZACZKES and GUGGENHEIM (*69*) found that rats maintained on a low-protein diet excrete so much riboflavin that they fail to maintain a functional level in their tissues, and ultimately die of ariboflavinosis despite the administration of comparatively large doses of this vitamin. An increase in the protein or fat content of such diets causes a reduction in the excretion of riboflavin, while decreasing the fat content raises the excretion. These effects are attributed to variations induced in the riboflavin levels of the tissue as influenced by the extent of synthesis of this vitamin by intestinal flora. Rats on high-protein and high-fat diets needed at least twice as much riboflavin as rats kept on a "normal" diet for the maintenance of the riboflavin level in the organs and in the urine. Unfortunately, neither gains-in-weight nor food consumptions were recorded in this experiment.

In an earlier report, POTTER *et al.* (*263*) were unable to show with puppies a similar increase in riboflavin requirements with high-fat diets.

Fat has a stimulatory effect on microbial growth as demonstrated in the microbiological methods for the determination of riboflavin (*21, 326*). In the case of the riboflavin assay, fat fails to act like the vitamin but its presence is responsible for greatly augmenting the microbiological response.

c) Pantothenic Acid and Fat. SALMON (*293*) observed that the activity of pantothenic acid is closely related to that of pyridoxine and the essential fatty acids. All three substances are required for the cure of the dermatitis caused by pantothenic acid deficiency. It has been suggested by STOTZ (*325*) that pantothenic acid in the form of a co-enzyme A, co-acetylase, may be involved in the fatty acid synthesis. LIPMANN and his co-workers (*215*) were the first to report the presence of such an enzyme which they called acetylase. It was suggested that acetylase plays a rôle in both pyruvate and acetate metabolism (*253, 254*). More recently, SOODAK and LIPMANN (*319*) have proved that co-enzyme A is concerned with the formation of acetoacetate from acetate. It would therefore appear probable that pantothenic acid is essential in the biosynthesis of fat. The requirements would then depend on the amount and type of fat in the diet.

Fat shows the same type of stimulating effect in the microbiological assay for pantothenic acid as is the case in the assay for riboflavin. Again, it acts atypically but it markedly increases the growth of microorganisms.

d) Pyridoxine (Vitamin B_6) and Fat. The sparing action of fat on the pyridoxine requirement has been demonstrated by many workers (*29, 176*). BIRCH and GYÖRGY (*29*) showed that a dermatitis, caused by a low pyridoxine intake on a 10% butter fat diet, could be cured by the addition of small amounts of lard. The addition of fat delayed or completely prevented the onset of the dermatitis caused by pyridoxine deficiency even to the time of death (*28*). In 1938, SALMON (*292*) reported that vitamin B_6 and fats supplement each other in curing dermatitis; the effectiveness of the fat was directly proportional to its degree of unsaturation.

Linoleic acid is the most effective unsaturated acid in curing pyridoxine deficiency, followed by arachidonic acid and then by linolenic acid. The same sequence is valid in curing uncomplicated scaly skin. SALMON (*292*) and others (*268, 304*) first reported the greater potency of linoleic acid over that of linolenic acid. Also, it was shown that the effectiveness of several natural fats was in proportion to their linoleic acid content. Thus, corn oil had greater activity than linseed oil; and other fats followed in the decreasing order of their linoleic acid content. Cod liver oil gave a negative result. RICHARDSON and his colleagues (*276*) reported that linoleic acid displays a greater curative action on dermatitis than arachidonic acid. The latter workers also suggest that the unsaturated acids do not replace pyridoxine but delay the onset of the skin symptoms. The decreasing antidermatitic effectiveness of butter fat when the fat becomes rancid is apparently associated with the destruction of its linoleic acid content (*303*).

The dermatitis produced by deficiency of the essential fatty acids is exaggerated when the non-essential fatty acids or fats are fed. Thus, SARMA *et al.* (*296*) showed that the addition of oleic acid to a ration containing suboptimal amounts of pyridoxal or pyridoxine resulted in growth inhibition. This could be counteracted by the administration of more vitamin B_6.

The interrelation of essential fatty acid and pyridoxine deficiencies has been emphasized recently by MEDES and co-workers (*237*). When the diet is lacking in both of these components, a relief from the deficiency can be obtained by the administration of either ethyl linoleate or pyridoxine. However, the growth response to pyridoxine alone was greater than that resulting from ethyl linoleate alone; the best results were obtained when both components were fed simultaneously.

McHENRY and GAVIN (*226*) have concluded that the role of pyridoxine in the mammalian organism includes that of fat synthesis from protein. Since UMBREIT and GUNSALUS (*343*) have proved that pyridoxine is involved in the enzymatic decarboxylation of amino acids, which is one of the essential steps in the production of fat from protein, the suggestion of McHENRY and GAVIN would seem to gain considerable support.

e) Niacin, Folic Acid, and Fat. According to ELVEHJEM (*107*), the intestinal synthesis of both niacin and folic acid is stimulated by the addition of fats to the diet. Butter fat gave somewhat better results than did corn oil.

Fat apparently also exerts a niacin-sparing action. The growth-promoting action of supplementary cystine when fed to rats on a casein diet as first observed by OSBORNE and MENDEL (*255 a*) only occurs on high-fat diets when adequate labile methyl groups are available. The effect of the fat is also partly ascribable to the fact that the nicotinic acid requirement is lower when fat is the chief source of calories than when the diet is predominantly a carbohydrate one (*293 a*). The necessity of dietary fat in order for the sparing effect of cystine on protein metabolism to be observed has been confirmed by GEIGER (*136 a*).

f) Biotin and Fat. In 1927, BOAS (*30*) reported that rats which had suffered severe deficiency in biotin, possess almost no stores of body fat. The administration of biotin to rats has been shown to increase the body fat as well as to cause a fatty infiltration in the liver (*136*). These fatty livers are characterized by their high cholesterol content and resemble those produced by feeding a fraction from beef liver (*226*). It was discovered by PAVCEK and SHULL (*260*) that rancid fats inactivate biotin but that tocopherol inhibits this process. Although such inactivated biotin is not utilizable for supporting the growth of *Lactobacillus casei*, it is still active for the growth of yeast. MELVILLE (*241*) believes that upon inactivation of biotin by rancid fats, biotin sulfoxide is formed.

There is much evidence from microbiological studies that biotin and fat are interrelated. A particular strain of *Lactobacillus* isolated from the caecal contents of rats was found to require oleic acid for growth (*353*). This organism could be transferred repeatedly on synthetic media but only when oleic acid was present. WILLIAMS and FIEGER (*357*) reported that oleic acid as well as lipid extracts of rice polishings possess a growth-stimulating action on *Lactobacillus casei* similar to biotin. Although a maximum acid production had taken place by the bacteria in the presence of lipids, no synthesis of biotin had occurred, as was demonstrated by the analysis of the acid-hydrolyzed cells. It is suggested that the bacteria are able to adapt themselves either to the biotin-free or to a biotin-low environment. These results would seem to contraindicate the idea that the fatty acids are concerned with biosynthesis of biotin in the intestine. This is in line with the results of OPPEL (*255*) who found that a relatively constant level of biotin excretion occurs in the urine (apparently from bacterial synthesis) and that superimposed on this are sudden variations due to biotin changes in the diet.

Although TRAGER (*336*) reported that the biotin-like activity of acid-hydrolyzed plasma could be traced to an ether-soluble component of the non-saponifiable residue, HOFMANN and AXELROD (*175*) were unable to confirm this claim. The biotin-like effect of human and beef plasma was found to occur exclusively in the saponifiable fraction. Esterification of the biologically active fraction with diazomethane, followed by distillation of the resulting methyl esters gave a fraction of high activity. Oleic acid on a weight basis was shown to possess about the same potency. The liquid acid fraction of plasma (containing oleic, linoleic, and arachidonic acids), obtained by lead soap separation, had a greater growth activity for the *Lactobacilli*, although the saturated acids were found to possess some synergistic action (*12*). More recently, TRAGER (*334*) has noted that the intramuscular injection of oleic acid does not have the same effect in reducing the severity of chick dermatitis (produced by raw egg white) as has biotin or the fat-soluble fraction from hydrolyzed plasma. However, WILLIAMS and FIEGER (*356*) have reported that the lipids from rice polishings which have a biotin-like activity in the case of *Lactobacillus casei* and *L. arabinosus*, will not reduce the biotin-deficiency symptoms in white leghorn chicks. Oleic acid can also substitute for biotin in the nutrient medium for mosquito larvae (*335*).

There have been several suggestions as to how biotin functions in fatty acid metabolism. TRAGER (*335*) concluded that biotin must serve in the synthesis of fatty compounds or of substances containing fatty acids, such as lecithin. The results of RUBIN and SCHEINER (*289*) indicate that the vitamin is primarily concerned with the synthesis of one or more fatty acids. In proof of this hypothesis, they have shown that such biotin antagonists as homobiotin, trishomobiotin, homobiotin sulfone, and trishomobiotin sulfone, have no effect on *Lactobacillus arabinosus* when oleic acid is used as a biotin substitute. This indicates that the function of biotin is to effect the synthesis of oleic acid since the biotin antagonists are inactive when the products of such synthesis are employed.

The biotin-like effect of oleic acid is shared by a number of related compounds. Thus, a number of configurational and structural isomers of oleic acid are relatively active. A decreased effectiveness occurs only when the fatty acid has a higher degree of unsaturation than that caused by the presence of one double bond. Hydroxylation or hydrogenation of the double bond results in complete inactivity. The relative biotin-like effect of a number of acids tested on *Lactobacillus arabinosus* by AXELROD *et al.* (*12*) is given in Table 12.

Table 12. The Biotin-like Effect of Oleic Acid Derivatives on the Growth of *Lactobacillus arabinosus* (*12*).

Compound	Formula	Activity in millimicrograms of biotin per mg
Oleic acid	$CH_3(CH_2)_7CH:CH(CH_2)_7COOH$ (*cis*)	6,0
Methyl oleate	$CH_3(CH_2)_7CH:CH(CH_2)_7COOCH_3$	3,2
Oleic acid amide	$CH_3(CH_2)_7CH:CH(CH_2)_7CONH_2$	6,3
Oleyl alcohol	$CH_3(CH_2)_7CH:CH(CH_2)_7CH_2OH$	$<0,05$
Linoleic acid	$CH_3(CH_2)_4CH:CHCH_2CH:CH(CH_2)_7COOH$	4,0
Linolenic acid	$CH_3CH_2CH:CHCH_2CH:CHCH_2CH:CH(CH_2)_7COOH$	1,7
Elaidic acid	$CH_3(CH_2)_7CH:CH(CH_2)_7COOH$ (*trans*)	1,0
Vaccenic acid	$CH_3(CH_2)_5CH:CH(CH_2)_9COOH$	1,2
Stearic acid	$CH_3(CH_2)_{16}COOH$	$<0,05$
Dihydroxystearic acid (94° C.)	$CH_3(CH_2)_7CHOHCHOH(CH_2)_7COOH$	$<0,05$
Dihydroxystearic acid (130,7° C.)	$CH_3(CH_2)_7CHOHCHOH(CH_2)_7COOH$	$<0,05$
Azelaic acid*	$HOOC(CH_2)_7COOH$	$<0,05$
Pelargonic acid	$CH_3(CH_2)_7COOH$	$<0,05$

g) Fat-soluble Vitamins and Fats. The quantity and type of the fat in the diet is of particular importance in assisting in the absorption of the fat-soluble vitamins from the gastro-intestinal tract. Thus, when an inert hydrocarbon such as mineral (paraffin) oil, which cannot be absorbed in the intestine, is fed, an interference in the utilization of critical doses of vitamin A (*288*), carotene (*68*), and vitamins D, E, and K (*7*) occurs. However, MELNICK and OSER (*240*) report that by means of a diet containing an inert hydrocarbon it was possible to add sufficient vitamins A and D to provide nutritionally adequate quantities of these vitamins.

MUELDER and KELLY (*247*) have shown that the absorption of vitamin A by depleted rats was aided by dietary fat, although its subsequent utilization was independent of the fat. However, it was reported by RUSSELL and his collaborators (*291*) that the absorption of preformed vitamin A by the chick is independent of the level of dietary fat, but that the absorption of carotene is markedly improved by the presence of fat in the ration. KRINGSTAD and LIE (*203*) noted that preformed vitamin A had about the same biological activity whether administered

* Aqueous solution employed for assay.

in alcohol or in ester form. However, there is some evidence that the biological value of the vitamin A alcohol and ester forms vary depending on the type and level of the fat in the diet or supplement (*349, 349 a*). Carotenoid utilization by the rat has been shown to depend upon the solvent in which it is administered (*211, 314*). FRAPS and MEINKE (*134*) in a comparative test based on liver storage and fecal excretion in rats, found that carotene, when fed in the form of vegetables, has considerably less biological value than when fed in butter fat. The same relative results were observed in man (*202*). GOTTLIEB *et al.* (*144*) found that on a diet rich in fat there was an increased requirement for tocopherols (vitamin E).

BOOTH *et al.* (*36*) reported that both vegetable and animal fats have some antirachitic effect on rats receiving a high-calcium-low-phosphorus rachitogenic diet, thus reducing the requirement for vitamin D. STRONG and CARPENTER (*326*) obtained avitaminosis K in rats when the animals were fed a diet high in dihydroxystearic acid (8%). The hydroxyacid may have interfered with the bacterial synthesis of vitamin K in the intestinal tract.

6. Relation of Fat to Protein Metabolism.

Until recently the general opinion has been prevalent that only carbohydrate can display a protein-sparing action while fat has been considered totally void of such activity. These conclusions have been based on the classical experiments of LUSK (*221*). Although there is still much evidence that carbohydrate plays a dominant rôle in controlling the level of protein metabolism, more recent work has indicated that fat may likewise take a specific part in preventing the breakdown of protein under certain conditions. SWANSON and her co-workers (*358*) were the first to demonstrate that nitrogen retention is better in rats on a protein-free diet when the calories are restricted, if fat is present in the diet, than when it is absent. Thus, the so-called "wear and tear quota"* of protein metabolism was maintained on a protein-free diet containing fat when the calories were restricted to 50% of that normally required. The level of nitrogen excretion was only moderately increased when the caloric intake was owered to 25%. On the other hand, when fat was absent from the diet, the nitrogen minimum was only maintained when the diet was fed at a level of 75% of the caloric requirement. At the 50% level, a marked increase of nitrogen excretion occurred and, when the diet was reduced to 25% of the caloric needs, the nitrogen excretion reached an extremely high level.

* "Wear and tear quota" or "nitrogen minimum" is the nitrogen excretion which is produced by ingestion of a protein-free, carbohydrate-rich diet over a prolonged period. The urinary nitrogen is reduced to the lowest possible value by this dietary procedure.

Similar conclusions may be deduced from other tests carried out by Swanson and her co-workers, in which methionine was introduced to produce a further sparing effect on protein metabolism. Their results are summarized in Table 13.

Table 13. Nitrogen Excretion of Rats Receiving Protein-free Diets which were Fat-free or Contained 10% Fat, Fed Alone or with Methionine at 100, 75, 50 and 25% of the Required Caloric Levels (358).

Per cent caloric requirement	Diet	Protein-free diet			Protein-free diet + methionine		
		Basal N-excretion (mg.)	Experimental N-excretion (mg.)	Difference (mg.)	Basal N-excretion (mg.)	Experimental N-excretion (mg.)	Difference (mg.)
100	Fat-free ...	280	185	— 95	295	177	—118
	10% fat...	295	210	— 85	315	188	—127
75	Fat-free ...	272	211	— 61	270	204	— 66
	10% fat...	310	263	— 47	310	215	— 95
50	Fat-free ...	283	594	+311	257	266	+ 9
	10% fat...	265	289	+ 24	296	256	— 42
25	Fat-free ...	249	818	+569	281	577	+296
	10% fat...	290	544	+254	329	523	+194

The recent report of Samuels, Gilmore, and Reinecke (295) who made use of an entirely different technique has offered confirmation of the sparing action of fat on protein metabolism. When rats were fed over a period of four weeks on diets containing 80% of the calories furnished by carbohydrate, fat, or protein, it was found that the lowest nitrogen excretion obtains in those animals on the high-fat diet. Moreover, in a subsequent fasting period which continued until the death of the animals, it was found that the longest survival occurred in the group which had received the fat diet. This is probably to be ascribed to the greater sparing action of fat on the nitrogen metabolism, and the prolongation of the period before the so-called pre-mortal rise in nitrogen appears. These results are summarized in Table 14.

Table 14. Average Nitrogen Excretion of Rats on Special Diets and During Subsequent Fasting Periods with Voluntary Activity and in Rats Undergoing Exhaustive Exercise (295).

Diet	Nitrogen excretion in gram/day/rat*				Nitrogen excreted during work mg./1000 revolutions
	On diet	0—4 days of fasting	6—11 days of fasting	13 day-death in fasting period	
Carbohydrate .	0,3375 ± 0,0053	0,1770 ± 0,0068	0,1318 ± 0,0111	0,2286 ± 0,0131	0,0465 ± 0,0058
Protein...	1,369 ± 0,038	0,2547 ± 0,0314	0,1427 ± 0,0392	—	0,0521 ± 0,0095
Fat......	0,3165 ± 0,0080	0,1104 ± 0,0057	0,1004 ± 0,0094	0,1407 ± 0,0272	0,0341 ± 0,0044

* Including the standard error of the mean.

The improvement in protein metabolism by the inclusion of fat in the diet is also emphasized in the earlier work of FRENCH, BLACK, and SWIFT (*135*). They demonstrated that more rapid and efficient gains-in-weight occurred in rats on low-protein diets when the diets contained 30% of fat than when only 2% was present.

That the protein-sparing action of fats may likewise have application to man is indicated by the recent work of SCHWIMMER and MCGAVACK (*308*). Low nitrogen excretion was maintained in human subjects only when the protein-free diets, fed at considerably below the levels required for caloric equilibrium, contained fat.

A possible explanation for the behavior of fat on protein metabolism may be obtained from the recent results of PEARSON and PANZER (*261*). These workers found that the loss of phenylalanine, valine, lysine, and methionine in the urine and feces of growing rats was less when the dietary regimen contained corn oil than when no fat was present. Although the variations in the quantities of the amino acids excreted were small, the differences in figures on the fat-free diets and fat-containing diets were statistically significant in a number of cases. Similar changes were shown to occur in mature rats when the fecal and urinary excretion of the four amino acids mentioned was compared on diets containing no fat and 8% corn oil; however, these differences are less marked in favor of the fat ration. We may expect further developments in this field in the near future.

7. Fats in Relation to Specific Dynamic Action of Foodstuffs.

Fats exert a peculiar effect on the specific dynamic action* of carbohydrate and protein, when fed with either or both of these foodstuffs. Previous to the recent studies (*130*), it had been generally assumed that the specific dynamic effects of foodstuffs when fed in combination with each other are strictly additive. Some years ago, MURLIN and LUSK (*248*) had demonstrated on dogs that the extra heat produced when glycine (20 grams), glucose (70 grams), and fat (75 grams) were fed together was practically identical with the value calculated from the sum of the three individual specific dynamic effects as determined separately. Although the results of MURLIN and LUSK are quite definite, it is possible that an erroneous conclusion may have been reached since the calculations were based only on a two-hour period (at which time the specific dynamic effects were believed to be at the maximum) rather than for the whole period during which the action had continued.

* "Specific dynamic action" refers to the extra heat which results over a period of hours following the ingestion of food over that produced in basal metabolism. The extent of increase varies with the type of foodstuff, the amount ingested, "nutritional state" and the environmental conditions.

FORBES and SWIFT (*130*) have, by the use of a new method for the study of specific dynamic effects, recently come to the opposite conclusion. They suggest that the advantage of their new procedure as compared with the measurement of dynamic effects of single test meals (fed to animals during a post-absorptive state) is that they are more representative of nutritive practice. The new method does not cause a confusion with the dynamic effects of catabolized body nutrients. Finally, the results are more significant and probably more accurate since they are measured as differences between the amounts of heat produced at established levels of metabolism, without need for concern as to the beginning, the maximum, or the end of specific dynamic effects of a single meal.

The determination were carried out for a period of seven and one-half hours. They were started one and one-half hours after the morning feeding. When beef protein in an amount corresponding to 1000 calories was fed separately, the specific dynamic effect amounted to 323 calories. With cerelose (commercial glucose) the specific dynamic effect was 202 calories and with lard it was only 160 calories when a quantity of either of these foodstuffs equivalent to 1000 calories was given.

When cerelose and protein were fed together, the response agreed quite closely with that expected as calculated from the specific dynamic effect of that quantity of each foodstuff when fed separately. On the other hand, whenever lard was included in the mixture tested, the calculated value was always considerably greater than that experimentally determined. For example, when beef protein and lard were fed together, only 113 extra calories were produced compared with a computed value of 244 calories. This sparing effect of fat on specific dynamic action was observed not only when it was combined with protein but also mixed with cerelose or with a mixture of cerelose and protein. No such deviation from the expected values was noted when carbohydrate and protein were combined. It is therefore evident that fat is the only foodstuff having the specific effect described. FORBES and SWIFT (*130*) termed these phenomena the "associative dynamic effect". A summary of their experiments is given in Figure 1.

It is therefore evident that the use of fat instead of carbohydrate results in a considerable economy of food utilization, since less calories are lost as a result of the specific dynamic action. According to FORBES and his co-workers, it is not necessary to diminish the protein content of a diet during hot weather in order to obtain a low-heat increment, but rather one needs only to substitute fat for a part of the carbohydrate. There is a physiological economy in the high-fat diet of working people in certain tropical countries such as Brazil.

These results have recently been reviewed by SWIFT and BLACK (*329*).

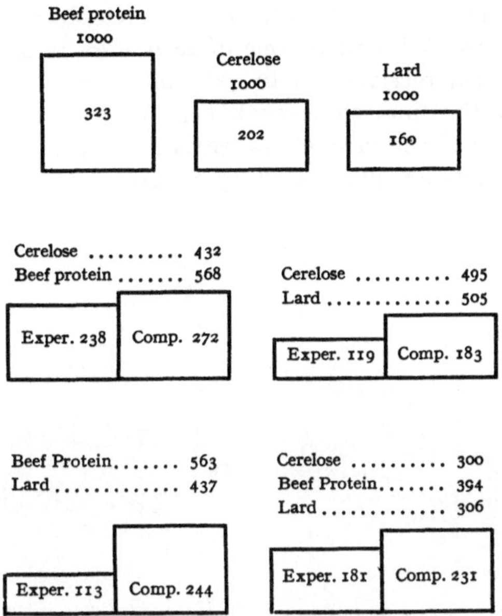

Fig. 1. Dynamic effects per 1000 calories of gross energy of nutrients as affected by nutrient combination. [J. Nutrit. 27, 453 (1944)] (*130*).

8. Fats in Relation to Growth.

The beneficial effects of diets containing generous quantities of fat are also apparent from the more usual procedures for determining nutritional values. For example, on the basis of growth tests on weanling rats, it has been shown by DEUEL *et al.* (*91*) that the feeding of diets containing 20 to 40% of fat results in superior growth over that obtained with rats, fed diets containing no appreciable quantities of fat, with or without the presence of essential fatty acids; even when 5 or 10% of fat was incorporated in the diet, inferior growth was observed. The superior growth on the high-fat diets is partly to be ascribed to the greater intake of energy. A similar result has also been reported by HOAGLAND and SNIDER (*174*).

The effect of fat on growth may be partly attributed to a greater efficiency of utilization, as has been reported by FORBES and his coworkers (*131, 132*). The comparative effects of different levels of fat in the diet is illustrated in Figure 2 which refers to an eighteen-week period with the cottonseed oil tests and to a twelve-week period with the margarine fat tests.

The improved nutrition which results from the inclusion of generous amounts of fat in the diet is likewise indicated in experiments on rats which were fed for a period of twelve weeks on calories so restricted that the weanling animals failed to gain any appreciable weight (*302*). This restricted period was followed by one of ***ad libitum*** feeding during which those animals which received 10 to 40% of fat in the diet grew best, especially during the final three-week period. The rats on the highest fat levels still showed the greatest potentiality in growth during the last period. These results are illustrated by Figure 3.

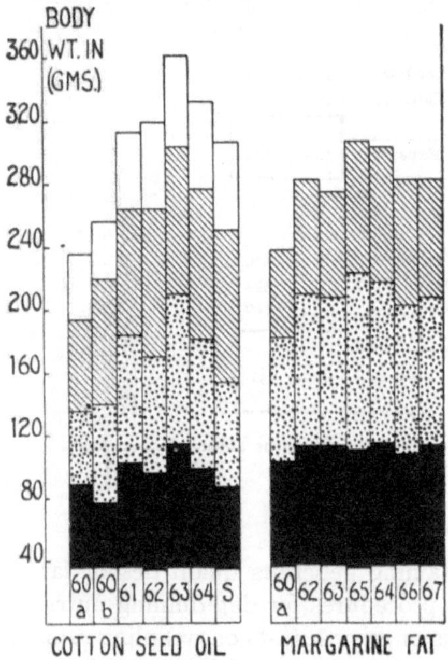

Figure 2. The average body weight of weanling male rats at the start (lower blank space), after 3 weeks (solid black), after 6 weeks (stippled), after 12 weeks (cross-line), and after 18 weeks (top blank space). The figures in the lower blank are the diet numbers. [J. Nutrit. *33*, 569 (1947).]

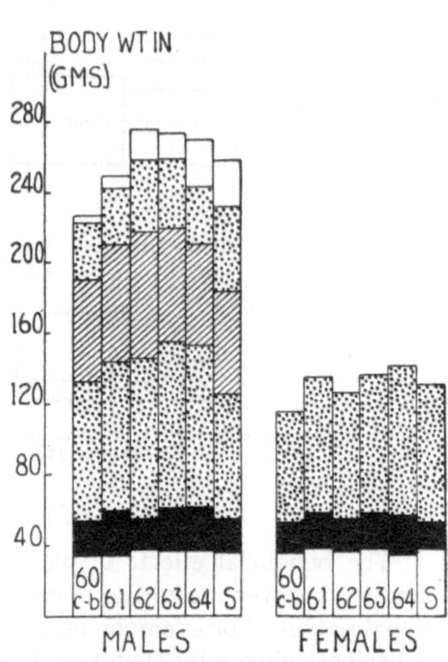

Figure 3. The average body weight of weanling rats at the start (lower blank space), after 12 weeks on restricted calories (solid black), after 3 weeks ad libitum (stippled), after 6 weeks ad libitum (lined), after 9 weeks (upper stippled), and after 12 weeks (upper blank space). [J. Nutrit. *33*, 583 (1947).]

Essentially similar data have been reported by SCHEER ***et al.*** (*300*) for high-fat diets when fed in conjunction with an experiment where young adult rats were first subjected to severe caloric restrictions. Under such conditions the loss-in-weight in the rats receiving the high-fat diet as compared with those receiving the fat-free diet was found to be less rapid. A smaller mortality likewise obtained when a liberal proportion of fat was present in the diet as contrasted with a fat-free diet. During the period following caloric restriction, when ***ad libitum*** feeding was employed, the recovery was more rapid on diets containing 20 to 40% of fat than on a fat-free regime.

A number of other investigators have confirmed the fact that the inclusion of a considerable quantity of fat in the diet improves growth.

These reports include a second one of FORBES *et al.* (*133*) as well as a study of FRENCH, BLACK, and SWIFT (*135*). In the latter paper, it was shown that more rapid and efficient gains in weight resulted on low-protein diets when 30% of the fat was included than when only 2% was present. Most recently, PEARSON and PANZER (*261*) have shown that rats fed an *ad libitum* diet containing 8% of corn oil or lard gained 29% more than those kept on a diet containing no added fats but supplemented with ethyl linoleate. However, there was no significant difference in the gains-in-weight of the rats on the several diets when the intake of protein and energy were equalized.

9. Fats in Relation to Sexual Maturity.

Although the time at which sexual maturity is reached has ordinarily not been considered as one of the indices of nutritional value, the results are indirectly a register of such activity since they are closely bound up with growth. The time at which the female rat becomes sexually mature can be determined with reasonable precision from the opening of the vagina. Whereas this phenomenon did not occur in the normal rat on the fat-free diet until it was eighty-one days of age, it took place as early as seventy days in rats receiving 5% fat, and at sixty-five days in the animals eating a 40% fat diet (*91*). On the other hand. female rats which had been so restricted in calories that they failed to grow appreciably for twelve weeks after weaning (105 days of age), were in no case sexually mature. When *ad libitum* feeding was instituted, the animals grew rapidly. The vaginas opened in fifteen days in the rats receiving the fat-free regime, while only eight days were needed for this physiological change in the rats on a 20% fat diet (*302*). These results are summarized in Table 15.

Table 15. The Effect of Fat Level of the Diet on the Time When Maturity Develops in the Female Rat (*302*).

Previous dietary regime	Average time in days to maturity after start of *ad libitum* feeding with diets containing				
	0% fat	5% fat	10% fat	20% fat	40% fat
Normal rats, 3 weeks old at start, on diets *ad libitum* from weaning (*91*)	60,2	49,0	45,5	44,8	44,1
Restricted rats, 15 weeks old at start, after 12 weeks on restricted calories (*302*)	14,7	11,9	10,5	7,7	9,8

10. Fats in Relation to Pregnancy and Lactation.

Nutritionists are unanimous in the opinion that pregnancy affords a better index of nutritional value than does growth alone. Lactation is known to be a still more critical method of establishing nutritive value than is pregnancy. Although no differences in reproductive and lactation performance could be noted in normal rats receiving diets which included 5 to 40% of fat, such responses were found in all cases to be markedly superior to those obtained on a fat-free diet (*91*). On the other hand, in the case of the female rats recovering from a period of undernutrition, the total weight of the litters was highest at twenty-one days in those females on the 40% fat diet. The results of the groups receiving 5 to 20% of fat did not show much variation but in these instances, the weight of the offspring was considerably higher than for the rats receiving a fat-free diet.

Only 22% of the females which were kept on a fat-free diet were successful in casting litters, contrasted with variations from 55 to 90% in the animals receiving the diet containing fat. Moreover, the individual baby rats weighed on an average only 17,3 grams, at twenty-one days of age in those cases where their mothers had been on a fat-free diet, while the weight of the young varied between 27,3 and 36,9 grams in the cases where their mothers had received appreciable amounts of fat during the pregnancy period.

However, in recent tests (*302*) in which the animals had been subjected to a twelve-week period of undernutrition followed by *ad libitum* feeding, the best reproductive and lactation performances were observed in those animals which received a diet containing 40% fat. Thus, the total litter weight at twenty-one days was only 152 grams in the animals receiving the fat-free diet containing methyl linoleate compared with a value of 283 grams for the animals receiving 40% fat in the diet. These data are summarized in Table 16.

Table 16. Reproductive Performance of Female Rats Recovering from a Period of Undernutrition (*302*).

Diet		Rats per litter	Average litter weight		Average rat weight	
Number	Fat content (per cent)		3 days (g.)	21 days (g.)	3 days (g.)	21 days (g.)
60b	0	5,3	47,6	152	8,9	26,1
61	5	6,8	47,7	219	7,1	32,4
62	10	7,7	64,3	167	8,4	27,8
Stock	14	7,0	53,8	175	7,7	23,3
63	20	8,5	60,8	230	7,2	31,8
64	40	8,9	72,3	283	8,1	30,6

SCHEER and co-workers (*300*) likewise offered confirmation to the better effect on pregnancy to be obtained with high-fat diets. These tests were carried out on female rats, following caloric restriction which was so severe that the animals lost approximately half of their weight during a period of twelve weeks. Breeding tests were conducted immediately following the recovery on *ad libitum* feeding. Whereas no litters were born to the rats which received 5% of fat in the diet, only one-fifth of those receiving no fat or 10% were fertile while four-fifths of those on the 20% and 40% levels cast litters. However, in none of these cases were the mothers able to raise their litters successfully to twenty-one days.

It would appear that the most satisfactory results on pregnancy and lactation performance are given by those animals which receive fairly generous amounts of fat in the diet.

11. Fats in Relation to Working Capacity and Survival.

Although capacity for work has not ordinarily been used as a method for evaluation of the nutritional value of diets, the length of survival under adverse nutritional conditions has sometimes been so employed. Both of these indices suggest that high-fat diets have a better nutritional status than do fat-free ones. Although KROGH and LINDHARD (*204*) reported some time ago that work could be accomplished approximately 10% more efficiently if carbohydrate were the source of calories than when fat served this same purpose, their results appeared to be contradictory to the exhaustive tests made earlier by ANDERSON and LUSK (*4*). The latter workers reported that, from a caloric standpoint, dogs ran equally efficiently in the respiration calorimeter irrespective of whether they had a respiratory quotient of 1,0, indicating the utilization of carbohydrate, or whether they had been previously fasting for approximately twenty days and showed from their respiratory quotient that they were using fat almost exclusively as a source of energy.

Table 17. The Effect of Fat Level in the Diet on Working Capacity of Rats as Determined by a Swimming Test (*301*).
(Figures in parentheses represent the average number of experiments.)

Sex	Weeks on diet	Duration of swim in seconds on different diets containing the following per cent. of fat								
		60 a 0%	60 b 0% *	61 5%	62 10%	Stock 14%	63 20%	65 30%	64 40%	66 50%
Cottonseed oil diets.										
Female	6	512 (5)	580 (5)	602 (6)	777 (5)	1025 (4)	1062 (5)	—	922 (3)	—
Male	6	805 (5)	692 (5)	1019 (5)	921 (5)	955 (4)	1223 (5)	—	1143 (5)	—
Male	12	645 (20)	637 (19)	769 (19)	850 (17)	1084 (19)	846 (19)	—	1068 (15)	—
Margarine fat diets.										
Male	12	485 (23)	—	—	806 (10)	1218 (8)	1174 (8)	1065 (25)	876 (4)	1122 (10)

* Contained linoleic acid.

More recently, DEUEL *et al.* (*91*) found that rats maintained on diets containing 5 to 50% fat were able to swim for markedly longer periods than those which had received a diet containing no fat. That this was not due to differences in buoyancy of their bodies resulting from variations in the fat content, was indicated by the fact that the animals previous to the tests had been adjusted to similar specific gravity. Table 17 which illustrates these results is given below.

The average duration of the swimming period for animals which received no fat in the diet varied from 485 seconds to 805 seconds, whereas the results on animals receiving 20% of fat in diet gave values varying from 846 to 1223 seconds. The tests were carried out by a special procedure worked out by SCHEER and his collaborators (*301*).

The greater physical capacity of rats held on a fat regime is beautifully illustrated in the tests of SAMUELS, GILMORE, and REINECKE (*295*). These investigators fed rats on diets containing 80% of the calories from fat, protein, or carbohydrate, for a period of four weeks prior to the experiments which were made during subsequent fasting. Not only was the nitrogen excretion lowest in the previously fat-fed animals, but the spontaneous and forced activity as well as the period of survival were also greatest in the same group.

Table 18. Effect of a Previous Diet Consisting Largely of Carbohydrate, Fat, or Protein on Spontaneous and Forced Activity, Nitrogen Excretion and Period of Survival of Rats During Fasting (*295*). (Figures in parentheses designate the number of rats in each group. The values include the standard error of the mean.)

Previous diet	Spontaneous activity for 2–4 days, 9–11 days and 16–18 days of fasting	Forced activity	Survival time
	Revolutions of cages	Revolutions of cages	days
Fat	6730 ± 915 (12) 8800 ± 1020 (13) 7960 ± 738 (7)	19700 ± 1780 (12)	18,3 ± 0,44 (12)
Carbohydrate	4040 ± 411 (16) 8660 ± 1017 (14) 8650 (3)	13800 ± 1630 (14)	15,5 ± 0,46 (13)
Protein	3600 ± 262 (7) 7220 ± 500 (4) (0)	6465 ± 1650 (9)	10,2 ± 1,68 (9)

GILMORE and SAMUELS (*142*) have shown that a difference in glucose utilization in the diaphragmatic muscle obtains in the tissues from those animals previously fed on carbohydrate and from those which had received fat. This is in line with an earlier report by LUNDBAEK and STEVENSON (*219*), in which it was demonstrated that the voluntary muscle of a carbohydrate-fed rat utilized glucose *in vitro* at twice the rate of that of the fat-fed animal, when both were exposed to the same environment.

The fact that the period of survival of rats previously fed on a high-fat diet is longer than the corresponding period is the case of those previously kept on a high-carbohydrate diet, is to be traced to a difference in the metabolic pattern; this was indicated by earlier work of SAMUELS and his collaborators (*280*). Thus, ROBERTS and SAMUELS (*281, 282*) found that rats which had been accustomed to the high-fat regimen had a markedly lower susceptibility to injected insulin than did those which had been on a high-carbohydrate diet. These authors ascribed the phenomenon to the more prolonged retention of liver glycogen in fat-fed animals. It was also suggested (*283*) that animals which had been fed on a diet in which 95% of the calories originate from fat, develop a sparing action for carbohydrate. This hypothesis is supported by the fact that when the livers are removed from such rats, the hepatectomized animals live longer and are free from hypoglycemic symptoms for a more prolonged interval than are rats previously given a high-carbohydrate diet.

12. Fats as Antithyrotoxic Agents.

When an excess of thyroid hormone is given to rats or other animals, the following readily reproducible toxic symptoms develop. Growth is suppressed below normal. An increased basal metabolism obtains, as well as an inhibition in ovarian development. The kidneys, adrenal glands and heart are increased to an abnormal size.

ABELIN and his co-workers (*1, 1 a, 1 b*) were the first to report the beneficial effects of dietary fat on hyperthyroid animals. High fat diets were shown to counteract the increased basal metabolic rate, as well as the reduction in liver glycogen which follows the administration of the thyroid hormone. BERG (*24 a*) has corroborated the protective action of fat in lowering the basal metabolic rate in dogs, while a number of other workers have confirmed ABELIN's observations on rats (*174 a, 247 a, 319 a*).

There is some evidence that the protective effect of fat may be related to the presence of essential fatty acids in this foodstuff. Sodium oleate and linoleic acid are effective in preventing a rise in metabolism (*201 a*) while GUERRA (*147 a*) correlated the unsaturation of the fat with such protective action. The results of ZAIN (*361, 362*) confirm this general claim. He found that linoleic acid but not stearic acid prevents losses of liver glycogen after massive doses of thyroid extracts. Oleic acid occupies an intermediate position.

ERSHOFF (*108 a*) recently demonstrated that fats also exert a marked beneficial influence in partially checking the growth inhibition caused by thyroid administration to rats. Olive oil, peanut oil, cottonseed oil, wheat germ oil, corn oil, soybean oil, hydrogenated cottonseed oil, and lard were all shown to cause a growth significantly greater than that of rats kept on the fat-free diet. The investigator mentioned failed to find any protective action of fat on the inhibition of ovarian development or on the hypertrophy of the kidneys, adrenal glands and heart.

In an analysis of the causes which might account for the protective action of fat in conditions of thyroid hyperactivity, ERSHOFF (*108 a*) considered the fact that the fat may furnish the essential fatty acids and thus alleviates a deficiency arising from a increased fatty acid requirement. Since a practically fat-free liver preparation also has a beneficial effect in thyrotoxicosis, ERSHOFF does not believe that the above fatty acid deficiency theory is a plausible one. Another suggestion is that fat may stimulate the synthesis of an antithyrotoxic factor by the intestinal flora. The favorable effect (*351 a*) on the intestinal thiamin synthesis in the rat would seem to support this theory.

13. Optimum Fat Levels in the Diet.

The beneficial effects of fat on nutrition are sufficiently varied in character to make one question the concept that fats are merely optional components of the diet. One is almost forced to the conclusion that this foodstuff should be regarded as an obligatory constituent.

The next question which arises is what level of the fat intake is to be considered as the optimum? In the case of the rat, most of the nutritional indices point to the fact that somewhat better results are obtained when the diet includes 20 to 40% fat than when smaller fractions are present. However, in many instances, the results obtained with diets containing only 10% fat are quite satisfactory.

Another question which may well be asked is whether all types of fats can fulfill this function equally well. At the present time there are not sufficient experimental data available to answer this question or to decide whether the optimum level varies with the specific composition of the fat ingested. The results of growth experiments on rats indicate that both cottonseed oil and margarine give the best results at levels of 20 to 40%. It also seems probable that the beneficial effect of fats is not exclusively a function of their content of essential fatty acids. In recent experiments (*74*) it has been demonstrated that female rats previously depleted on a fat-free diet and then given 20 mgs. of linoleic acid per day, along with the fat-free diet, until their growth curve plateaued, showed no further increase in growth when the daily linoleic acid supplement was raised to 60 mgs. However, when the animals were transferred to a diet containing 10% cottonseed oil, there appeared a marked growth-stimulating effect which lasted over a number of weeks. These results might indicate that the fat possesses some beneficial action which is apparently not immediately traceable to the unsaturated fatty acids or to tocopherol.

There is no definite evidence as to what extent the results obtained in rats may be translated into terms of human nutrition. It has been estimated on the basis of dietary studies that the average consumption of fats in the United States in approximately 125 grams daily, representing 33% of the total calories [BRANDT (*42*)]. The same author who made these estimates indicates that the daily fat intake of the British people was 106 grams before the first World War, while that of the Japanese prior to 1930 amounted to only 29 grams. Fat consumption in England had risen to 124 grams per day in 1934. On the other hand, SHEN (*309*) estimates that the soldiers of South China ate fat to the extent of only 3% of their total caloric intake, which is less than one-tenth of the 40% which has been calculated by HOWE (*188*) as the consumption of the average American soldier in World War II. Although the normal height and weight relationships of the Chinese soldier were maintained on the low-fat ration over several years, this period is obviously too short for any critical evaluation of the effect of a possible defective diet on human subjects.

Much experimental work still needs to be done on man to determine the tolerance for fat, as well as factors which make it possible for human subjects to consume large quantities of this foodstuff without the usual gastric distress. Neither do we have any information on man as to the possible effect of other substances, fed concomitantly with the fat, on the utilization of the latter.

III. Comparison of the Nutritional Value of Animal and Vegetable Fats.

1. General Remarks.

Since fats appear to have a prime nutritional function, it is of considerable importance to ascertain whether the need for them can be met equally as well by vegetable as by animal fats. From the chemical standpoint, vegetable and animal fats contain in general the same types of fatty acids. However, stearic acid is found in larger proportion in the animal fats; and the unsaturated acids, such as linoleic and linolenic, more frequently make up considerable portions of the vegetable fats. There is some indication, also, that the manner of distribution of the fatty acids in the triglyceride molecules shows a variation between fats originating from the vegetable and animal kingdoms. Thus, HILDITCH (*160*) has pointed out that in most vegetable fats, the acids are distributed according to the rule of "even distribution" in contrast to a random arrangement. On the other hand, the saturated fatty acids occur in many of the animal fats according to a chance or random distribution (p. 18), although HILDITCH calls attention to the fact that such an arrangement is not valid for the unsaturated acids. In view of the fact that the animal body is able to hydrolyze the triglyceride molecules and rearrange the fatty acids in any of the triglycerides which are resynthesized, no profound differences in the metabolism of vegetable and animal fats would be expected.

Differences in nutritive value of the respective fats may be attributed to variations in their content of the fat-soluble vitamins. Whereas vitamin A as such is never found in vegetable fats, it is present in some animal fats including butter, although many animal fats such as lard are completely devoid of this component. On the other hand, the provitamin A, β-carotene is present in some animal fats (*363a, 365*) and butter fat (*141c*) while a second provitamin A, cryptoxanthin has been shown to occur in egg yolk (*141b*) and butter (*141a*). However, in general, the provitamins A are found in animal fats less frequently, while carotene may make up a marked portion of some vegetable fats such as palm oil. Vitamin D is found exclusively in animal fats but it is only present in extremely small amounts in all sources except in fish liver oils. Insofar as the important fat-soluble vitamin, vitamin E (tocopherol) is concerned, its several forms are present in much greater amounts in vegetable than in animal fats. The same statement is valid concerning the distribution of the "essential" unsaturated acids.

2. The Comparative Amounts of Vitamins and Essential Fatty Acids in Animal and Vegetable Fats.

a) Vitamin A and Provitamin A. The commonest source of preformed vitamin A in foods is butter. Although the A-content in this fat varies considerably with season, breed of the animal, and diet, it has recently been found on the basis of a cooperative study (*8*) that the average vitamin A content of butter produced in the United States somewhat exceeds 15000 I. U. per pound. This figure, however, includes both vitamin A as such and the varying amounts of carotene which are present in milk fat. The highest vitamin A concentrations are found in fish liver oils.

Cod liver oil *(Gadus morrhua)* which has always been regarded as an excellent source of this constituent contains only a relatively small concentration, namely, about 600 I. U. per gram, whereas many of the common fish liver oils such as those from the mackerel *(Scomber scombrus)*, barracuda *(Sphyraena argentea)*, yellowtail *(Seriola dorsalis)*, and the yellowfin tuna *(Neothunnus macropterus)* and others contain anywhere between 30000 and 80000 units of vitamin A per gram (*287*). Certain other fishes carry especially high quantities of vitamin A in the liver oils.

These are the bonito *(Sarda chiliensis)*, 120000 I. U.; cabrilla pinta *(Epinephelus analogus)*, 170000 I. U.; swordfish *(Xiphias gladius)*, 250000 I. U.; ishinagi *(Stereolepis ishinagi)*, 300000 I. U.; Black Sea bass *(Stereolepis gigas)*, 600000 I. U.; and several varieties of shark, especially the soupfin, *Galeorhinus zyopterus* whose vitamin A content may be even higher [640000 I. U. (*279*) to 800000 I. U. (*328*)].

Vitamin A as such is found also in egg yolk as well as in the lungs and kidneys, although that contained in the corpora lutea is probably largely present in the form of carotene. In the retina, vitamin A occurs in an especially high concentration presumably as vitamin A aldehyde, which can be reduced to the vitamin proper.

Although vitamin A as such is not found in vegetables, provitamin A is present in an extremely large proportion in such sources as turnip greens, spinach, broccoli, as well as in such yellow, orange, or red colored foods as squash, carrots, apricots, persimmons, and also in tomatoes. Ordinarily, provitamin A is accompanied in these sources by a minimum quantity of fat.

β-Carotene is biologically the most effective of the provitamins A, while α-carotene, and cryptoxanthin are only about half as active (*90*, *96*). Although two samples of γ-carotene prepared from *Mimulus* and from *Pyracantha angustifolia* both had potencies of approximately 25% of that of β-carotene (*86*, *89*), a recent sample of all-*trans*-γ-carotene, prepared by isomerization of pro-γ-carotene, was found to have a biological value of almost half that of all-*trans*-β-carotene (*144 a*, *364*).

Pro-γ-carotene, a naturally occuring poly*cis* compound not used for human food but consumed by birds and some other animals (porcupine) has about 40% the potency of β-carotene in rats (*89*).

In commercial vegetable oils, red palm oil has the highest content of β-carotene. Buckley (*47*) has reported carotene levels equivalent to 1900 I. U. of vitamin A per gram of red palm oil obtained from the ripe fruits, while Kaufmann (*199*) found carotene values varying between 0,20 and 0,05%, in which case the maximum level corresponds to a vitamin A potency of 3000 I. U. per gram. Since it was formerly believed that the average vitamin A-content in butter approximated 9000 I. U. per pound (20 I. U. per gram), the standard for fortification of margarine in the United States as set by the Food and Drug Administration in 1941 was based on this figure (*244*). However, because of the recent survey in the United States in which 15000 I. U. of vitamin A per pound (33 I. U. per gram) was indicated as the average level in butter (*8*), practically all margarine sold in this country now contains that higher quantity.

b) Vitamin D. Vitamin D is found only rarely in common animal fats and never in the vegetable fats. Small quantities, however, may be present in milk fats and in butter. On the other hand, exceedingly concentrated sources of this vitamin are present in some fish liver oils which are widely employed as practical sources of this material for therapeutic use. The vitamin D contents of fish liver oils vary from a value of zero in the sturgeon liver oil to approximately 40000 I. U. per gram in the liver oil of the bluefin tuna *(Thunnus thynnus)*. Swordfish liver oil *(Xiphias gladius)* and yellowfin tuna liver oil *(Neothunnus macropterus)* containing 10000 I. U. per gram are good sources of vitamin D. Cod liver oil *(Gadus morrhua)*, which has long been regarded as the standard medicine for rickets, contains only 100 I. U. per gram and is twenty-third in a list of fish liver oils arranged in the decreasing order of the respective vitamin D potencies (*27*).

However, since it had been demonstrated recently that vitamin D can be formed by irradiation of the provitamins D which are present in various foodstuffs, the occurrence of preformed vitamin D has become of less importance. Thus, 7-dehydrocholesterol in animal fats and ergosterol, and dihydroergosterol in yeast, as well as dihydrositosterol in numerous vegetable fats, are active as provitamins D.

c) Vitamin E. A number of compounds display vitamin E activity. From the standpoint of the anti-sterility effect, α-tocopherol (5,7,8-trimethyltocol), β-tocopherol (5,8-dimethyltocol), γ-tocopherol (7,8-dimethyltocol) and possibly also δ-tocopherol (8-methyltocol) are effective in that decreasing order, while their potency in terms of anti-oxidant activity runs in the reverse direction. The most effective source of tocopherol is wheat germ oil, although a number of other common vegetable oils contain relatively high quantities. However, cod liver oil, which contains the highest quantity observed in any animal fat so far [26 mg. per 100 grams (*266*)] contains less than one-twentieth of the value reported for solvent-extracted wheat germ oil [550 mg. per 100 grams (*13*)].

The comparative amounts of tocopherol occurring in some vegetable and animal fats and fatty oils are given in Table 19.

While it is uncertain whether or not vitamin E plays a rôle as an anti-sterility agent in human subjects, there is considerable evidence that it may have other important functions in man. Moreover, it is of great importance as an anti-oxidant, protecting fats. Thus, it must be present in the fat if vitamin A is to be effectively utilized. Vitamin E is not destroyed by the usual commercial processes of fat hydrogenation (*243*). In fact, several workers have demonstrated that hydrogenated cottonseed oil is a more reliable vitamin E source than is the unhydrogenated material (*67, 115, 348*).

d) Unsaturated Fatty Acid Content in Vegetable and Animal Fats. Vegetable fats are much more frequently sources of the essential fatty acids than are the animal fats. Moreover, the proportion of the essential fatty acids found in vegetable fats is usually much higher than has been noted in the animal fats. Linoleic and linolenic acids are the only forms of these essential acids which have been reported in the vegetable fats, while these two acids and also arachidonic acid have been detected in the animal fats.

Table 19. Vitamin E (Tocopherol) Content of Some Vegetable and Animal Fats (*71*).

Oil or fat	Reference	Tocopherol (mg. per 100 g.)
Vegetable Fats:		
Babassu, crude	(*13*)	3
Castor	(*13*)	50
Coconut, refined	(*13*)	3
„ , hydrogenated	(*266*)	3
Corn, crude	(*266*)	119
„ , refined	(*13*)	95
Cottonseed, crude	(*266*)	100
„ , refined	(*13*)	90
Olive	(*13, 266*)	20 25
Palm, crude	(*13*)	50
Peanut, crude	(*13*)	52
„ , refined	(*13*)	48
Pecan, refined	(*13*)	45
Rice bran, crude	(*13*)	100
„ „ , refined	(*13*)	90
Safflower, crude	(*13*)	80
Sesame, refined	(*13*)	50
Soybean, crude	(*266*)	212, 152
„ , refined	(*13, 266*)	175, 125 110
„ , phosphatide	(*195*)	200
Wheat germ, crude	(*13, 266*)	400 380, 310
„ „ oil, solvent extracted	(*13*)	550
Animal Fats:		
Cod liver oil	(*266*)	26
Lard, prime steam	(*13, 58, 197*)	2,8–0,5 0,5 0,2
Mangona shark liver oil	(*284*)	10
Oleo oil	(*13*)	2
Soupfin shark liver oil	(*284*)	4

The highest linoleic acid content has been found in safflower seed oil (*194, 245*), hackberry tree seed oil (*307*), and walnut oil (*149, 157*) while several other seed fats have a content in excess of 70%. As far as the results of total essential fatty acid analyses (linoleic plus linolenic) are concerned, linseed oil (*286*) and hempseed oil (*145, 200*) are at the top of the list. High proportions of poly-unsaturated acids occur in animal fats only when they have made up a considerable proportion of the diet. The highest value which has been reported in animals is that for the egg yolk fat of the hen (*66*) which had previously been fed on a hempseed oil

Table 20. Relative Maximum Amounts of Essential Fatty Acids in Various Vegetables and Animal Fats (71).

Vegetable fat	Essential fatty acid		Animal fat	Essential fatty acid		
	Linoleic (%)	Linolenic (%)		Linoleic (%)	Linolenic (%)	Arachidonic (%)
Almond	19,9	—	Egg yolk (hen)	—	—	—
Avocado	10,3	—	Hempseed oil diet	41,9	10,0	—
Brazil nut	22,8	—	Linseed oil diet	24,9	17,4	—
Cocoa butter	21,1	—	Various diets	21,7	2,9	2,3
Coconut	2,6	—	Phosphatides	8,2	—	—
Corn	39,1	—	Goose fat (body)	19,3	—	—
Cottonseed	50,4	—	Hen fat (body)	21,3	—	0,6
Grapeseed	73	—	Human fat	11,0	—	1,0
Grapefruit seed	51,4	—	Ox depot fat	5,3	—	0,5
Hackberry tree seed	76,7	—	Pig depot fat	15,6	—	2,1
Hempseed	68,8	24,4	Soybean oil diet	38,9	0,5	—
Linseed	23,3	60,9	Peanut diet	19,7	—	—
Olive	15,0	—	Cottonseed oil diet (12%)	26,8	—	—
Palm	10,9	—	Cottonseed oil diet (8%)	18,2	—	—
Peanut	27,4	—	Cottonseed oil diet (4%)	13,3	—	—
Perilla seed	33,6	—	Sheep depot fat	5	—	—
Pistachio nut	20,0	—	Milk fat (cow)	5,8	—	0,4
Poppyseed	62,2	—	Milk fat (goat)	1,5	—	—
Rapeseed	49	3,5	Milk fat (human)	7,9	5,4	—
Safflower seed	78	—				
Sesame seed	40,4	—				
Soybean	58,8	8,1				
Sunflower seed	6,8	—				
Tobacco seed	74,5	—				
Tomato seed	38,2	—				
Thom apple	74,5	—				
Walnut	75,5	10				
Watermelon seed	65,8	—				

rich diet. A comparable value was noted in pig depot fats for animals previously on a soybean oil diet (*104*). Values for the unsaturated fatty acids in vegetable and animal fats are summarized in Table 20.

3. Comparative Digestibility and Absorption of Vegetable and Animal Fats.

The digestibility of a fat is a very important index in establishing its nutritional value. No obvious differences in the digestibility of vegetable and animal fats as recorded in Tables 3 and 4 (p. 15, 16) for human subjects are apparent. Thus, in the case of the thirty-four vegetable fats examined, only two were found to be utilized to an extent of less than 94%. Avocado fat had a coefficient of digestibility of 88%, while a single experiment on tea seed oil gave 91%. On the other hand, all of the eighteen animal fats investigated yielded values of 94% or higher. In the case of the fats melting above 50° C., the digestibility figures are considerably lower for both the vegetable and the animal fats. In the tests made on rats (Table 7, p. 20) the completeness of utilization of hydrogenated cottonseed oil is practically identical with that for hydrogenated lard melting at corresponding temperatures. In the latter tests account is taken of the amount of the fat which has been lost in the form of fecal soaps.

Moreover, no appreciable differences in the rate of absorption appeared between vegetable and animal fats (Table 10, p. 28—29). One must conclude that from the standpoint of digestibility and absorption there is little to choose between vegetable and animal fats.

4. Growth Tests with Vegetable and Animal Fats.

Growth is the index of nutritional value which has been most widely employed and has been widely used for comparing vegetable and animal fats.

a) Experiments with Normal Rats. It has been shown in the previous section that better growth occurs on diets containing fat than on a fat-free regime. There has been considerable discussion as to whether all readily-digestible fats can function equally well in supplying this need, provided, of course, that fat-soluble vitamins and unsaturated fatty acids are available in the required quantities.

Several reports stated that butter fat possesses a specific growth effect. Thus, Schantz, Elvehjem, and Hart (*299*) observed that weanling rats grew best over a six-week period when they were fed on skimmed milk, homogenized with butter fat. When corn oil, cottonseed oil, coconut oil, or soybean oil were used instead, the growth was markedly inferior, although the differences between the several fat diets largely disappeared after a six-week period. In later assays, it was noted that the saturated acid fraction of butter produced a growth rate superior to that of an equal quantity of whole butter, while the volatile fraction was ineffective (*298*). The unsaturated residue was likewise ineffective unless it had been hydrogenated (*38*). SCHANTZ and his co-workers concluded that the superior growth observed on butter diet was to be traced to the presence of certain saturated long-chain acids which are not contained in other fats.

In contrast, two groups of investigators have recently failed to confirm that the saturated fatty acid fraction of butter has any specific effect on growth (*158, 192*). More recently, GEYER *et al.* (*138*) have reported a superior growth-promoting action of the liquid fraction of butter which remained, after separation from the solid glycerides, in acetone solution at -4° C. (Some doubt is cast on this result by the authors as they were able to obtain the fraction in only one instance and failed to duplicate the results with other samples of butter.) JACK *et al.* (*193*), likewise, have found better growth on a liquid fraction of butter provided that care was taken to prevent oxidative destruction. It is obvious that these results are contradictory. Although there seems to be some evidence that the liquid fraction may act more strongly than the others, the results obtained by DEUEL *et al.* (*95*) on whole butter failed to bear out this finding.

BOER (*31*) is another investigator who has reported, on the basis of growth experiments, that butter fat possesses a nutritive value superior to that of olive oil. The present reviewer believes, however, that the choice of olive oil as a representative vegetable fat is an unfortunate one since this oil contains very little tocopherol, and readily becomes rancid. DEUEL *et al.* (*95*) observed that growth on an olive oil sample which had become rancid was inferior to that obtained when other vegetable fats or butter were used. BOER, JANSEN, and KENTIE (*34*), reported that rapeseed oil is a poorer material than butter for producing growth in rats. Their results have been confirmed by DEUEL *et al.* (*78*) as well as by EULER, EULER, and LINDEMAN (*111*). The less satisfactory performance of rapeseed oil is apparently related to its low digestibility in the rat (*73*).

On the other hand, there are a number of reports which have failed to confirm the results of the Wisconsin group (*299*) or of the Dutch investigators (*31–34*). Thus, DEUEL and his associates (*95*) found that when skimmed milk powder was employed instead of liquid skimmed milk, and the fats to be tested were added to the skimmed milk powder in the same proportion found in whole milk powder, no differences could be observed in the rate of growth of rats at any time over a twelve-week period, regardless of whether the fat supplied in the diet was a butter, a margarine, corn oil, cottonseed oil, olive oil, peanut oil, or soybean oil. The identity in growth was established not only by the increase-in-weight but also by the rate of bone growth.

Not only was a similar efficiency in the utilization of the food observed in the experiments with the different fats, but likewise, analyses of the tissues of rats raised on the different diets, indicated that the same distribution of protein, fat, carbohydrate, and ash obtained (*82*). Considering the similarity in body weight, tibia length, and tissue composition, one is forced to the conclusion that the extent of growth obtained with the different fats (Figures 4, 5) is identical.

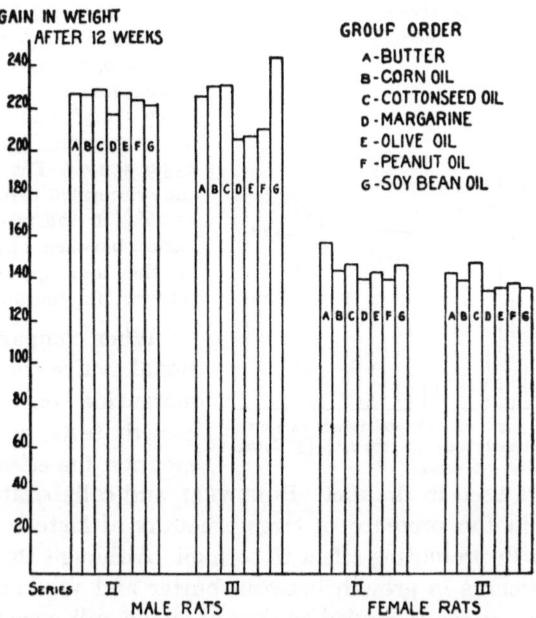

Figure 4. The total gain in weight of male and female rats on diets containing the different fats over a 12 week period. [J. Nutrit. 27, 107 (1944).]

DEUEL and MOVITT (*92*) have suggested that the more extensive growth reported by SCHANTZ *et al.* for butter diets (*299*) may have been caused by greater food consumption. Failure to observe such differences in the tests of DEUEL *et al.* (*95*) is probably to be ascribed to the inclusion of diacetyl or butter flavor in all of the diets. It was shown that rats prefer a diet of bland fat flavored with diacetyl to one which is unflavored (*92*). Although BOUTWELL *et al.* (*40*) did not succeed in bringing about a sufficient increase in food consumption by the inclusion of diacetyl to make the growth on the corn oil diet identical with that obtained on the butter diet, they did find some improvement in the food consumption.

EULER, EULER, and RÖNNESTAM-SÖBERG (*113*) have obtained better growth in rats receiving oleomargarine than in those fed butter. Recently,

EULER and co-workers (*112*) have found as satisfactory growth of rats on a margarine-fat mixture as on butter, over a 700-day period. DEUEL *et al.* (*77, 78*) likewise have observed identical growth on butter and margarine diets. Other workers who have also demonstrated similarity in growth response between butter and vegetable fats are ZIALCITI and MITCHELL (*366*), HENRY *et al.* (*158*), and LASSEN and BACON (*209*).

The latter two investigators showed that the mean daily weight gain (in grams) for 70 days (30 days to 100 days) and the average body length (in millimeters) at 100 days were, respectively, the following: summer butter fat, 4,28 ± 0,14 g. and 243,0 ± 1,36 mm.; oleomargarine, 4,27 ± 0,10 and 245,0 ± 1,14; olive oil, 3,99 ± 0,095 and 241,4 ± 1,31; cottonseed oil, 4,42 ± 0,12 and 246,7 ± 1,78; and for controls kept on fat-free diet, 3,63 ± 0,089 and 235,9 ± 1,74. The fat content of the eviscerated carcasses averaged 12,47% in the fat-free group and varied between 14,26 and 15,89% in the four groups which had received fat-containing diets.

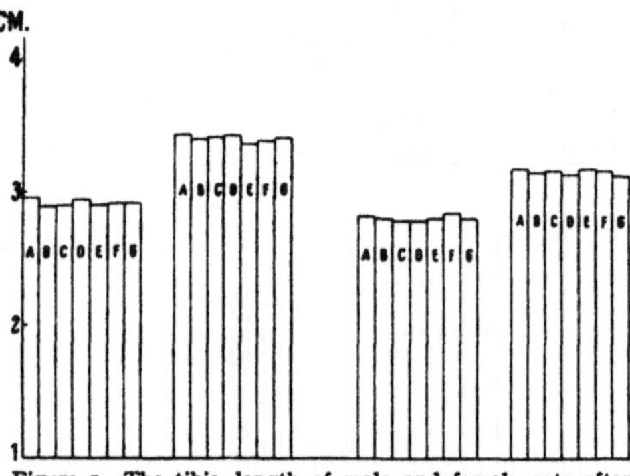

Figure 5. The tibia length of male and female rats after 3 and 6 weeks on diets containing different fats. [J. Nutrit. 27, 107 (1944).]

When comparing the nutritional value of butter and margarine on the basis of growth tests, one should also mention the effect of the type of carbohydrate used in the diet. BOUTWELL and collaborators (*37*) have reported that in the presence of large amounts of lactose, butter gives a superior growth-promoting effect to corn oil. Although DEUEL *et al.* (*95*) found no differences in growth between butter and vegetable fats when the lactose content corresponded to that of whole milk powder, *i.e.* 32%, the superior growth of butter-fed rats reported by BOUTWELL *et al.* occurred when lactose comprised 48% of the diet. However, this variation was noted only in those cases where the vitamin B intake was restricted. In contrast, no such differences were obtained between corn oil and butter when a galactose-glucose mixture was fed or when glucose alone was the carbohydrate present. It must be emphasized that the tests with lactose were made with a highly artificial diet. One is not so much impressed with the difference in response to the two fats mentioned on a high-lactose diet, as one is with the poor showing of all animals on a high-lactose diet compared with those which received other carbohydrates. ERSHOFF and DEUEL (*110*) have recently reported that, when lactose was fed at a level of 75%, death of young rats resulted, while some other sugars were well tolerated under similar conditions. The toxic effect of lactose has recently been confirmed by other investigators (*150, 186*).

Vaccenic acid (11-octadecenoic acid) has recently been proposed as an acid of high nutritional value to which the supposed superior growth-promoting effects of butter can be ascribed. BOER, JANSEN, and KENTIE (*34*) have reported that summer butter contains significant amounts of this acid while it is absent from winter butter and vegetable fats. The presence of vaccenic acid in butter and animal fats was first established by BERTRAM (*25*), GROSSFELD and SIMMER (*147*), and confirmed by GEYER *et al.* (*140*). The latter workers likewise have shown the absence of vaccenic acid in vegetable oils. On the other hand, the presence of considerable amounts of this acid has been reported in hydrogenated fat (*25, 147*).

The suggestion that vaccenic acid possesses special growth-promoting powers was supported by BOER *et al.* in a later report (*35*), by the demonstration that vaccenic acid prepared from China wood oil had a growth-promoting effect on rats receiving a diet which contained rapeseed oil as the fat. These workers attribute the failure of the EULER group (*113, 114*), in 1942 and 1943, to substantiate the earlier report of BOER (*31*) and of BOER and JANSEN (*33*) on the superior growth-promoting effect of butter fat (as compared with olive oil) to the fact that winter butter produced in Sweden probably does not contain vaccenic acid.

However, a number of investigators have failed to confirm BOER's results on the importance of vaccenic acid as a dietary adjunct. DEUEL *et al.* (*78*), using similar technique to BOER *et al.* (*34*), found that rapeseed oil did give poorer growth than butter or cottonseed oil; however, the addition of vaccenic acid to the rapeseed oil or cottonseed oil diet in no way improved the growth response. Similar negative results with vaccenic acid have been recorded by EULER, EULER, and LINDEMAN (*111*) as well as by NATH and associates (*250*). In fact, BOER and co-workers (*32*) have recently retracted their earlier report.

There is a possibility that BOER *et al.* (*35*) may have mistaken 12-octadecenoic acid for vaccenic acid since both isomers have quite similar properties. However, tests by the present reviewers (*77*) of 12-octadecenoic acid show that it has no specific nutritional value.

The evidence seems to be preponderant that, on complete diets, vegetable fats and margarine possess as high a nutritional value as an animal fat such as butter. There is no evidence that butter contains such fatty acids specifically required for growth which are not found in vegetable fats. Vaccenic acid certainly has no special merit from a nutritional standpoint.

b) Experiments on Prematurely-Weaned Rats. BOUTWELL and collaborators (*39*) have stated that the particular fatty acids present in butter which are required by the young rat are no longer necessary if the animal is kept on the stock diet until thirty days of age. However, these investigators report that the need for butter is accentuated in prematurely-weaned rats. These results are denied by ZIALCITA and MITCHELL (*366*) who were able to raise rats weaned at seven days of age, instead of fourteen days, as used by BOUTWELL *et al.* (*39*) or at the usual twenty-one day interval, on artificial mixtures containing either butter fat or corn oil, without observing any superiority of one fat over the other. In tests carried out in the writers' laboratory (*93*), it was found that rats weaned at fourteen days grew as satisfactorily on diets of corn, cottonseed, peanut, or soybean oils, or on a margarine fat, as on a diet containing a butter fat. The weight of evidence would, therefore, seem to discount this type of experiment for the demonstration of a superior growth-promoting action of butter.

c) Comparative Effectiveness of Animals Receiving Fats in a Restricted Diet. Another means of evaluation of the growth-promoting activity of a fat is to test it when the animal is receiving a restricted caloric intake. If a specific fat is needed for growth, then the requirement for it may not be satisfied during a period of limited caloric intake. However, when such tests were made, no differences could be found in the rate of growth of rats receiving vegetable oils, a margarine, a commercial hydrogenated fat, or a butter, when given in connection with a complete basal diet fed at such a level that growth occurred at a slower than normal rate (*85*). That no permanent damage resulted from the lack of animal fat in this diet was indicated by the fact that the increased growth response during the period of *ad libitum* feeding, following the twelve-week period of restricted caloric intake, was similar in the butter-receiving group and in the various vegetable fat groups.

d) Experiments on Rats Using Growth Hormones. A further possible method for the estimation of nutritional value is to determine whether diets will be satisfactory when growth requirement is increased above normal. It is conceivable that a substance might be able to promote normal growth but that it would fail in its purpose when an increased growth rate was required. ERSHOFF and DEUEL (*109*) have shown that such a deficiency as that caused by the lack of vitamin A can be accentuated when growth hormone is injected. Under such conditions, an earlier death of the injected rats ensued. However, when a pituitary-growth hormone was injected parenterally in normal rats, it was found that diets containing such vegetable fats as corn, cottonseed, peanut, or soybean oil, or a vegetable margarine, supported the growth as effectively as did a butter fat sample (*85*).

It thus appears that the vegetable fats are able to be as readily utilized as animal fats and serve equally well for their nutritional functions in cases of a restricted diet or when an augmented metabolism obtains because of the presence of a growth hormone.

e) Growth and Nutrition Experiments on Children. While most of the evidence on the nutritive value of fats has been deduced from tests on rats, there is one report in which this experimental method has been used with children. LEICHENGER, EISENBERG, and CARLSON (*213*) recently have reported a two-year test which was carried out on the children of two orphanages. In one case margarine served exclusively as a source of table spread, while the second group received a comparable diet in which butter was employed. At the conclusion of the tests, it was shown that all children which had received margarine had grown as well as the children who had received the butter diet. The results of these tests would indicate that some data obtained on rats may also have application for man.

5. Pregnancy and Lactation Performance of Rats Receiving Animal and Vegetable Fats.

The studies on pregnancy and lactation performance for animals receiving vegetable and animal fat diets likewise can be interpreted as indicating the equivalence of these two types of fat. As has been indicated earlier, nutritionists in general agree that pregnancy affords a more critical index of the nutritional value of a diet than does growth, while satisfactory lactation can be produced only if the diet is still better from a nutritional standpoint.

ANDERSON and WILLIAMS (*6*) as well as MEIGS (*238*) have emphasized the well-known fact that fat-low or fat-free diets are especially poor for lactation. However, as pointed out by SURE (*327*), satisfactory lactation cannot be promoted on otherwise *inadequate* diets by butter fat, lard, hydrogenated oil, olive oil, or wheat germ oil. On *adequate* diets, it was shown by MAYNARD and RASMUSSEN (*236*) that an improvement obtains in rats when corn or coconut oil is added to a fat-free diet, although LOOSLI and collaborators (*218*) later reported that hydrogenated coconut oil was ineffective in promoting lactation.

QUACKENBUSH, KUMMEROW, and STEENBOCK (*267*) have shown that linoleic acid is necessary for satisfactory milk production, a fact which would seem to indicate that vegetable fats should be preferable to animal fats since they generally contain a larger proportion of the essential unsaturated acids. However, other workers (*218*) have been unable to demonstrate an improvement in lactation of rats (kept on a fat-free diet) when 125 mg. of methyl linoleate were administered daily.

More recently, DEUEL and collaborators (*94*) have shown that in a series of vegetable fats including a margarine fat and a butter sample, all act similarly in allowing a normal pregnancy and lactation period. Although the results on olive oil were somewhat less satisfactory than those obtained with the other fats tested, it is believed that this is partly to be ascribed to the poor quality of the olive oil. Some data for the pregnancy and lactation tests are summarized in Table 21.

Table 21. Comparative Nutritional Value of Vegetable Fats, a Vegetable Oil Margarine and a Butter as Determined by Ability to Promote Successful Pregnancy and Lactation (*94*).

(The figures in parentheses indicate the number of rats weaned.)

Skimmed milk powder diet mixed with one of the following fats	Pregnancy tests			Lactation tests	
	Animals bred, number	Litters cast, number	Average 3-day weight per rat (g.)	Rats died during 21 days, number	Average 21-day weight per rat (g.)
Butter fat	29	29	9,07	23	33,4 (84)
Corn oil	31	28	9,06	1	32,9 (112)
Cottonseed oil	31	30	9,15	2	34,2 (98)
Margarine fat	35	33	7,57	8	34,8 (112)
Olive oil	23	18	7,88	1	35,2 (70)
Peanut oil	32	31	8,03	4	33,6 (84)
Soybean oil	28	26	7,60	1	32,4 (98)

EULER, EULER, and RÖNNESTAM-SÖBERG (*112*) have presented data which indicates that margarine exerts a considerable superiority to butter as demonstrated by its performance in reproduction tests.

In their first series of experiments, the total weight of all offspring at twenty-eight days produced over an eighteen-month period was 8508 grams compared with a total of 15956 grams for the progeny of a similar number of females which had received a diet containing a margarine fat in place of butter fat. Although vitamin E supplementation improved the response on the butter diet in a second series, it did not equal that of the unsupplemented margarine group.

6. Growth and Reproduction of Rats over Many Generations on a Diet Containing a Vegetable Margarine.

The so-called "multi-generation" test is a procedure which combines most of the other methods for evaluating the nutritional value of a foodstuff. If a particular diet is the only food consumed over a number of generations, then such a regime must be satisfactory from the standpoint of growth, pregnancy, and lactation. A prolonged test of this kind is more conclusive than the comparison of growth, reproduction, and lactation in a single generation. It is conceivable that a diet might support growth exceedingly well over one generation. It might also allow for a perfectly normal pregnancy and lactation, but on subsequent trials deficiencies might accumulate and nutritional failure result after several generations.

SHERMAN and CAMPBELL (*310*, *311*) devised a simple diet which has enabled rats to continue growth, reproduction, and lactation satisfactorily over a period of sixty-seven generations (*313*). This basal diet "A" consists of whole milk powder ($^1/_6$), ground whole wheat ($^5/_6$) and sodium chloride equal to 2% of the weight of the wheat. That this diet, however, is a suboptimal one, was shown by the fact that when fortieth-generation rats on diet *A* were transferred to a diet *B* (in which the proportion of whole milk powder was increased from $^1/_6$ to $^1/_3$), a marked improvement in a number of biometric measurements which are generally regarded as indices of nutritive value obtained. The rats grew larger on diet *B*, growth took place at a faster rate, and the duration over which reproduction was possible in the females was practically doubled. The number

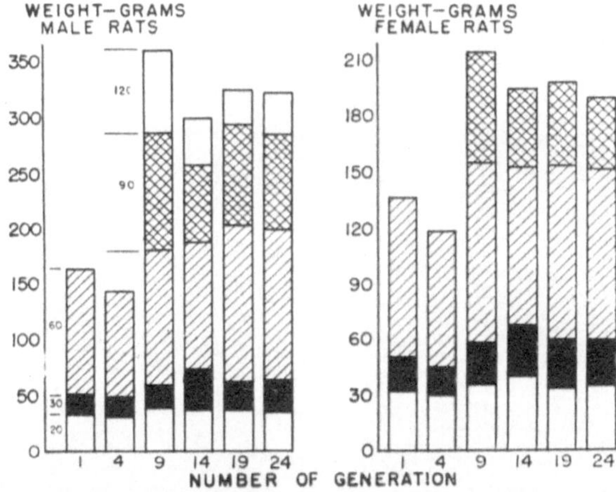

Figure 6. The average body weights of weanling rats at 20 days (lower blank space), after 30 days (solid black), after 60 days (lined), after 90 days (cross-hatched), after 120 days (top blank space).

of rats weaned was increased. In a later study SHERMAN and CAMPBELL (*312*) showed that the life span of animals fed on diet *B* was prolonged by a statistically significant period over that with the less satisfactory diet *A*.

DEUEL *et al.* (*76, 81*) have employed the SHERMAN diet *B* (which is the better one of the two diets tested by SHERMAN and CAMPBELL) for demonstrating whether or not butter fat is essential for continued fertility and reproduction. In these tests, rats were maintained on diet *B* with the exception that skimmed milk powder was employed in place of whole milk powder; and an amount of a vegetable margarine fat corresponding to the butter fat content of whole milk powder was added to the diet. At the end of ten generations, the rate of growth of the male and female rats was considerably higher than at the start of the test (*81*). Moreover, fertility was maintained and the average weight of the rat at three days of age and at weaning was considerably higher than had obtained in the early generations. In a later report (*76*) the data were reviewed again, after the completion of twenty-five generations. Although the growth rates were slightly less than that reached by the tenth generation, they maintained a level considerably higher than had been noted at the start of the test. Moreover, fertility and lactation performance continued at a high level. There is no indication yet that the experiment cannot continue indefinitely with the present diet. The comparative rates of growth of the male and female rats in representative generations are represented in Figure 6, while Figure 7 gives the data on fertility and growth of newborn rats at various representative times during the experiment.

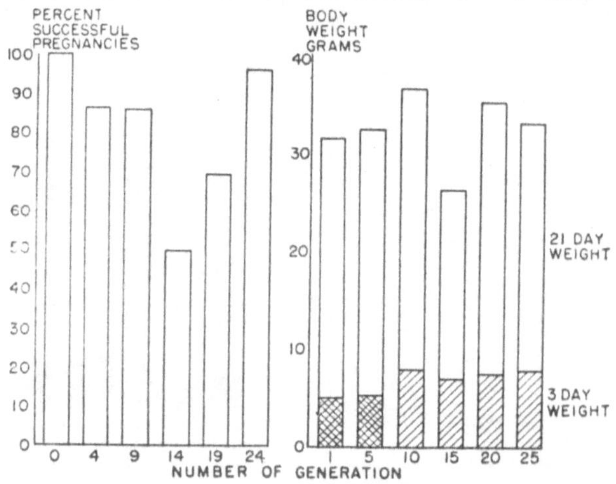

Figure 7. (Left) Average per cent successful pregnancies. (Right) Average body weight for progeny. Weight at one day of age (cross-hatched), weight at 3 days of age (lined). Total column represents 21 day weight.

The tests of the authors were started in April 1940 and have now continued for over ten years. If it is possible to translate these values into comparable experiments on human subjects, with an average of thirty years for each generation, then this would account for a period of 750 years.

IV. Conclusions.

A considerable amount of experimental evidence has accumulated to indicate that fats possess a definite rôle in animal nutrition. Not only are they the foodstuffs having the highest caloric density, but likewise they are the sources of certain essential unsaturated fatty acids which cannot be synthesized in the animal organism. Recently, it has become evident that fats possess considerable importance in the metabolism of some of the water-soluble as well as the fat-soluble vitamins. Although it was formerly believed that carbohydrates alone display a protein-sparing action, it now becomes evident that fat and fat alone under certain conditions can exhibit this same property. Thus, in cases where animals have been reduced to the "wear and tear quota" on protein-free diets, only fat is able to maintain this low level when the total caloric intake is reduced.

Fat likewise has an interesting rôle in relation to the specific dynamic action of foodstuffs. When it is fed in combination with protein or carbohydrate, the resultant specific dynamic action is less than would be expected if these foodstuffs were fed separately. No such effect is exhibited when protein and carbohydrate are combined. Other indices ordinarily employed to establish nutritive values, which indicate that fat is essential in the diet, are those of growth, pregnancy, and lactation. Moreover, animals on diets containing generous amounts of fat are able to carry out more work and to survive during periods of subsequent fasting for longer intervals than animals which have received fat-low diets. Best results seem to be obtained on diets containing 20 to 40% fat.

While these data are quite clear-cut for the rat, there is as yet no cogent evidence that analogous conditions necessarily exist in man.

There are a number of facts which indicate that vegetable and animal fats have similar nutritive value. Thus, while animal fats more frequently possess vitamin A, provitamin A (carotene) and tocopherol are more frequently found in large proportions in vegetable fats. The essential fatty acids occur in a much higher proportion in vegetable than in animal fats. In growth tests on normal rats, on animals which have been prematurely weaned, on rats receiving restricted diets, or rats receiving injected growth hormone, the results are uniform in indicating that corn oil, cottonseed oil, peanut oil, soybean oil, a vegetable margarine, and a butter possess similar nutritive properties. An experiment performed on young children over a two-year period showed that the growth of the children receiving a vegetable margarine is similar to, if not better than that of a group receiving butter. The results from such critical indices of nutrition as pregnancy and lactation performance of rats are as satisfactory for vegetable as for animal fats. Finally, it has been demonstrated that rats have been kept in excellent condition over twenty-five generations on a diet containing a vegetable margarine as the sole fat. CAMPBELL and SHERMAN arrived at the same conclusion using a similar diet containing butter fat. Not only has growth been maintained at a high level, but reproduction and lactation remained excellent. This would seem to indicate that margarine fat can be used in place of butter fat over long periods of time and that it is able to maintain a high nutritional state.

References.

1. ABELIN, I.: Über Fett- und Schilddrüsenwirkung. Klin. Wschr. 5, 367 (1926).

1a. ABELIN, I., M. KNUCHEL u. W. SPICHTIN: Ernährung und Schilddrüsenwirkung. II. Über die Bedeutung der Vitamine für den Verlauf der experimentellen Hyperthyreose. Biochem. Z. **228**, 189 (1930).

1b. Abelin, I. u. P. Kürsteiner: Über den Einfluß der Schilddrüsensubstanzen auf den Fettstoffwechsel. Biochem. Z. **198**, 19 (1928).
1c. Achaya, K. T. and B. N. Banerjee: Characteristics of Indian Animal Fats. Current Sci. **15**, 23 (1946).
2. Adlersberg, D. and H. Sobotka: Influence of Lecithin Feeding on Fat and Vitamin A Absorption in Man. J. Nutrit. **25**, 255 (1943).
3. Anderson, R. J.: The Chemistry of the Lipoids of the Tubercle Bacillus and Certain Other Microorganisms. Fortschr. Chem. organ. Naturstoffe **3**, 145 (1939).
4. Anderson, R. J. and G. Lusk: Animal Calorimetry. XIII. The Interrelation Between Diet and Body Condition and the Energy Production During Mechanical Work. J. biol. Chemistry **32**, 421 (1917).
5. Anderson, W. E. and L. B. Mendel: The Relation of Diet to the Quality of Fat Produced in the Animal Body. J. biol. Chemistry **76**, 729 (1928).
6. Anderson, W. E. and H. H. Williams: The Rôle of Fat in the Diet. Physiologic. Rev. **17**, 335 (1937).
7. Anonymous: Mineral Oil in Foods. Current Comment in J. Amer. med. Assoc. **131**, 1426 (1946).
8. Anonymous: (Bureau Dairy Indust.) Vitamin A in Butter. U. S. Dept. Agric. Miscell. Pub. No. 571 (1945).
9. Arnold, A. and C. A. Elvehjem: Influence of the Composition of the Diet on the Thiamin Requirement of Dogs. Amer. J. Physiol. **126**, 289 (1939).
10. Arnschink, L.: Versuche über die Resorption verschiedener Fette aus dem Darmkanale. Z. Biol. **26**, 434 (1890).
11. Augur, V., H. S. Rollman and H. J. Deuel, Jr.: The Effect of Crude Lecithin on the Coefficient of Digestibility and the Rate of Absorption of Fat. J. Nutrit. **33**, 177 (1947).
12. Axelrod, A. E., M. Mitz and K. Hofmann: The Chemical Nature of Fat-soluble Materials with Biotin Activity in Human Plasma. Additional Studies on Lipide Stimulation of Microbial Growth. J. biol. Chemistry **175**, 265 (1948).
12a. Baer, E. and H. O. L. Fischer: Studies in Acetone-Glycerylaldehyde and Optically Active Glycerides. IX. Configuration of the Natural Batyl, Chimyl and Selachyl Alcohols. J. biol. Chemistry **140**, 397 (1941).
12b. Baer, E., H. O. L. Fischer and L. J. Rubin: Naturally Occuring Glycerol Ethers III. Selachyl Alcohol and Its Geometrical Isomers. J. biol. Chemistry **170**, 337 (1947).
13. Bailey, A. E.: Industrial Oil and Fat Products. New York: Interscience Publishers, Inc. 1945.
14. Banerji, G. G.: The Effect of a High-fat Diet on the Excretion of Bisulfite-binding Substances in the Urine of Rats Deficient in Vitamin B_1. Biochemic. J. **34**, 1329 (1940).
15. Banks, A. and T. P. Hilditch: The Glyceride Structure of Beef Tallows. Biochemic. J. **25**, 1168 (1931).
16. — — The Body Fats of the Pig. II. Some Aspects of the Formation of Animal Depot Fats Suggested by the Composition of their Glycerides and Fatty Acids. Biochemic. J. **26**, 298 (1932).
17. Banks, A., T. P. Hilditch and E. C. Jones: The Component Fatty Acids of Rat Body Fats. Biochemic. J. **27**, 1375 (1933).
18. Barnes, R. H., E. S. Miller and G. O. Burr: The Adrenals and Fat Absorption. J. biol. Chemistry **140**, 241 (1941).
19. Barnes, R. H., M. F. Primrose and G. O. Burr: The Influence of the Protein Content of the Diet upon Fat Digestibility. J. Nutrit. **27**, 179 (1944).

20. Barnes, R. H., I. I. Rusoff and G. O. Burr: Adrenalectomy and the Absorption of Different Fats. Proc. Soc. exp. Biol. Med. **49**, 84 (1942).

21. Bauernfeind, J. C., A. L. Sotier and C. S. Boruff: Growth Stimulants in the Microbiological Assay for Riboflavin and Pantothenic Acid. Ind. Engng. Chem., Analyt. Edit. **14**, 666 (1942).

22. Bavetta, L. A.: The Effect of Adrenalectomy on the Absorption of the Short-chain Fatty Acids and their Triglycerides. Amer. J. Physiol. **140**, 44 (1943).

23. Bavetta, L. A. and H. J. Deuel, Jr.: The Effect of Adrenalectomy on the Absorption of Hydrogenated Cottonseed Oil, Corn Oil, Tributyrin and Sodium Butyrate. Amer. J. Physiol. **136**, 712 (1942).

24. Bavetta, L. A., L. Hallman, H. J. Deuel, Jr. and P. O. Greeley: The Effect of Adrenalectomy on Fat Absorption. Amer. J. Physiol. **134**, 619 (1941).

24a. Berg, J.: Untersuchungen über die Heilwirkung von Lipoiden auf den experimentellen Hyperthyreodismus. Z. ges. exp. Med. **93**, 143 (1934).

25. Bertram, S. H.: Die Vaccensäure (eine neue Fettsäure aus Rinder-, Schafs- und Butterfett). Biochem. Z. **197**, 433 (1928).

26. Bhattacharya, R. and T. P. Hilditch: Structure of Synthetic Mixed Triglycerides. Proc. Roy. Soc. (London), Ser. A **129**, 468 (1930).

27. Bills, C. E.: Physiology of the Sterols, Including Vitamin D. Physiologic. Rev. **15**, 1 (1935).

28. Birch, T. W.: The Relation Between Vitamin B_6 and the Unsaturated Fatty Acid Factors. J. biol. Chemistry **124**, 775 (1938).

29. Birch, T. W. and P. György: A Study of the Chemical Nature of Vitamin B_6 and Methods for its Preparation in a Concentrated State. Biochemic. J. **30**, 304 (1936).

30. Boas, M. A.: The Effect of Desiccation Upon the Nutritive Properties of Egg White. Biochemic. J. **21**, 712 (1927).

31. Boer, J.: A New Growth Factor in Butter Fat. Acta brevia neerl. Physiol., Pharmacol., Microbiol. **11**, 180 (1941).

32. Boer, J., E. H. Groot and B. C. P. Jansen: Further Investigations on the Growth-promoting Factor in Butter. Voeding **9**, 60 (1948); Chem. Abstr. **42**, 7847 (1948).

33. Boer, J. and B. C. P. Jansen: Over die voedingswaarde van boter, vergeleken met die andere vetten, waaraan de vitamin A en D zijn toegevoegd. Voeding **2**, 204 (1941); cited in (*35*).

34. Boer, J., B. C. P. Jansen and A. Kentie: On the Growth-promoting Factor for Rats Present in Summer Butter. J. Nutrit. **33**, 339 (1947).

35. Boer, J., B. C. P. Jansen, A. Kentie and H. W. Knol: The Growth-promoting Action of Vaccenic Acid. J. Nutrit. **33**, 359 (1947).

35a. Boldingh, J.: Fatty Acid Analysis by Partition Chromatography Rec. Trav. chim. Pays-Bas **69**, 247 (1950).

36. Booth, R. G., K. M. Henry and S. K. Kon: A Study of the Antirachitic Effect of Fat on Rats Receiving High-calcium, Low-phosphorus Rachitogenic Diets. Biochemic. J. **36**, 445 (1942).

37. Boutwell, R. K., R. P. Geyer, C. A. Elvehjem and E. B. Hart: Studies on the Interrelation of Fats, Carbohydrates, and B-Vitamins in Rat Nutrition. Arch. Biochemistry **7**, 143 (1945).

38. — — — — The Effect of Hydrogenation on the Nutritive Value of the Fatty Acid Fractions of Butter Fat and of Certain Vegetable Oils. J. Dairy Sci. **24**, 1027 (1941).

39. — — — — Further Studies on the Growth-promoting Value of Butter Fat. J. Dairy Sci. **26**, 429 (1943).

40. BOUTWELL, R. K., R. P. GEYER, C. A. ELVEHJEM and E. B. HART: Effect of Flavor on the Nutritive Value of Fats. Proc. Soc. exp. Biol. Med. **55**, 153 (1944).
41. BRADLEY, T. F.: Drying Oils and Resins. Ind. Engng. Chem., Analyt. Edit. **29**, 440 (1937).
42. BRANDT, K.: Production and Consumption of Fats and Oils. Ann. Amer. Acad. Polit. Soc. Sci. **225**, 210 (1943).
43. BROCKLESBY, H. N.: The Chemistry and Technology of Marine Animal Oils with Particular Reference to those of Canada. Fisheries Research Bd. Canada, Bull. **59**, 107 (1941).
44. BROWN, J. B.: Low Temperature Crystallization of Fatty Acids and Glycerides. Chem. Reviews **29**, 333 (1941).
45. BROWN, W. R. and A. E. HANSEN: Arachidonic and Linoleic Acid of the Serum in Normal and Eczematous Human Subjects. Proc. Soc. exp. Biol. Med. **36**, 113 (1937).
46. BROWN, W. R., A. E. HANSEN, G. O. BURR and I. MCQUARRIE: Effects of Prolonged Use of Extremely Low-fat Diet on an Adult Human Subject. J. Nutrit. **16**, 511 (1938).
47. BUCKLEY, T. A.: The Dietetic Value of Palm Oil. Malayan agric. J **24**, 485 (1936).
48. BURR, G. O.: Chemistry and Medicine, p. 101. Univ. Minnesota Press. 1940; cited by G. O. BURR, Federat. Proc. **1**, 224 (1942), pp. 224, 225.
49. — Significance of the Essential Fatty Acids. Federat. Proc. **1**, 224 (1942).
50. BURR, G. O., J. B. BROWN, S. P. KASS and W. O. LUNDBERG: Comparative Curative Values of Unsaturated Fatty Acids in Fat Deficiency. Proc. Soc. exp. Biol. Med. **44**, 242 (1940).
51. BURR, G. O. and M. M. BURR: A New Deficiency Disease Produced by the Rigid Exclusion of Fat From the Diet. J. biol. Chemistry **82**, 345 (1929).
52. — — On the Nature and Rôle of the Fatty Acids Essential in Nutrition. J. biol. Chemistry **86**, 587 (1930).
53. BURR, G. O., M. M. BURR and E. S. MILLER: On the Fatty Acids Essential in Nutrition. III. J. biol. Chemistry **97**, 1 (1932).
54. CAHILL, W. M.: Urinary Excretion of Thiamine on High-fat and High-carbohydrate Diets. J. Nutrit. **21**, 411 (1941).
55. CASSIDY, H. G.: Adsorption Analysis. II. The Adsorption of Higher Fatty Acids. J. Amer. chem. Soc. **62**, 3073 (1940).
56. — Adsorption Analysis. IV. Separation of Mixtures of Higher Saturated Fatty Acids. J. Amer. chem. Soc. **63**, 2735 (1941).
57. CHENG, A. L. S., M. G. MOREHOUSE and H. J. DEUEL, Jr.: The Effect of the Level of Dietary Calcium and Magnesium on the Digestibility of Fatty Acids, Simple Triglycerides, and Some Natural and Hydrogenated Fats. J. Nutrit. **37**, 237 (1949).
58. CHIPAULT, J. R., W. O. LUNDBERG and G. O. BURR: The Chemical Determination of Tocopherols in Animal Fats: The Stability of Hog Fats in Relation to Fatty Acid Composition and Tocopherol Contents. Arch. Biochemistry **8**, 321 (1945).
59. CLAESSON, S.: A Study on the Adsorption of Normal Branched and Unsaturated Fatty Acids and Related Compounds by Means of Adsorption Analysis. Recueil Trav. chim. Pays-Bas **65**, 571 (1946).
60. COLLINS, H. S., J. M. KRAFT, T. D. KINNEY, C. S. DAVIDSON, J. YOUNG and F. J. STARE: Parenteral Nutrition. III. Studies on the Tolerance of Dogs to Intravenous Administration of Fat Emulsions. J. Lab. clin. Med. **33**, 143 (1948).

61. CORNBLEET, T. and E. R. PACE: Use of Maize Oil (Unsaturated Fatty Acids) in the Treatment of Eczema. Arch. Dermatol. Syphilology **31**, 224 (1935).
62. CRAIG, L. C.: Identification of Small Amounts of Organic Compounds by Distribution Studies. II. Separation by Countercurrent Distribution. J. biol. Chemistry **155**, 519 (1944).
63. CRAIG, L. C., C. GOLUMBIC, H. MIGHTON and E. TITUS: Identification of Small Amounts of Organic Compounds by Distribution Studies. III. The Use of Buffers in Counter-current Distribution. J. biol. Chemistry **161**, 321 (1945).
64. CRAMPTON, E. W.: Presented at the Gordon Conference, New London, N. H., Aug. 1948.
65. CROCKETT, M. E. and H. J. DEUEL, Jr.: A Comparison of the Coefficient of Digestibility and the Rate of Absorption of Several Natural and Artificial Fats as Influenced by Melting Point. J. Nutrit. **33**, 187 (1947).
66. CRUICKSHANK, E. M.: Studies in Fat Metabolism in the Fowl. I. The Composition of the Egg Fat and Depot Fat of the Fowl as Affected by the Ingestion of Large Amounts of Different Fats. Biochemic. J. **28**, 965 (1934).
67. CUMMINGS, M. J. and H. A. MATTILL: The Auto-oxidation of Fats with Reference to their Destructive Effect on Vitamin E. J. Nutrit. **3**, 421 (1931).
68. CURTIS, A. C. and E. M. KLINE: Influence of Liquid Petrolatum on the Blood Content of Carotene in Human Feces. Arch. intern. Med. **63**, 54 (1939).
69. CZACZKES, J. W. and K. GUGGENHEIM: The Influence of Diet on the Riboflavin Metabolism of the Rat. J. biol. Chemistry **162**, 267 (1946).
70. DAUBERT, B. F.: The Composition of Fats. J. Amer. Oil chem. Soc. **26**, 556 (1949).
71. DEUEL, H. J., Jr.: In A. E. BAILEY: Cottonseed and Cottonseed Products, p. 763. New York: Interscience Publishers, Inc. 1948.
72. — Studies on the Comparative Nutritive Value of Fats. IX. The Digestibility of Margarine Fat in Human Subjects. J. Nutrit. **32**, 69 (1946).
73. DEUEL, H. J., Jr., A. L. S. CHENG and M. G. MOREHOUSE: The Digestibility of Rapeseed Oil in the Rat. J. Nutrit. **35**, 295 (1948).
74. DEUEL, H. J., Jr., S. M. GREENBERG, C. E. CALBERT, E. E. SAVAGE and T. FUKUI: The Effect of Fat Level of the Diet on General Nutrition. V. The Relation of Linoleic Acid Requirement to the Optimum Fat Level of the Diet. J. Nutrit. **40**, 351 (1950).
75. DEUEL, H. J., Jr., S. M. GREENBERG and E. E. SAVAGE: Unpublished.
76. DEUEL, H. J., Jr., S. M. GREENBERG, and E. E. SAVAGE: Studies on the Comparative Nutritive Value of Fats. XIII. Growth and Reproduction Over 25 Generations on SHERMAN Diet B where Butter Fat was Replaced by Margarine Fat, Including a Study of Calcium Metabolism. J. Nutrit. (in press).
77. DEUEL, H. J., Jr., S. M. GREENBERG, E. STRAUB, T. FUKUI, C. M. GOODING and C. F. BROWN: Studies on the Comparative Nutritive Value of Fats. XI. On the Possible Growth-promoting Activity of Δ^{12}-octadecenoic Acid. J. Nutrit. **38**, 361 (1949).
78. DEUEL, H. J., Jr., S. M. GREENBERG, E. STRAUB, D. JUE, C. M. GOODING and C. F. BROWN: Studies on the Comparative Nutritive Value of Fats. X. On the Reported Growth-promoting Action of Vaccenic Acid. J. Nutrit. **35**, 301 (1948).
79. DEUEL, H. J., Jr. and L. HALLMAN: The Rate of Absorption of Synthetic Triglycerides in the Rat. J. Nutrit. **20**, 227 (1940).
80. DEUEL, H. J., Jr., L. HALLMAN and A. LEONARD: The Comparative Rate of Absorption of Some Natural Fats. J. Nutrit. **20**, 215 (1940).

81. DEUEL, H. J., Jr., L. HALLMAN and E. MOVITT: Studies on the Comparative Nutritive Value of Fats. VI. Growth and Reproduction Over 10 Generations on Sherman Diet B where Butter Fat was Replaced by a Margarine Fat. J. Nutrit. **29**, 309 (1945).
82. DEUEL, H. J., Jr., L. HALLMAN, E. MOVITT, F. H. MATTSON and E. WU: Studies on the Comparative Nutritive Value of Fat. II. The Comparative Composition of Rats Fed Different Diets. J. Nutrit. **27**, 335 (1944).
83. DEUEL, H. J., Jr., L. HALLMAN and S. QUON: A Simple Procedure for the Study of Fat Absorption. J. biol. Chemistry **128**, **XIX** (1939).
84. DEUEL, H. J., Jr., L. HALLMAN and A. REIFMAN: The Rate of Absorption of Various Fatty Acids by the Rat. J. Nutrit. **21**, 373 (1941).
85. DEUEL, H. J., Jr., C. HENDRICK and M. E. CROCKETT: Studies on the Comparative Nutritive Value of Fat. VII. Growth Rate with Restricted Calories and on Injection of the Growth Hormone. J. Nutrit. **31**, 737 (1946).
86. DEUEL, H. J., Jr., C. HENDRICK, E. STRAUB, A. SANDOVAL, J. H. PINCKARD and L. ZECHMEISTER: Stereochemical Configuration and Provitamin A Activity. VI. Some *cis-trans* Isomers of γ-Carotene. Arch. Biochemistry **14**, 97 (1947).
87. DEUEL, H. J., Jr. and A. D. HOLMES: Digestibility of Cod Liver, Java Almond, Tea Seed, and Watermelon Seed Oils, Deer Fat, and Some Blended Hydrogenated Fats. U. S. Dept. Agric. Bull. No. 1033 (1922).
88. DEUEL, H. J., Jr., R. M. JOHNSON, C. E. CALBERT, J. GARDNER and B. THOMAS: Studies on the Comparative Nutritive Value of Fats. XII. The Digestibility of Rapeseed and Cottonseed Oils in Human Subjects. J. Nutrit. **38**, 369 (1949).
89. DEUEL, H. J., Jr., C. JOHNSTON, E. SUMNER, A. POLGÁR, W. A. SCHROEDER and L. ZECHMEISTER: Stereochemical Configuration and Provitamin A Activity. II. All-*trans*-γ-Carotene and Pro-γ-Carotene. Arch. Biochemistry **5**, 365 (1944).
90. DEUEL, H. J., Jr., E. R. MESERVE, C. JOHNSTON, A. POLGÁR and L. ZECHMEISTER: Re-investigation of the Relative Provitamin A Potencies of Cryptoxanthin and β-Carotene. Arch. Biochemistry **7**, 447 (1945).
91. DEUEL, H. J., Jr., E. R. MESERVE, E. STRAUB, C. HENDRICK and B. T. SCHEER: The Effect of Fat Level of the Diet on General Nutrition. I. Growth, Reproduction and Physical Capacity of Rats Receiving Diets Containing Various Levels of Cottonseed Oil or Margarine Fat *ad libitum*. J. Nutrit. **33**, 569 (1947).
92. DEUEL, H. J., Jr. and E. MOVITT: Studies on the Comparative Nutritive Value of Fats. III. The Effect of Flavor on Food Preference. J. Nutrit. **27**, 339 (1944).
93. — — Studies on the Comparative Nutritive Value of Fat. V. The Growth Rate and Efficiency of Conversion of Various Diets to Tissues in Rats Weaned at 14 Days. J. Nutrit. **29**, 237 (1945).
94. DEUEL, H. J., Jr., E. MOVITT and L. HALLMAN: Studies on the Comparative Nutritive Value of Fats. IV. The Negative Effect of Different Fats on Fertility and Lactation in the Rat. J. Nutrit. **27**, 509 (1944).
95. DEUEL, H. J., Jr., E. MOVITT, L. HALLMAN and F. H. MATTSON: Studies on the Comparative Nutritive Value of Fats. I. Growth Rate and Efficiency of Conversion of Various Diets to Tissue. J. Nutrit. **27**, 107 (1944).
96. DEUEL, H. J., Jr., E. SUMNER, C. JOHNSTON, A. POLGÁR and L. ZECHMEISTER: Stereochemical Configuration and Provitamin A Activity. III. All-*trans*-α-Carotene and neo-α-Carotene U. Arch. Biochemistry **6**, 157 (1945).
97. DOERSCHUK, A. P. and B. F. DAUBERT: Low-temperature Fractionation of Corn Oil and Calculation of its Glyceride Structure. J. Amer. Oil chem. Soc. **25**, 425 (1948).

98. DUNHAM, L. J. and A. BRUNSCHWIG: Intravenous Administration of Fat for Nutritional Purposes. Arch. Surgery **48**, 395 (1944).

99. DUNN, H. C., T. P. HILDITCH and J. P. RILEY: The Composition of Seed Fats of West Indian *Citrus* Fruits. J. Soc. chem. Ind. **67**, 199 (1948).

100. ECKSTEIN, H. C.: The Influence of Diet on the Body Fat of the White Rat. J. biol. Chemistry **81**, 613 (1929).

101. — The Influence of the Ingestion of Tricaprin on the Body Fat of the White Rat. J. biol. Chemistry **84**, 353 (1929).

102. ELLIS, N. R. and O. G. HANKINS: Soft Pork Studies. I. Formation of Fat in the Pig on a Ration Moderately Low in Fat. J. biol. Chemistry **66**, 101 (1925).

103. ELLIS, N. R. and H. S. ISBELL: Soft Pork Studies. II. The Influence of the Character of the Ration Upon the Composition of the Body Fat of Hogs. J. biol. Chemistry **69**, 219 (1926).

104. — — Soft Pork Studies. III. The Effect of Food Fat Upon Body Fat, as Shown by the Separation of the Individual Fatty Acids of the Body Fat. J. biol. Chemistry **69**, 239 (1926).

105. ELLIS, N. R. and J. H. ZELLER: Soft Pork Studies. IV. The Influence of a Ration Low in Fat upon the Composition of the Body Fat of Hogs. J. biol. Chemistry **89**, 185 (1930).

106. ELSDEN, S. R.: The Application of Silica Gel Partition Chromatogram to the Estimation of Volatile Fatty Acids. Biochemic. J. **40**, 252 (1946).

107. ELVEHJEM, C. A.: The Rôle of Intestinal Bacteria in Nutrition. J. Amer. Dietet. Assoc. **22**, 959 (1946).

108. ENGEL, R. W.: The Relation of B-Vitamins and Dietary Fat to the Lipotropic Action of Choline. J. Nutrit. **24**, 175 (1942).

108a. ERSHOFF, B. H.: Protective Effects of Soybean Meal for the Immature Hyperthyroid Rat. J. Nutrit. **39**, 259 (1949).

109. ERSHOFF, B. H. and H. J. DEUEL, Jr.: The Effect of Growth Hormone on the Vitamin A-deficient Rat. Endocrinology **36**, 280 (1945).

110. — — Inadequacy of Lactose and β-Lactose as Dietary Carbohydrates for the Rat. J. Nutrit. **28**, 225 (1944).

111. EULER, B. v., H. v. EULER and G. LINDEMAN: Biologic Facts on Vaccenic Acid. II. Ark. Kem., Mineral. Geol. **26**, No. 3, 1 (1948).

112. EULER, B. v., H. v. EULER and I. RÖNNENSTAM-SÖBERG: Influence of Fat Nutrition Upon the Growth, the Fertility, and the Longevity of Rats. I. Ark. Kem., Mineral. Geol., Ser. A **22**, No. 8, 1 (1946).

113. — — — The Nutritional Value of Different Fats. Ernährung **7**, 65 (1942).

114. — — — Versuche über die Nährungsfaktoren der Butter. Ernährung **8**, 257 (1943).

115. EVANS, H. M. and G. O. BURR: Vitamin E. II. The Destructive Effect of Certain Fats and Fractions Thereof on the Antisterility Vitamin in Wheat Germ and in Wheat Germ Oil. J. Amer. med. Assoc. **89**, 1587 (1927).

116. — — A New Dietary Deficiency with Highly Purified Diets. Proc. Soc. exp. Biol. Med. **24**, 740 (1926/27).

117. EVANS, H. M. and S. LEPKOVSKY: Sparing Action of Fat on the Antineuritic Vitamin B. J. biol. Chemistry **83**, 269 (1929).

118. — — Beneficial Effects of Fat in High Sucrose Diets when the Requirements for Antineuritic Vitamin B and the Fat-soluble Vitamin are Fully Satisfied. J. biol. Chemistry **92**, 615 (1931).

119. — — Vital Need of the Body for Certain Unsaturated Fatty Acids. I. Experiments with Fat-free Diets in Which Sucrose Furnishes the Sole Source of Energy. J. biol. Chemistry **96**, 143 (1932).

120. EVANS. H. M. and S LEPKOVSKY: Vital Needs of the Body for Certain Unsaturated Fatty Acids. II. Experiments with High-fat Diets in which Saturated Fatty Acids Furnish the Sole Source of Energy. J. biol. Chemistry **96**, 157 (1932).

121. — — The Sparing Action of Fat on Vitamin B. II. The Rôle Played by the Melting Point and the Degree of Unsaturation of Various Fats. J. biol. Chemistry **96**, 165 (1932).

122. — — The Sparing Action of Fat on Vitamin B. III. The Rôle Played by Glycerides of Single Fatty Acids. J. biol. Chemistry **96**, 179 (1932).

123. — — Vital Need of the Body for Certain Unsaturated Fatty Acids. J. biol. Chemistry **99**, 231 (1932).

124. — — The Sparing Action of Fat on Vitamin B. IV. Is it Necessary for Fat to Interact with Vitamin B in the Alimentary Canal to Exert its Sparing Effect? J. biol. Chemistry **99**, 235 (1932).

125. — — Sparing Action of Fat on the Antineuritic Vitamin. Science (New York) **68**, 298 (1928).

126. EVANS, H. M., S. LEPKOVSKY and E. A. MURPHY: The Sparing Action of Fat on Vitamin G. J. biol. Chemistry **107**, 443 (1934).

127. FABER, H. K. and D. B. ROBERTS: Studies in Infantile Allergic Eczema. II. Serum Lipids with Special Reference to Saturation of the Fatty Acids. J. Pediatrics **6**, 490 (1935).

128. FINNERUD, C. W., R. L. KESLER and H. F. WIESE: Ingestion of Lard in the Treatment of Eczema and Allied Dermatoses. Arch. Dermatol. Syphilology **44**, 849 (1941).

129. FOLCH, J.: Cephalin, a Mixture of Phosphatides. Separation from it of Phosphatidyl Serine, Phosphatidyl Ethanolamine, and a Fraction Containing an Inositol Phosphatide. J. biol. Chemistry **146**, 35 (1942).

130. FORBES, E. B. and R. W. SWIFT: Associative Dynamic Effects of Protein, Carbohydrate and Fat. J. Nutrit. **27**, 453 (1944).

131. FORBES, E. B., R. W. SWIFT, R. F. ELLIOTT and W. H. JAMES: Relation of Fat to Economy of Food Utilization. I. By the Growing Albino Rat. J. Nutrit. **31**, 203 (1946).

132. — — — — Relation of Fat to Economy of Food Utilization. II. By the Mature Albino Rat. J. Nutrit. **31**, 213 (1946).

133. FORBES, E. B., R. W. SWIFT, W. H. JAMES, J. W. BRATZLER and A. BLACK: Further Experiments on the Relation of Fat to Economy of Food Utilization. I. By the Growing Albino Rat. J. Nutrit. **32**, 387 (1946).

134. FRAPS, G. S. and W. W. MEINKE: Relative Values of Carotenes in Foods as Measured by Storage of Vitamin A in Livers of Rats. Food Res. **10**, 187 (1945).

135. FRENCH, C. E., A. BLACK and R. W. SWIFT: Further Experiments on the Relation of Fat to Economy of Food Utilization. III. Low Protein Intake. J. Nutrit. **35**, 83 (1948).

136. GAVIN, G. and E. W. MCHENRY: The Effects of Biotin Upon Fat Synthesis and Metabolism. J. biol. Chemistry **141**, 619 (1941).

136a. GEIGER, E.: Personal Communication to the Authors.

137. GEYER, R. P., J. CHIPMAN and F. J. STARE: Oxidation *in vivo* of Emulsified Radioactive Trilaurin Administered Intravenously. J. biol. Chemistry **176**, 1469 (1948).

138. GEYER, R. P., B. R. GEYER, P. H. DERSE, H. NATH, V. H. BARKI, C. A. ELVEHJEM and E. B. HART: The Nutritive Value of Fractions of Butter Fat Prepared by Cold Crystallization. J. Dairy Sci. **30**, 299 (1947).

139. GEYER, R. P., G. V. MANN, J. YOUNG, T. D. KINNEY and F. J. STARE: Parenteral Nutrition. V. Studies on Soybean Phosphatides as Emulsifiers for Intravenous Fat Emulsions. J. Lab. clin. Med. **33**, 163 (1948).

140. GEYER, R. P., H. NATH, V. H. BARKI, C. A. ELVEHJEM and E. B. HART: The Vaccenic Acid Content of Various Fats and Oils. J. biol. Chemistry **169**, 227 (1947).

141. GIBSON, G. and C. F. HUFFMAN: Influence of Different Levels of Fat in the Ration Upon Milk and Fat Secretion. Michigan State Coll. Agric. appl. Sci., agric. Exp. Sta. quart. Bull. **21**, 258 (1939).

141a. GILLAM, A. E. and I. M. HEILBRON: Cryptoxanthine in Butter. Biochemic. J. **29**, 834 (1935).

141b. — — Cryptoxanthine in Egg Yolk. Biochemic. J. **29**, 1064 (1935).

141c. GILLAM, A. E. and M. S. EL RIDI: Carotene in Milk Fat. Biochemic. J. **31**, 251 (1937).

142. GILMORE, R. C. and L. T. SAMUELS: The Effect of Previous Diet on the Metabolic Activity of the Isolated Rat Diaphragm. J. biol. Chemistry **181**, 813 (1949).

143. GINSBERG, J. E., C. BERNSTEIN, Jr. and L. V. IOB: Effect of Oils Containing Unsaturated Fatty Acids on Patients with Dermatitis. Arch. Dermatol. Syphilology **36**, 1033 (1937).

144. GOTTLIEB, H. L., F. W. QUACKENBUSH and H. STEENBOCK: The Biological Determination of Vitamin E. J. Nutrit. **25**, 433 (1943).

144a. GREENBERG, S M., C. E. CALBERT, J. H. PINCKARD, H. J. DEUEL, Jr. and L. ZECHMEISTER: Stereochemical Configuration and Provitamin A Activity. IX. A Comparison of all-*trans*-γ-Carotene and pro-γ-Carotene with all-*trans*-β-Carotene in the Chick. Arch. Biochemistry **24**, 31 (1949).

145. GRIFFITHS, H. N. and T. P. HILDITCH: The Oleic-elaidic Acid Transformation as an Aid in the Analysis of Mixtures of Oleic, Linoleic and Linolenic Acid. J. Soc. chem. Ind. **53**, 75 T (1934).

146. GROER, F. VON: Zur Frage der praktischen Bedeutung des Nährwertbegriffes nebst einigen Bemerkungen über das Fett-Minimum des menschlichen Säuglings. Biochem. Z. **97**, 311 (1919).

147. GROSSFELD, J. and A. SIMMER: Separation and Determination of Solid Fatty Acids in Edible Fats. Z. Unters. Lebensmittel **59**, 237 (1930); Chem. Abstr. **24**, 5173 (1930).

147a. GUERRA, E. Z.: Accion inhibidora de diversas grasas dieteticas (saturadas y no saturadas) sobre elevacion del consumo de oxígeno producida por la tiroxina. Tesis Medico-Cirujano., Univ. de Chile, Inst. Fisiol. (1947).

148. GULLICKSON, T. W., F. C. FOUNTAINE and J. B. FITCH: Various Oils and Fats as Substitutes for Butter Fat in the Ration of Young Calves. J. Dairy Sci. **24**, 315 (1941).

149. GUNDE, B. G. and T. P. HILDITCH: The Mixed Unsaturated Glycerides of Liquid Seed Fats. I. Some "Non-drying" Oils. J. Soc. chem. Ind. **59**, 47 T (1940).

150. HANDLER, P.: The Biochemical Defect Underlying the Nutritional Failure of Young Rats on Diets Containing Excessive Quantities of Lactose or Galactose. J. Nutrit. **33**, 221 (1947).

151. HANKINS, O. G. and N. R. ELLIS: Some Results of Soft Pork Investigations. U. S. Dept. Agric., Dept. Bull. No. 1407 (April 1926).

152. HANKINS, O. G., N. R. ELLIS and J. H. ZELLER: Some Results of Soft Pork Investigations. II. U. S. Dept. Agric., Dept. Bull. No. 1492 (Feb. 1928).

153. Hansen, A. E.: Serum Lipids in Eczema and in Other Pathologic Conditions. Amer. J. Diseases Children 53, 933 (1937).

154. — Serum Lipid Changes and Therapeutic Effects of Various Oils in Infantile Eczema. Proc. Soc. exp. Biol. Med. 31, 160 (1933).

155. Hansen, A. E. and G. O. Burr: Iodine Numbers of Serum Lipids in Rats Fed on Fat-free Diets. Proc. Soc. exp. Biol. Med. 30, 1201 (1933).

156. Hansen, A. E. and H. F. Wiese: Studies with Dogs Maintained on Diets Low in Fat. Proc. Soc. exp. Biol. Med. 52, 205 (1943).

157. Heiduschka, A. and C. Wiesemann: Composition of Almond Oil; Comparison of Almond Oil with Oil of Apricot Kernels. J. prakt. Chem. 124, 240 (1930); Chem. Abstr. 24, 2907 (1930).

158. Henry, K. M., S. K. Kon, T. P. Hilditch and M. L. Meara: Comparison of the Growth-promoting Value for Rats of Butter Fat, of Margarine Fat and of Vegetable Oils. J. Dairy Res. 14, 45 (1945).

159. Hilditch, T. P.: Recent Advances in the Study of Component Acids and Component Glycerides of Natural Fats. Fortschr. Chem. organ. Naturstoffe 5, 72 (1948).

160. — The Chemical Constitution of Natural Fats, 2nd. ed. London: Chapman and Hall; and New York: J. Wiley. 1947.

161. — The Component Glycerides of Vegetable Fats. Fortschr. Chem. organ. Naturstoffe 1, 24 (1938).

162. Hilditch, T. P. and C. H. Lea: Investigation of the Constitution of Glycerides in Natural Fats. A Preliminary Outline of Two New Methods. J. chem. Soc. (London) 1927, 3106.

163. Hilditch, T. P. and J. A. Lovern: The Head and Blubber Oils of the Sperm Whale. I. Quantitative Determinations of the Mixed Fatty Acids Present. J. Soc. chem. Ind. 47, 105 (1928).

164. — — The Head and Blubber Oils of the Sperm Whale. II. Investigation of Some of the Component Wax Esters and the General Structure of the Oils. J. Soc. chem. Ind. 48, 359 T (1929).

165. — — Head and Blubber Oils of the Sperm Whale. III. Quantitative Determinations of the Higher Fatty Alcohols Present. J. Soc. chem. Ind. 48, 365 T (1929).

166. — — The Evolution of Natural Fats: a General Survey. Nature (London) 137, 478 (1936).

167. Hilditch, T. P. and L. Maddison: The Mixed Unsaturated Glycerides of Liquid Seed Fats. II. Low Temperature Crystallization of Cottonseed Oil. J. Soc. chem. Ind. 59, 162 T (1940).

168. Hilditch, T. P. and K. S. Murti: The Component Acids and Glycerides of Some Indian Ox Depot Fats. Biochemic. J. 34, 1301 (1940).

169. Hilditch, T. P. and S. Paul: The Component Glycerides of an Ox Depot Fat. Biochemic. J. 32, 1775 (1938).

170. Hilditch, T. P. and W. H. Pedelty: Component Glycerides of Perinephric and Outer Back Pig Fats from the Same Animal. Biochemic. J. 34, 971 (1940).

171. Hilditch, T. P. and Y. A. H. Zaky: Sheep Body Fats. II. Component Glycerides of Perinephric and External Tissue Fats from the Same Animal. Biochemic. J. 35, 940 (1941).

172. Hoagland, R. and G. G. Snider: Digestibility of Some Animal and Vegetable Fats. J. Nutrit. 25, 295 (1943).

173. — — Digestibility of Certain Higher Saturated Fatty Acids and Triglycerides. J. Nutrit. 26, 219 (1943).

174. — — Nutritive Properties of Certain Animal and Vegetable Fats. U. S. Dept. Agric. Tech. Bull. No. 725 (Jan. 1940).

174a. HOFFMANN, F., E. J. DE HOFFMANN and J. TALESNIK: Influencia de dietas rica en grasa sobre la elevacion del consumo de oxígeno inducida por la tiroxina. Rev. Med. Chile, Ano LXXIII, No. 5, 392 (1945).

175. HOFMANN, K. and A. E. AXELROD: The Biotin Activity of Fat-soluble Materials from Plasma. Arch. Biochemistry **14**, 482 (1947).

176. HOGAN, A. G. and L. R. RICHARDSON: Plural Nature of Vitamin B. Nature (London) **136**, 186 (1935).

177. HOLMAN, R. T. and L. HAGDAHL: Displacement Analysis of Lipids. Preliminary Studies with Normal Saturated Fatty Acids. Arch. Biochemistry **17**, 301 (1948).

178. — — Separation of Fatty Acids by Displacement Chromatography and its Application to the Analysis of Butter Fat. J. Dairy Sci. **32**, 700 (1949).

179. HOLMES, A. D.: Digestibility of Oleomargarine. Boston Med. and Surg. J. **192**, 1210 (1925).

180. — Digestibility of Certain Miscellaneous Animal Fats. U. S. Dept. Agric. Bull. No. 613 (Apr. 25, 1919).

181. — Studies on the Digestibility of Some Nut Oils. U. S. Dept. Agric. Bull. No. 630 (Apr. 16, 1916).

182. — Digestibility of Some Seed Oils. U. S. Dept. Agric. Bull. No. 687 (June 28, 1918).

183. — Digestibility of Some By-product Oils. U. S. Dept. Agric. Bull. No. 781 (May 31, 1919).

184. HOLMES, A. D. and H. J. DEUEL, Jr.: Digestibility of Some Hydrogenated Oils. Amer. J. Physiol. **54**, 479 (1921).

185. — — Digestibility of Certain Miscellaneous Vegetable Fats. J. biol. Chemistry **41**, 227 (1920).

186. HOLT, L. E. and C. N. KAJDI: The Nutritional Requirements in Inanition. I. Observations on the Ability of Single Foodstuffs to Prolong Survival. Bull. Johns Hopkins Hosp. **74**, 121 (1944).

187. HOLT, L. R., Jr., H. C. TIDWELL, C. M. KIRK, D. M. CROSS and S. NEALE: Studies in Fat Metabolism. J. Pediatrics **6**, 427 (1935).

188. HOWE, P. E.: The Dietaries of our Military Forces. Ann. Amer. Acad. Polit. Sci. **225**, 72 (1943).

189. HUME, E. M., L. C. A. NUNN, I. SMEDLEY-MACLEAN and H. H. SMITH: Studies of the Essential Unsaturated Fatty Acids in their Relation to the Fat-deficiency Disease of Rats. Biochemic. J. **32**, 2162 (1938).

190. — — — — Fat-deficiency Disease of Rats. The Relative Curative Potencies of Methyl Linoleate and Methyl Arachidonate with a Note on the Action of the Methyl Esters of Fat Acids from Cod Liver Oil. Biochemic. J. **34**, 879 (1940).

191. IRWIN, M. H., H. STEENBOCK and V. M. TEMPLIN: A Technic for Determining the Rate of Absorption of Fats. J. Nutrit. **12**, 85 (1936).

192. JACK, E. L., J. L. HENDERSON, D. F. REID and S. LEPKOVSKY: Nutritional Studies on Milk Fat. II. The Growth of Young Rats Fed Glyceride Fractions Separated from Milk Fat. J. Nutrit. **30**, 175 (1945).

193. JACK, E. L. and E. B. HINSHAW: Nutritional Studies on Milk Fat. III. The Effect of the Treatment of Milk Fat with Certain Solvents on the Growth of Young Rats. J. Nutrit **34**, 715 (1947).

194. JAMIESON, G. S., W. F. BAUGHMAN and R. M. HANN: Avocado Oil. Oil Fat Ind. **5**, 202 (1928).

195. JENSEN, J. L., K. C. D. HICKMAN and P. L. HARRIS: Effect of Tocopherols and Soybean Phosphatides on the Utilization of Carotene. Proc. Soc. exp. Biol. Med. **54**, 294 (1943).

196. JONES, C. M., P. J. CULVER, G. D. DRUMMEY and A. E. RYAN: Modification of Fat Absorption in the Digestive Tract by the Use of an Emulsifying Agent. Ann. intern. Med. **29**, 1 (1948).

197. KARRER, P. u. H. KELLER: Quantitative Bestimmung der Tocopherole in verschiedenen Ausgangsmaterialien. Helv. chim. Acta **21**, 1161 (1938).

198. KARRER, P. u. H. KOENIG: Beitrag zur Kenntnis der essentiellen ungesättigten Fettsäuren. Helv. chim. Acta **26**, 619 (1943).

199. KAUFMANN, H. P.: The Importance of Accompanying Substances in Natural Fats and their Fate During Refining. I. The Chemistry and Biology of Accompanying Substances. Fette u. Seifen **48**, 53 (1941); Chem. Abstr. **36**, 1202 (1942).

200. KAUFMANN, H. P. u. S. JUSCHKEWITSCH: Quantitative Analyse des Hanföles. Z. angew. Chem. **43**, 90 (1930).

201. KAUFMANN, H. P. and W. WOLF: Adsorption Separation in the Field of Fats. V. The Separation of *cis-trans* Isomers. Fette u. Seifen **50**, 519 (1943); Chem. Abstr. **39**, 205 (1945).

201a. KEESER, E.: Untersuchungen über antithyreoidwirksame Stoffe. Klin. Wschr. **17**, 1100 (1938).

202. KREULA, M. and A. I. VIRTANEN: Absorption of Carotene from Carrots in Human Subjects. Upsala Läkareфören. Förh. **45**, 355 (1939); Chem. Abstr. **34**, 5897 (1940); cited by D. MELNICK and B. L. OSER: Physiological Availability of Vitamins. Vitamins and Hormones **5**, 39 (1947), p. 59.

203. KRINGSTAD, H. and J. LIE: Biologic Effect of Vitamin A as Ester and as Free Alcohol (Axerophthol). Tidsskr. Kjemi Bergves. **2**, 57 (1942); Chem. Abstr. **28**, 2367 (1944).

204. KROGH, A. and J. LINDHARD: The Relative Values of Fat and Carbohydrate as Sources of Muscular Energy with Appendices on the Correlation Between Standard Metabolism and the Respiratory Quotient During Rest and Work. Biochemic. J. **14**, 290 (1920).

205. LANGWORTHY, C. F.: The Digestibility of Fats. Ind. Engng. Chem. **15**, 276 (1923).

206. LANGWORTHY, C. F. and A. D. HOLMES: Digestibility of Some Animal Fats. U. S. Dept. Agric. Bull. No. 310 (Nov. 9, 1915).

207. — — Digestibility of Some Vegetable Fats. U. S. Dept. Agric. Bull. No. 505 (Feb. 13, 1917).

208. — — Studies on the Digestibility of Some Animal Fats. U. S. Dept. Agric Bull. No. 507 (May 24, 1917).

209. LASSEN, S. and E. K. BACON: The Growth-promoting Effect on the Rat of Summer Butter and Other Fats. J. Nutrit. **39**, 83 (1949).

210. LASSEN, S., E. K. BACON and H. J. DUNN: The Digestibility of Polymerized Oils. Arch. Biochemistry **23**, 1 (1949).

211. LEASE, E. J., J. G. LEASE, H. STEENBOCK and C. A. BAUMANN: The Biological Value of Carotene in Various Fats. J. Nutrit. **17**, 91 (1939).

212. LEBEDEFF, A.: Woraus bildet sich das Fett in Fällen der akuten Fettbildung? Experimenteller Beitrag zur Kenntnis der Leber- und Milchfette. Pflügers Arch. ges. Physiol. Menschen Tiere **31**, 11 (1883).

213. LEICHENGER, H., G. EISENBERG and A. J. CARLSON: Margarine and the Growth of Children. J. Amer. med. Assoc. **136**, 368 (1948).

214. LERNER, S. R., I. L. CHAIKOFF, C. ENTENMAN and W. G. DAUBEN: Oxidation of Parenterally Administered Carbon 14-labeled Tripalmitin Emulsion. Science (New York) **109**, 13 (1949).

215. LIPMANN, F., N. O. KAPLAN, G. D. NOVELLI, L. C. TUTTLE and B. M. GUIRARD: Coenzyme for Acetylation, a Pantothenic Acid Derivative. J. biol. Chemistry **167**, 869 (1947).
216. LONGENECKER, H.: Composition and Structural Characteristics of Glycerides in Relation to Classification and Environment. Chem. Reviews **29**, 201 (1941).
217. — Deposition and Utilization of Fatty Acids. I. Fat Synthesis from High-carbohydrate and High-protein Diets in Fasted Rats. J. biol. Chemistry **128**, 645 (1939).
218. LOOSLI, J. K., J. F. LINGENFELTER, J. W. THOMAS and L. A. MAYNARD: The Rôle of Dietary Fat and Linoleic Acid in the Lactation of the Rat. J. Nutrit. **28**, 81 (1944).
219. LUNDBAEK, K. and J. A. F. STEVENSON: The Effect of Previous Carbohydrate Deprivation on the Carbohydrate Metabolism of Isolated Muscle. Federat. Proc. **7**, 75 (1948).
220. LUSK, G.: Elements of the Science of Nutrition. 4th ed., p. 186. Philadelphia: Saunders. 1928.
221. — Über den Einfluß der Kohlehydrate auf den Eiweißzerfall. Z. Biol. **27**, 459 (1890).
222. LYMAN, J. F.: Metabolism of Fats. Utilization of Palmitic Acid, Glyceryl Palmitate, and Ethyl Palmitate by the Dog. J. biol. Chemistry **32**, 7 (1917).
223. McAMIS, A. J., W. E. ANDERSON and L. B. MENDEL: Growth of Rats on "Fat-free" Diets. J. biol. Chemistry **82**, 247 (1929).
224. McCAY, C. M. and H. PAUL: The Effect of Melting Point of Fat Upon its Utilization by Guinea Pigs. J. Nutrit. **15**, 377 (1938).
225. McHENRY, E. W. and M. L. CORNETT: The Rôle of Vitamins in the Anabolism of Fats. Vitamins and Hormones **2**, 1 (1944).
226. McHENRY, E. W. and G. GAVIN: The Effects of Liver and Pancreas Extracts upon Fat Synthesis and Metabolism. J. biol. Chemistry **134**, 683 (1940).
227. — — The B Vitamins and Fat Metabolism. IV. The Synthesis of Fat From Protein. J. biol. Chemistry **138**, 471 (1941).
228. McKIBBEN, J. M., R. M. FERRY and F. J. STARE: Parenteral Nutrition. II. The Utilization of Emulsified Fat Given Intravenously. J. clin. Invest. **25**, 679 (1946).
229. McKIBBEN, J. M., A. POPE, S. THAYER, R. M. FERRY and F. J. STARE: Parenteral Nutrition. I. Studies on Fat Emulsions for Intravenous Alimentation. J. Lab. clin. Med. **30**, 488 (1945).
230. MACDONALD, D. G. H. and E. W. McHENRY: Studies on Rat Bradycardia. Amer. J. Physiol. **128**, 608 (1940).
231. MANNERING, G. J., M. A. LIPTON and C. A. ELVEHJEM: Relation of Dietary Fat to Riboflavin Requirement of Growing Rats. Proc. Soc. exp. Biol. Med. **46**, 100 (1941).
232. MANNERING, G. J., D. ORSINI and C. A. ELVEHJEM: Effect of the Composition of the Diet on the Riboflavin Requirement of the Rat. J. Nutrit. **28**, 141 (1944).
233. MARTIN, G. J.: Studies of Fat-free Diets. J. Nutrit. **17**, 127 (1939).
234. MATTIL, K. F. and J. W. HIGGINS: The Relationship of Glyceride Structure to Fat Digestibility. I. Synthetic Glycerides of Stearic and Oleic Acids. J. Nutrit. **29**, 255 (1945).
235. MAYNARD, L. A., K. E. GARDNER and A. HODSON: Soybeans as a Source of Fat in the Dairy Ration. Cornell Univ. Agric. Exp. Sta., Bull. No. 7, 22 (Mar. 10, 1939).
236. MAYNARD, L. A. and E. RASMUSSEN: The Influence of Dietary Fat on Lactation Performance in Rats. J. Nutrit. **23**, 385 (1942).

237. MEDES, G., D. C. KELLER and A. KURKJIAN: Essential Fatty Acid Metabolism. I. Essential Fatty Acid Content of Rats on Fat-free and Pyridoxine-free Diets. Arch. Biochemistry **15**, 19 (1947).

238. MEIGS, E. B.: Milk Secretion as Related to Diet. Physiologic. Rev. **2**, 204 (1922).

239. MELNICK, D. and H. FIELD, Jr.: A Note on the Vitamin B_1-sparing Action of Fat. J. Nutrit. **17**, 223 (1939).

240. MELNICK, D. and B. L. OSER: Physiological Availability of Vitamins. Vitamins and Hormones **5**, 39 (1947).

241. MELVILLE, D. B.: The Chemistry of Biotin. Vitamins and Hormones **2**, 29 (1944).

242. MENG, H. C. and S. FREEMAN: Experimental Studies on the Intravenous Injection of a Fat Emulsion into Dogs. J. Lab. clin. Med. **33**, 689 (1948).

243. MILLER, H. G.: Nutritive Differences in Rations Containing Unhydrogenated or Hydrogenated Fats as Shown by Rearing Successive Generations of Rats. J. Nutrit. **26**, 43 (1943).

244. MILLER, W. B.: Oleomargarine; Definition and Standard of Identity. Federal Register **6**, 2761 (1941).

245. MILNER, R. T., J. E. HUBBARD and M. B. WIELE: Sunflower and Safflower Seeds and Oils. Oil and Soap **22**, 304 (1945).

246. MOYLE, V., E. BALDWIN and R. SCARISBRICK: Separation and Estimation of Saturated C_2—C_8 Fatty Acids by Buffered Partition Columns. Biochemic. J. **43**, 308 (1948).

247. MUELDER, K. D. and E. KELLY: The Effect of the Level of Fat in the Diet upon Utilization of Vitamin A. J. Nutrit. **23**, 335 (1942).

247a. MUNOZ ORTIZ, E.: Acción de la tiroxina sobre la cuantía del metabolismo en ratas sometidas a dietas ricas en grasa y relativamente pobres en colina. Tesis Circujano-Dentista, Univ. de Chile, Inst. Fisiol. (1944).

248. MURLIN, J. R. and G. LUSK: Animal Calorimetry. XII. The Influence of the Ingestion of Fat. J. biol. Chemistry **22**, 15 (1915).

249. MUNK, I.: Zur Lehre von der Resorption, Bildung und Ablagerung der Fette im Tierkörper. Virchow's Arch. pathol. Anatom. Physiol. klin. Med. **95**, 407 (1884).

250. NATH, H., V. H. BARKI, C. A. ELVEHJEM and E. B. HART: Studies on the Alleged Growth-promoting Property of Vaccenic Acid. J. Nutrit. **36**, 761 (1948).

251. NORRIS, F. A. and K. F. MATTIL: A New Approach to the Glyceride Structure of Natural Fats. J. Amer. Oil chem. Soc. **24**, 274 (1947).

252. NORRIS, F. A. and D. E. TERRY: Precise Laboratory Fractional Distillation of Fatty Acid Esters. Oil and Soap **22**, 41 (1945).

253. NOVELLI, G. D. and F. LIPMANN: Bacterial Conversion of Pantothenic Acid Into Coenzyme A (Acetylation) and its Relation to Pyruvic Oxidation. Arch. Biochemistry **14**, 23 (1947).

254. — — The Involvement of Coenzyme A in Acetate Oxidation in Yeast. J. biol. Chemistry **171**, 833 (1947).

255. OPPEL, T. W.: Studies of Biotin Metabolism in Man. J. clin. Invest. **21**, 630 (1942).

255a. OSBORNE, T. B. and L. B. MENDEL: The Comparative Nutritive Value of Certain Proteins in Growth and the Problem of Protein Minimum. J. biol. Chemistry **20**, 351 (1915).

256. PAINTER, E. P.: Some Relationships Between Fat Acid Composition and the Iodine Number of Linseed Oil. Oil and Soap **21**, 343 (1944).

257. PAINTER, E. P. and L. L. NESBITT: Thiocyanogen Absorption of Linseed Oils. Thiocyanogen Absorption of Linoleic and Linolenic Acids and the Composition of Linseed Oils. Ind. Engng. Chem., Analyt. Edit. **15**, 123 (1943).
258. — — Fat acid Composition of Linseed Oil from Different Varieties of Flaxseed. Oil and Soap **20**, 208 (1943).
259. PAUL, H. and C. M. McCAY: Utilization of Fats by Herbivora. Arch. Biochemistry **1**, 247 (1942).
260. PAVCEK, P. L. and G. M. SHULL: Inactivation of Biotin by Rancid Fats. J. biol. Chemistry **146**, 351 (1942).
261. PEARSON, P. B. and F. PANZER: Effect of Fat in the Diet of Rats on their Growth and their Excretion of Amino Acids. J. Nutrit. **38**, 257 (1949).
262. PETERSON, M. H. and M. S. JOHNSON: The Estimation of Fatty Acids of Intermediate Chain Length by Partition Chromatography. J. biol. Chemistry **174**, 775 (1948).
263. POTTER, R. L., A. E. AXELROD and C. A. ELVEHJEM: The Riboflavin Requirement of the Dog. J. Nutrit. **24**, 449 (1942).
264. POWELL, M.: The Metabolism of Tricaprylin and Trilaurin. J. biol. Chemistry **89**, 547 (1930).
265. — The Metabolism of Tricaprin. J. biol. Chemistry **95**, 43 (1932).
266. QUACKENBUSH, F. W., H. L. GOTTLIEB and H. STEENBOCK: Distillation of Tocopherols from Soybean Oil. Ind. Engng. Chem. **33**, 1276 (1941).
267. QUACKENBUSH, F. W., F. A. KUMMEROW and H. STEENBOCK: The Effectiveness of Linoleic, Arachidonic, and Linolenic Acids in Reproduction and Lactation. J. Nutrit. **24**, 213 (1942).
268. QUACKENBUSH, F. W., H. STEENBOCK, F. A. KUMMEROW and B. R. PLATZ: Linoleic Acids, Pyridoxine and Pantothenic Acid in Rat Dermatitis. J. Nutrit. **24**, 225 (1942).
269. RALSTON, A. W.: Fatty Acids and their Derivatives, p. 289. New York: J. Wiley. 1948.
270. RAMSEY, L. L. and W. I. PATTERSON: Separation and Identification of the Volatile Saturated Fatty Acids (C_1—C_4). J. Assoc. off. agric. Chemists **28**, 644 (1945).
271. — — Separation and Determination of the Straight-chain Saturated Fatty Acids C_5—C_{10} by Partition Chromatography. J. Assoc. off. agric. Chemists **31**, 139 (1948).
272. — — Separation of the Saturated Straight-chain Fatty Acids C_{11}—C_{19}. J. Assoc. off. agric. Chemists **31**, 441 (1948).
273. REINBOLD, C. L. and H. J. DUTTON: Adsorption Analysis of Lipids. II. The Fractionation of Soybean Oil and Derived Ethyl Esters. J. Amer. Oil chem. Soc. **25**, 117 (1948).
274. — — Adsorption Analysis of Lipids. III. Synthetic Mixtures of Ethyl Stearate, Oleate, Linoleate and Linolenate. J. Amer. Oil chem. Soc. **25**, 120 (1948).
275. REINHOLD, J. G., J. T. L. NICHOLSON and K. O'SHEA-ELSOM: The Utilization of Thiamine in the Human Subject; the Effect of High Intake of Carbohydrate or of Fat. J. Nutrit. **28**, 51 (1944).
276. RICHARDSON, L. R., A. G. HOGAN and K. F. ITSCHNER: Vitamin B_6, Pantothenic Acid, and the Unsaturated Fatty Acids as they Affect Dermatitis in Rats. Univ. Missouri Agric. Exper. Sta. Res. Bull. 333 (1941).
277. RIECKEHOFF, I. G., R. T. HOLMAN and G. O. BURR: Polyethenoid Fatty Acid Metabolism. Effect of Dietary Fat on Polyethenoid Fatty Acids of Rat Tissues. Arch. Biochemistry **20**, 331 (1949).

278. RIEMENSCHNEIDER, R. W., S. F. HERB and P. L. NICHOLS, Jr.: Isolation of Pure Natural Linoleic and Linolenic Acids as their Methyl Esters by Adsorption Fractionation on Silicic Acid. J. Amer. Oil chem. Soc. **26**, 371 (1949).
279. RIPLEY, W. E. and R. A. BOLOMEY: The Relation of the Biology of the Soupfin Shark to the Liver Yield of Vitamin A. Calif. Dept. Nat. Resources, Div. Fish and Game, Bu. Marine Fisheries, Fish Bull. No. *64*, 39 (1946).
280. ROBERTS, S. and L. T. SAMUELS: Influence of Previous Diet on Metabolism During Fasting. Amer. J. Physiol. **158**, 57 (1949).
281. — — The Influence of Previous Diet on the Preferential Utilization of Foodstuffs. Bull. Minnesota Med. Foundation **4**, 55 (Feb. 1944).
282. — — Influence of Previous Diet on Insulin Tolerance. Proc. Soc. exp. Biol. Med. **53**, 207 (1943).
283. ROBERTS, S., L. T. SAMUELS and R. M. REINECKE: Previous Diet and the Apparent Utilization of Fat in the Absence of the Liver. Amer. J. Physiol. **140**, 639 (1944).
284. ROBESON, C. D. and J. G. BAXTER: α-Tocopherol, a Natural Antioxidant in a Fish Liver Oil. J. Amer. chem. Soc. **65**, 940 (1943).
285. ROCKWOOD, E. W. and P. B. SIVICKES: Relative Digestibility of Maize Oil (Corn Oil), Cottonseed Oil and Lard. J. Amer. med. Assoc. **71**, 1649 (1918).
286. ROSE, W. G. and G. S. JAMIESON: The Composition of Seven American Linseed Oils. Oil and Soap **18**, 173 (1941).
287. ROSENBERG, H. R.: Chemistry and Physiology of the Vitamins. New York: Interscience Publishers, Inc. 1942.
288. ROWNTREE, J. I.: The Effect of the Use of Mineral Oil Upon the Absorption of Vitamin A. J. Nutrit. **3**, 345 (1931).
289. RUBIN, S. H. and J. SCHEINER: Antibiotin Effect of Homologs of Biotin and Biotin Sulfone. Arch. Biochemistry **23**, 400 (1949).
290. RUSSELL, W. C., M. W. TAYLOR and L. J. POLSKIN: Fat Requirements of the Growing Chick. J. Nutrit. **17**, 555 (1940).
291. RUSSELL, W. C., M. W. TAYLOR, H. A. WALKER and L. J. POLSKIN: The Absorption and Retention of Carotene and Vitamin A by Hens on Normal and Low-fat Rations. J. Nutrit. **24**, 199 (1942).
292. SALMON, W. D.: The Effect of certain Oils in Alleviating Localized Erythematous Dermatitis (Acrodynia or Vitamin B_6 Deficiency) in Rats. J. biol. Chemistry **123**, CIV (1938).
293. — The Relation of Pantothenic Acid, Pyridoxine, and Linoleic Acid to the Cure of Rat Acrodynia. J. biol. Chemistry **140**, CIX (1941).
293a. — Some Physiological Relationships of Protein, Fat, Choline, Methionine, Cystine, Nicotinic Acid and Tryptophane. J. Nutrit. **33**, 155 (1947).
294. SALMON, W. D. and J. G. GOODMAN: Alleviation of Vitamin B Deficiency in the Rat by Certain Natural Fats and Synthetic Esters. J. Nutrit. **13**, 477 (1937).
295. SAMUELS, L. T., R. C. GILMORE and R. M. REINECKE: The Effect of Previous Diet on the Ability of Animals to do Work During Subsequent Fasting. J. Nutrit. **36**, 639 (1948).
296. SARMA, P. S., E. E. SNELL and C. A. ELVEHJEM: The Bioassay of Vitamin B_6 in Natural Materials. J. Nutrit. **33**, 121 (1947).
297. SATO, Y., G. T. BARRY and L. C. CRAIG: Identification of Small Amounts of Organic Compounds by Distribution Studies. VII. Separation and Estimation of Normal Fatty Acids. J. biol. Chemistry **170**, 501 (1947).
298. SCHANTZ, E. J., R. K. BOUTWELL, C. A. ELVEHJEM and E. B. HART: The Nutritive Value of the Fatty Acid Fractions of Butter Fat. J. Dairy Sci. **23**, 1205 (1940).

299. SCHANTZ, E. J., C. A. ELVEHJEM and E. B. HART: The Comparative Nutritive Value of Butter Fat and Certain Vegetable Oils. J. Dairy Sci. **23**, 181 (1940).
300. SCHEER, B. T., J. F. CODIE and H. J. DEUEL, Jr.: The Effect of Fat Level of the Diet on General Nutrition. III. Weight Loss, Mortality and Recovery in Young Adult Rats Maintained on Restricted Calories. J. Nutrit. **33**, 641 (1947).
301. SCHEER, B. T., S. DORST, J. F. CODIE and D. F. SOULE: Physical Capacity in Rats in Relation to Energy and Fat Content of the Diet. Amer. J. Physiol. **149**, 194 (1947).
302. SCHEER, B. T., D. F. SOULE, M. FIELDS and H. J. DEUEL, Jr.: The Effect of Fat Level of the Diet on General Nutrition. II. Growth, Mortality and Recovery in Weanling Rats Maintained on Restricted Calories. J. Nutrit. **33**, 583 (1947).
303. SCHNEIDER, H. A.: Butter Fat in Dermatitis-producing Diets. Proc. Soc. exp. Biol. Med. **44**, 266 (1940).
304. SCHNEIDER, H. A., H. STEENBOCK and B. R. PLATZ: Essential Fatty Acids, Vitamin B_6, and Other Factors in the Cure of Rat Acrodynia. J. biol. Chemistry **132**, 539 (1940).
305. SCHOENHEIMER, R. and D. RITTENBERG: Deuterium as an Indicator in the Study of Intermediary Metabolism. J. biol. Chemistry **114**, 381 (1936).
306. — — The Study of Intermediary Metabolism of Animals with the Aid of Isotopes. Physiologic. Rev. **20**, 218 (1940).
307. SCHUETTE, H. A. and R. G. ZENNPFENNIG: The Seed Oil of the Hackberry. Oil and Soap **14**, 269 (1937).
308. SCHWIMMER, D. and T. H. MCGAVACK: Some Newer Aspects of Protein Metabolism. I. Resumé of Experimental Data. New York State J. Med. **48**, 1797 (1948).
309. SHEN, T.: The Diet of Chinese Soldiers and College Students in Wartime. Science (New York) **98**, 302 (1943).
310. SHERMAN, H. C. and H. L. CAMPBELL: Growth and Reproduction upon Simplified Food Supply. IV. Improvement in Nutrition Resulting from an Increased Proportion of Milk in the Diet. J. biol. Chemistry **60**, 5 (1924).
311. — — Further Experiments on the Influence of Food upon Longevity. J. Nutrit. **2**, 415 (1929/30).
312. — — Nutritional Well-being and Length of Life as Influenced by Different Enrichments of an Already Adequate Diet. J. Nutrit. **14**, 609 (1937).
313. SHERMAN, H. C. and H. Y. TRUPP: Further Experiments with Vitamin A in Relation to Aging and to Length of Life. Proc. nat. Acad. Sci. USA **35**, 90 (1949).
314. SHERMAN, W. C.: The Effect of Certain Fats and Unsaturated Fatty Acids Upon the Utilization of Carotene. J. Nutrit. **22**, 153 (1941).
315. SINCLAIR, R. G.: Growth of Rats on High-fat and Low-fat Diets, Deficient in the Essential Unsaturated Fatty Acids. J. Nutrit. **19**, 131 (1940).
316. SMEDLEY-MACLEAN, I. and E. M. HUME: Fat-Deficiency Disease of Rats. The Influence of Tumor Growth on the Storage of Fat and of Polyunsaturated Acids in the Fat-starved Rat. Biochemic. J. **35**, 996 (1941).
317. SMEDLEY-MACLEAN, I. and L. C. A. NUNN: Fat-deficiency Disease of Rats. The Effect of Doses of Methyl Arachidonate and Linoleate on Fat Metabolism, with a Note on the Estimation of Arachidonic Acid. Biochemic. J. **34**, 884 (1940).
318. SMITH, E. L.: Comments on the Partition Chromatogram of MARTIN and SYNGE. Biochemic. J. **36**, xxii (1942).

319. SOODAK, M. and F. LIPMANN: Enzymatic Condensation of Acetate to Acetoacetate in Liver Extracts. J. biol. Chemistry **175**, 999 (1948).

319a. SPALLONI CIALDEA, W.: Influencia de dietas grasas sobre la elevacion del consumo de oxígeno en ratas tratadas con tiroxina. Tesis Quimico Farmaceutico, Univ. de Chile, Inst. Fisiol. (1945).

320. STARE, F. J., G. V. MANN, R. P. GEYER and D. M. WATKIN: The Need of Fat in Intravenous Feeding. J. Amer. Oil chem. Soc. **26**, No. 4, 145 (1949).

321. STARLING, E. H.: The Significance of Fats in the Diet. Brit. med. J. **2**, 105 (1918).

322. STEENBOCK, H., M. H. IRWIN and J. WEBER: The Comparative Rate of Absorption of Different Fats. J. Nutrit. **12**, 103 (1936).

323. STEWARD, W. C. and R. G. SINCLAIR: The Absence of Ricinoleic Acid from the Phospholipids of Rats Fed Castor Oil. Arch. Biochemistry **8**, 7 (1945).

324. STIRN, F. E., A. ARNOLD and C. A. ELVEHJEM: The Relation of Dietary Fat to the Thiamin Requirements of Growing Rats. J. Nutrit. **17**, 485 (1939).

325. STOTZ, E.: Biological Synthesis of Fatty Acids. J. Amer. Oil chem. Soc. **26**, 341 (1949).

326. STRONG, F. M. and L. E. CARPENTER: Preparation of Samples for Microbiological Determination of Riboflavin. Ind. Engng. Chem., Analyt. Edit. **14**, 909 (1942).

327. SURE, B.: Dietary Requirements for Fertility and Lactation. J. Nutrit. **22**, 499 (1941).

328. SWAIN, L. A. and B. H. MCKERCHER: Examination of Unsaponifiable Matter of Marine Oils. Fisheries Research Bd., Canada, Prog. Repts., Pac. Coast Stas., No. **65**, 67 (1945); Chem. Abstr. **40**, 1681 (1946).

329. SWIFT, R. W. and A. BLACK: Fats in Relation to Caloric Efficiency. J. Amer. Oil chem. Soc. **26**, 171 (1949).

330. TANGE, U.: A Study on the Effects of Fatty Acid on Nutrition. Sci. Pap. Inst. physic. chem. Res. **20**, 13 (1932).

331. — Vitamin B_2 Complex. VI. Rat Acrodynia and Fat Acids. Sci. Pap. Inst. physic. chem. Res. **36**, 2 (1939).

332. TAUB, S. J. and S. J. ZAKON: The Use of Unsaturated Fatty Acids in the Treatment of Eczema. J. Amer. med. Assoc. **105**, 1675 (1935).

333. TOYAMA, Y.: The Unsaponifiable Constituents (Higher Alcohols) of Shark and Torpedo Liver Oils. Chem. Umschau Gebiete Fette, Öle, Wachse, Harze **31**, 61, 153 (1924); Chem. Abstr. **18**, 2613, 3733 (1924).

334. TRAGER, W.: Further Studies on a Fat-soluble Material from Plasma Having Biotin Activity. J. biol. Chemistry **176**, 133 (1948).

335. — Biotin and Fat-soluble Materials with Biotin Activity in the Nutrition of Mosquito Larvae. J. biol. Chemistry **176**, 1211 (1948).

336. — A Fat-soluble Material from Plasma Having the Biological Activities of Biotin. Proc. Soc. exp. Biol. Med. **64**, 129 (1947).

337. TSUJIMOTO, M.: Saturated Hydrocarbons in Basking Shark Liver Oil. Ind. Engng. Chem. **9**, 1098 (1917).

338. — Squalene. A highly Unsaturated Hydrocarbon in Shark Liver Oil. Ind. Engng. Chem. **12**, 63 (1920).

339. — Occurrence of Squalene in the Egg Oil from Shark. Ind. Engng. Chem. **12**, 73 (1920).

340. TSUJIMOTO, M. and Y. TOYAMA: The Unsaponifiable Constituents (Higher Alcohols) of the Liver Oils of Sharks and Rays. Chem. Umschau Gebiete Fette, Öle, Wachse, Harze **29**, 27, 43, 237, 245 (1922); Chem. Abstr. **16**, 1512 (1922); **17**, 892 (1923).

341. TURPEINEN, O.: Further Studies on the Unsaturated Fatty Acids Essential in Nutrition. J. Nutrit. 15, 351 (1938).
342. — Effectiveness of Arachidonic Acid in Curing "Fat-deficiency" Disease. Proc. Soc. exp. Biol. Med. 37, 37 (1937).
343. UMBREIT, W. W. and I. C. GUNSALUS: The Function of Pyridoxine Derivatives: Arginine and Glutamic Acid Decarboxylases. J. biol. Chemistry 159, 333 (1945).
344. VAN DEN HENDE, A.: The Raman Spectrograms and the Structure of Oleic and Elaidic Acids. Bull. Soc. chim. Belgique 56, 328 (1947).
345. VERZÁR, F. u. L. LASZT: Die Hemmung der Fettresorption nach Exstirpation der Nebennieren. Biochem. Z. 276, 11 (1935).
346. — — Nebennierenrinde und Fettresorption. Biochem. Z. 278, 396 (1935).
347. VISSCHER, F. E.: Storage of Hendecanoic Acid in the White Rat. J. biol. Chemistry 162, 129 (1946).
348. WEBER, J., M. H. IRWIN and H. STEENBOCK: Vitamin E Destruction by Rancid Fats. Amer. J. Physiol. 125, 593 (1939).
349. WEEK, E. F. and F. J. SEVIGNE: Vitamin A Utilization Studies. I. The Utilization of Vitamin A Alcohol, Vitamin A Acetate and Vitamin A Natural Esters by the Chick. J. Nutrit. 39, 219 (1949).
349a. — — Vitamin A Utilization Studies. II. The Utilization of Vitamin A Alcohol, Vitamin A Acetate and Vitamin A Natural Esters by the Rat. J. Nutrit 39, 233 (1949).
350. WHIPPLE, D. V. and C. F. CHURCH: Relation of Vitamin B (B_1) to Fat Metabolism. I. The Rôle of Fat in the Refection Phenomenon. J. biol. Chemistry 109, XCVIII (1935).
351. — — The Composition of Growth Induced by Vitamin B (B_1). J. biol. Chemistry 114, CVII (1936).
351a. — — Relation of Vitamin B (B_1) to Fat Metabolism. I. The Role of Fat in the Refection Phenomenon. J. biol. Chemistry 109, Proc. Amer. Soc. Biol. Chem. XCVIII (1935).
352. WHITE, E. A., J. R. FOY and L. R. CERECEDO: Essential Fatty Acid Deficiency in the Mouse. Proc. Soc. exp. Biol. Med. 54, 301 (1943).
353. WHITEHILL, A. R., J. J. OLESON and Y. SUBBAROW: A *Lactobacillus* of Cecal Origin Requiring Oleic Acid. Arch. Biochemistry 15, 31 (1947).
354. WIDMER, C. and R. T. HOLMAN: Polyethenoid Fatty Acid Metabolism. Deposition of Polyunsaturated Acids in Fat-deficient Rats upon Single Fatty Acid Supplementation. Abstract 5 C (10) Amer. Chem. Soc. 115th meeting, San Francisco (Mar.-Apr. 1949).
355. WILLIAMS, F. C. and J. O. OSBURN: Fatty Acid Distillation. Comparison of Two Column Types. J. Amer. Oil chem. Soc. 26, 663 (1949).
356. WILLIAMS, V. R. and E. A. FIEGER: Further Studies on Lipide Stimulation of *Lactobacillus casei*. II. J. biol. Chemistry 177, 739 (1949).
357. — — Growth Stimulants for Microbiological Biotin Assay. Ind. Engng. Chem., Analyt. Edit. 17, 127 (1945).
358. WILLMAN, W., M. BRUSH, H. CLARK and P. SWANSON: Dietary Fat and the Nitrogen Metabolism of Rats Fed protein-free Rations. Federat. Proc. 6, 423 (1947).
359. WOLLAEGER, E. E., M. W. COMFORT and A. E. OSTERBERG: Total Solids, Fat and Nitrogen in the Feces. III. A Study of Normal Persons Taking a Test Diet Containing a Moderate Amount of Fat; Comparison with Results Obtained with Normal Persons Taking a Test Diet Containing a Large Amount of Fat. Gastroenterology 9, 272 (1947).

360. WOLLAEGER, E. E., M. W. COMFORT, J. F. WEIR and A. E. OSTERBERG: The Total solids, Fat and Nitrogen in the Feces. II. A Study of Persons who had Undergone Partial Gastrectomy with Anastomosis of the Entire Cut End of the Stomach and the Jejunum *(Polya anastomosis)*. Gastroenterology **6**, 93 (1946).
361. ZAIN, H.: Zur antithyreoidalen Wirkung einiger ungesättigter Fettsäuren. Klin. Wschr. **15**, 1722 (1936).
362. — Der Einfluß ungesättigter Fettsäuren mit 1, 2 und 3 Doppelbindungen auf die experimentell erzeugte Rattenhyperthyreose. Arch. exp. Pathol. (Leipzig) **187**, 302 (1937).
363. ZECHMEISTER, L.: Progress in Chromatography 1938—1947. London: Chapman and Hall; New York: J. Wiley. 1949.
363 a. — Die Carotinoide im tierischen Stoffwechsel. Ergebn. Physiol. **39**, 117 (1937).
364. ZECHMEISTER, L., J. H. PINCKARD, S. M. GREENBERG, E. STRAUB, T. FUKUI and H. J. DEUEL, Jr.: Stereochemical Configuration and Vitamin A Activity. VIII. Pro-γ-carotene (a Poly-*cis* Compound) and its All-*trans* Isomer in the Rat. Arch. Biochemistry **23**, 242 (1949).
365. ZECHMEISTER, L. and P. TUZSON: Zur Kenntnis der tierischen Fettfarbstoffe. Ber. dtsch. chem. Ges. **67**, 154 (1934).
366. ZIALCITA, L. P., Jr. and H. H. MITCHELL: Corn Oil and Butter Fat Essentially Equal in Growth-promoting Value. Science (New York) **100**, 60 (1944).

(Received, December 16, 1949.)

Odeurs et parfums des animaux.

Par E. LEDERER, Paris.

Avec 10 figures.

Sommaire.

Introduction.

En essayant de résumer nos connaissances actuelles de la chimie des substances odorantes élaborées par les animaux, nous rencontrons, d'une part, de vastes domaines inconnus où tout reste à défricher (nous ne savons rien, par exemple, des substances odorantes si variées des Invertébrés) et, d'autre part, cet édifice imposant qu'est aujourd'hui la chimie des substances macrocycliques, chimie des corps à odeur musquée qui s'est développée si rapidement au cours des dernières 25 années, grâce aux travaux de L. RUZICKA et de son école. La chimie des drogues animales utilisées en parfumerie est actuellement assez bien connue; les résultats scientifiques acquis dans ce domaine nous font espérer que, bientôt, des progrès semblables seront réalisés dans les domaines voisins, encore incultes.

Rôle de l'odeur dans le règne animal.

La perception de substances odorantes semble universellement répandue dans le règne animal. MONCRIEFF (*88*) ainsi que LE MAGNEN (*82*) ont résumé nos connaissances à ce sujet.

Invertébrés. Chez les Invertébrés, ce sont surtout les Insectes qui sont doués d'un odorat extrêmement développé. Citons LE MAGNEN:

«FABRE, le premier, avait expérimenté avec soin les faits troublants de l'attirance sexuelle chez les papillons. Il constate que la femelle d'une espèce rare dans la région, attire en quelques heures un grand nombre de mâles, ce qui laisse supposer que son odeur a permis de la déceler à grande distance. Il démontre l'origine olfactive de la stimulation. Les mâles, amputés de leurs antennes, sièges des récepteurs olfactifs, ne reviennent pas. Il montre aussi l'existence au moins d'une «source» matérielle d'où émane l'odeur. La femelle placée sous une cloche de verre n'attire pas les mâles. Enfin, si l'on place durant un certain temps des femelles dans une enceinte fermée en présence de matières poreuses, telles que la terre d'infusoire ou d'argile, ces corps adsorbants, imprégnés de l'odeur des femelles, se montrent capables comme elles d'attirer les mâles. STANDFUSS, plus tard, précise l'ordre de grandeur des distances parcourues. Il retient captif dans un jardin une femelle de *Saturnia pavonia*; en six heures, cent vingt sept mâles sont attirés et capturés. Ces papillons sont amenés à une distance de quatre kilomètres, 40% d'entre eux retrouvent la femelle. Amenés à une distance de 12 km., 26% perçoivent encore l'appel sexuel et reviennent à la femelle ...» (*82*, p. 14.)

La production, par les Lépidoptères, de corps odorants à rôle sexuel est un fait bien connu. Les glandes qui produisent ces corps volatils sont extrêmement variés par leur structure et leur emplacement, surtout chez les mâles. On en a trouvé sur presque toutes les parties de l'Insecte, y compris les pattes et les ailes. Chez les femelles, les variations sont moins grandes et les organes, généralement rétractiles, se rencontrent surtout à l'abdomen [voir BOURGOGNE (*10*)].

D'après GOETZ (*31*), on peut se servir de pièges garnis de débris de Papillons femelles pour la lutte contre certaines espèces nuisibles (*Clysia ambiguella, Polychrosis botrana*, etc.).

HALLER, ACREE et POTTS (*36*) ont décrit les efforts faits en Amérique pour lutter par de tels pièges contre le Lepidoptère *Porthetria dispar* L. qui cause de grands dégats dans les vergers. La femelle, non ailée, secrète au bout de l'abdomen une substance odorante qui attire les mâles à de longues distances. La substance odorante peut être extraite au benzène; les essais de purification de HALLER, ACREE et POTTS (*36*) montrent qu'elle est insaponifiable et de nature alcoolique. Cette substance ne peut être remplacée, pour attirer des mâles de *Porthetria dispar*, ni par des extraits de femelles d'autres espèces voisines, ni par les substances suivantes: cantharidine; acides butyrique, caproïque, iso-caproïque, salicylique; aldéhyde salicylique; iso-amylamine; exaltone et civettone.

Dans une note préliminaire (*13 a*), BUTENANDT a décrit quelques propriétés d'une substance sécrétée par la femelle de *Bombyx mori* et qui attire les mâles par son odeur. Les abdomens de 1500 femelles ont donné 160 mg de substance soluble dans l'éther de pétrole. De là on a isolé, en faible quantité, la substance active (non-cétonique, neutre et exempte d'azote) dont 0,01 γ exercent une action nette sur les mâles du Ver à soie.

FLASCHENTRAEGER et AMIN (*28 a*) ont étudié récemment la substance odorante sécrétée par l'abdomen des femelles de *Prodenia litura* FAB. et *Agrotis ypsilon* ROTT., Lépidoptères qui ravagent les plantations de coton. La substance qui attire les mâles est soluble dans l'éther, stable envers l'action d'acides et de bases dilués et contient un groupe CO.

FABRE (*23*) a aussi étudié le comportement d'un petit coléoptère *Bolbocera gallicus* qui se nourrit d'une certaine espèce de Truffe (*Hydnocystis arenaria*). L'insecte, attiré par l'odeur, trouve sa nourriture, enfouie jusqu'à 25 cm. sous la terre et creuse ensuite un canal vertical pour y arriver; il n'est pas attiré par d'autres espèces de Truffes.

Les Insectes vivant en colonies (Fourmis, Termites, Abeilles) reconnaissent à l'odeur les membres de la même colonie; des Fourmis ou Abeilles étrangères pénétrant dans un nid sont immédiatement mises à mort. L'odorat, localisé dans les antennes, sert aussi aux Fourmis à trouver leur chemin de retour; si l'on enlève à une Fourmi ses antennes, elle est incapable de retrouver son nid et mord indifféremment ses compagnons de nid et des étrangères [voir par exemple LE MAGNEN (*82*)].

Les exemples cités montrent le grand rôle que joue chez ces animaux la perception des odeurs, pour la recherche de la nourriture et pour leur vie sexuelle et sociale.

On trouvera des détails biologiques sur les substances odorantes des Insectes dans les livres de FORD (*29*) et de WIGGLESWORTH (*171*); voir aussi les articles de HARDY (*41*) et de KETTLEWELL (*60*).

Vertébrés. Chez les Vertébrés à mesure que l'on s'élève depuis les Prochordés jusqu'aux Mammifères, les récepteurs olfactifs sont de plus en plus étroitement localisés dans les cavités respiratoires. L'odorat ne semble jouer qu'un rôle secondaire pour les Poissons et les Oiseaux. Chez certains Batraciens, nous trouvons, au contraire des glandes à parfum assez développées, ce qui fait penser que leur odorat l'est aussi; c'est le cas des Crocodiles, Alligators, Lézards et Géckos (voir le chapitre «yacarol», p. 119). C'est surtout chez les Mammifères que l'odorat joue de nouveau un rôle considérable. Reconnaissance alimentaire, recherche et poursuite de la nourriture, reconnaissance des sites et orientation par les traces olfactives, la reconnaissance des individus, et l'attirance des sexes font intervenir essentiellement des stimuli olfactifs.

Rien n'est encore connu au sujet de la *chimie de l'olfaction*; la région olfactive de la muqueuse nasale des Vertébrés est pigmentée en jaune et l'intensité du pigment semble être proportionnelle à l'acuité olfactive des animaux [voir par exemple LE MAGNEN (*82*)].

MILAS, POSTMAN et HEGGIE (*87a*) ont trouvé que la muqueuse nasalle du Boeuf contient de la *vitamine A* et des *caroténoïdes.* Cette observation importante semble être passée inaperçue et attend d'être généralisée.

Drogues animales utilisées en parfumerie.

Certains Mammifères possèdent des glandes spéciales produisant des substances odorantes et jouant un rôle important dans la vie sexuelle. L'homme n'échappe pas à l'attrait de ces odeurs; déjà dans l'antiquité et jusqu'à nos jours il s'est servi de différentes glandes et excrétions animales comme source de parfum. La valeur de ces drogues est dûe à la note animale et à la ténacité de leur odeur.

Les principales drogues animales utilisées en parfumerie sont: le *musc*, la *civette*, le *castoréum* et l'*ambre gris*.

L'analyse de ces drogues a été poussée très loin et nous donnerons dans la suite tous les détails sur leur chimie; nous insisterons non seulement sur les substances odorantes, mais aussi sur les corps inodores, car ces derniers ont aussi leur intérêt biochimique et le Parfumeur doit savoir quelles sont les substances présentes dans les extraits qu'il utilise. Souvent même, les substances odorantes prennent naissance par transformation de corps inodores (voir le cas de l'ambréine, p. 123).

Quelquefois, la nature s'est ingéniée à doter des animaux de sécrétions particulièrement malodorantes servant de moyens de défense; c'est le cas des Mustelidès, dont nous parlerons plus loin.

Parmi les mises au point récentes mentionnant la chimie des odeurs et parfums d'origine animale citons celles de SABETAY (*131*, *132*) notre article de 1946 (*69*), la revue de FESTER (*23a*), les livres de MONCRIEFF (*88*, *89*) et de NAVES et MAZUYER (*92*). On trouvera des indications utiles sur les odeurs et l'hygiène dans le livre de McCORD et WITHERIDGE (*84*).

Nous avons classé la matière du présent article d'après les chapitres suivants:

I. Produits d'excrétion de *glandes sébacées* (musc et civette, musc américain, mustélidès, yacarol, castoréum, suintine);

II. *concrétions intestinales* (ambre gris, bézoaires);

III. substances odorantes de l'*urine* (phénols, acides, dérivés d'ionone, stéroïdes odorants, divers);

IV. *fèces* (hyracéum).

I. Produits d'excrétion des glandes sébacées.

Généralement disséminées sur toute la surface du corps et de taille microscopique, les glandes sébacées prennent, chez certaines espèces animales, le plus souvent à proximité des glandes sexuelles, un développement considérable et forment de véritables *glandes à parfum*. On signale généralement que leur sécrétion augmente parallèlement avec l'activité sexuelle.

1° Musc et civette.

a) Le Musc.

Le chevrotain ou bouquetin porte-musc (*Moschus moschiferus*, famille des Moschidés) (Fig. 1) est un ruminant de la taille d'une chèvre que l'on trouve en Asie centrale. Chez le mâle la glande à parfum atteint la taille d'une orange, visible à l'abdomen. Sa sécrétion est brun-foncé et de consistance semiliquide. La poche dans laquelle se forme la sécrétion a une ouverture par où le musc s'épanche au dehors. Le musc ne commence à se former que chez les adultes; lorsqu'il est trop abondant, surtout à l'époque du rut, l'animal se frotte contre les arbres pour s'en débarrasser, et dépose ainsi des traces qui trahissent sa présence par leur odeur pénétrante (*40*). L'animal est capté dans des pièges et la glande enlevée avec la peau qui l'entoure; on la sèche au soleil et c'est dans cet état qu'on la trouve dans le commerce sous le nom de «musc en poches». Le contenu enlevé du sac est appelé «musc en grains».

Constituants des glandes à musc. En 1906, WALBAUM (*165*) a isolé la substance odorante du musc naturel, sous forme d'une cétone $C_{16}H_{30}O$ qu'il a appelée *muscone*. La muscone fait de 0,5 à 2% du musc naturel.

SANO (*136*) a trouvé, en 1938, que la glande à musc contient des hormones sexuelles mâles. Plus récemment, SCHINZ, RUZICKA, GEYER et PRELOG (*137*) ont isolé des glandes à musc une base $C_{16}H_{25}N$ qu'ils ont appelé «*muscopyridine*» (huile, éb. 155—160°; $[\alpha]_D = +17{,}4°$).

L'oxydation de cette base par le permanganate donne l'acide pyridine-α,α-dicarboxylique, ce qui prouve que la muscopyridine est une pyridine α,α-disubstituée; les deux substituants sont saturés, l'un d'eux doit être monocyclique.

Fig. 1. Chevrotain porte-musc (*Moschus moschiferus*, Musc deer, Moschustier.)

b) La Civette.

Le mot *civette* désigne à la fois l'animal (*Viverra civetta*) (Sig. 2) et la sécrétion de ses glandes à parfum. L'animal, Mammifère carnivore de la famille des Viverridés est répandu en Afrique et en Asie. Il atteint une longueur d'environ 1 m. et une hauteur de 25 à 30 cm. La majeure partie de la drogue importée en Europe provient de civettes élevées en Abyssinie; la sécrétion se forme dans une poche profonde, à proximité des organes sexuels, et se présente sous forme d'une graisse de couleur jaune. Elle est prélevée par des curettages hebdomadaires ou mensuels, suivant la saison et l'âge de l'animal.

La civette vient sur le marché dans des cornes de buffles contenant jusqu'à 1 kg. de substance. Elle est presque entièrement soluble dans l'éther de pétrole.

HARDY (*39*) donne les indications suivantes sur d'autres animaux, proches parents de la civette, et ayant des glandes à parfum analogues:

"There is no doubt that the production of civetone is by no means confined to the African or Abyssinian civet; in point of fact these two anal glandular pouches occur frequently in other members of the viverrine family. For instance the odorous secretion from the glands of the rasse *(Viverra malaccensis)* smallest of the civets, is known to the Javanese as 'dedes' and to the Malays as 'jibet', and is quite a favourite perfume amongst the Javanese. They apply it to their dresses or, mixed with ointments to their bodies until their appartments and their festivities reek with it ... The secretion of the Indian civet *(Viverra zibetha)* is similar in odour and equally extolled by the natives although proving so repulsive to Europeans ... The Javan civet or so called 'wild cat' *(Viverra tangalunga)* produces a secretion similar to that of the Indian civet and the rasse ... There is an allied European and African genera of small mammals very similar to civets and called genets.

Fig. 2. Civette (*Viverra zibetha*, Civet, Zibetkatze). (Photo New York Zoological Society.)

In the case of the cat-like common genet *(Genetta vulgaris)* which inhabits Spain and Southern France ... the odoriferous anal pouches are reduced to a mere depression in the skin, with very little secretion, but nevertheless sufficient to produce a very perceptible odour ... In the common palm-civet or tree-cat *(Paradoxurus typus)* once known as the 'pougonne' ... the scent gland is represented by a superficial glandular space, placed a little below the anal opening ... There is a solitary glandular pouch in the anal region of the cusimanse or mangue *(Crossarchus obscurus)* a smooth-nosed mongoose of West Africa."

Constituants de la civette. En 1915, SACK (*133*) a trouvé que le principe odorant de la civette est une cétone $C_{17}H_{30}O$ qu'il appela *civettone.* La teneur de la civette en civettone est de 2,5 à 3% [RUZICKA (*110*)].

Le carbinol correspondant, le *civettol* est également présent dans la glande [RUZICKA, SCHINZ et SEIDEL (*124*)]; cet alcool est inodore; on le transforme en civettone par oxydation [NAEF & Co. (*90*)].

La note animale de l'odeur de la drogue naturelle est, en partie aussi, due à d'autres substances odorantes telles que le scatol et des acides gras (*110*, *133*).

c) La chimie des grands cycles.

Nos premières notions précises sur la nature des principes odorants du musc et de la civette sont dues aux travaux classiques de RUZICKA et de son école; ils ont donné un essor considérable à la chimie synthétique et de nombreuses revues bibliographiques en font état [par exemple MONCRIEFF (*88, 89*), RUZICKA (*113*), SABETAY (*131, 132*), M. STOLL (*147, 149*)].

RUZICKA (*110, 111*) a établi en 1926, que la muscone et la civettone sont des substances macrocycliques, c'est-à-dire que leurs molécules sont composées de cycles contenant un grand nombre d'atomes de carbone. La constitution de la *civettone* a été établie la première (*110*).

L'hydrogénation catalytique en dihydrocivettone $C_{17}H_{32}O$ (I) montrait la présence d'une double liaison et d'un cycle. Les résultats d'une transposition de BECKMANN ainsi que de l'oxydation permanganique de la dihydrocivettone montraient que le groupement cétonique se trouvait dans ce cycle; la première réaction donne une iso-oxime (II) qui est transformée en acide aminé (III), par saponification. L'oxydation chromique fournit un acide dicarboxylique (IV), ayant comme l'acide aminé (III) le même nombre de carbones que la civettone.

$$C_{15}H_{30}\left\{\begin{array}{l}-CO\\ \;|\\ -CH_2\end{array}\right. \qquad C_{16}H_{32}\left\{\begin{array}{l}-CO\\ \;|\\ -NH\end{array}\right. \qquad C_{16}H_{32}\left\{\begin{array}{l}-COOH\\ \\ -NH_2\end{array}\right. \qquad C_{15}H_{30}\left\{\begin{array}{l}-COOH\\ \\ -COOH\end{array}\right.$$

(I.) Dihydrocivettone. (II.) (III.) (IV.)

L'oxydation de la civettone par le permanganate, à froid, attaque la molécule au niveau de la double liaison et donne un acide céto-dioïque $C_{16}H_{28}O_5$ (V), ce qui prouve que la double liaison se trouve également dans le cycle. Au cours de cette oxydation, il y a perte d'un atome de carbone, phénomène observé souvent au cours de l'oxydation permanganique de substances aliphatiques ayant un nombre impair d'atomes de carbone. L'oxydation permanganique énergique de l'acide (V) a donné un mélange d'acides dicarboxyliques, principalement de l'acide subérique (VI), puis aussi les acides succinique, pimélique et azélaïque (VII). La production de ces acides ne pouvait s'expliquer que par la formule (V) de l'acide cétodicarboxylique.

$$\begin{array}{l}HOOC-(CH_2)_6\diagdown\\ \qquad\qquad\qquad CO-\\ HOOC-(CH_2)_7\diagup\end{array} \begin{array}{l}\longrightarrow HOOC-(CH_2)_6-COOH\\ \\ \longrightarrow HOOC-(CH_2)_7-COOH\end{array} \qquad \begin{array}{l}CH_2-(CH_2)_7\diagdown\\ |\qquad\qquad\quad CO\\ CH_2-(CH_2)_7\diagup\end{array}$$

(V.) (VI.) Acide subérique. (VII.) Acide azélaïque. (VIII.) Dihydrocivettone.

L'exactitude de cette interprétation a été vérifiée par l'identification de l'acide dicarboxylique (IV) avec l'acide pentadécaméthylène-1,15 dioïque préparé synthétiquement par CHUIT (*15*), ainsi que par la synthèse de la dihydrocivettone (VIII) par RUZICKA, STOLL et SCHINZ (*129*).

La formation de l'acide (V) prouve la position de la double liaison et la civettone doit avoir la formule (IX).

La *muscone*, $C_{16}H_{30}O$ (X) était, de toute évidence, une méthylcyclopentadécanone; sa réduction d'après CLEMMENSEN donna l'hydrocarbure $C_{16}H_{32}$ qui a pu être identifié avec le méthylcyclo-pentadécane synthétique.

L'oxydation chromique de la muscone donne un mélange d'acides, dont deux acides $C_{16}H_{30}O_4$ sont les principaux et dans lequel RUZICKA (*III*) a pu identifier tous les homologues supérieurs de l'acide succinique jusqu'à l'acide décane-1,10-dioïque.

Ceci permettait d'exclure les positions 5, 6 et 7 comme point d'attache du groupement méthyle. La position 1 a pu être exclue, parce que les 1-méthyl-cétones ne donnent pas d'acides dicarboxyliques à nombre égal d'atomes de carbone, mais un acide possèdant un atome de carbone de moins.

```
CH—(CH2)7                        CH3                          CH3
 ‖        \                       |                            |
 ‖         CO      CH2—(CH2)5—CH—CH2           CH2—(CH2)5—CH—COOH
 ‖        /        |               /           |
CH—(CH2)7          CH2—(CH2)5—CO               CH2—(CH2)5—COOH
```

(IX.) Civettone. (X.) Muscone. (XI.)

Les semicarbazones des cétones méthylées correspondant aux positions 2, 3 et 4 ont été synthétisées et comparées avec celle de la muscone, mais cette comparaison ne pouvait naturellement avoir qu'une valeur limitée, puisque les cétones synthétiques sont optiquement inactives; ces semicarbazones fondent environ à 30° plus haut que celle de la muscone. On prit cependant des point de fusion des mélanges; seul un mélange de la semicarbazone de la 3-méthylcétone avec celle de la muscone provoquait une dépression. Ceci exclut la position 3 et il ne resta donc plus qu'à décider entre les positions 2 et 4.

En recristallisant bien des fois les acides dicarboniques $C_{16}H_{30}O_4$ mentionnés plus haut, on arriva à faire monter le point de fusion jusqu'à 78°. On péprara alors les acides 1-, 2-, 3- ou 4-méthyl-tridécane 1,13-dioïques et il ressortit que seul l'acide 1-méthylé possède un F. supérieur à 78° (94°). Il ne peut donc y avoir dans le produit fondant à 78° que l'acide méthyl-1-tridécane 1,13-dicarbonique (XI) comme produit principal et celui-ci est probablement accompagné d'un peu d'acide 2-méthylé.

```
               CH3
               |                       CH3
CH2—(CH2)5—CH—CH2                      |                     CH—(CH2)5—CO
|               |         CH2—(CH2)5—CH—CH2                  ‖             \
CH2—(CH2)4—C——CO          |               \                  ‖              O
           ‖              |                COOH              ‖             /
           CH—C6H5        CH2—(CH2)4—COOH                    CH—(CH2)7—CH2
```

(XII.) Benzylidène-muscone. (XIII.) Acide méthyl-2-tridécane-1,13-dicarboxylique. (XIV.) Ambrettolide.

La muscone fut ensuite condensée avec l'aldéhyde benzoïque, opération dans laquelle il était à croire que la condensation serait empêchée sur le groupe méthylénique se trouvant entre le carbone méthylé et le groupe cétonique et que seul l'autre méthylène, voisin du groupe cétonique réagirait. Il se forma, par traitement de la benzylidène-muscone (XII) avec l'ozone et ensuite avec l'acide chromique, un acide homogène, F. 69°, dextrogyre (XIII). L'acide *d,l*-méthyl-2-tridécane-1,13-dicarboxylique synthétique (XIII) fond à 76° et mélangé avec l'acide droit ne donne pas de dépression du point de fusion.

La formule (X) de la muscone était ainsi rendue extrêmement probable (*112*).

Jusque là, on ne connaissait pas de substances analogues et leur existence semblait même impossible, d'après la théorie de BAEYER qui a montré que les cycles en C_5 et C_6 sont les plus stables et ont le plus de tendance à se former. Des cycles plus grands ne pourraient plus se former qu'avec grande difficulté par suite de la forte tension que provoquerait la déviation des angles valentiels du carbone. Nous savons maintenant que de grands cycles peuvent être parfaitement stables et sans tension grâce à une configuration spaciale appropriée (voir cidessous p. 98).

RUZICKA et ses collaborateurs, après avoir établi la structure de la muscone et de la civettone, ont développé d'une façon magistrale la chimie des corps à grand cycle (voir plus de 50 mémoires „Zur Kenntnis des Kohlenstoffringes" dans les Helvetica chimica Acta de 1926 à 1949). Nous connaissons aujourd'hui des centaines de substances de cette catégorie (alcools, aldéhydes, cétones, lactones, oxydes, anhydrides, carbonates, imines, sulfones etc.) dont les membres ayant des cycles de 14 à 18 chaînons présentent généralement une odeur plus ou moins musquée (voir p. 106).

Pour des revues détaillées de ces corps, voir: MONCRIEFF (*88, 89*), SABETAY (*131, 132*) et M. STOLL (*147, 149*).

Le règne animal n'a, du reste, pas le monopole des substances macrocycliques à odeur musquée; KERSCHBAUM (*59*) a isolé de l'huile des semences d'ambrette (*Hibiscus abelmoschus*) une lactone, l'*ambrettolide*, qu'il identifia comme étant une hexadécène-7,8-olide (XIV), et il isola la *pentadécanolactone* (« *exaltolide* ») (XV) de l'huile des semences d'angélique (*Archangelica officinalis*).

Rappelons que l'odeur de musc se trouve aussi dans une série de corps aromatiques synthétiques, nitrés (musc Baur, musc xylène, etc.).

Dans la suite, nous exposerons surtout les travaux chimiques ayant un intérêt immédiat pour la chimie des *cétones macrocycliques naturelles* d'origine animale. Les *stéroïdes* à odeur de musc sont traités p. 137.

d) Premières synthèses de cétones macrocycliques.

RUZICKA, STOLL et SCHINZ (*129*) ont décrit en 1926 la première de ces synthèses: la décomposition thermique de sels d'acides α,ω-dicarboxyliques (XVI), spécialement de sels de thorium et de quelques terres rares (cérium, ytterbium). Les rendements en cétones cycliques sont de 45% pour la cycloheptanone (subérone), de 25% pour la cyclooctanone,

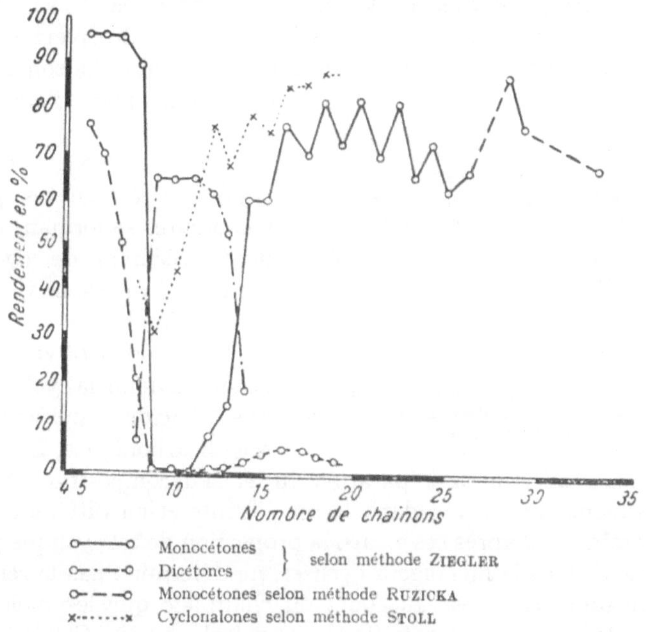

Fig. 3. Rendement en cétones macrocyclique d'après diverses méthodes (M. STOLL *149*).

mais de quelques dixièmes de pour cent seulement pour les cycles en C_{10} et C_{11}. Ils augmentent ensuite pour atteindre 5% pour les cycles en C_{17} et C_{18} (Fig. 3). Ces difficultés de synthèse de cycles en C_{10} à C_{12} sont dues à des empêchements stériques par suite de la forme de la chaîne carbonée. Les grands cycles, au-dessus de C_{17}, se présentent

CH_2—$(CH_2)_5$—CO
| O
CH_2—$(CH_2)_6$—CH_2
(XV.) Exaltolide.

CH_2—$(CH_2)_5$—CH_2—COO
| Me
CH_2—$(CH_2)_5$—CH_2—COO
(XVI.)

→

CH_2—$(CH_2)_5$—CH_2
| CO
CH_2—$(CH_2)_5$—CH_2
(XVII.) Cyclopentadécanone (exaltone).

comme deux longues chaînes parallèles reliées aux deux bouts par une partie courbée (Fig. 4); dans ces molécules, les espaces de vibration des atomes d'hydrogène se trouvent placés à moitié vers l'extérieur, à moitié vers l'interieur du cycle. Dans les cycles de C_7 à C_{16}, l'espace à l'intérieur du cycle est trop petit, les espaces de vibration sont restreints, ce qui amène une augmentation de la densité et des difficultés de cyclisation [voir STOLL et STOLL-COMTE (*161*)].

Parmi les cétones cycliques de C_{10} à C_{18} synthétisées par RUZICKA, STOLL et SCHINZ (*129*), la cyclopentadécanone (XVII) présente l'odeur de musc la plus pure et se trouve dans le commerce sous le nom d'*exaltone*; la lactone correspondante, également très appréciée (XV), s'appelle *exaltolide*. L'odeur de la cycloheptadécanone (dihydrocivettone, VIII) ressemble beaucoup à celle de la civettone; la double liaison n'est pas
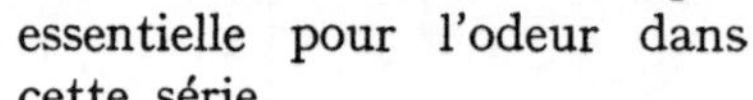
essentielle pour l'odeur dans cette série.

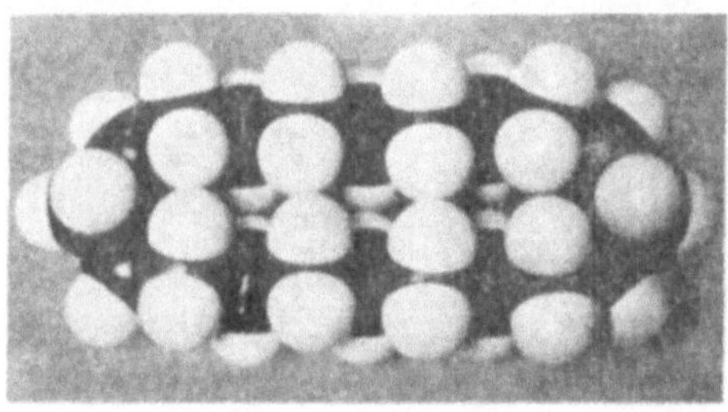

Fig. 4. Modèle moléculaire du cyclo-eicosane, d'après RUZICKA. Le modèle montre les deux chaînes parallèles. (Photo M. STOLL, Genève.)

RUZICKA, STOLL et SCHINZ (*130*) ont étudié les produits secondaires se formant au cours de la réaction de cyclisation (cétones inférieures, cétones non cycliques).

STOLL et ROUVE (*158*) ont étudié en détail la cyclisation des sels d'acides polyméthylène-α,ω-dicarboniques. Leur travail a porté en particulier sur la préparation du sel, la décomposition thermique et sur l'influence du cation dans le sel. Trente-et-un différents cations ont été examinés. D'après ces essais, la proportion de sel cyclique présente primitivement dans le mélange à cycliser, ne détermine pas le rendement de la cyclisation, car c'est au cours du chauffage que les molécules se cyclisent grâce à l'action spécifique du métal utilisé. Celui-ci a deux fonctions distinctes à remplir: «d'une part, il doit maintenir les bouts de la chaîne afin que le mouvement moléculaire active davantage le milieu de la chaîne que ses deux extrémités, d'autre part, il doit faciliter la transposition moléculaire des deux bouts de la chaîne, d'abord en sel cyclique, puis en cétone, acide carbonique et oxyde du métal.»

La cyclisation des sels de terres rares ne permet pas, cependant, la synthèse de la *muscone*, car la présence d'un ou de plusieurs groupes méthyles au voisinage du carbonyle empêche toute cyclisation.

e) Synthèses de la muscone.

ZIEGLER, EBERLE et OHLINGER (*175*) ont élaboré, en 1933, une nouvelle méthode de cyclisation permettant des rendements satisfaisants. Ils appliquent *le principe des grandes dilutions*, qui permet de diminuer la fréquence des réactions entre deux ou plusieurs molécules dicarboniques et d'augmenter ainsi le nombre de cyclisations monomoléculaires. En utilisant ce principe pour la cyclisation des dinitriles d'acides polyméthylène dicarboniques (XVIII) en présence d'un amidure substitué du lithium [par exemple du lithium-éthyl-aniline, $C_6H_5N(C_2H_5)Li$], ils obtiennent d'abord un nitrile de cétimide (XIX) qu'ils saponifient en cétonitrile (XX); ce dernier est ensuite scindé en cétone cyclique (XXI).

$$\begin{array}{l} \ulcorner\!-\!-\!C{\equiv}N \\ (CH_2)_n \\ \llcorner\!-\!-\!CH_2{-}C{\equiv}N \end{array} \quad \begin{array}{l} \ulcorner\!-\!-\!C{=}NH \\ (CH_2)_n \\ \llcorner\!-\!-\!CH{-}C{\equiv}N \end{array} \quad \begin{array}{l} \ulcorner\!-\!-\!CO \\ (CH_2)_n \\ \llcorner\!-\!-\!CH{-}C{\equiv}N \end{array} \quad \begin{array}{l} \ulcorner\!-\!-\!CO \\ (CH_2)_n \\ \llcorner\!-\!-\!CH_2 \end{array}$$

(XVIII.) (XIX.) (XX.) (XXI.)

ZIEGLER et WEBER (*176*) ont ainsi effectué la *synthèse de la muscone racémique*; 5 gr. de dinitrile (XXII) ont donné 2,5 gr. de muscone (X).

ZIEGLER et AURNHAMMER (*174*) ont étudié en détail le problème de la tendance de cyclisation en fonction de la longueur de la chaîne.

RUZICKA et STOLL (*128*) ont repris le problème de la synthèse de la muscone en introduisant le groupe méthyle dans un cycle déjà formé. La débromuration de la bromoexaltone (XXIII) donne une cétone insaturée (XXIV) qui est transformée en dérivé malonique (XXV) que

$$\begin{array}{l} \ulcorner\!-\!-\!CH_2{-}C{\equiv}N \\ (CH_2)_{11} \quad CH_2C{\equiv}N \\ \llcorner\!-\!-\!-\!-\!CH{-}CH_3 \end{array} \quad \begin{array}{l} CO{-}CHBr \\ (CH_2)_{12} \\ \llcorner\!-\!-\!CH_2 \end{array} \quad \begin{array}{l} CO{-}CH \\ \quad\quad \| \\ (CH_2)_{12}{-}CH \end{array} \quad \begin{array}{l} CO{-}CH_2 \\ (CH_2)_{12}{-}CH{-}CH \\ \quad\quad\quad\quad \| \\ \quad\quad\quad (COOR)_2 \end{array}$$

(XXII.) (XXIII.) Bromoexaltone. (XXIV.) (XXV.)

l'on décarboxyle en (XXVI); une nouvelle décarboxylation conduit au composé vinylé (XXVII) qui donne la muscone (X) par hydrogénation.

L'odeur de la méthyl-1-cyclopentadécanone-2 est analogue à celle de la muscone, mais moins intense. La diméthyl-1,3-cyclotridécanone-2 n'a cependant pas d'odeur de musc.

$$\begin{array}{l} CO{-}CH_2 \\ (CH_2)_{12}{-}CH{-}CH_2{-}COOH \end{array} \quad \begin{array}{l} CO{-}CH_2 \\ (CH_2)_{12}{-}C{=}CH_2 \end{array} \quad HOOC[(CH_2)_nOOC(CH_2)_n]_zOH$$

(XXVI.) (XXVII.) (XXVIII.)

SPANAGEL et CAROTHERS (*140*) ont étudié un autre type de cyclisation, qui est basé sur une *dépolymérisation* en milieu dilué. L'estolide de l'ω-hydroxypentadécanoïque ($n = 14$) (XXVIII) donne avec du $Cl_2Mg \cdot 6\,H_2O$ comme catalyseur et au-dessus de 270°, 70% de pentadécanolide (XV).

De même, les polyesters (XXIX, p. 100) donnent facilement des diesters (XXX). Malheureusement, ces méthodes de cyclisation conduisent à des produits hétérocycliques, tandis que les réactions de condensation qui pourraient donner des produits isocycliques sont le plus souvent trop lentes, ou trop compliquées pour être utilisées en très grande dilution.

$$HOOC(CH_2)_{n-x}-COO[(CH_2)_xOOC(CH_2)_{n-x}-COO]_z(CH_2)_xOH \longrightarrow \begin{matrix} \ulcorner CO-O \\ (CH_2)_{n-x} \quad (CH_2)_x \\ \llcorner CO-O \end{matrix}$$

(XXIX.) (XXX.)

HUNSDIECKER (*48*) a publié en 1942 une nouvelle méthode, consistant à *cycliser* des composés métalliques d'esters halogène acyl-acétiques avec *scission cétonique* consécutive. Il condense un chlorure d'acide polyméthylène-ω-halogéné (XXXI) avec l'acétoacétate d'éthyle; le produit de condensation (XXXII) donne, par traitement au méthylate de soude, l'ester ω-bromo-β-cétonique correspondant (XXXIII) dont le sel sodique (XXXIV) est cyclisé en cétoester (XXXV). Les rendements sont de 40 à 75% pour les cycles en C_{14} à C_{17}. HUNSDIECKER (*48*) a ainsi synthétisé la *muscone racémique* (X, p. 95).

$$\underset{Br}{CH_2}-(CH_2)_x-COOR + \underset{CO-CH_3}{CH_2}-COOR \longrightarrow \underset{Br}{CH_2}-(CH_2)_x-CO-\underset{COCH_3}{CH}-COOR$$

(XXXI.) (XXXII.)

$$\underset{Br}{CH_2}-(CH_2)_x-CO-CH_2-COOR \longrightarrow \underset{Br}{CH_2}-(CH_2)_x-CO-\underset{Na}{CH}-COOR$$

(XXXIII.) Ester ω-bromo-β-cétonique. (XXXIV.)

Le traitement des esters ω-bromés par le méthylate de soude peut provoquer différentes réactions secondaires (par exemple $Br-CH_2-$ en CH_3O-CH_2- etc.).

Une autre synthèse de la muscone, décrite par STOLL et ROUVÉ (*160*) se sert de la dodécaméthylène-α,ω-diméthyl-dicétone (ou dioxo-2,15-hexadécane, XXXVIII) comme matière de départ. Ce produit est facilement accessible, soit par une condensation de l'α,ω-dibromodécane avec de l'ester acétylacétique, suivie d'une scission en dicétone, soit par l'électrolyse d'un mélange de oxo-6-heptanoïque (XXXVI) et de oxo-10-undécanoïque (XXXVII).

$$\underbrace{CH_2-(CH_2)_x-CO-CH}-COOR \qquad CH_3-CO-(CH_2)_8-COOH + CH_3-CO-(CH_2)_4COOH$$

(XXXV.) (XXXVI.) Acide oxo-6-heptanoïque. (XXXVII.) Acide oxo-10-undécanoïque.

En appliquant la méthode de GRIGNARD et COLONGE (*17, 34*) à la condensation interne de la dicétone (XXXVIII) et en choisissant les conditions les plus favorables, STOLL et ROUVE (*160*) ont obtenu 17% de β-méthyl-cyclopentadécénone (muscénone, XL) par deshydratation de (XXXIX). Par réduction catalytique,. celle-ci est facilement transformée en muscone (X). Cette synthèse rappelle, du reste, la biosynthèse de la muscone, telle qu'elle a été discutée par STEVENS (*144*) (voir p. 109).

$$CH_3{-}CO{-}(CH_2)_{12}{-}CO{-}CH_3 \longrightarrow$$

(XXXVIII.) 2,15-Dioxo-hexadécane.

$$\longrightarrow \begin{array}{l} \ulcorner\!-\!-\!-\!C(CH_3){-}OH \\ (CH_2)_{12} \quad\quad CH_2 \\ \llcorner\!-\!-\!-\!CO \end{array} \xrightarrow[-H_2O]{} \begin{array}{l} \ulcorner\!-\!-\!-\!C{-}CH_3 \\ (CH_2)_{12} \quad\quad \| CH \\ \llcorner\!-\!-\!-\!CO \end{array} + H_2 \longrightarrow (X)$$

(XXXIX.) (XL.) β-Méthyl-cyclopenta-décénone.

Blomquist et Spencer (*9*) ont étudié la polymérisation de dicétènes (XLII) obtenus par déshydrohalogénation de dichlorures d'acides dicarboxyliques (XLI) par l'action d'amines aliphatiques. Ces dicétènes (XLII) se condensent sous forme linéaire pour donner des cétènes polymères, mais en travaillant à de très grandes dilutions, on obtient aussi une certaine quantité de composés macrocycliques. L'hydrolyse et la décarboxylation de ces cétènes cycliques (probablement XLII a) donne des cétones (XLIII) et des dicétones (XLIV) macrocycliques.

$$\begin{array}{c} COCl \\ | \\ (CH_2)_n \\ | \\ COCl \end{array} \longrightarrow \begin{array}{c} CH{=}CO \\ | \\ (CH_2)_{n-2} \\ | \\ CH{=}CO \end{array} \longrightarrow$$

(XLI.) (XLII.)

$$\longrightarrow \begin{array}{l} \ulcorner\!-\!-\!-\!C{=}C{=}O \\ (CH_2)_{n-1} \;| \\ \llcorner\!-\!-\!-\!C{=}O \end{array} \longrightarrow \begin{array}{l} \ulcorner\!-\!-\!-\!\urcorner \\ (CH_2)_{n-1}\; CO \\ \llcorner\!-\!-\!-\!\lrcorner \end{array} + \begin{array}{c} CO \\ (CH_2)_n \quad (CH_2)_n \\ CO \end{array}$$

(XLII a.) (XLIII.) (XLIV.)

Ici encore les rendements sont faibles parce que la transformation des deux fonctions du chlorure d'acide n'a pas lieu simultanément. En outre, le produit primaire de cyclisation (XLII a) semble réagir avec le produit initial puisque le rendement fléchit lorsqu'on réduit la vitesse d'introduction.

Blomquist, Holley et Spencer (*8*) ont effectué la synthèse de la *muscone racémique* (X) en utilisant l'acide bromo-14, méthyl-3, tétradécanoïque (XLV) comme point de départ. Par une synthèse malonique effectuée sur son ester, ils obtiennent le diacide (XLVI) dont le dichlorure donne la muscone (X, p. 95) après transformation en dicétène.

$$Br(CH_2)_{11}{-}\underset{\displaystyle CH_3}{\underset{|}{CH}}{-}CH_2COOH \longrightarrow HOOC(CH_2)_{12}{-}\underset{\displaystyle CH_3}{\underset{|}{CH}}{-}CH_2{-}COOH$$

(XLV.) Acide bromo-14-méthyl-3-tétradécanoïque. (XLVI.)

Toutes les réactions de cyclisation, décrites jusqu'ici, ont cependant l'inconvénient de s'effectuer seulement en milieu très dilué et avec des rendements loins d'être satisfaisants.

A l'heure actuelle, une seule réaction de condensation permet la préparation rapide des produits mégacycliques avec d'excellents rendements: c'est la méthode de condensation acyloïnique de BOUVEAULT et LOCQUIN (*11*) que HANSLEY (*37*) avait déjà mise au point dans la série aliphatique. Cette condensation consiste dans l'action de 4 atomes de sodium finement pulvérisés et fondus sur une molécule de diester (XLVII). Elle aboutit à un énolate sodé (XLVIII) et deux molécules d'alcoolate de sodium. Par hydrolyse, l'énolate se transforme en une cyclanolone (IL). La cyclisation acyloïnique est environ 300 fois plus rapide que la plus rapide des autres cyclisations. C'est la raison pour laquelle on obtient par cette méthode, pour les produits musqués, des rendements dépassant 80%, même en maintenant une vitesse d'introduction de 100 g. de produit par litre et par heure. Les cycles moyens se préparent avec des rendements allant jusqu'à 40% [STOLL et coll. (*156*, *159*); PRELOG et coll. (*97*)]. Voir Figure 3; p. 97 [d'après STOLL (*149*)].

Ayant d'abord mis au point la déshydratation des cyclo-polyméthylène-cétols-1,2 (IL) en cyclo-polyméthylène-cétones non saturées en α, β (L) (*148*), STOLL et COMMARMONT (*152*) ont pu effectuer la synthèse de la *muscone* en fixant, sur la cyclopentadécène-2-one-1 (L), en position 1,4, du système conjugué, le magnésien du bromure de méthyle, selon la méthode de KHARASCH et LAMBERT (*61*). Après hydrolyse des produits

$(CH_2)_n$(COOR)(COOR) (XLVII.) $\xrightarrow{4\,Na}$ $(CH_2)_n$(CONa=CONa) (XLVIII.) + 2 NaOR ⟶ $(CH_2)_n$(C=O)(CHOH) (IL.) $\xrightarrow{Al_2O_3}$

⟶ $(CH_2)_{12}$(CO)(CH=CH) (L.) Cyclopentadécène-2-one-1. + $BrMgCH_3$ ⟶ $(CH_2)_{12}$(C(OMgBr)(CH_3))(HC=CH) (LI.) + $(CH_2)_{12}$(COMgBr=CH)(HC(CH_3)) (LII.)

(LI.) ↓ $(CH_2)_{12}$(C(OH)(CH_3))(HC=CH) (LIII.) Cyclopentadécène-2-méthyl-1-ol-1.

(LII.) ↓ (X.) Muscone (p. 95).

d'addition (LI et LII), on obtient un mélange de ***muscone racémique*** (X), à côté du cyclopentadécène-2-méthyl-1-ol-1 (LIII). Le rendement en produit provenant de la fixation en position 1,4 n'était toutefois que le cinquième environ de celui provenant de la fixation en position 1,2 sur le groupe carbonyle.

Il était, d'autre part, naturel d'essayer des cyclisations de diesters portant déjà le groupe méthyle de la muscone. Mais, la cyclisation d'un ester méthylé en position α (LIV) mene exclusivement à une cyclanolone ayant le goupe carbonyle en position α du groupement méthyle (LV). Par réduction, on obtient donc exclusivement l'α-méthyl-cyclopentadécanone (LVI) [STOLL (*151*)]. Si le groupe méthyle se trouve

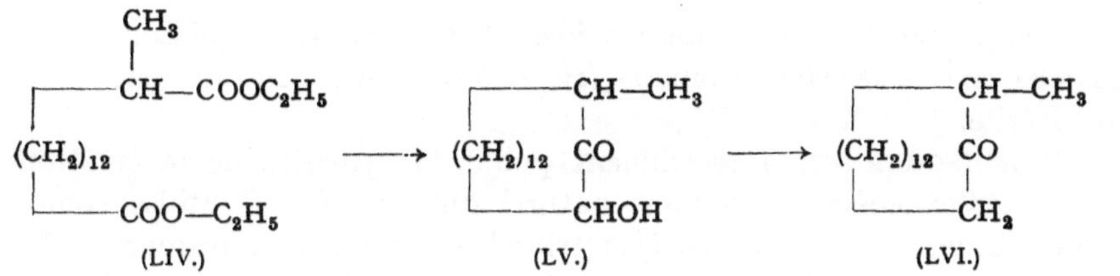

CH_3 | $CH—COOC_2H_5$; $(CH_2)_{12}$; $COO—C_2H_5$ (LIV.) ⟶ $CH—CH_3$; $(CH_2)_{12}$ CO ; $CHOH$ (LV.) ⟶ $CH—CH_3$; $(CH_2)_{12}$ CO ; CH_2 (LVI.)

en position β (LVII), la synthèse réussit partiellement [STOLL et COMMARMONT (*153*)]. On obtient un mélange de deux cyclopentadécanolones (LVIII et LIX) qui donne par réduction un mélange de γ-méthylcyclopentadécanone (LX) et de ***muscone*** (X, p. 95), dont l'odeur est identique à celle de la muscone naturelle.

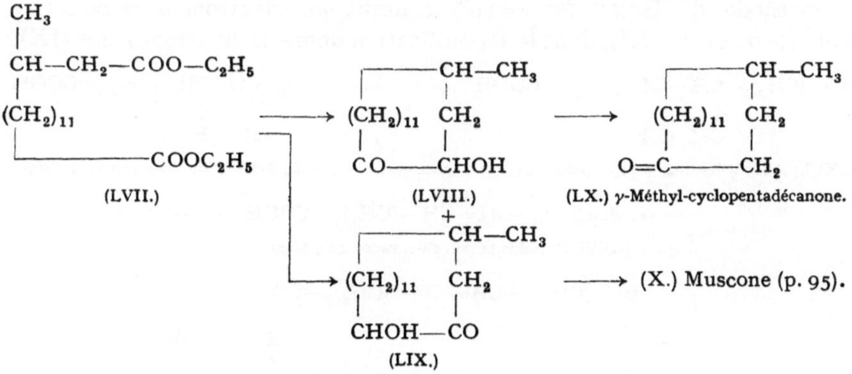

CH_3 | $CH—CH_2—COO—C_2H_5$; $(CH_2)_{11}$; $COOC_2H_5$ (LVII.) ⟶ $CH—CH_3$; $(CH_2)_{11}$ CH_2 ; $CO—CHOH$ (LVIII.) ⟶ $CH—CH_3$; $(CH_2)_{11}$ CH_2 ; $O{=}C—CH_2$ (LX.) γ-Méthyl-cyclopentadécanone.

\+ $CH—CH_3$; $(CH_2)_{11}$ CH_2 ; $CHOH—CO$ (LIX.) ⟶ (X.) Muscone (p. 95).

f) Synthèses de la civettone.

La difficulté de cette synthèse réside dans l'obtention de l'acide dicarbonique correspondant, ayant la double liaison à la place voulue. RUZICKA, PLATTNER et WIDMER (*120*) ont récemment synthétisé les formes *cis* et *trans* de cet acide octadécène-9-dicarbonique-1,18 (LXV) par la méthode suivante: l'acide acétylénique (LXI) est condensé en acyloïne (LXII), celle-ci réduite en glycol (LXIII) que l'on oxyde en acide dicarbonique (LXIV); on remplacé ensuite les deux —OH par une double liaison et obtient (LXV). HUNSDIECKER (50) a également

$$CH_3-C\equiv C-(CH_2)_7-COOH \longrightarrow CH_3-C\equiv C-(CH_2)_7-CO-CHOH-(CH_2)_7-C\equiv C-CH_3$$

(LXI.) (LXII.)

$$CH_3-C\equiv C-(CH_2)_7-CHOH-CHOH-(CH_2)_7-C\equiv C-CH_3 \longrightarrow$$

(LXIII.)

$$\longrightarrow HOOC-(CH_2)_7-CHOH-CHOH-(CH_2)_7-COOH \longrightarrow$$

(LXIV.)

$$\longrightarrow HOOC-(CH_2)_7-CH=CH-(CH_2)_7-COOH$$

(LXV.) Acide octadécène-9-dicarbonique-1,18.

décrit la préparation de ces deux acides. Leur cyclisation qui doit donner les deux civettones stéréoisomères (*cis* et *trans*, ou α et β) n'a pas encore été décrite.

HUNSDIECKER (49) a récemment publié la synthèse de la civettone *trans*, stéréoisomère du produit naturel qui est *cis*. Il utilise comme matière de départ l'*acide aleuritique* (trihydroxy-9,10,16-palmitique, LXVI) de la gomme-laque. La bromuration de cet acide donne l'acide tribromo-9,10,16-palmitique (LXVII) qui fournit, après débromuration partielle, l'acide bromo-16-hexadécène-9-oïque-1 (LXVIII), dont le chlorure est condensé avec l'acétoacétate d'éthyle. Une saponification partielle du produit de condensation (LXIX) mène au bromo-18-octadécène-11-céto-3-oate d'éthyle *trans* (LXX). La cyclisation de ce céto-β-ester selon la méthode de HUNSDIECKER (48) conduit au civettone-α-carboxylate d'éthyle *trans* (LXXI), dont la saponification donne la civettone *trans* (IX).

$$HO-(CH_2)_6-\underset{OH}{CH}-\underset{OH}{CH}-(CH_2)_7-COOH \longrightarrow Br-(CH_2)_6-\underset{Br}{CH}-\underset{Br}{CH}-(CH_2)_7-COOH$$

(LXVI.) Acide trihydroxy-9.10,16-palmitique (ac. aleuritique). (LXVII.) Acide tribromo-9,10,16-palmitique

$$\longrightarrow Br(CH_2)_6-CH=CH-(CH_2)_7-COOH \longrightarrow$$

(LXVIII.) Acide bromo-16-hexadécène-9-oïque-1.

$$\longrightarrow \begin{array}{l} Br-(CH_2)_6-CH=CH-(CH_2)_7-CO \\ \qquad\qquad\qquad\qquad\qquad\qquad\qquad \diagdown \\ \qquad\qquad\qquad\qquad\qquad\qquad\qquad HC-COOR \\ \qquad\qquad\qquad\qquad\qquad\qquad\qquad \diagup \\ \qquad\qquad\qquad\qquad\qquad\quad H_3C-CO \end{array}$$

(LXIX.)

$$\longrightarrow Br(CH_2)_6-CH=CH-(CH_2)_7-CO-CH_2-COOC_2H_5 \longrightarrow$$

(LXX.) Bromo-18-octadécène-11-céto-3-oate d'éthyle.

$$\longrightarrow \begin{array}{l} CH-(CH_2)_6-CH-COOC_2H_2 \\ \| \qquad\qquad\qquad\quad | \\ CH-(CH_2)_7-CO \end{array} \longrightarrow \text{(IX.) Civettone (p. 95).}$$

(LXXI.) Civettone-α-carboxylate d'éthyle.

BLOMQUIST et HOLLEY (*7*) ont amélioré les rendements de cette synthèse. En appliquant leur méthode de cyclisation de dicétènes au dichlorure de l'acide préparé d'après HUNSDIECKER (*48*), BLOMQUIST, HOLLEY et SPENCER (*8*) ont obtenu la civettone *trans* en un rendement de 33% à partir du diacide.

STOLL, HULSTKAMP et ROUVE (*157*), ont récemment décrit la première synthèse de la *civettone naturelle* cis.

Le céto-9-heptadécane-dioate d'éthyle (LXXII), préparé selon STOLL (*150*), cétalisé par du glycol d'éthylène (LXXIII) est cyclisé selon STOLL et HULSTKAMP (*156*) en éthylène-cétal de la cyclocéto-9-heptadécanolone-1,17 (LXXIV). Par réduction catalytique, cette dernière est transformée en éthylène-cétal du cyclocéto-9-heptadécanediol-1,17 (LXXV). Un traitement à l'acide bromhydrique dans de l'acide acétique le transforme en bromo-acétate cyclique (LXXVI) qui donne par réduction au zinc la *civettone* (IX, p. 95). La séparation des civettones α et β a présenté quelques difficultés; la cristallisation directe des cétones et la cristallisation des semicarbazones ayant échoué, les auteurs ont finalement réussi la séparation par cristallisation fractionnée des éthylène-cétals (LXXVII).

$O{=}C[(CH_2)_7{-}COOC_2H_5]_2$ ⟶ $(CH_2{-}O)_2C[(CH_2)_7{-}COOC_2H_5]_2$ ⟶

(LXXII.) Céto-9-héptadécane-dioate d'éthyle. (LXXIII.)

⟶ $(CH_2{-}O)_2C\langle(CH_2)_7{-}C{=}O,\ (CH_2)_7{-}CHOH\rangle$ ⟶ $(CH_2{-}O)_2C\langle(CH_2)_7{-}CHOH,\ (CH_2)_7{-}CHOH\rangle$ ⟶

(LXXIV.) Éthylène-cétal de la cyclocéto-9-héptadécanolone-1,17. (LXXV.) Éthylène-cétal du cyclocéto-9-héptadécanediol-1,17.

⟶ $O{=}C\langle(CH_2)_7{-}CHBr,\ (CH_2)_7{-}CHOOCCH_3\rangle$ ⟶ (IX) $(CH_2{-}O)_2C\langle(CH_2)_7{-}CH{=}CH{-}(CH_2)_7\rangle$ ⟶

(LXXVI.) (LXXVII.)

⟶ $(CH_2{-}O)_2C\langle(CH_2)_7{-}CHBr,\ (CH_2)_7{-}CHBr\rangle$ $\xrightarrow{-2\,HBr}$

(LXXVIII.)

⟶ $(CH_2{-}O)_2C\langle(CH_2)_7{-}C{\equiv}C{-}(CH_2)_7\rangle$ ⟶ (LXXVII) $(CH_2)_{16}\ NH$

(LXXIX.) (LXXIX a.) Héxadéca-méthylène-imine.

A cette occasion, ils ont aussi purifié la civettone naturelle par son éthylène-cétal et mis en évidence qu'elle contient, à côté de la cyclo-heptadécène-9-one-1 (IX) déterminée par Ruzicka (*110*), de petites quantités d'autres isomères, mais pas de civettone β *(trans)*.

La comparaison de l'éthylène-cétal et de la semicarbazone de la civettone naturelle avec les dérivés synthétiques α et β a permis d'*identifier la civettone naturelle avec la civettone α (cis)*.

Pour déterminer la *stéréoisomérie des civettones* α et β, Stoll, Hulstkamp et Rouve (*157*) ont transformé le cétal de la civettone (LXXVII) en son dibromure (LXXVIII) puis en cétal de déhydrocivettone (LXXIX). Par réduction catalytique de cette dernière, ils ont obtenu le cétal de la civettone α. Puisque la réduction catalytique d'une triple liaison mène presque toujours à l'isomère *cis*, ils attribuent à l'isomère α identique au produit naturel, la forme *cis* et à l'isomère β la forme *trans*. Des mesures de spectres Raman de la civettone naturelle confirment cette conclusion.

2° Constitution chimique et odeur musquée des substances macrocycliques.

L'influence de la grandeur des mégacycles sur l'odeur a été étudiée en détail par Hill et Carothers (*44—46*). Ils ont trouvé que les cycles avec 9 à 12 membres avaient une odeur camphrée ou menthée, ceux avec 13 membres une odeur de bois ou de cèdres et que l'odeur musquée était propre aux cycles à 14, 15 ou 16 membres; les cycles avec 17 ou 18 membres ont quelquefois une odeur de civette; à partir de 18 membres, l'odeur devient faible et disparaît pour des cycles plus grands. Le tableau 1 illustre cette dépendance de l'odeur de la taille des cycles.

Tous ces corps contiennent des groupes carbonyles, mais Ruzicka, Salomon et Meyer (*123*) ont montré que des imines cycliques (par exemple l'hexadéca-méthylène-imine, LXXIXa) ont aussi une odeur musquée. Nous voyons donc que c'est moins la composition qualitative de la molécule que sa *forme* qui est responsable de l'odeur musquée.

3° Musc américain.

Stevens et Erickson (*145*) ont étudié en 1942 le parfum des glandes odorantes du Rat musqué d'Amérique *(Ondatra zibethicus rivalicius)* (Fig. 5) et ont ainsi découvert la source naturelle la plus abondante de substances macrocycliques, car cet animal est extrêmement répandu à certains endroits (6 millions sont capturés annuellement dans le seul État de Louisiana).

Les glandes à parfum se trouvent chez les animaux des deux sexes et pèsent en moyenne 1,25 g. à l'état frais. 2,650 glandes fraîches (3,320 g.) ont fourni 71 g. d'une huile insaponifiable distillant de 130 à

Tableau 1. Odeurs de substances macrocycliques, d'après HILL et CAROTHERS (*45*).

Nombre total de membres du cycle (CO et O compris)	Cétones $(CH_2)_n$ CO	Lactones $(CH_2)_n$ O—CO	Carbonates $(CH_2)_n$ O—CO—O	Anhydrides $(CH_2)_n$ CO—O—CO
5	amandes amères, menthée	fruitée	faible	néant
6	amandes amères, menthée	faible	»	»
7	intermédiaire	faible épicée florale	—	faible
8	»	—	—	—
9	camphrée	—	menthée empyreumatique	—
10	»	—	menthée, moins empyreumatique	épicée
11	»	—	terreuse, camphrée	»
12	»	—	terreuse, camphrée	»
13	bois de cèdre, puis musc, surtout en solution diluée	—	cèdre faiblement rosée et camphrée	—
14	musc	musc	musc	aromatique
15	»	musc et ambre gris	»	musc
16	intermédiaire	musc	»	—
17	civette	civette	»	—
18	faiblement civette	faible	—	—
19	faible	—	—	musc
21	—	—	faible	—

170° sous 1 mm. Par distillation fractionnée on la sépare en deux parties principales, contenant chacune un *carbinol macrocyclique inodore* accompagné de petites quantités de la *cétone* correspondante, *odorante.*

Le carbinol de la partie la plus volatile (éb. 136 à 138°; 1 mm., F. 77 à 78°) est le *cyclo-pentadécanol* [nor-muscol, exaltol (LXXX); la cétone qui l'accompagne est la *cyclo-pentadécanone* [nor-muscone, exaltone (XVII) p. 97].

La partie moins volatile (éb. 155 à 157°) contient surtout le *cyclo-heptadécanol* [dihydrocivettol (LXXXI)] et de petites quantités de *cyclo-heptadécanone* [*dihydrocivettone* (VIII)]. Toutes ces substances ont été identifiées par comparaison avec des produits synthétiques.

La faible activité optique des cétones naturelles serait due à des traces d'isomères et expliquerait la différence des nuances d'odeur des cétones naturelles et synthétiques.

CH_2——CHOH	CH_2——CHOH	CH_2——CO	CH_2——CO
$(CH_2)_{12}$—CH_2	$(CH_2)_{14}$—CH_2	$(CH_2)_{10}$—CH_2	$(CH_2)_{16}$—CH_2
(LXXX.) *Nor*-muscol (exaltol).	(LXXXI.) Dihydrocivettol.	(LXXXII.) Cyclo-tridécanone.	(LXXXIII.) Cyclo-nonadécanone.

Le musc distillé du Rat musqué de Louisiana contient, 58% de dihydrocivettol, 40% de nor-muscol et 2% des deux cétones odorantes. Un brevet de STEVENS et ERICKSON (*146*) décrit l'oxydation des carbinols en cétones odorantes [voir aussi le brevet de SPARHAWK (*141*)].

Fig. 5. Rat d'Amérique (*Ondatra zibethica*, Musc rat, Moschusratte). [Photo New York Zoological Society.]

Dans un mémoire suivant, STEVENS (*144*) rapporte l'isolement de deux nouvelles cétones macrocycliques, la *cyclo-tridécanone* (LXXXII) et la *cyclo-nonadécanone* (LXXXIII) obtenues par oxydation des carbinols naturels des glandes à parfum du Rat musqué.

STEVENS (*144*) a été frappé par le parallélisme entre les quantités de cétones et carbinols macrocycliques et la teneur des glandes en acides gras ayant un atome de carbone de plus (Tableau 2).

Tableau 2. Teneur des glandes à parfum du Rat musqué en acides gras et en substances macrocycliques [STEVENS (*144*)].

Acides gras %	Substances macrocycliques %
C_{10} 0,3	C_9 —
C_{12} 0,7	C_{11} —
C_{14} 8,1	C_{13} 1,0
C_{16} 29,8	C_{15} 40,0
C_{18} 59,3	C_{17} 58,0
C_{20} 1,5	C_{19} 0,7
C_{22} 0,2	C_{21} ?

Il en tire la conclusion que les cétones sont formées à partir des acides ayant un atome de carbone de plus, et propose le mécanisme suivant:

$$\begin{array}{l} \diagup CH_2{-}CH_3 \\ (CH_2)_x \\ \diagdown CH_2{-}CH_2{-}COOH \end{array} \xrightarrow[\text{partielle}]{\omega\text{-oxydation}} \begin{array}{l} \diagup CH_2{-}CHO \\ (CH_2)_x \\ \diagdown CH_2{-}CH_2{-}COOH \end{array} \xrightarrow{\text{cyclisation}}$$

$x = 9, 11, 13, 15$
(pour la civettone prendre l'acide oléique).

$$\longrightarrow \begin{array}{l} \diagup CH_2{-}CHOH \\ (CH_2)_x \quad | \\ \diagdown CH_2{-}CH\cdot COOH \end{array} \xrightarrow{-\,CO_2} \begin{array}{l} \diagup CH_2{-}CHOH \\ (CH_2)_x \quad | \\ \diagdown CH_2{-}CH_2 \end{array}$$

$$\downarrow -\,2\,H \qquad\qquad\qquad \downarrow \text{oxydation}$$

$$\begin{array}{l} \diagup CH_2{-}CO \\ (CH_2)_x \quad | \\ \diagdown CH_2{-}CH{-}COOH \end{array} \xrightarrow{-\,CO_2} \begin{array}{l} \diagup CH_2{-}CO \\ (CH_2)_x \quad | \\ \diagdown CH_2{-}CH_2 \end{array}$$

La *synthèse biologique de la muscone* s'expliquerait par les formules suivantes:

$$\begin{array}{l} \diagup CH_2{-}CH_2{-}CH_3 \\ (CH_2)_{11} \\ \diagdown CH_2{-}CH_2{-}CH_2{-}COOH \end{array} \xrightarrow{\omega\text{-oxydation}} \begin{array}{l} \diagup CH_2{-}CH_2{-}COOH \\ (CH_2)_{11} \\ \diagdown CH_2{-}CH_2{-}CH_2{-}COOH \end{array} \longrightarrow$$

Acide stéarique.

$$\xrightarrow[\text{bilaterale}]{\beta\text{-oxydation}} \begin{array}{l} \diagup CO{-}CH_2{-}COOH \\ (CH_2)_{11} \\ \diagdown CH_2{-}CO{-}CH_2{-}COOH \end{array} \xrightarrow{-\,2\,CO_2} \begin{array}{l} \diagup CO{-}CH_3 \\ (CH_2)_{11} \\ \diagdown CH_2{-}CO{-}CH_3 \end{array} \longrightarrow$$

$$\xrightarrow{\text{condensation}} \begin{array}{l} \diagup CO{-}CH_2 \\ (CH_2)_{11} \quad | \\ \diagdown CH_2{-}C{-}CH_3 \\ \qquad\quad | \\ \qquad\quad OH \end{array} \xrightarrow{-\,H_2O} \begin{array}{l} \diagup CO{-}CH \\ (CH_2)_{11} \quad \| \\ \diagdown CH_2{-}C{-}CH_3 \end{array} \xrightarrow{\text{réduction}}$$

$$\longrightarrow \begin{array}{l} \diagup CO{-}CH_2 \\ (CH_2)_{11} \quad | \\ \diagdown CH_2{-}CH{-}CH_3 \end{array}$$

Muscone.

Les carbinols du Rat musqué pourraient se former par réduction des cétones, ou plutôt, puisqu'ils constituent 98% de la fraction macrocyclique en être les précurseurs (formés par décarboxylation de l'acide β hydroxylé).

Ainsi s'expliquerait l'existence du civettol à côté de la civettone; le mécanisme de formation de la muscone esquissé ci-dessus ne laisse pas, d'autre part, prévoir de formation intermédiaire du carbinol muscol. Ruzicka (*111*) n'a pas, en effet, trouvé de muscol dans l'extrait de musc. Des théories analogues à celle de Stevens (*144*) avaient, du reste, déjà été émises par Ruzicka (*111*) en 1926, qui avait fait dériver la civettone de l'acide oléique et la muscone de l'acide palmitique.

Dans une fraction de tête des cétones macrocycliques, Stevens (*144*) a trouvé une cétone, probablement identique à la *décanone-2*, trouvée par Pfau (*96*) dans l'huile de Rue (*Ruta graveolens*).

4° Autres animaux à sécrétion musquée.

Le tableau 3 énumère d'autres animaux ayant des sécrétions musquées, mais dont nous ne possèdons encore aucun détail chimique.

Tableau 3. Animaux à sécrétion musquée [en partie d'après Hardy (*43*)].

	Nom	Organe ou partie du corps ayant l'odeur musquée
	I. Mammifères.	
Bœuf musqué	*Ovibos moschatus*	Toute la chair
Zébu	*Bos indicus*	
Chamois	*Capra ibex*	Le sang séché
Pécari	*Dicotyles torquatus*	Glandes inguinales
Desman	*Myogal moschatus*	Glandes caudales
Fourmilier	*Myrmecophaga tetradactyla*	La chair
Blaireau	*Meles meles*	Glandes caudales
Furet-blaireau	*Helictis moschata*	Glandes caudales
Martre	*Mustela foina*	Les fèces
Kangourou musqué	*Potorus*	
	II. Oiseaux.	
Canard musqué	*Ana moschata*	La chair
Canard musqué d'Australie	*Biziura lobata*	
Vautour	*Gyps fulvus*	L'œuf
	III. Reptiles.	
Tortue	*Cinosternon pennsylvanicum*	
Crocodile	*Crocodilus vulgaris*	4 paires de glandes
	IV. Insectes.	
Coléoptères	*Aromia moschata*	
»	*Callichromides*	
	V. Mollusques.	
Escargot	*Fasciolaria trapezium*	
Céphalopode	*Eledone moschata*	Glandes à parfum

Des glandes odorantes bien développées se trouvent en outre chez le Renard (glandes caudales et plantaires), chez le Cochon et l'Hippopotame (glandes caudales et inguinales), chez les Lézards et Geckos (glandes sous-maxillaires et autres).

L'influence des substances macrocycliques à odeur de musc sur la *psychologie humaine* est certaine. LE MAGNEN (*82*) écrit que «l'odeur de musc provoque des images sexuelles et de plaisir physique»; il a trouvé, d'autre part, que certaines sensibilités étaient en relation avec l'état hormonal. «L'odeur de musc d'une lactone à grand cycle n'était perçue de façon intense que par les femmes adultes. Les mâles et les impubères des deux sexes ne perçoivent pas, ou perçoivent faiblement l'odeur. L'intensité perçue par la femme semble évoluer nettement avec le cycle génital; l'intensité maxima de la perception étant atteinte au moment où la sécrétion endocrine des hormones sexuelles est maximale» (*81*).

5° Substances soufrées des Mustélidés et d'autres animaux.

Les sécrétions odorantes des Mustélidés semblent surtout servir à la défense, car ce sont des substances soufrées généralement très malodorantes.

Le Mydaus des Philippines (*Mydaus marchei* Huet) éjecte la sécrétion de ses deux glandes anales contre son ennemi à une distance de presqu'un

Fig. 6. Moufette (*Mephitis nigra*, Skunk, Skunks). [Photo New York Zoological Society.]

mètre. ALDRICH (*1*) et BECKMANN (*4*) ont déjà montré en 1896 que leur «parfum» contenait du butyl-mercaptan et ils avaient soupçonné la présence du disulfure correspondant. En 1937, FESTER et BERTUZZI (*25*) ont étudié l'excrétion malodorante d'une Moufette (du «zorrino») d'Amérique du Sud (*Conepatus suffocans*) et constaté la présence d'un mercaptan en C_4 (probablement *n*-butyl-mercaptan) à côté d'un mercaptan en C_4 insaturé et du disulfure correspondant. En 1945, STEVENS (*143*) a examiné le «parfum» de la Moufette d'Amérique du Nord (*Mephitis mephitis nigra*) (Fig. 6) HARDY (*43*) dit de cet animal "...from two anal glands the animals squirt for ten to twelve feet a yellow fluid that is so pungent, nauseous and foetid that no enemy—unless it be the great horned owl—can withstand it. The smell often pervades the atmosphere for a quarter of a mile". STEVENS (*143*) a isolé de cette secrétion, à côté de mercaptans très volatils, une substance insaturée de formule $C_8H_{14}S$ qu'il a identifié avec le *sulfure de dicrotyle* $CH_3 \cdot CH{=}CH{-}CH_2{-}S{-}CH_2{-}CH{=}CH \cdot CH_3$. Il n'y avait pas de substances macrocycliques.

Des substances soufrées malodorantes se forment quelquefois aussi dans l'*organisme humain* et se font sentir dans l'haleine. C'est le cas par exemple des malades traités par la thiourée, contre l'hyperthyroidisme; il s'agit probablement d'un mercaptan formé à partir de la thiourée [CHALLENGER (*14*)]. Des personnes recevant des préparations de bismuth étaient autrefois connues pour leur «haleine de bismuth»; là il s'agissait de dérivés méthylés du tellurium présents comme impuretés dans le bismuth [CHALLENGER (*14*)]. L'odeur désagréable de l'haleine après absorption d'ail est certainement aussi dûe à des dérivés alcoylés du soufre.

On trouve aussi des mercaptans chez les Invertébrés, par exemple dans les glandes à pourpre de *Murex trunculus* [JULLIEN, GARABEDIAN et GUIBAULT (*57*)].

6° Castoréum.

Les glandes à parfum du Castor Canadien (*Castor fiber* L.) (Fig. 7) se trouvent dans le commerce sous le nom de *castoréum*. Ces glandes, piriformes et paires, existent dans la région anale des animaux de deux sexes et pèsent

Fig. 7. Castor (*Castor fiber*, Beaver, Biber).

à l'état sec, de 50 à 100 g. par animal (Fig. 8). En marchant, le Castor laisse une trace odorante due à la sécretion de ces glandes; c'est cette odeur, phénolique et animale, qui semble jouer un rôle dans l'attraction sexuelle et qui est encore aujourd'hui appréciée en parfumerie. A l'état frais la sécrétion se présenterait sous forme d'un liquide blanchâtre, crémeux; à l'état séché elle devient brune, foncée, dure et cassante (*42*).

Walbaum et Rosenthal (*167*) ainsi que Pfau (*95*) ont étudié plus en détail les constituants du castoréum et y ont trouvé l'*acide benzoïque*, l'*acétophénone*, l'*alcool benzylique*, le *p-éthylphénol*, le *bornéol* et deux phénols représentant soit l'*o*-crésol et le gaïacol, soit l'*o*-éthylphénol et le créosol. Pfau (*95*) mentionne également la présence d'une *lactone* très odorante.

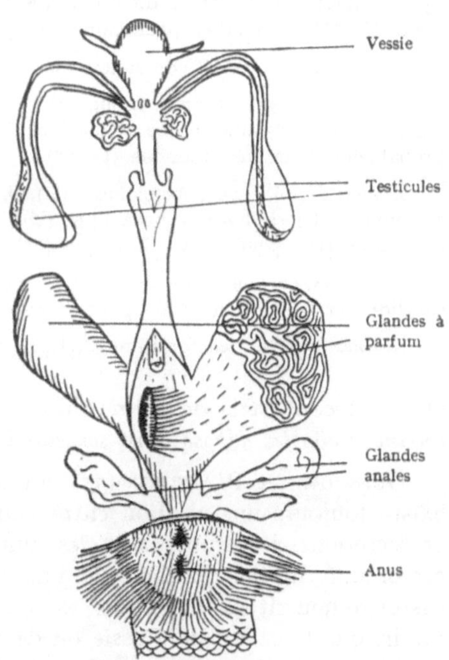

Fig. 8. Organes sexuels du Castor (d'après Guibourt et Plauchon).

Nous y avons trouvé, en plus des substances isolées par les auteurs précédents, une trentaine de composés nouveaux (*70*); Tableau 4, p. 114.

Stevens (*142*) a récemment cherché des substances macrocycliques dans le castoréum, sans en trouver; il rapporte l'isolement d'acide anisique; son «acide castorique» n'est certainement pas un produit homogène, de même la «castorine» d'auteurs anciens (voir *62*, *67*); dans ce dernier cas il s'agit certainement de cholestérol accompagné de faibles quantités d'autres stéroïdes (*67*).

En ajoutant aux 45 substances identifiées avec certitude, autant d'autres dont la purification n'a pas encore été possible, nous arrivons au total à une centaine de corps chimiques dont se composerait la sécrétion des glandes à parfum du Castor. La présence d'un tel nombre de substances plus ou moins étranges dans une glande animale nous a paru très surprenante; nous croyons cependant avoir trouvé les deux principales raisons qui expliquent ce phénomène. Tout d'abord, le Castor se nourrit principalement d'*écorces* et de *bourgeons* d'arbres, donc de parties végétales très riches en glycosides, phénols, terpènes, etc. Mais l'absorption de ces substances par d'autres Vertébrés entraînerait simplement leur élimination par l'urine, le plus souvent après conjugaison appropriée.

Tableau 4. Constituants du castoréum (glandes à parfum du Castor Canadien *70*).
(Les substances marqués d'un U se trouvent dans l'urine de Vertébrés.)

Alcools (*67, 68, 70, 78*): alcool benzylique (libre et estérifié), cholestérol, β-cholestanol, *cis*-hydroxy-5-tétrahydro-ionol (U), mannitol (U).

Phénols (*63, 68, 70, 75*): *p*-éthylphénol (U, **), *p*-propyl-phénol (**), pyrocatéchol (U), hydroquinone (U), éther monométhylique d'hydroquinone (*), chavicol (*), éthyl-gaïacol (*), méthyl-pyrocatéchol (*), éthyl-pyrocatéchol (*), bétuligénol (*), dihydroxy-2,4'-diphénylméthane (**), dihydroxy-4,4'-dibenzo-α-pyrone (pigment I, ***), dilactone de l'acide dihydroxy-4,4'-diphénique (pigment II, ***); un éther phénolique, F. 83° (***).

Aldéhydes (*64*): aldéhyde salicylique (*).

Cétones (*64, 78*): acétophénone (*), *p*-hydroxy-acétophénone (*), *p*-méthoxy-acétophénone (**), une cétone aromatique $C_{10}H_{12}O$, deux cétones isomères $C_{13}H_{24}O_2$, probablement dérivés d'ionone (U, **).

Acides (*62, 68, 70*): benzoïque (U), hydrocinnamique (*), cinnamique (U), salicylique (U), *m*-hydroxybenzoïque (U), *p*-hydroxybenzoïque (U), anisique (*), gentisique (U), 5-méthoxy-salicylique (**).

Esters (*68, 70*): oléate de cholestérol, un ester phénolique (F. 170°), esters d'alcool cérylique et benzylique et de phénols avec l'acide gentisique.

Amines (*67, 70*): castoramine $C_{15}H_{23}O_2N$, F. 66° (***).

Or — et ceci est la deuxième raison — le castor ne *les élimine pas par l'urine*, mais *les dépose dans ses glandes à parfum* et les accumule.

GUIBOURT et PLANCHON (*35*) ont déjà en 1876 émis l'opinion «qu'il existe toujours une relation entre l'odeur que présentent les excrétions et sécrétions des animaux et les aliments dont ils se nourrissent. Le castoréum du Canada possède une odeur térébinthinée parce que le castor se nourrit surtout d'écorces de conifères si communes en Amérique tandis que le castor de Russie ou de Sibérie fournit le castoréum de ce nom, caractérisé par l'odeur de cuir de Russie ou d'écorce de bouleau, qui sert tout à la fois à nourrir les castors et à tanner les cuirs».

Les substances suivantes, isolées du castoréum se trouvent dans les *écorces* d'arbres: *bornéol, pyrocatéchol, bétuligénol, p-hydroxy-acétophénone, acides benzoïque et salicylique. L'acide ellagique* des écorces d'arbres est un proche parent des pigments du castoréum (voir ci-dessous). De la *lignine* dérivent certainement les phénols méthoxylés *gaïacol, créosol* et *ethyl-gaïacol*. Les *bourgeons* d'arbres contiennent: de l'*hydroquinone* et son *ether monométhylique*, de l'*aldéhyde salicylique*, de l'*acétophénone*, du *salicylate de benzyle*, etc.

(*) Substances isolées pour la première fois d'un organe animal.
(**) Substances isolées pour la première fois d'une source naturelle.
(***) Substances non encore décrites précédemment.

On a isolé, d'autre part, les substances suivantes de l'*urine* de Vertébrés: le *mannitol*, le *cis-hydroxy-5-tétrahydro-ionol*, le *p-éthylphénol*, le *pyrocatechol*, l'*hydroquinone*, les *acides benzoïque*, *cinnamique*, *salicylique*, *m-* et *p-hydroxybenzoïque*, etc. (voir au Tableau 4 toutes les substances marquées d'un U, p. 114).

Quelques unes des substances de la nourriture sont déposées inchangées, d'autres après *hydrolyse* d'une liaison glycosidique (*picéol* → *p-hydroxyacétophénone*; *bétuloside* → *bétuligénol*), d'autres après *oxydation* (*acide salicylique* → *acide gentisique*) ou réduction (*xanthophylles* → *tétrahydroionols*; *acide cinnamique* → *acide hydrocinnamique*).

Passons maintenant à une discussion plus détaillée de la chimie et de la biochimie des constituants les plus intéressants du castoréum.

Acétophénone. La présence dans le castoréum d'assez grandes quantités de cette cétone qui est assez rare dans le règne végétal peut s'expliquer par un mécanisme décrit par DAKIN (*18*); cet auteur a trouvé que l'acide hydrocinnamique, administré à des Chiens ou des Chats conduit à l'excrétion d'acétophénone dans l'urine. Le mécanisme de cette réaction est probablement le suivant:

$$C_6H_5\cdot CH_2\cdot CH_2\cdot COOH \rightarrow C_6H_5\cdot CHOH\cdot CH_2\cdot COOH \rightarrow$$
$$\rightarrow C_6H_5\cdot CO\cdot CH_2\cdot COOH \rightarrow C_6H_5\cdot CO\cdot CH_3 + CO_2.$$

Phénols volatils. Parmi les phénols volatils du castoréum, le *p*-éthylphénol et le *p*-propylphénol sont les plus importants; on ne les a jamais trouvé dans le règne végétal. Ceci conduit à penser qu'ils sont formés dans l'organisme même du Castor, par dégradation d'autres substances phénoliques, dont la principale pourrait être la tyrosine. Nous allons revenir plus tard sur cette question (p. 136).

Bétuligénol. Cette substance est un phénol $C_{10}H_{14}O_2$ (F. 80°), dont nous avons établi la constitution en 1941, indépendamment de SOSA (*139*) qui l'avait isolé comme aglycone du glycoside betuloside, extrait de l'écorce de Bouleau. L'oxydation de l'éther monométhylique du bétuligénol (LXXXIV) donne l'acide anisique, son oxydation chromique fournit une cétone qui s'est avérée identique à l'anisyl-1-butanone-3 synthétique (LXXXV) (*75*).

$$H_3CO-C_6H_4-CH_2-CH_2-CHOH-CH_3 \xrightarrow{C_2O_3}$$

(LXXXIV.) Éther monométhylique du bétuligénol.

$$\longrightarrow H_3CO-C_6H_4-CH_2-CH_2-CO.CH_3 \qquad HO-C_6H_4-CH_2-CH_2-CHOH-CH_3$$

LXXXV.) Anisyl-1 butanone-3.

(LXXXVI.) Bétuligénol.

Le bétuligénol est donc le *p*-hydroxyphényl-1-butan-3-ol (LXXXVI). DELEPINE et SOSA (*19*) ont décrit le dédoublement du *d,l-p*-méthoxyphenyl-1-butanol-3 en ses composants optiquement actifs; l'antipode lévogyre a été identifié avec le méthyl-bétuligénol.

Dihydroxy-2,4'-diphénylméthane. Cette substance (F. 120°) (LXXXVII) a été identifiée par oxydation chromique ménagée en 2,4'-dihydroxybenzophénone (LXXXVIII). On ne l'avait pas encore trouvée dans la nature (*75*).

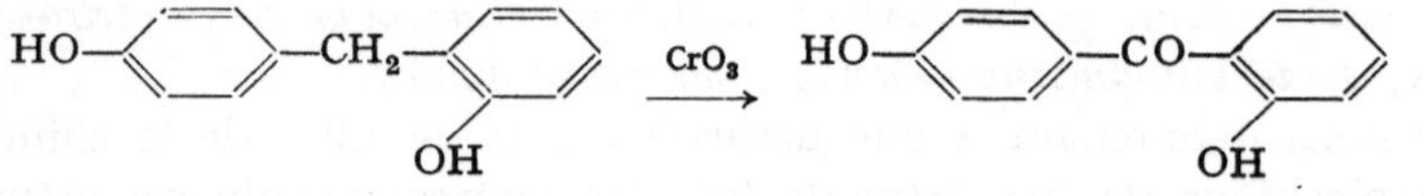

HO—C₆H₄—CH₂—C₆H₄(OH) —CrO₃→ HO—C₆H₄—CO—C₆H₄(OH)

(LXXXVII.) Dihydroxy-2,4'-diphénylméthane. (LXXXVIII.) Dihydroxy-2,4'-benzophénone.

Pigments jaunes du castoréum (*63*). Les glandes à parfum du Castor contiennent deux nouveaux pigments jaunes qui sont solubles dans le carbonate aqueux avec une couleur jaune intense et que nous avons séparé par cristallisation fractionnée de leurs acétates incolores.

Le pigment «I», $C_{13}H_8O_4$ (LXXXIX) peut être obtenu à partir de son diacétate incolore (F. 210°) sous forme d'aiguilles jaunes fondant au dessus de 360° (avec décomposition). Il contient deux hydroxyles phénoliques et un groupement lactonique. Par distillation sur poudre de zinc on obtient du *fluorène*, $C_{13}H_{10}$ (XC), ce qui fixe la structure de tout le squelette carboné. La fusion alcaline donne de l'*acide m-hydroxybenzoïque.* Ces résultats, ainsi que la réaction de GRIESSMAYER positive (voir ci-dessous) nous ont conduit à formuler le pigment «I» comme dihydroxy-4,4'-dibenzo-α-pyrone (LXXXIX). Nous avons récemment con-

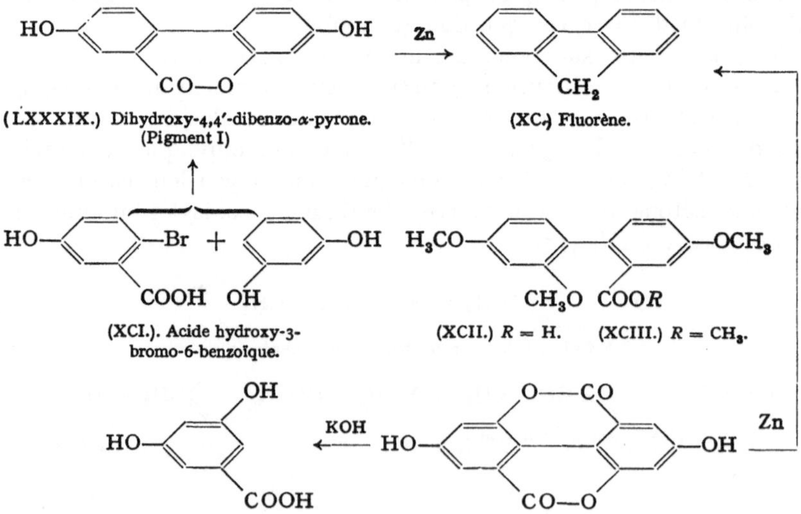

(LXXXIX.) Dihydroxy-4,4'-dibenzo-α-pyrone. (Pigment I) —Zn→ (XC.) Fluorène.

(XCI.). Acide hydroxy-3-bromo-6-benzoïque. + HO—C₆H₄(OH) → (LXXXIX)

(XCII.) R = H. (XCIII.) R = CH_3.

(XCV.) Acide β-résorcylique. ←KOH— (XCIV.) Pigment II. —Zn→ (XC)

firmé cette formule par synthèse d'après la méthode de HURTLEY (*51*). En condensant l'acide hydroxy-3-bromo-6-benzoïque (XCI) avec le résorcinol en présence de traces de sulfate de cuivre comme catalyseur, nous avons obtenu un pigment jaune, dont le diacétate, l'éther diméthylique et les deux dérivés (XCII) et (XCIII) étaient identiques avec les substances correspondantes dérivées du pigment naturel. La même méthode peut servir à la synthèse de divers analogues du pigment «I» (*76*).

Le pigment «II» (XCIV) du castoréum est obtenu par hydrolyse de son diacétate incolore (F. 320°); il se présente sous forme d'aiguilles vert-jaunâtre qui ne fondent pas au dessus de 360°. Sa formule brute est $C_{14}H_6O_6$ et il contient deux hydroxyles phénoliques et deux groupements lactoniques. La distillation sur poudre de zinc donne également du fluorène (XC) et la fusion alcaline l'acide β-résorcylique (XCV). Ces résultats conduisent à la formule (XCIV) pour ce pigment. Une formule asymétrique, avec un seul hydroxyle en 4 n'était cependant pas exclue à priori. Ceci nous a amenés à étudier la réaction de GRIESSMAYER (*33*); cette réaction consiste en une coloration rouge obtenue par chauffage de l'acide ellagique (XCVI) et de pigments analogues

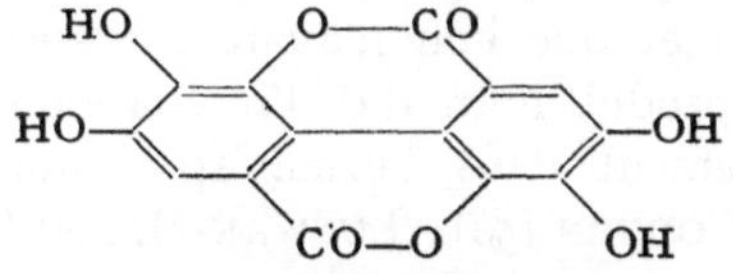

(XCVI.) Acide ellagique.

avec de l'acide nitrique contenant de l'acide nitreux; la coloration est stable après dilution par l'eau. Nous avons trouvé que seuls les pigments ayant deux hydroxyles libres en position 4,4' donnaient la réaction de GRIESSMAYER (*33*); ceci a été également confirmé avec des dérivés synthétiques du pigment «I». Etant donné que les deux pigments du castoréum donnent une réaction positive, le pigment «II» doit bien avoir la formule (XCIV).

Les deux pigments du castoréum sont de proches parents de l'acide ellagique (XCVI), substance très répandue dans les écorces d'arbres. Nous ne savons pas encore si les pigments du castoréum doivent être considérés comme produits de réduction de l'acide ellagique ou s'ils existent déjà préformés dans le règne végétal.

Dérivés d'ionone. En 1943, nous avons décrit un glycol amorphe isolé du castoréum, pour lequel nous avons proposé la formule $C_{12}H_{24}O_2$ d'un glycol monocyclique; l'oxydation chromique de cette substance nous a donné une dicétone (caractérisée par sa *bis*-2,4-dinitrophénylhydrazone) (F. 200°) qui donnait la réaction de LÉGAL caractéristique des méthylcétones (*67*). Au cours des années suivantes, notre glycol a cristallisé (F. 100°) et une comparaison de ce glycol, de la dicétone et de deux de ses dérivés cristallisés a finalement montré son identité avec le mélange de diols stéréoisomères dérivés de l'ionone que PRELOG et coll. (*101*) avaient isolé de l'urine de Jument gravide et dont nous parlerons plus en détail ci-dessous (p. 133). Il est très probable que les deux céto-alcools $C_{13}H_{24}O_2$ isomères que nous avons isolés du castoréum (*64*) sont aussi identiques aux dérivés correspondants isolés par PRELOG et ses collaborateurs (*101*).

Nous avons ici une nouvelle illustration du parallélisme entre les constituants de l'urine et ceux du castoréum et une indication sur l'importance générale de ces dérivés d'ionone comme produits de dégradation des xanthophylles (*78*).

Acides du castoréum (*62*). Ici, l'analogie entre les constituants du castoréum et ceux de l'urine des Vertébrés est particulièrement frappante; dans les deux cas, l'acide benzoïque est l'acide prépondérant; dans les deux cas, nous trouvons les trois acides hydroxy-benzoïques. Notons cependant l'absence de l'acide phénylacétique dans le castoréum; il y est remplacé par l'acide phénylpropionique.

L'acide gentisique (dihydroxy-2,5-benzoïque) qui est un des constituants majeurs du castoréum et que l'on n'avait pas encore trouvé dans un organe animal, est un produit normal de l'oxydation de l'acide salicylique et se trouve généralement dans l'urine après administration d'acide salicylique [KAPP et COBURN (*58*); LUTWAK-MANN (*83*)].

Amines du castoréum (*67*). La partie basique d'un extrait de castoréum contient un mélange de bases assez volatiles et malodorantes; la moitié de cette fraction est constituée par une base cristallisée, F. 66°, (sulfate: F. 216°) que nous avons appelée *castoramine*; sa formule est $C_{15}H_{23}O_2N$; elle ne contient ni méthoxyle, ni hydrogène mobile. SCHINZ, RUZICKA, GEYER et PRELOG (*137*) ont récemment isolé une base analogue de la glande de musc du chevrotain porte-musc qu'ils ont appelée «*muscopyridine*». Cette base a la formule $C_{16}H_{25}N$; malgré la similitude des formules de ces deux «alcaloïdes animaux», les deux substances ne semblent pas être apparentées chimiquement, car la castoramine ne donne pas, comme la muscopyridine, le spectre d'absorption des dérivés de pyridine (*70*).

7° Suintine.

La suintine, ou graisse de laine est la substance cireuse excrétée sur la laine par les glandes sébacées du Mouton. Elle a une odeur particulière et qui est différente suivant les races de Mouton; elle permet aux spécialistes de distinguer les différentes sortes de laine.

La suintine est un mélange très complexe d'esters d'acides gras ramifiés avec des alcools aliphatiques et cycliques. Nous n'entrerons pas en détail dans sa composition chimique que nous avons exposée ailleurs (*80*).

Il n'y a pas, à proprement parler, de travaux chimiques sur l'odeur de la suintine; elle est probablement dûe, en partie, à des acides gras libres. A la saponification, on obtient un mélange fortement odorant d'acides gras très interessants dont WEITKAMP (*168*) a precisé la nature. Après distillation fractionnée des esters méthyliques, il a caractérisé 32 acides gras appartenant à quatre séries homologues: 1° Acides à chaîne normale, 2° acides α-hydroxylés, 3° acides «iso» ayant un groupe méthyle attaché sur l'avant dernier atome de carbone (XCVII) et 4° acides

$$H_3C-CH(CH_3)-(CH_2)_{2n}-COOH \qquad n = 3 \text{ à } 11.$$

(XCVII.)

$$H_3C-CH_2-CH(CH_3)-(CH_2)_{2n}-COOH \qquad n = 2 \text{ à } 13.$$

(XCVIII.)

«antéiso» dextrogyres, ayant un groupe méthyle attaché au carbone antépénultième (XCVIII). Les acides «anteiso» sont ainsi les premiers acides gras naturels à nombre impair d'atomes de carbone. Dans la série «iso» il y a des représentants de 10 à 28 atomes de carbone, dans la série «antéiso» de 9 à 31 atomes de carbone. Les membres inférieurs de ces séries contribuent fortement à l'odeur des acides isolés de la suintine.

WEITKAMP, SMILJANIC et ROTHMAN (*169*) ont trouvé récemment que la graisse de cheveux d'adultes contient des acides libres comprenant des acides gras saturés et insaturés de 7 à 22 atomes de carbone. C'est la première fois que l'on ait trouvé dans une source naturelle des acides gras normaux à 7, 9, 11, 13, 15 et 17 atomes de carbone. La double liaison des acides insaturès est généralement en 6, 7, quelquefois en 8, 9. Ces acides gras sont fungicides contre *Microsporon andouini* (champignon pathogène causant la teigne, *Tinea capitis*, des enfants et contre laquelle les adultes s'avèrent résistants).

Nous avons récemment insisté sur l'intérêt biochimique de ces résultats et avons conclu que la composition particulière de la graisse de laine (présence d'alcools triterpéniques, d'acides ramifiés et à nombre impair d'atomes de carbone) reflète les capacités synthétiques des glandes sébacées, différentes de celles du foie (*73*).

8° Yacarol.

La seule sécrétion odorante des Vertébrés inférieurs ayant été étudiée plus en détail, est celle des Alligators d'Amérique du Sud (*Alligator sclerops* et *A. latirostris*). Ces animaux ont quatre paires de «glandes à musc» qui sont utilisées par les indigènes comme source de parfum, comme celles du Crocodile d'Egypte.

Une paire de glandes se trouve près des yeux, une autre sous la mâchoire inférieure, une troisième sur les pattes de devant et la quatrième, la plus importante, près des glandes sexuelles. Ces dernières pèsent de 4 à 35 g. par animal et contiennent une huile jaune de laquelle FESTER et BERTUZZI (24) ont isolé les substances suivantes: les *acides isovalérique, myristique* et *palmitique*, du *cholestérol*, de l'*alcool cétylique* et une substance liquide à odeur douce et rosée qu'ils ont appelée *yacarol* (d'après *yacare*: alligator en espagnol). L'extrait éthéré des glandes contient 80% d'acides gras et 20% d'insaponifiable, dont un cinquième de yacarol.

Constitution du yacarol. L'oxydation du yacarol a donné de l'*acétone* et de l'*acide lévulique*, ce qui a conduit les auteurs à attribuer à cette

$$CH_3{-}\underset{\displaystyle CH_3}{\underset{|}{C}}{=}CH{-}CH_2{-}CH_2{-}\underset{\displaystyle CH_3}{\underset{|}{C}H}{-}CH_2OH$$

(IC.) Diméthyl-2,6-heptène-2-ol-7.

substance la structure d'un diméthyl-2,6-heptène-2-ol-7 (IC). Mais FESTER et PUCCI (27) ont ensuite trouvé que le yacarol est différent de ce corps, obtenu par synthèse. FESTER, BERTUZZI et PUCCI (26) ont enfin montré que le yacarol n'est pas un corps homogène; la partie principale en est le *d-citronellol* (C), identifié par trois dérivés cristallisés

$$CH_3{-}\underset{\displaystyle CH_3}{\underset{|}{C}}{=}CH{-}CH_2{-}CH_2{-}\underset{\displaystyle CH_3}{\underset{|}{C}H}{-}CH_2{-}CH_2OH$$

(C.) Citronellol.

avec un produit authentique; les têtes de distillation contiennent peut-être un alcool saturé $C_8H_{17}OH$. Ils pensent que le yacarol provient de l'oxydation d'une chaîne latérale d'un stérol. Cette hypothèse paraît intéressante, mais il faudrait aussi tenir compte de la possibilité de l'origine alimentaire du *d*-citronellol.

II. Concrétions intestinales.

1° Ambre gris.

L'ambre gris est une concrétion intestinale du Cachalot (*Physeter macrocephalus* L.), (Fig. 9) dont les conditions de formation sont encore très peu connues. L'ambre gris trouvé dans l'intestin du Cachalot se présente sous forme d'amas foncés, presque noirs, de taille très variable et pouvant atteindre 100 kgs. Leur odeur est plus ou moins caractéristique, fécale et animale, avec des notes de tabac, de marais et de musc. Après la mort des animaux, les morceaux d'ambre gris flottent à la surface de la mer.

le soleil les blanchit et leur odeur s'affine. C'est à cet état que l'ambre gris est le plus apprécié en parfumerie, à la fois pour son odeur et son pouvoir fixateur.

Composition chimique de l'ambre gris.

Jusqu'en 1946, nos connaissances sur la composition de l'ambre gris étaient assez maigres. Des auteurs anciens [PELLETIER et CAVENTOU (*94*), RIBAN (*107*)] puis, plus récemment SUZUKI (*162*), avaient isolé une substance cristallisée appelée ***ambréine*** et ressemblant d'assez près au

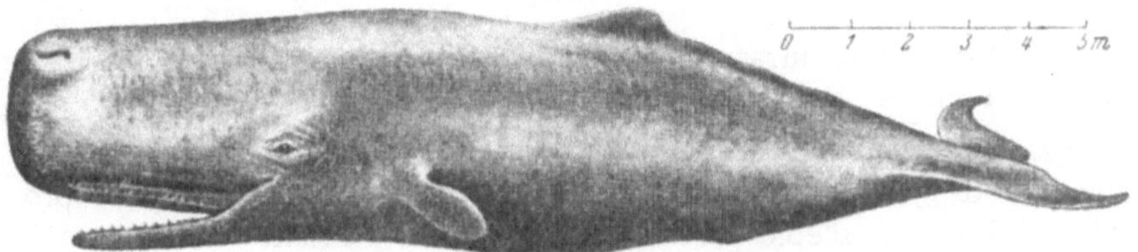

Fig. 9. Cachalot (*Physeter* ***macrocephalus***, Sperm whale, Pottwal).

cholestérol. ISIGURO et WATANABE (*52*) y ont trouvé de l'acide ***arachidique*** et de l'***épicoprostérol***. JANISTYN (*53*) a étudié de près les stérols de l'ambre gris et y a constaté aussi la présence de ***coprostérol***. OKAHARA (*93*) a décrit l'isolement d'une ***porphyrine*** et SANDULESCO et SABETAY (*135*) mentionnent l'isolement d'une partie ***cétonique*** odorante.

Pristane. Dans une première note sur l'ambre gris, nous avons décrit (*66*) l'isolement d'une ***paraffine liquide*** qui fait environ 4% de l'ambre gris. Cette substance est, d'après toutes ses propriétés physiques et chimiques, identique au ***pristane*** que TSUJIMOTO (*163*) et d'autres auteurs japonais (*165*) avaient trouvé dans des foies de Requins et d'autres huiles de Poissons et pour lequel la formule $C_{18}H_{38}$ semblait la plus probable. Le pristane de l'ambre gris pourrait donc simplement provenir de la nourriture du Cachalot.

Récemment, SÖRENSEN et MEHLUM (*138 c*) ont trouvé que les propriétés physiques du pristane étaient plutôt en accord avec celles d'une nonadecane $C_{19}H_{40}$. La synthèse des huit monométhyl-octadécanes (*138 a*) fondant entre —16,5° et 13,2° a montré que le pristane (F. —100°) devait être plus ramifié. SÖRENSEN et SÖRENSEN (*138 b*) ont finalement ***identifié le pristane avec le nor-phytane*** ($C_{19}H_{40}$, tétraméthyl-2, 6, 10, 14-pentadécane) obtenu par synthèse à partir du phytol. Cette identité vient d'être confirmée par la mesure des spectres infrarouges (PLIVA, sous presse).

SÖRENSEN et SÖRENSEN (*138 b*) pensent que le pristane = norphytane se forme par action enzymatique (des bactéries intestinales?)

à partir du phytol, par décarboxylation de l'acide correspondant et hydrogénation ou par décarboxylation de l'acide saturé.

$$\begin{array}{ll} CH_3\cdot \underset{\displaystyle CH_3}{CH}\cdot CH_2CH_2CH_2\cdot \underset{\displaystyle CH_3}{CH}\cdot CH_2CH_2CH_2\underset{\displaystyle CH_3}{CH}\cdot CH_2CH_2CH_2\cdot \underset{\displaystyle CH_3}{C}{=}CH\cdot CH_2OH & \text{phytol} \\ CH_3\cdot \underset{\displaystyle CH_3}{CH}\cdot CH_2CH_2CH_2\cdot \underset{\displaystyle CH_3}{CH}\cdot CH_2CH_2CH_2\cdot \underset{\displaystyle CH_3}{CH}\cdot CH_2CH_2CH_2\underset{\displaystyle CH_3}{C}{=}CH_2 & \text{zamène} \\ CH_3\cdot \underset{\displaystyle CH_3}{CH}\cdot CH_2CH_2CH_2\cdot \underset{\displaystyle CH_3}{CH}\cdot CH_2CH_2CH_2\cdot \underset{\displaystyle CH_3}{CH}\cdot CH_2CH_2CH_2\underset{\displaystyle CH_3}{CH}\cdot CH_3 & \text{norphytane=} \\ & \text{=pristane} \end{array}$$

L'hydrocarbure insaturé $C_{19}H_{38}$, appelé *zamène*, accompagne le pristane dans les huiles de Poissons (*163, 138 b*)

Stérols de l'ambre gris. Nous avons confirmé les résultats de SUZUKI (*162*) ainsi que de JANISTYN (*53*) sur la présence, dans l'ambre gris, de coprostérol (CIIIb) et d'épicoprostérol (CIIIa). En traitant l'insaponifiable d'ambre gris avec le réactif *T* de GIRARD et SANDULESCO (*30*), nous avons

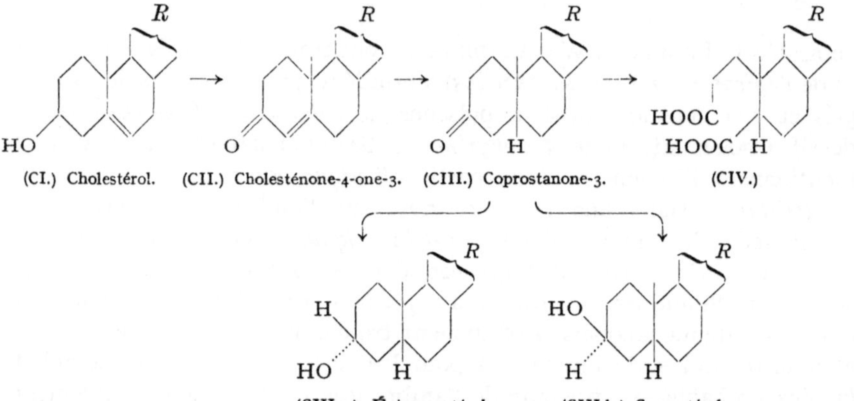

(CI.) Cholestérol. → (CII.) Cholesténone-4-one-3. → (CIII.) Coprostanone-3. → (CIV.)

(CIII a.) Épicoprostérol. (CIII b.) Coprostérol.

isolé une cétone cristallisée que nous avons identifiée avec la coprostanone-3 (CVI). C'est la première fois que cette substance a été isolée d'une source naturelle (*71*). Sa présence dans l'ambre gris, à côté de l'épicoprostérol et du coprostérol est une nouvelle preuve de l'exactitude de la théorie de ROSENHEIM et STARLING (*108, 109 a*) et de SCHOENHEIMER et RITTENBERG (*138*) sur la formation du coprostérol à partir du cholestérol. D'après cette théorie, le cholestérol (CI) est d'abord oxydé en cholestène-4-one-3 (CII) qui est ensuite réduite en coprostanone-3 (CIII). La réduction de cette cétone donne les deux alcools épimères (CIIIa et CIIIb). Récemment, ROSENHEIM et WEBSTER (*109*) ont isolé la cétone (CIII) des fèces de Chiens et de Rats; MARKER, WITTBECKER, WAGNER et

TURNER (*85*) ont trouvé l'épicoprostérol (CIIIb) dans les fèces de Chiens. La coprostanone-3 (CIII) a été le dernier intermédiaire non encore isolé.

L'acide dicarboxylique $C_{27}H_{46}O_4$ (CIV), F. 245°, que WINDAUS (*173*) avait obtenu par oxydation chromique de la coprostanone-3 se trouve également dans l'ambre gris. Etant donné le mémoire de TURFITT (*164*) sur l'oxydation microbienne de stéroïdes, il est assez probable que cet acide se forme dans l'intestin du Cachalot par oxydation microbienne de la coprostanone-3.

Le Tableau 5 indique les pourcentages approximatifs des constituants principaux de l'ambre gris.

Tableau 5. Constituants de l'ambre gris (*71*).

Insoluble dans l'éther	10—16%
Pristane	2—4%
Ambréine	25—45%
Epicoprostérol, libre et estérifié	30—40%
Coprostérol	1—5%
Cholestérol	moins de 0,1%
Cétones, dont plus de 50% de coprostanone-3	6—8%
Acides libres	5%
Acides estérifiés	5—8%

Ambréine.

Cette substance, la plus intéressante de l'ambre gris, a été déjà isolée en 1820 par PELLETIER et CAVENTOU (*94*). Ils l'ont décrite comme substance cireuse incolore fortement odorante fondant à 36°. RIBAN (*107*) l'a purifiée et a proposé la formule $C_{23}H_{40}O$. SUZUKI (*162*) l'a enfin obtenue à l'état pur (F. 83°, $[\alpha]_D = +20{,}5°$; alcool) et lui a attribué la formule $C_{26}H_{44}O$. La formule brute exacte et la constitution chimique de l'ambréine sont restées inconnues jusqu'en 1946. Grâce à des travaux sortis du laboratoire de RUZICKA et du nôtre, la astrucure de l'ambréine a été rapidement éclaircie jusque dans les moindres détails.

L'ambréine (CV) est un triterpène tricyclique de formule $C_{30}H_{52}O$ contenant deux doubles liaisons et un hydroxyle tertiaire (*118*, *71*). C'est ainsi le premier triterpène *tricyclique* connu et un représentant de la petite classe des triterpènes du règne animal, dont les autres sont: le *squalène* $C_{30}H_{50}$ aliphatique, du foie des Squalidés, et les quatre alcools triterpéniques, tétracycliques de la graisse de laine: *lanostérol* $C_{30}H_{50}O$, *dihydrolanostérol* $C_{30}H_{52}O$, *agnostérol* $C_{30}H_{48}O$ et *dihydroagnostérol* $C_{30}H_{50}O$ (*122*).

Hydrogénation et deshydratation de l'ambréine. L'ambréine peut être hydrogénée en tétrahydroambréine, $C_{30}H_{56}O$, ce qui indique la présence de deux doubles liaisons; l'ambréine et la tétrahydroambréine sont facilement deshydratées par ébullition avec l'anhydride acétique ou l'acide formique. L'hydroxyle de l'ambréine est ainsi caractérisé comme

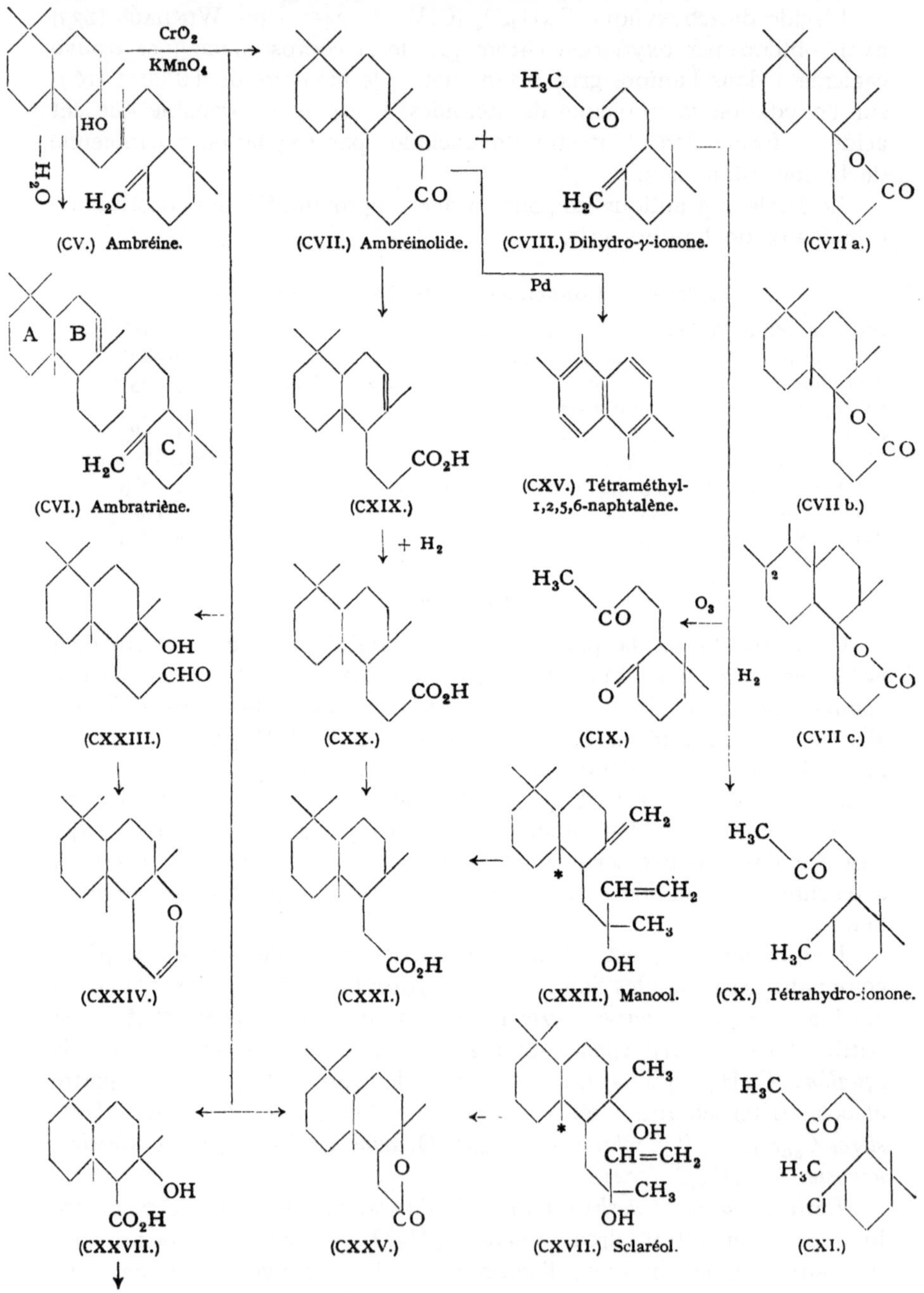

CrO_2 $KMnO_4$ $-H_2O$ HO H_2C

(CV.) Ambréine.

(CVII.) Ambréinolide.

(CVIII.) Dihydro-γ-ionone.

(CVII a.)

Pd

A B C H_2C

(CVI.) Ambratriène.

(CXIX.)

(CXV.) Tétraméthyl-1,2,5,6-naphtalène.

(CVII b.)

$+ H_2$ O_3 H_2

(CXXIII.)

(CXX.)

(CIX.)

(CVII c.)

(CXXIV.)

(CXXI.)

(CXXII.) Manool.

(CX.) Tétrahydro-ionone.

(CXXVII.)

(CXXV.)

(CXVII.) Sclaréol.

(CXI.)

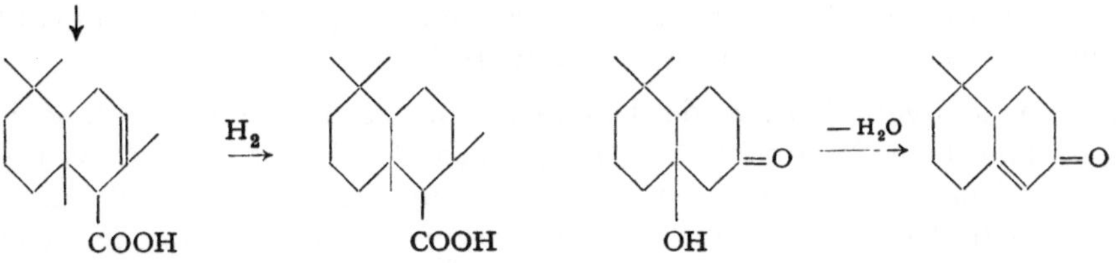

H_2 → ; $-H_2O$ →

COOH (CXXVIII.) COOH (CXXIX.) OH (CXII.) =O (CXIII.) =O

tertiaire. On ne connaît aucun ester de cet hydroxyle. La nouvelle double liaison formée par deshydratation n'est pas hydrogénable. La tétrahydro-ambréine $C_{30}H_{56}O$ et l'ambratriène $C_{30}H_{50}$ (CVI) sont huileux; ce dernier peut être caractérisé par son trichlorhydrate, F. 143° (*71, 118*).

Oxydation de l'ambréine. L'oxydation de l'ambréine (CV) par le permanganate ou l'acide chromique coupe la molécule en deux parties au niveau d'une des doubles liaisons; on obtient ainsi une lactone $C_{17}H_{28}O_2$ bicyclique saturée, appelée ambréinolide, (CVII) et une cétone $C_{13}H_{22}O$ monocyclique ayant une double liaison (CVIII) (*71*). RUZICKA et LARDON (*118*) ont montré que cette cétone donne du formol à l'ozonisation (comme aussi l'ambréine) ce qui montre qu'il y a une double liaison exocyclique, sous forme d'un groupement méthylénique. Le reste de la molécule est isolé sous forme d'une dicétone $C_{12}H_{20}O_2$ (CIX); cette même dicétone est aussi obtenue au cours de l'oxydation de l'ambréine par le permanganate (*71*).

Dihydro-γ-ionone. L'hydrogénation catalytique de la cétone $C_{13}H_{22}O$ (CVIII) ayant donné de la tétrahydro-ionone (CX), la structure doit être celle d'une dihydro-γ-ionone (CVIII) (*118*). Elle a une odeur ambrée très prononcée et se trouve aussi dans les parties volatiles de l'ambre gris (*126*).

La *synthèse de la dihydro-γ-ionone* vient d'être accomplie par RUZICKA, BÜCHI et JEGER (*114*) à partir de la dihydro-α-ionone. Par addition d'acide chlorhydrique à celle-ci on obtient un chlorure assez instable, ayant probablement la structure (CXI). En déchlorurant cette substance avec du stéarate d'argent dans le xylène bouillant, les auteurs suisses ont obtenu un mélange de (CVIII) et de dihydro-α-ionone qu'ils ont pu séparer par cristallisation fractionnée ou par chromatographie des semicarbazones.

La dicétone $C_{12}H_{20}O_2$ (CIX) a été cyclisée par RUZICKA et LARDON (*118*) en une hydroxy-cétone bicyclique $C_{12}H_{20}O_2$ (CXII); cette substance se deshydrate en solution acide en une cétone insaturée $C_{12}H_{18}O$ (CXIII) (λ_{max} 243 mμ, log $\varepsilon = 4{,}2$), la $\Delta^{5,10}$-diméthyl-1,1-octalone-6 dont la synthèse a été effectuée par BÜCHI, JEGER et RUZICKA (*12*).

Structure de l'ambréinolide (CVII). La première formule de l'ambreinolide, proposée par RUZICKA et LARDON (*118*) était celle d'une γ-lactone de structure (CVIIa); notre première formule (CVIIb) n'en différait que par la position d'un groupe méthyle (*71*).

La deshydrogénation de l'ambréine ou de l'ambreinolide par le sélénium donne le triméthyl-1,5,6-naphtalène (CXIV) et le diméthyl-1,6-naphtalène (*118*); celle par le palladium sur charbon nous a donné le tétraméthyl-1,2,5,6-naphtalène (CXV) (*71*). Ce résultat a induit RUZICKA et LARDON (*118*) à changer la formule de l'ambreinolide de (CVIIa) en (CVIIc), pour tenir compte du fait que le tétraméthyl-1,2,5,6-naphtalène avait jusque là été obtenu seulement à partir de triterpènes hydroxylés en 2, où sa formation pouvait être le résultat d'une transposition rétro-pinacolique. L'ambréine n'ayant pas d'hydroxyle en 2, ne pourrait donner le tétraméthyl-1,2,5,6-naphtalène qu'à condition d'avoir ses groupes méthyles comme dans (CVIIc, p. 124).

Nous avons ensuite montré que cette conclusion n'était pas justifiée, car notre catalyseur donnait le même tétraméthyl-naphtalène (CXV) aussi à partir du tétracyclosqualène (CXVI) et du diterpène scIaréol (CXVII), substances ayant les deux groupements méthyls géminaux (*74*). Notre catalyseur (15% de Pd sur charbon *végétal,* au lieu du charbon *animal* employé d'habitude) avait donc bien la capacité de faire migrer l'un de ces groupes méthyles. Il n'y avait plus d'objection à formuler l'ambreinolide et l'ambréine d'après (CVII) et (CV) et de les faire dériver ainsi du tricyclosqualène hypothétique (CXVIII), comme nous l'avions fait déjà dans notre premier mémoire sur l'ambréine (*71*).

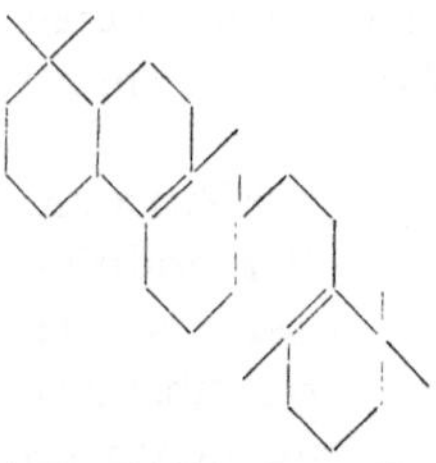

COOH

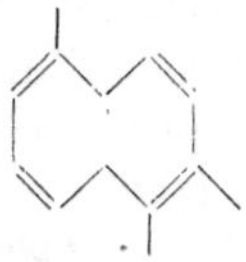

(CXIV.) Triméthyl-1,5,6-naphtalène.

(CXVIII.) Tricyclo-squalène. 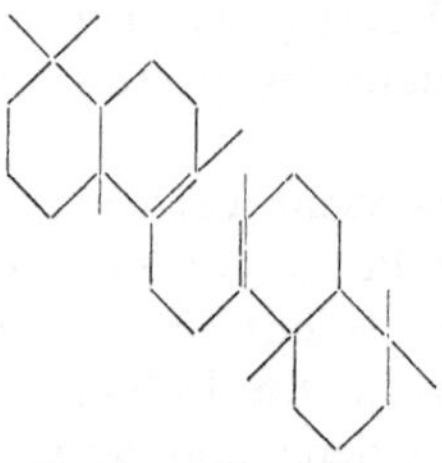 (CXVI.) Tétracyclo-squalène. 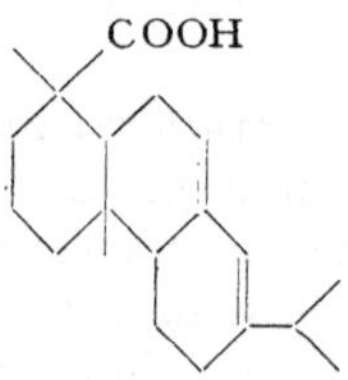 (CXXVI.)

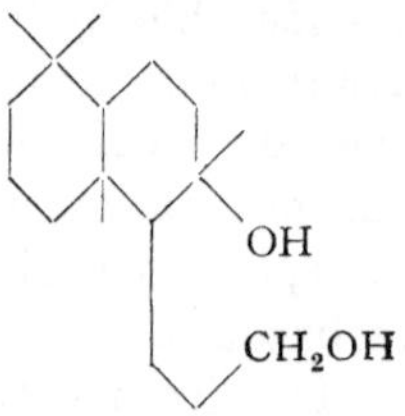

OH

CH_2OH

(CXXVI a.)

HO

CO_2H

O

O·CO

CH_3O_2C

CH_3O_2C

→ → → (CXXIX)

O

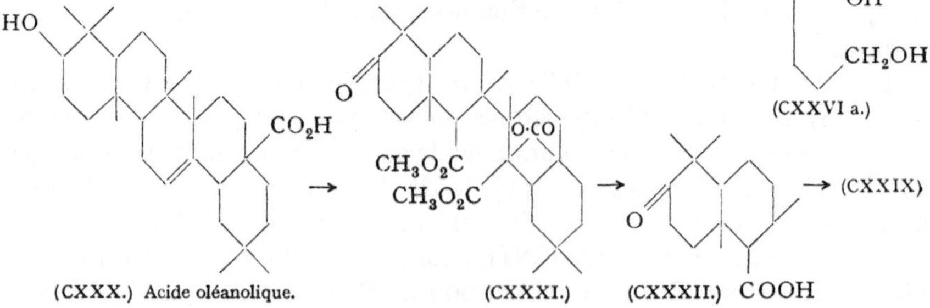

(CXXX.) Acide oléanolique. (CXXXI.) (CXXXII.) COOH

Entre temps, RUZICKA, DÜRST et JEGER (*115*) avaient réussi à établir une importante liaison entre l'ambréine et les *diterpènes des plantes*. En dégradant, d'après WIELAND, SCHLICHTING et JAKOBI (*170*), l'acide $C_{17}H_{30}O_2$ (CXX) obtenu par RUZICKA et LARDON (*118*) à partir de l'ambreinolide (via l'acide insaturé CXIX), ils ont préparé l'acide $C_{16}H_{28}O_2$ (CXXI) qui s'est avéré être identique à l'acide obtenu par HOSKING et BRANDT (*47*) à partir du diterpène manool (CXXII). Ceci prouvait que les parties bicycliques de l'ambréine et du manool étaient identiques. Comme, d'autre part, le manool (du bois de *Dacrydium biformae*) et le sclaréol (de *Salvia sclarea*) donnent le même trichlorhydrate, cette identité s'étend aussi au sclaréol (CXVII).

En isolant des produits d'oxydation de l'ambréine une lactone $C_{16}H_{26}O_2$ (F. 121°) (CXXV), identique à une lactone obtenue par RUZICKA et JANOT (*117*) [voir aussi RUZICKA, SEIDEL et ENGEL (*125*)] à partir du sclaréol (CXVII), nous avons prouvé que l'hydroxyle de l'ambréine se trouve bien au même carbone que l'un des hydroxyles du sclaréol. L'ambréinolide est donc une δ-lactone (CVII) et non une γ-lactone (comme dans CVII a, b et c, p. 124).

JEGER, DÜRST et BÜCHI (*55*) ont ensuite transformé le *manool* (CXXII) en un produit de dégradation de l'*acide abiétique* (CXXVI), prouvant ainsi que les parties bicycliques de ces corps sont identiques et fixant, en particulier, l'emplacement du dernier méthyle du manool et du sclaréol qui n'avait pas encore pu être précisé. (Marqué d'un * dans les formules CXVII et CXXII.) Etant donné l'identité de structure des parties bicycliques du groupe sclaréol-manool avec celle de l'ambréine, la structure exacte de ce dernier était ainsi connue et l'ambréine est le premier triterpène cyclique dont tous les détails de structure soient élucidés (*56*).

Dans un mémoire récent, COLLIN-ASSELINEAU, LEDERER, MERCIER et POLONSKY (*16*) décrivent la préparation et les propriétés de quelques produits d'oxydation et de dégradation de l'ambréine:

Un *hexol* $C_{30}H_{56}O_6$, F. 200°, obtenu par hydroxylation de l'ambratriène (CVI) par OsO_4 (analogue au pentol $C_{30}H_{56}O_5$, F. 203°, obtenu par RUZICKA, DÜRST et JEGER (*115*) par un traitement analogue de l'ambréine;

un *hydroxy-oxyde* $C_{30}H_{52}O_2$, F. 143°, formé au cours de l'oxydation chromique de l'ambréine;

un *trioxyde* $C_{30}H_{52}O_3$, F. 178°, présent dans les eaux-mères de la substance précédente;

trois *lactones isomères*, $C_{17}H_{28}O_2$, F. 143°, F. 96°, la troisième liquide, obtenues par traitement de l'ambréinolide (CVII) par l'acide sulfurique à 80%, à des températures de 20°, 60° et 90°;

quatre *diols* $C_{17}H_{32}O_2$, (CXXVI b), F. 133°, F. 93°, F. 89°, le quatrième huileux, obtenus par réduction de l'ambréinolide et de ses isomères par $LiAlH_4$.

Dans ce mémoire, nous décrivons également une hydroxy-aldéhyde $C_{17}H_{30}O_2$, F. 192° (CXXIII) qui se trouve aussi dans les parties volatiles de l'ambre gris; elle peut être oxydée par CrO_3 en ambréinolide (CVII) et réduite par $LiAlH_4$ en le diol F. 133° (CXXVI a). L'hydroxyaldéhyde (CXXIII) peut être deshydratée en un oxyde insaturé $C_{17}H_{28}O$, F. 85° (CXXIV). Dans un mémoire antérieur (*74*) nous avions, par erreur, attribué à ces deux substances des formules à 16 atomes de carbone.

Rapport de l'ambréine avec les triterpènes et diterpènes des Plantes. En fractionnant les produits d'oxydation de l'ambréine, nous avons isolé un hydroxy-acide $C_{15}H_{26}O_3$, F. 182° qui devait avoir la formule (CXXVII); sa deshydratation a donné l'acide insaturé $C_{15}H_{24}O_2$, F. 95° (CXXVIII) dont la double liaison n'est pas en conjugaison avec le carboxyle (pas de bande d'absorption correspondante). L'hydrogénation de cet acide a donné un acide saturé $C_{15}H_{26}O_2$, F. 135° (CXXIX) qui s'est avéré être identique avec un acide isolé à Zürich par réduction d'après WOLFF-KISHNER de l'acide cétonique (CXXXII), obtenu lui-même après pyrolyse d'un produit de dégradation (CXXXI) de l'acide oléanolique (CXXX), acide triterpénique le plus répandu des Plantes, appartenant au groupe de la β-amyrine (*116*). MEISELS, JEGER et RUZICKA (*87*) viennent d'isoler le même acide aussi à partir de l'α-amyrine. Ceci montre que la partie bicyclique de l'ambréine a la même structure et même configuration que la partie correspondante des triterpènes du groupe des amyrines et a permis de relier entre eux, par l'intermédiaire de l'ambréine, les triterpènes pentacycliques du groupe des amyrines, avec les diterpènes bi- et tri-cycliques mentionnés plus haut.

Biosynthèse de l'ambréine et formation de l'ambre gris.

Nous ne savons rien encore sur l'origine biochimique de l'ambréine; une formation par cyclisation et hydratation du squalène serait évidemment à envisager, mais des essais *in vitro* ont échoué. Il est probable que la présence d'ambréine et la formation de la concrétion intestinale qu'est l'ambre gris sont en rapport, direct ou indirect avec la nourriture du Cachalot. Seul parmi les Baleines, le Cachalot a des dents et se nourrit d'animaux de mer de taille plus importante, parmi lesquels des Céphalopodes prendraient une place prépondérante. On trouve, en effet, dans presque tous les morceaux d'ambre gris des becs très pointus de Céphalopodes. Nous avons discuté, avec TIXIER (*79*), le rôle que pourraient jouer ces becs dans la formation de l'ambre gris. OKAHARA (*93*) avait isolé une porphyrine de l'ambre gris qu'il avait appelée «ambraporphyrine». Nous avons montré qu'il s'agissait simplement d'un mélange de proto- et mésoporphyrine, facile à séparer par chromatographie. Or, il est connu que de tels mélanges se trouvent dans les excréments de Vertébrés après hémorrhagies dans l'intestin. L'hémoglobine est dégradée par les Bactéries intestinales en protoporphyrine, dont une partie est réduite en mésoporphyrine. La présence de becs de Céphalopodes très pointus et de ce mélange de deux porphyrines est donc une assez forte indication que la formation d'ambre gris est dûe à un processus pathologique.

Quant à l'origine des matières odorantes mêmes de l'ambre gris, nous avons vu que des produits d'oxydation de l'ambréine (dihydro-γ-ionone CVIII, peut-être aussi la dicétone CIX) en font partie; une partie des substances odorantes pourrait provenir de la nourriture, étant donné que le Céphalopode *Elledone moschata* a lui-même une glande à parfum et se trouve dans la nourriture du Cachalot. La persistance de l'odeur des extraits d'ambre gris pourrait être dûe à une lente autoxydation de l'ambréine catalysée peut-être par le cuivre présent dans l'ambre gris et provenant de l'hémocyanine des Céphalopodes de la nourriture du Cachalot (70).

Essais synthétiques dans la série de l'ambréine.

Synthèse de l'ambratriène. La deshydratation de l'ambréine conduit à l'ambratriène, hydrocarbure tricyclique à trois doubles liaisons (CVI). DÜRST, JEGER et RUZICKA (20) viennent d'en faire une synthèse partielle en condensant le magnésien du bromure (CXXXIII) (dérivé de l'acide

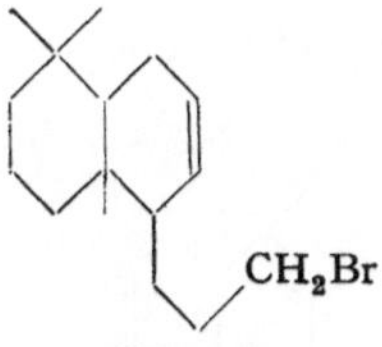

CH_2Br (CXXXIII.) OH H_2C (CXXXIV.) Isomère d'ambréine. CHO (CXXXV.) CHO (CXXXVI.) COOH (CXXXVII.)

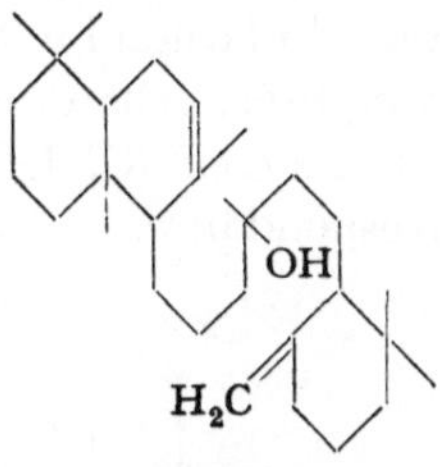

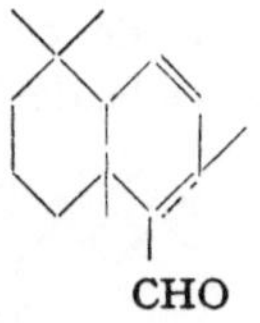

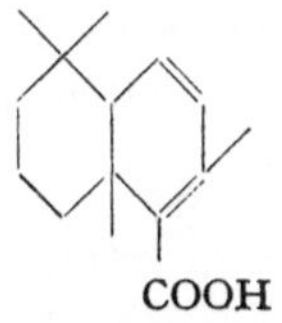

insaturé CXIX par réduction par $LiAlH_4$ en alcool et bromuration de celui-ci) avec la dihydro-γ-ionone (CVIII). L'alcool (CXXXIV) ainsi obtenu, isomère de l'ambréine, a été deshydraté en ambratriène (CVI).

La position de la double liaison dans le noyau *B* de l'ambratriène (CVI) et dans l'acide $C_{17}H_{28}O_2$ (CXIX) n'est pas encore prouvée sans équivoque; DÜRST, JEGER et RUZICKA (20) lui attribuent celle indiquée dans les formules (CVI et CXIX) et promettent d'en fournir la preuve dans une prochaine communication.

Synthèses du système bicyclique de l'ambréine. La première synthèse du système bicyclique de l'ambréine (dérivé de la tétraméthyl-1,1,6,10-décaline) a été effectué par BATTY, HEILBRON et JONES (2) par cyclisation de la ψ-ionylidène-acetaldéhyde (CXXXV), produit de condensation du citral avec l'aldéhyde β-méthyl-crotonique. Ils ont ainsi obtenu une aldéhyde à deux doubles liaisons ayant la formule (CXXXVI).

L'oxydation argentique de cette aldéhyde nous a donné un acide, F. 170° (CXXXVII) dont l'hydrogénation a donné un acide F. 145°, stéréoisomère de l'acide naturel (CXXIX) (*16 a*).

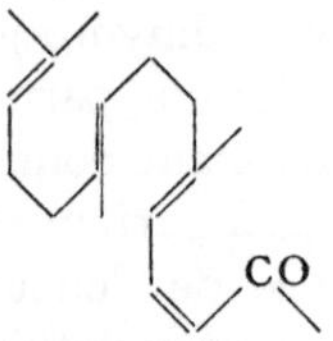

(CXXXVIII.) Farnésylidène-acétone. 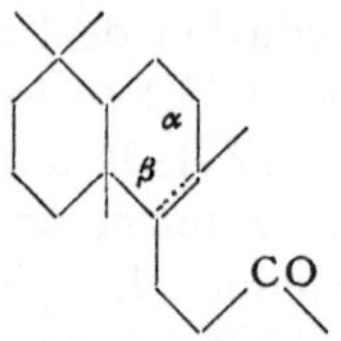 (CXXXIX.) Bicyclo-farnésylidène-acétone. 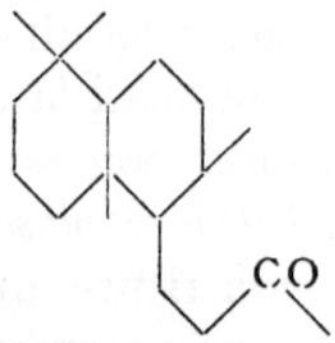 (CXL.) Tétraméthyl-(1,1,6,10)-(oxo-3'-butyl)-5-décaline.  (CXLI.) Farnésal.

ZOBRIST et SCHINZ (*177*) ainsi que NAVES (*91*) ont obtenu la bicyclo-farnésylidène-acétone (CXXXIX) par cyclisation de la farnésylidène-acétone (CXXXVIII). Le produit cyclique est un mélange de formes α et β. Par hydrogénation il donne la tétraméthyl (1,1,6,10)-(oxo-3'-butyl)-5-décaline (CXL).

STOLL et COMMARMONT (*154*) viennent de décrire la cyclisation du farnésal (CXLI) en α-bicyclo-farnésal (CXLII) par l'acide phosphorique et en β-bicyclo-farnésal (CXLIII) par l'acide sulfurique. La réduction de ce dernier par le $LiAlH_4$ donne le β-bicyclo-farnésol (CXLIV). L'odeur de la première aldéhyde (CXLII) est camphrée, celle de son isomère β a une note de bois de cèdre. L'odeur de l'alcool (CXLIV) était encore plus faible que celle de l'aldéhyde correspondante.

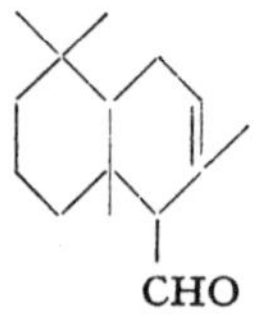

(CXLII.) α-Bicyclo-farnésal. 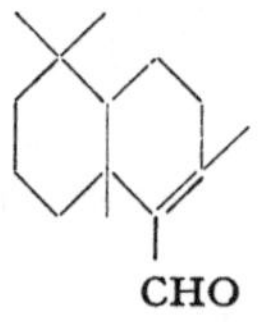 (CXLIII.) β-Bicyclo-farnésal. 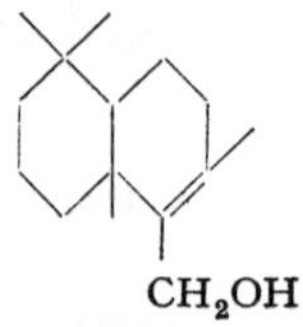 (CXLIV.) β-Bicyclo-farnésol. 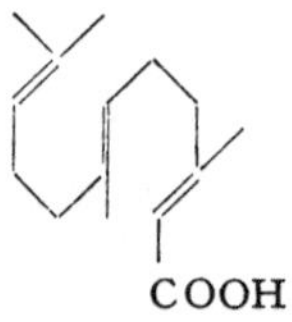 (CXLIV a.) Acide farnésique. 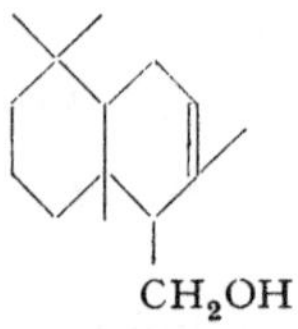 (CXLIV b.) α-Bicyclo-farnésol.

La cyclisation de l'acide farnésique (CXLIV a) en un mélange de deux acides bicyclo-farnésiques (CXXVIII) vient d'être décrite par CALIEZI et SCHINZ (*13 b*). L'un de ces acides est cristallisé (F. 131°), l'autre liquide; ils paraissent être des stéréoisomères de l'acide F. 95° (CXXVIII) obtenu à partir de l'ambréine (*116*). La réduction de l'acide F. 131° par $LiAlH_4$ donne un-α-bicyclo-farnésol F. 65° (CXLIV b), celle de l'acide liquide donne un alcool liquide. Ces deux alcools ont une odeur de bois de cèdre.

Récemment, nous avons préparé des acides stéréoisomères de (CXXVII) et de (CXXIX) en cyclysant par l'acide phosphorique des acides monocycliques monoinsaturés obtenus par condensation des dihydro-α- et dihydro-β-ionones avec le bromacétate d'éthyle, d'après REFORMATZKY (*16 a*).

2° Bézoards.

Les bézoards sont des concrétions de l'estomac de Ruminants; on les appelle aussi égagropiles (du grec: aigagros = Chèvre sauvage et pilos = feutre, laine pressée); ce sont des formations dures, légères, sphériques ou ovoïdes, constituées de poils enchevetrés et de substances organiques d'origine alimentaire. On les trouve surtout chez les animaux sauvages et rarement chez des animaux domestiqués ou tenus en captivité.

SALGUES (*134*) a consacré un article détaillé aux bézoards des Ruminants et à ceux des Poissons. D'après cet auteur «à l'état frais, et sensiblement durant les deux années qui suivent leur élimination par voie nécropsique, tous les bézoards, après section, présentent une odeur très spéciale de musc. Il est facile de caractériser et de doser les amides: indol et scatol, après isolement, par leurs points de fusion et d'ébullition, pour le second de ces corps par son chlorhydrate et sa combinaison avec l'acide picrique. Mais c'est surtout la présence d'une cétone cyclique, la muscone, qui donne à ces concrétions leur véritable relief.»

«Caractérisée par son oxime et sa semicarbazone, nous avons trouvé cette cétone, avec un pourcentage de: égagropiles de chèvre 0,56 à 0,91; bouquetin 0,76; chamois 0,38; girafe 0,17; *Addax* 0,38; *Gazella* 0,52; *Bubalis* 0,86; Cerf 0,74; *Capreolus* 0,86.»

Les constituants des concrétions de Poissons comprendraient, d'après SALGUES (*134*): l'indol et le scatol, une cétone odorante, des stéroïdes odorants, du *p*-crésol, une substance aromatique non phénolique de formule $C_{10}H_{16}O$, isomère du camphre et, enfin des acides benzoïque et *p*-hydroxybenzoïque.

Il serait intéressant de vérifier les données de SALGUES présentées sans détails expérimentaux.

III. Substances odorantes de l'urine de Vertébrés.

La plupart des substances odorantes normalement excrétées par l'urine des Vertébrés, est conjuguée sous une forme ou une autre [voir à ce sujet WILLIAMS (*172*)]. Des fermentations ou l'hydrolyse chimique les libèrent. Quatre groupes de substances chimiques odorantes ont été récemment étudiées plus en détail: des substances dérivées de l'ionone, les phénols, les acides et les stéroïdes.

1° Dérivés de l'ionone.

a) De l'urine de Lapin.

L'étude de FISCHER et BIELIG (*28*) sur les transformations de l'ionone β dans l'organisme animal, montre l'intérêt que peut présenter la synthèse biologique de substances odorantes. Ayant administré à des Lapins de l'ionone β, ces auteurs ont isolé de l'urine les substances suivantes, dont certaines n'avaient pas encore été obtenues par voie de synthèse chimique: l'ionol-β (CXLV), le dihydro-ionol-β (CXLVI), l'hydroxy-ionone-β (CXLVII), l'hydroxy-ionol-β (CXLVIII), l'hydroxy-dihydro-ionone-β (CXLIX) et l'hydroxy-dihydro-ionol-β (CL).

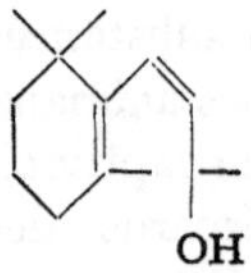

(CXLV.) Ionol-β. (CXLVI.) Dihydro-ionol-β. (CXLVII.) Hydroxy-ionone-β. (CXLVIII.) Hydroxy-ionol-β.

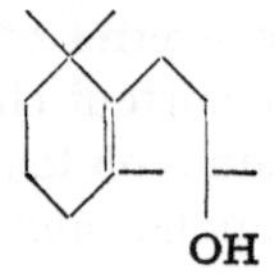

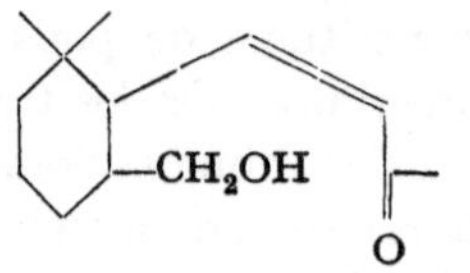

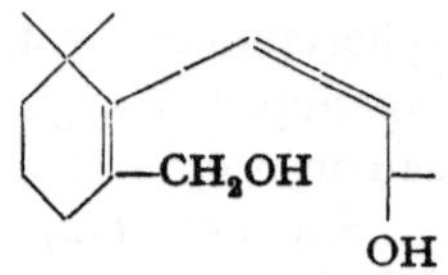

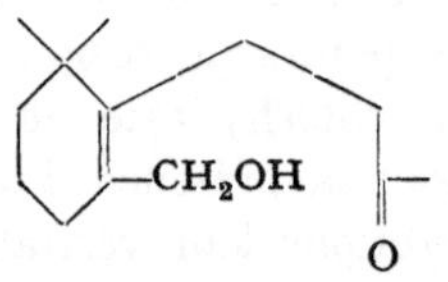

(CXLIX.) Hydroxy-dihydro-ionone-β. (CL.) Hydroxy-dihydro-ionol-β. (CLII.)

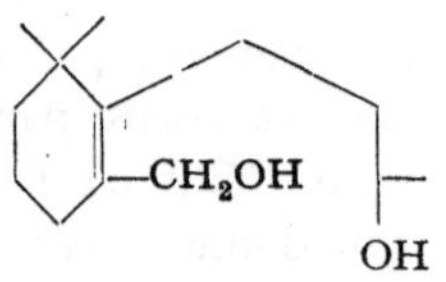

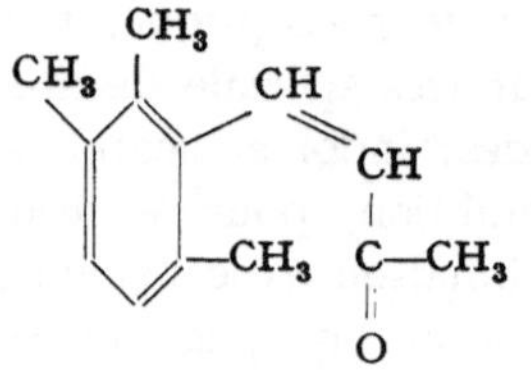

Ces quatre dernières substances porteraient leur oxhydryle sur un des groupes méthyles du noyau d'ionone; les auteurs n'ont pas pu préciser ce point davantage; de plus, il nous semble que la structure des corps isolés n'a pas été élucidée de façon suffisamment certaine.

Ces corps se présentent sous forme d'huiles très visqueuses, à odeur de bois de cèdre. Fischer et Bielig (*28*) pensent qu'une partie des dérivés hydroxylés est oxydée *in vivo* en aldéhyde et finalement en acide.

b) De l'urine de Jument gravide.

En fractionnant la partie neutre d'un extrait d'urine de Jument gravide Prelog et Führer (*99*) isolèrent, entre autres, deux substances à 13 atomes de carbone, une cétone insaturée $C_{13}H_{18}O$ (F. 94,5°) et un alcool $C_{13}H_{20}O$ (F. 90,5°), ce dernier présentant le spectre d'absorption des benzènes polysubstitués. La constitution de l'alcool (CLI)

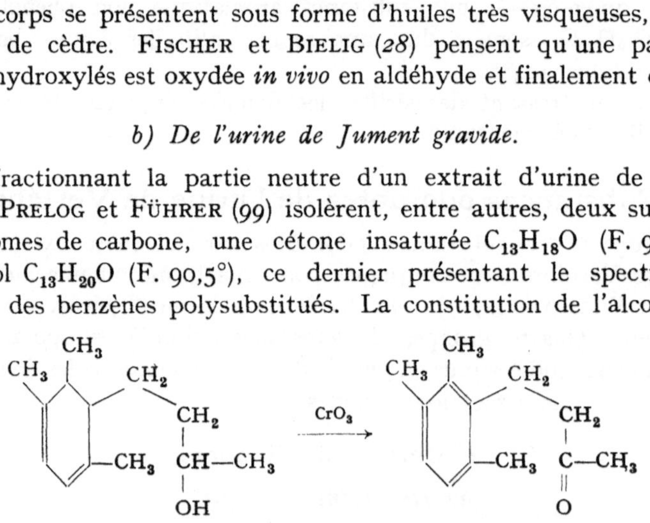

$\xrightarrow{CrO_3}$

(CLI.) (CLIa.) Triméthyl-2',3',6'-phényl-1-butanone-3.

a été établie par Prelog, Führer, Hagenbach et Frick (*100*). L'oxydation chromique de l'alcool donne une cétone liquide $C_{13}H_{18}O$ dans laquelle la présence d'un groupement $—COCH_3$ est indiquée par la réaction au iodoforme. Ayant synthétisé quelques cétones isomères, dérivées de benzènes alcoylés, les auteurs constatèrent, d'après les spectres d'absorption, que le groupement cétonique ne devait pas être en position β du noyau benzénique. La (triméthyl-2',3',6'-phényl)-1-butanone-3 (CLI a) synthétique s'est finalement trouvée être identique à la cétone obtenue par oxydation chromique de l'alcool naturel. Celui-ci est donc le (—)-(triméthyl-2,3,6)-phényl -1-butanol-3 (CLI).

L'isolement de la cétone insaturée correspondante (CLII) a été rapporté par PRELOG, FÜHRER, HAGENBACH et SCHNEIDER (*101*). Ces deux substances peuvent être dérivées de l'ionone par migration d'un groupe méthyle, accompagnée d'aromatisation du noyau. De telles réactions peuvent être accomplies assez facilement, comme l'ont montré BÜCHI, SEITZ et JEGER (*13*) en isolant la (triméthyl-2',3',6'-phényl)-1-butanone-3 (CLI a) des produits de débromuration d'un bromure d'ionone β.

PRELOG et coll. (*101*) ont en outre décrit l'isolement, à partir d'urine de Jument gravide, des dérivés d'ionone suivants:

deux diols isomères $C_{13}H_{26}O_2$ (*A* et *B*) (CLV),
une cétone $C_{13}H_{22}O_2$ (*C*),
deux dicétones isomères $C_{13}H_{22}O_2$ (*D* et *F*) (CLIII),
deux oxycétones isomères $C_{13}H_{24}O_2$ (*E* et *G*) (CLVI).

Les deux diols *A* et *B* ainsi que les oxycétones *E* et *G* ont pu être transformés en la dicétone *D* (CLIII), par oxydation chromique. Il s'agit donc de substances ayant le même squelette carboné et les deux atomes d'oxygènes fixés sur les mêmes atomes de carbone. Elles sont

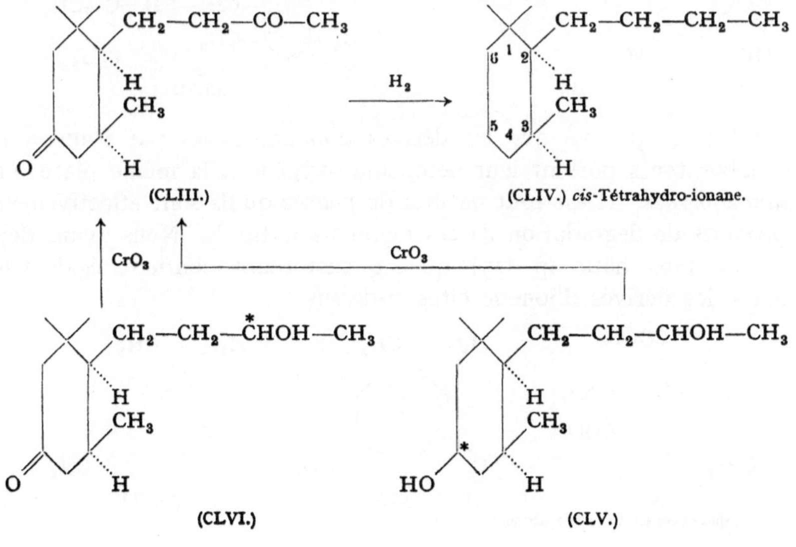

$CH_2—CH_2—CO—CH_3$; H ; CH_3 ; O ; H — $\xrightarrow{H_2}$ — $CH_2—CH_2—CH_2—CH_3$; H ; CH_3 ; H

(CLIII.) (CLIV.) *cis*-Tétrahydro-ionane.

CrO_3 CrO_3

$CH_2—CH_2—\overset{*}{C}HOH—CH_3$; H ; CH_3 ; O ; H — $CH_2—CH_2—CHOH—CH_3$; H ; CH_3 ; HO ; H

(CLVI.) (CLV.)

toutes saturées, donc monocycliques. La réduction d'après WOLFF-KISHNER de la dicétone *D* (CLIII) a donné un hydrocarbure liquide $C_{13}H_{26}$ qui a été identifié par son spectre infrarouge avec le *cis*-tétrahydroionane (CLIV). [Pour la préparation et les propriétés des deux tétrahydro-ionanes, voir PRELOG et FRICK (*98*).]

La dicétone *D* donne la réaction d'iodoforme et une réaction colorée des méthyl-cétones. Il s'ensuit que l'un des oxygènes des substances *A*, *B*, *D*, *E* et *G* a la même position que dans les ionones. La dicétone *D* est, de ce fait, une céto-*cis*-tétrahydro-ionone. Les deux diols (CLV) sont deux des quatre hydroxy-*cis*-tétrahydro-ionols stéréoisomères correspondants. Les deux hydroxy-cétones ne donnent pas la réaction colorée des methylcétones; elles seraient donc des céto-*cis*-tétrahydro-ionols (CLVI). La cétone *C* est probablement une epoxycétone de la même série. Elle donne, après ébullition avec de la dinitrophénylhydrazine dans l'HCl, la même dinitrophénylhydrazone que la dicétone *D*.

La position en 5 du 2ème carbonyle de la dicétone *D* (CLIII) vient d'être prouvée par PRELOG et SCHNEIDER (*105*). La dicétone *D* a été condensée avec le magnésien de bromobenzène et le produit de réaction (CLVII) deshydrogéné par le sélénium. L'oxydation du produit de deshydrogénation (CLVIII) par le permanganate a donné un acide, F. 230°, qui a été identifié par synthèse avec l'acide diphényl-3,4,5-tricarboxylique (CLIX).

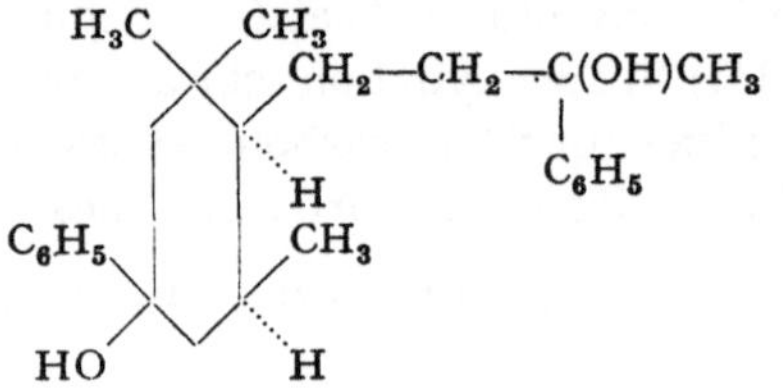

(CLVII.) 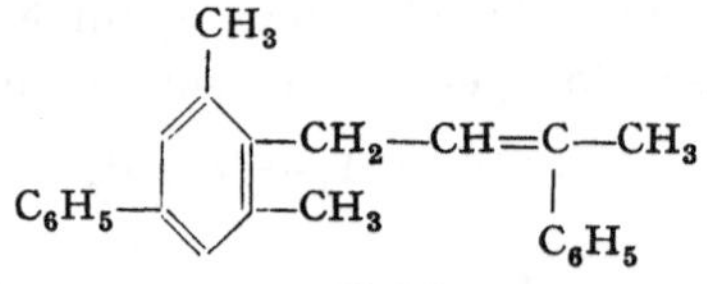 (CLVIII.)

Il est donc prouvé que les dérivés d'ionone isolés par PRELOG et ses collaborateurs portent leur deuxième oxygène à la même place que les *xanthophylles*. Il est tout naturel de penser qu'ils sont effectivement des produits de dégradation de ces pigments naturels. Nous avons déjà mentionné plus haut (p. 117) que le castoréum contient également plusieurs des dérivés d'ionone cités ci-dessus.

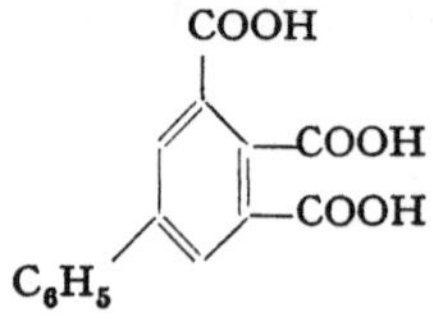

(CLIX.) Acide diphényl-3,4,5-tricarboxylique.

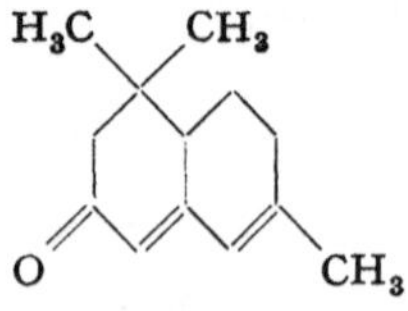

(CLX.)

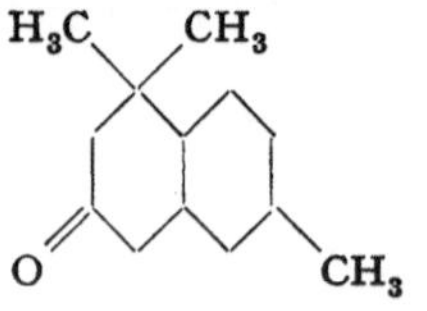

(CLXI.)

Prelog et Vaterlaus (*106*) viennent d'éclaircir la constitution chimique de la cétone insatureé $C_{13}H_{18}O$, F. 94,5° (CLX), décrite précédemment (*99*). Le spectre d'absorption de cette cétone montre la présence du système C=C—C=C—O (bande d'absorption à 293 mμ, log ε = 4,5). La réduction catalytique donne la tétrahydrocétone (CLXI) qui est réduite en hydrocarbure (CLXII) par réduction d'après Wolff-Kishner. La tétrahydrocétone (CLXI) réagit d'autre part avec le magnésien du bromure de méthyle en donnant le carbinol (CLXIII) dont la deshydrogénation par le sélénium fournit le triméthyl-1,3,6-

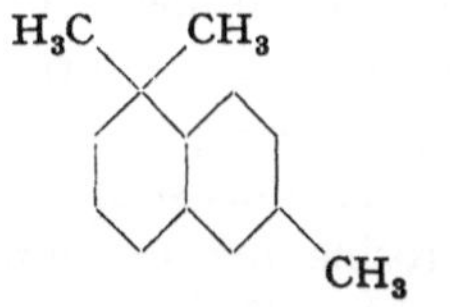

(CLXII.) 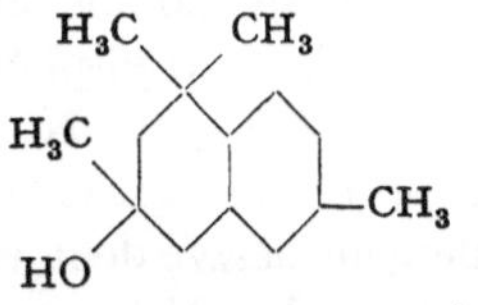 (CLXIII.) 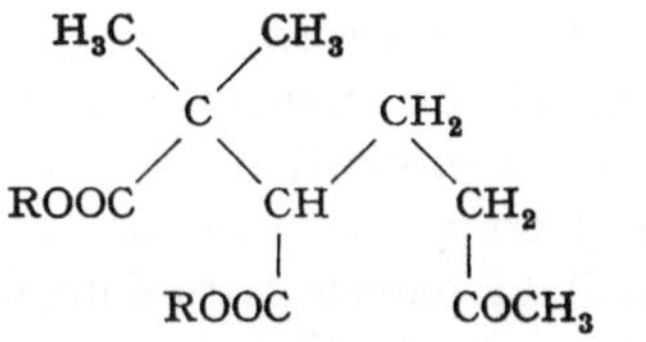 (CLXIV.)

naphtalène, ce qui fixe le position du groupe CO de la cétone (CLX). Son oxydation conduit à l'acide (CLXIV), ce qui prouve la position des doubles liaisons. Cette cétone est également à considérer comme un dérivé de l'ionone, formé par cyclisation entre le méthyle du noyau et le groupe carbonyle de la chaîne latérale.

2° Phénols de l'urine de Jument gravide.

L'analogie entre les constituants des glandes à parfum du Castor et ceux de l'urine des Vertébrés, mentionnée déjà plus haut, nous a incité à étudier les phénols volatils de l'urine de Jument gravide (qui représentent un produit de déchet de la fabrication des hormones sexuelles femelles). Marshall (*86*) a trouvé que cette partie phénolique est essentiellement constituée de *p*-crésol.

p-Ethyl-phénol. On sait depuis plus de 50 ans que l'urine de Vertébrés contient les sulphates éthérés de différentes substances phénoliques dérivées de la tyrosine par dégradation bactérienne. Baumann (*3*) a déjà en 1879 proposé le schéma suivant de la dégradation bactérienne de la tyrosine: tyrosine → acide *p*-hydroxy-phényl-propionique → *p*-éthyl-phénol → acide *p*-hydroxyphényl-acétique → *p*-crésol → acide *p*-hydroxy-benzoïque → phénol. Toutes ces substances avaient déjà été isolées de l'urine des Vertébrés, sauf le *p*-éthylphénol.

Ayant confirmé les rapports de Walbaum et Rosenthal (*167*) ainsi que de Pfau (*95*) sur la présence d'assez fortes quantités de *p*-éthyl-phénol dans le castoréum, nous nous sommes demandé si ce phénol n'était pas un constituant normal de l'urine des Vertébrés. Une distillation fractionnée des phénols volatils de l'urine de Jument gravide nous a conduit facilement à l'isolement d'un phénol dont le phényluréthane et l'ester glycolique ne donnait pas de dépression de point de fusion avec les dérivés correspondants du *p*-éthyl-phénol (*65*).

Dans l'urine de Jument gravide, le *p*-éthyl-phénol ne constitue que 4 à 10% par rapport au *p*-crésol; dans le castoréum, au contraire il n'y a pas de *p*-crésol. L'existence de différences assez marquées dans les phénols d'urine de différentes espèces animales est démontrée aussi par le travail récent de GRANT (*32*) qui a isolé le sulphate éthéré du *p*-éthyl-phénol de l'urine de Chèvre et qui rapporte que le *p*-crésol ne s'y trouve qu'à l'état de trace.

p-Propyl-phénol. Ce phénol n'avait jamais été isolé d'une source végétale ou animale; il accompagne le *p*-éthyl-phénol dans le castoréum dans la proportion de 1 : 10 environ (*68*, *70*). Nous n'avons pas trouvé de *p*-propyl-phénol dans l'urine de Jument gravide, mais nous supposons qu'il est, dans le castoréum, également dérivé de la tyrosine, par desamination, et réduction du groupe carboxyle. Il est assez significatif, que ni le castoréum, ni l'urine de Jument gravide ne contiennent de *p*-butyl-phénol ou d'autres *p*-alcoyl-phénols supérieurs (dont la formation, à partir de la tyrosine, serait impossible).

3° Acides de l'urine de Jument gravide.

La présence de l'acide *m*-hydroxy-benzoïque dans le castoréum nous avait surprise; cet acide n'avait été isolé qu'une fois d'une source naturelle: BIELIG et HAYASIDA (*5*) l'avaient trouvé dans l'urine de Lapin après administration d'ionone β. Ces auteurs avaient évidemment reconnu que cet acide ne pouvait pas être dérivé de l'ionone β, mais ils n'avaient pas étudié la possibilité d'une présence naturelle de cet acide dans l'urine du Lapin.

Les nombreuses analogies, déjà citées, entre les constituants du castoréum et ceux de l'urine de Vertébrés, nous ont conduit à supposer que l'acide *m*-hydroxy-benzoïque était, en effet, un constituant normal de l'urine de Vertébrés; ceci nous a incité à analyser en détail, les acides de l'urine de Jument gravide (qui constituent également un déchet de la fabrication industrielle des hormones sexuelles femelles).

Après distillation fractionnée des esters ethyliques de 900 g. d'acides d'urine de Jument gravide, nous avons isolé 15 acides à l'état pur, dont 13 ont été identifiés; 4 seulement avaient été considérés comme constituants normaux de l'urine des Vertébrés (les acides benzoïque, phénylacétique, *p*-hydroxybenzoïque et azélaïque). Les neuf acides nouveaux sont: les acides salicylique, vanillique, *m*-hydroxybenzoïque (qui est donc effectivement un constituant normal de l'urine) trois acides insaturés: cinnamique, *p*-hydroxy-cinnamique (*p*-coumarique) et férulique méthoxy-3-hydroxy-4-cinnamique) et trois acides isolés pour la première fois d'une source naturelle: *m*-méthoxy-benzoïque, dihydroférulique et decane-1,10-dicarboxylique (*77*).

Le tableau 6 indique le pourcentage de ces acides dans l'urine.

Tableau 6. Acides d'urine de Jument Gravide (*77*).

Acide	F.	% d'acides totaux	mg/l d'urine
Benzoïque	122°	74	2460
Phénylacétique	76°	14	460
p-Hydroxybenzoïque	214°	1,3	43
Salicylique	156°	0,7	23
Azélaïque	107°	0,4	13
Dihydroférulique	90°	0,2	6,6
Décane-1,10-dicarboxylique	128°	0,1	3,3
m-Hydroxy-benzoïque	197°	0,1	3,3
Salicylique substitué (non identifié)	118°	0,07	2,3
Férulique	170°	0,04	1,3
Vanillique	205°	0,03	0,9
m-Méthoxy-benzoïque	105°	0,025	0,7
Cinnamique	134°	0,02	0,6
Phénolique (non identifié)	208°	0,01	0,3
p-Coumarique	210°	0,002	0,06

L'odeur de la fraction acide brute, qui est très forte, d'urine et vanillée à la fois, est surtout dûe à l'acide phénylacétique.

Les acides ci-dessus comprennent onze acides aromatiques et deux acides aliphatiques dicarboxyliques. Il est possible que d'autres acides, plus solubles dans l'eau aient été perdus au cours de la préparation industrielle. Le matériel étudié ayant été extrait d'urine hydrolysée à l'acide, nous ne pouvons rien dire sur la forme de conjugaison de ces acides dans l'urine.

Nous avons discuté précédemment l'intérêt biochimique de nos résultats (*70*, *77*).

4° Les stéroïdes odorants.

En 1944, PRELOG et RUZICKA (*102*) ont isolé d'un extrait de testicules de porc, deux stéroïdes $C_{19}H_{30}O$ (F. 143,5° et F. 123°), les Δ^{16}-androstène-ol-3 épimères (CLXV et CLXVI), qui présentent une odeur musquée très nette. C'est la première fois que l'on a observé l'odeur d'un stéroïde.

PRELOG et RUZICKA (*102*) font remarquer, cependant, qu'il y a une forte analogie de structure entre les deux androsténols odorants et la civettone (CLXVII = IX) dont on pourrait les faire dériver par des cyclisations appropriées et par l'addition de deux groupes méthyles.

Rappelons à ce sujet qu'une hypothèse de WINDAUS avait fait considérer la civettone comme précurseur biochimique des stérols; cette hypothèse n'est cependant plus soutenable à la lumière de nos connaissances actuelles sur la biogénèse des stérols [formation à partir de petites molécules, en C_2; voir BLOCH et RITTENBERG (*6*)].

PRELOG, RUZICKA et WIELAND (*104*) ont ensuite synthétisé les deux stéroïdes odorants, ainsi que leurs dérivés dihydrogénés et la cétone correspondante (CLXIX). La pyrolyse à 300° de l'hexahydrobenzoate d'androstanol-(17-β)-one-3 (CLXVIII) a donné la cétone (CLXIX) dont la réduction par l'isopropylate d'aluminium a donné les 2 alcools épimères

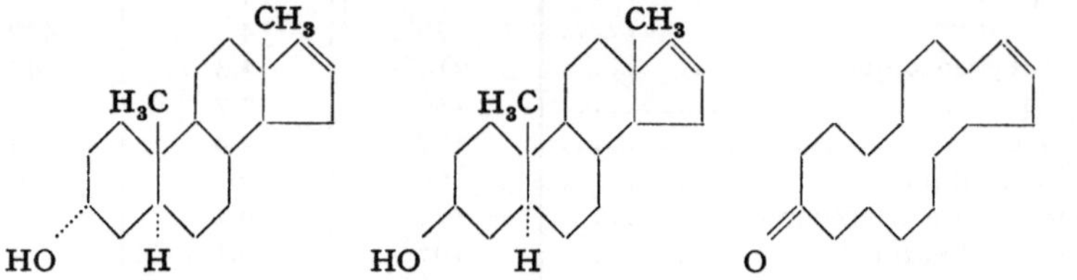

(CLXV.) Δ^{16}-*epi*-Androstène-ol-3.

(CLXVI.) Δ^{16}-Androstène-ol-3.

(CLXVII.) Civettone.

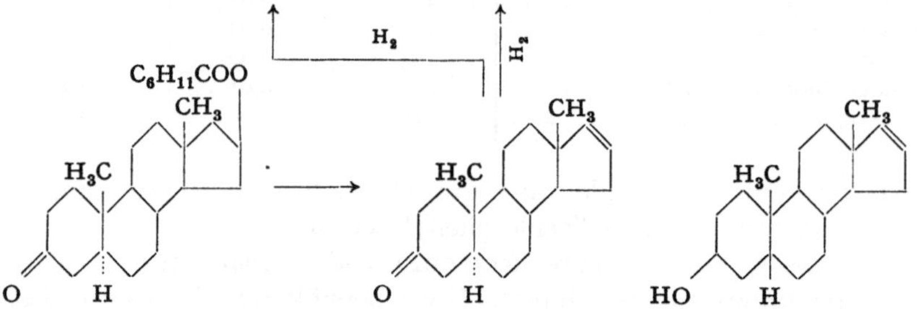

(CLXVIII.)

(CLXIX.) Δ^{16}-Androstène-one-3.

(CLXX.) Δ^{16}-Etio-cholène-ol-3-β.

(CLXV) et (CLXVI). L'odeur de la cétone est beaucoup plus intense que celle des alcools et «rappelle celle de récipients ayant servi longtemps à la conservation d'urine». Les auteurs pensent que la Δ^{16}-androstène-one-3 pourrait s'y former effectivement à partir de stéroïdes naturels excrétés dans l'urine. L'odeur des alcools est musquée, mais avec une nuance nettement plus animale et plus proche de celle du musc naturel que celle de la civettone et de la muscone pures, qui ont une nuance plutôt fleurie.

La configuration stérique de l'hydroxyle en C_3 a une forte influence sur l'odeur; le dérivé 3 α (CLXV) a une odeur beaucoup plus forte que l'épimère 3 β (CLXVI) dont l'odeur n'est décelable que par un nez exercé.

PRELOG, RUZICKA, MEISTER et WIELAND (*103*) ont consacré une étude détaillée aux *rapports entre la constitution et l'odeur des stéroïdes*. L'odeur musquée semble être assez rare, dans cette série, car elle ne se trouve dans aucune des nouvelles substances étudiées. Le simple changement de configuration au carbone 5 transforme les Δ^{16}-androstène-ol-3 odorants en les Δ^{16}-étio-cholène-ol-3 (CLXX et CLXXI) inodores. Leurs dérivés dihydrogénés sont également inodores.

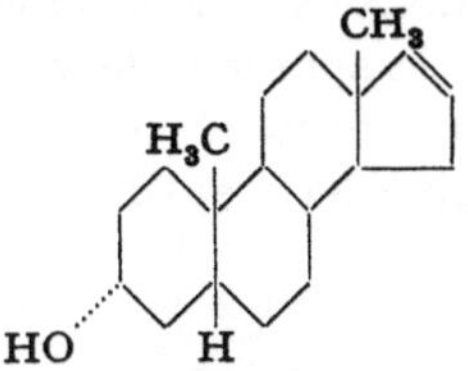

(CLXXI.) Δ^{16}-Etio-cholène-ol-3-α.

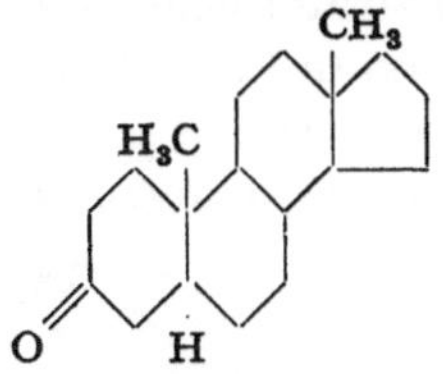

(CLXXII.) Androstan-one-3.

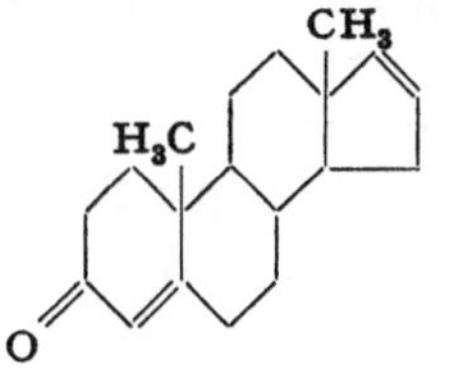

(CLXXIII.) $\Delta^{4,16}$-Androstadiène-one-3.

L'odeur d'«urine» appartient cependant à plusieurs nouvelles substances: l'androstan-one-3 (CLXXII) et la $\Delta^{4,16}$-androstadiène-one-3 (CLXXIII). L'odeur de la Δ^{16}-étio-cholène-one-3 (CLXXIV) et de l'étio-cholan-one-3 (CLXXV) est plus faible que celle des substances précitées. Il est interessant de constater que la Δ^{2}-androstène-one-17 (CLXXVI), isolée récemment par plusieurs auteurs américains à partir d'urine humaine (pour les références voir *103*), a, surtout à chaud, une très nette odeur «d'urine».

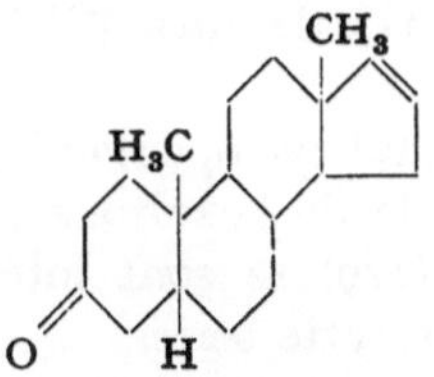

(CLXXIV.) Δ^{16}-Etio-cholène-one-3.

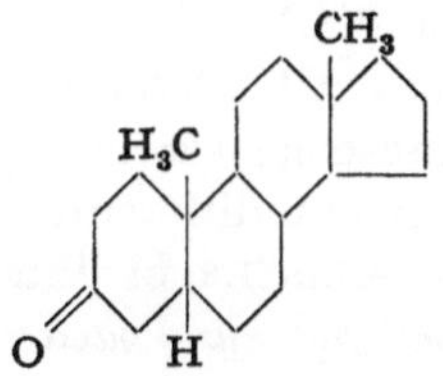

(CLXXV.) Etio-cholan-one-3.

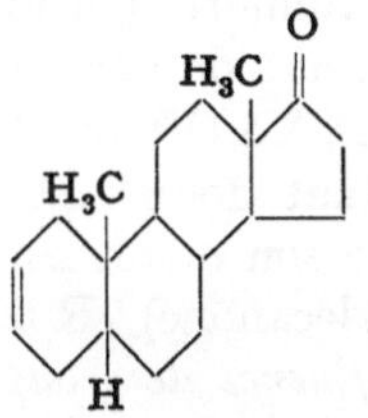

(CLXXVI.) Δ^{2}-Androstène-one-17.

Les lactones (par exemple CLXXVII) dérivées des cétones fortement odorantes ont également une nette odeur d'«urine».

Ainsi se confirme dans cette série l'observation faite pour les grands cycles, sur la conservation de l'odeur après passage d'une cétone à la lactone correspondante. Il est interessant aussi de constater que les changements de configuration qui augmentent ou diminuent l'intensité de

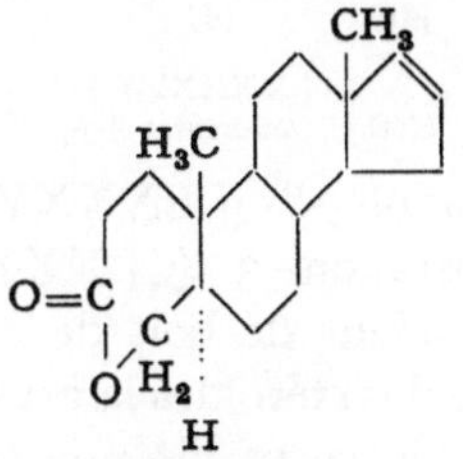

(CLXXVII.)

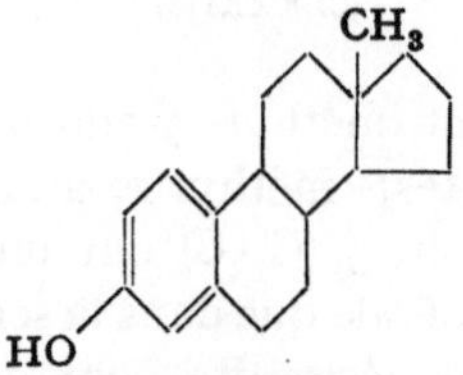

(CLXXVIII.) $\Delta^{1,3,5}$-Oestratriène-ol-3.

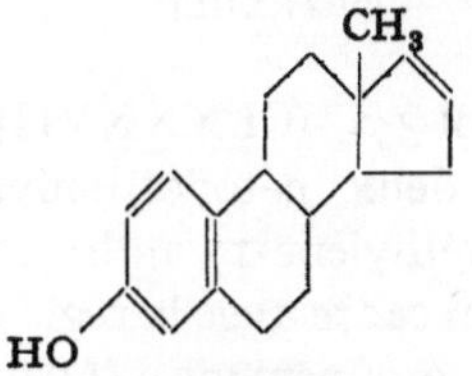

(CLXXIX.) $\Delta^{1,3,5,16}$-Oestratétraène-ol-3.

l'odeur, influencent dans le même sens l'activité hormonale mâle de ces stéroïdes.

Des stéroïdes ayant un noyau aromatique, $\Delta^{1,3,5}$-oestratriène-ol-3 (CLXXVIII), et $\Delta^{1,3,5,16}$-oestratétraène-ol-3 (CLXXIX), sont inodores.

RUZICKA, PRELOG et MEISTER (*121*) décrivent la synthèse de stéroïdes dérivant de l'androstane par rétrécissement du noyau *A* (CLXXX à CLXXXII) ou par élargissement du noyau *D* (CLXXXIII et CLXXXIV).

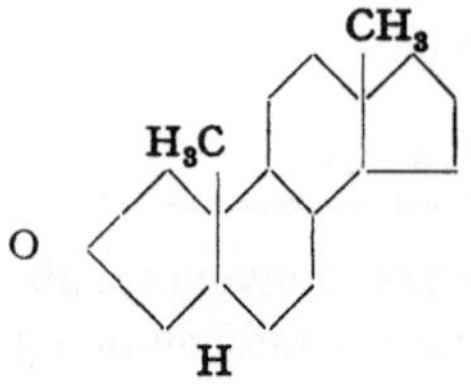

(CLXXX.) 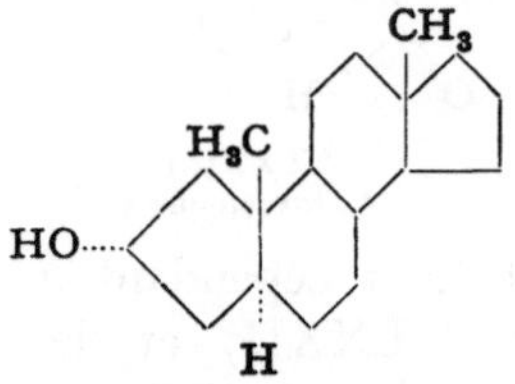 (CLXXXI.) 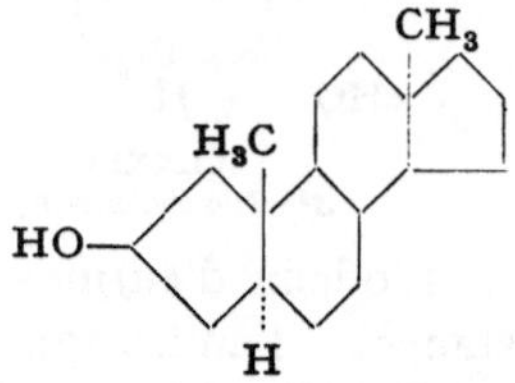 (CLXXXII.)

Ils constatent que seuls les dérivés de l'androstane dont la structure se rapproche le plus de celle des stéroïdes naturels ont des propriétés odorantes très nettes. Les nouveaux dérivés synthétiques (avec un cycle *A* en C^5 ou un cycle *D* en C^6) n'ont pas d'odeur notable, tout au moins à température ordinaire. Quelques-unes de ces substances ont, cependant, à chaud, des odeurs de bois de cèdre (CLXXX et CLXXXII), de musc (CLXXXI et CLXXXIV) ou d'«urine» (CLXXXIII).

Etant donné que la muscone est une 3-méthyl-cyclopentadecanone et que son odeur est nettement différente de celle de l'exaltone (cyclopentadecanone), RUZICKA, MEISTER et PRELOG (*119*) se sont intéressés à *l'influence de groupes méthyles sur l'odeur*, dans cette série.

Ils ont trouvé que les corps synthétiques suivants avaient, surtout à chaud, une très nette odeur de musc: méthyl-3-androstanol-3-α (CLXXXV), méthyl-3-androstanol-3-β (CLXXXVI), méthyl-17-andro-

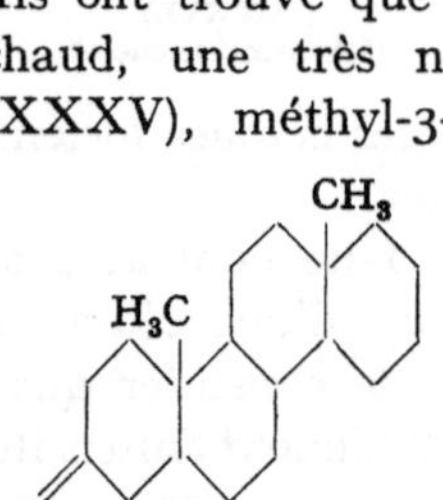

(CLXXXIII.) 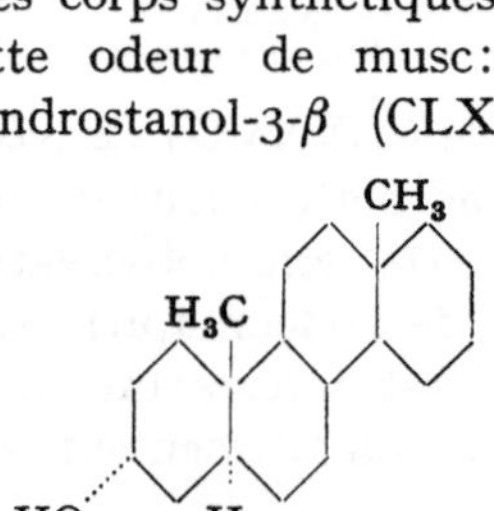 (CLXXXIV.) 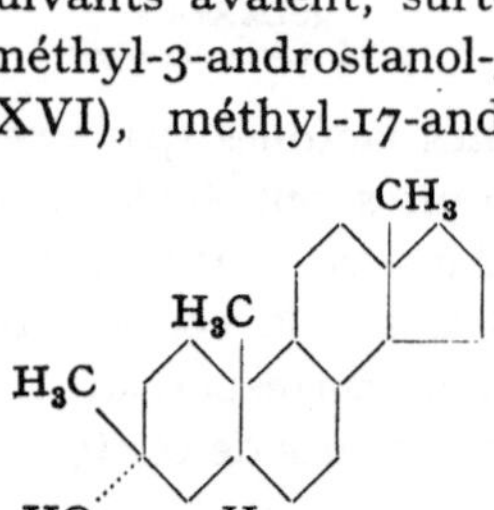 (CLXXXV.) Méthyl-3-androstanol-3-α.

stanol-3-α (CLXXXVII) et méthyl-17-androstanol-3-β (CLXXXVIII). Les deux cétones suivantes: méthyl-17-androstanone-3 (CLXXXIX) et méthylène-17-androstanone-3 (CXC) ont une odeur de bois de cèdre, ce qui cadre avec la règle générale que dans la série des stéroïdes la présence d'un groupement —OH est nécessaire pour que le corps ait une odeur musquée. Parmi les deux dérivés de l'allopregnane, allopregnanol-3-α

(CXCI) et $\Delta^{17,20}$-allopregnène-one-3 (CXCII), c'est de nouveau l'alcool qui a l'odeur de musc et non la cétone.

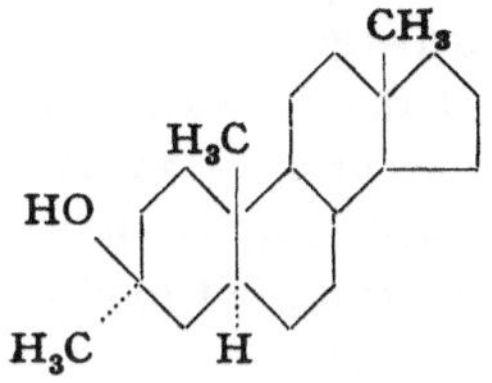

(CLXXXVI.) Méthyl-3-androstanol-3-β.

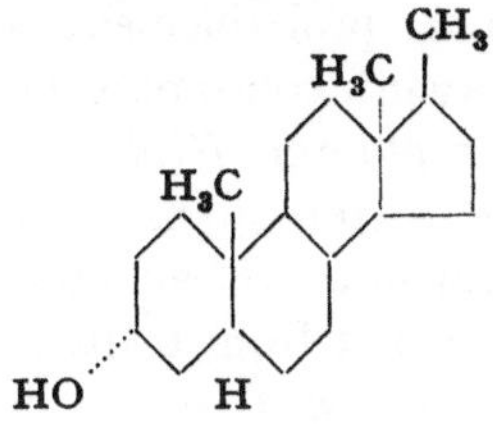

(CLXXXVII.) Méthyl-17-androstanol-3-α.

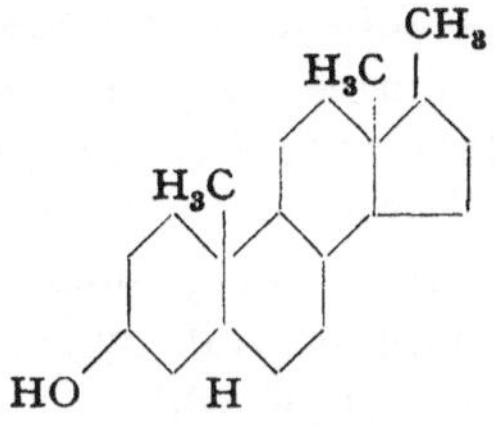

(CLXXXVIII.) Méthyl-17-androstanol-3-β.

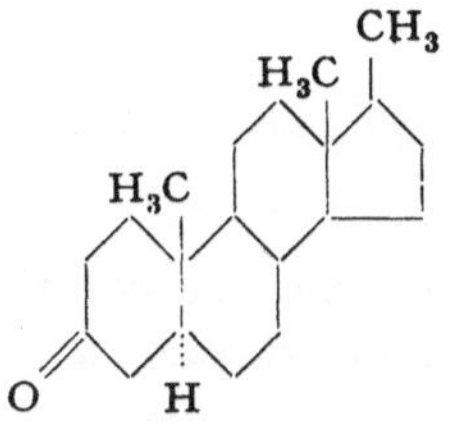

(CLXXXIX.) Méthyl-17-androstanone-3.

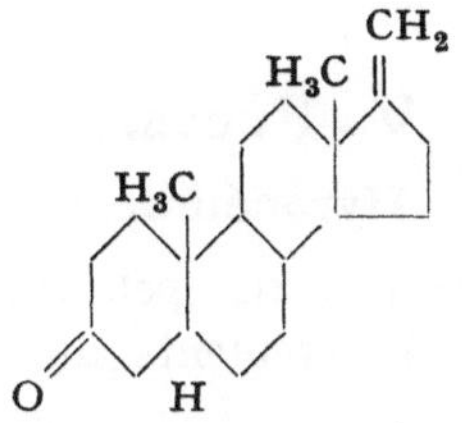

(CXC.) Méthylène-17-androstanone-3.

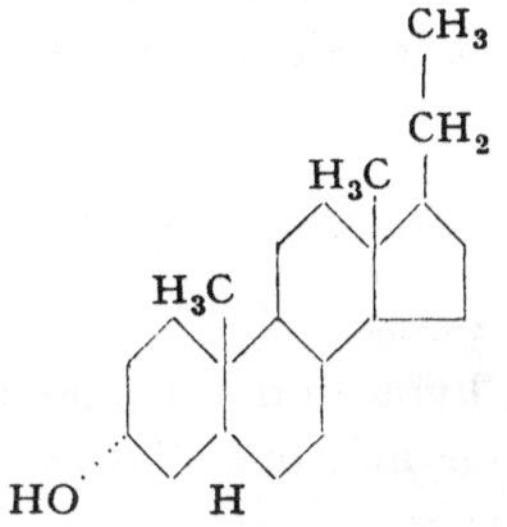

(CXCI.) Allopregnanol-3-α.

Mais, s'il y a des stéroïdes odorants à activité hormonale, il n'y a pas pour autant de macrocycles à activité hormonale; c'est ce que STOLL, HINDER et RUZICKA (*155*) ont montré en transformant la civettone en époxyde (CXCIII), puis en cyclo-heptadécane-dione-1,9 (CXCIV); ce dernier dont la position des atomes d'oxygène est calquée sur celle de nombreuses hormones sexuelles, s'est montrée tout à fait inactive dans différents tests physiologiques d'hormones sexuelles.

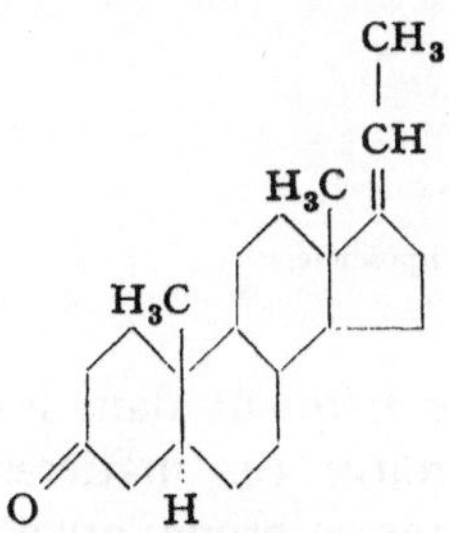

(CXCII.) $\Delta^{17,20}$-Allopregnène-one-3.

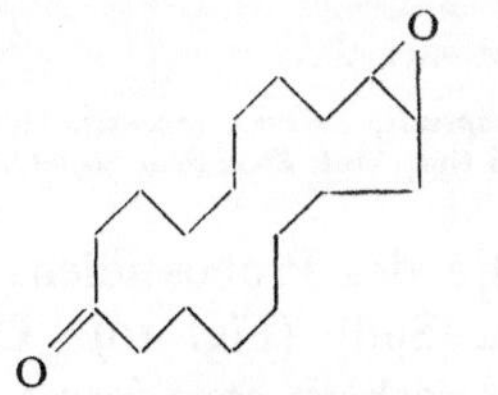

(CXCIII.)

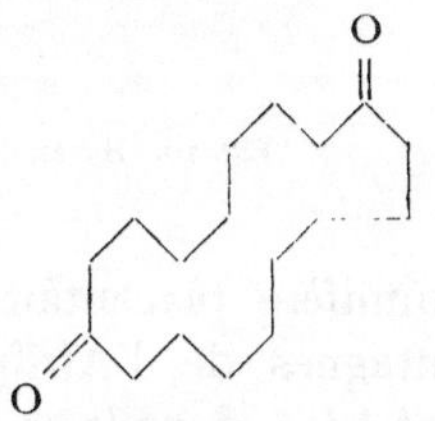

(CXCIV.) Cyclo-heptadécane-dione-1,9.

5° Divers.

EULER (*21, 22*) a trouvé qu'une partie de l'action hypertensive d'urine était due à la présence de ***pipéridine***. L'urine normale de non-fumeurs contient 3 à 20 mg. de pipéridine par litre.

JANSEN (*54*) a récemment rapporté l'isolement de l'acide dithiol-2,2'-isobutyrique à partir d'asperges; cette substance n'est pas le précurseur de la substance malodorante, probablement soufrée qui se manifeste dans l'urine de personnes ayant consommé des asperges.

HARDY (*43*) relate que le glouton *(Gulo luscus)* des forêts aretiques de l'Europe et d'Amérique secrète dans l'urine le produit fortement odorant de ses glandes à parfum. «Even now it still troubles folks as far south as Utah by stealing bait from traps and carrying off food from cabins, all of which it despoils with its foetid urine.»

L'hyracéum que nous traiterons dans le chapitre suivant, consiste en partie d'urine séchée du Daman du Cap *(Hyrax capensis)* et peut avoir une odeur très nette de castoréum.

V. Fèces.

Hyracéum.

L'hyracéum est une drogue assez peu connue; elle est constituée par un mélange d'urine et d'excréments du Daman (*Hyrax capensis,*

Fig. 10. Hyrax (*Hyrax capensis, Procaria capensis,* Hyrax, Klippschliefer).
[Photo New York Zoological Society.]

Mammifère placentaire, ordre des Proboscidiens, très fréquent dans les montagnes de l'Afrique du Sud) (Fig. 10). On trouve ces matières desséchées, dans le creux des rochers, sous forme de masses brunes ayant une odeur fécale, scatolique et phénolique [voir HARDY (*38*)].

Nous avons récemment examiné deux échantillons d'hyracéum fort différents l'un de l'autre.

Le premier, un morceau gris pesant 60 g. avait l'aspect d'une pierre et une odeur animale et phénolique (*69*); 12% seulement en étaient solubles dans l'éther. La partie acide contenait une forte proportion d'*acide benzoïque*, comme celle de toutes les urines des Herbivores. Nous en avons isolé en outre une *substance acide* bien cristallisée, F. 195°, ayant probablement la formule $C_{14}H_{14}O_6$ [cette substance avait été, par erreur, désignée comme neutre (*69*)]. L'odeur de la plupart des fractions rappelait celles du castoréum. Nous pensons que cet échantillon représentait de l'urine desséchée.

Le 2ème échantillon, de plusieurs kgs., avait l'aspect de crottin de lapin, consistant en petites boules sèches contenant des brins de cellulose. L'odeur de cet échantillon était fécale avec une note très nette d'ambre. Nous y avons identifié les substances suivantes: une *paraffine* (F. 60 à 62°), un *alcool saturé* en C_{20}, de l'*alcool cérylique* $C_{26}H_{54}O$, les *acides palmitique* et *stéarique*, un mélange de *stérols* et trois *triterpènes*, dont l'*acide oléanolique* $C_{30}H_{48}O_3$ et l'*acide ursolique* $C_{30}H_{48}O_3$ (LEDERER et POLONSKY, observations inédites).

Conclusions.

Les principaux progrès de nos connaissances de la *chimie des substances odorantes des animaux* concernent des sécrétions et excrétions de *Mammifères*. Cinq grands groupes de travaux sont à signaler:

1° *Chimie macrocyclique*: L'école de RUZICKA a continué à développer la chimie des substances macrocycliques; ces travaux ont abouti à des synthèses de la muscone et de la civettone, principales substances odorantes naturelles, de structure macrocyclique. STEVENS a isolé une série de corps macrocycliques nouveaux à partir des glandes à parfum du Rat musqué d'Amérique.

2° *Castoréum*: Les travaux de LEDERER et coll. sur le castoréum ont montré que les glandes à parfum du Castor représentent un dépôt très complexe d'une multitude de substances organiques, provenant pour la plupart directement de la nourriture végétale. La chimie et la biochimie de ces substances a été étudiée de près.

3° *Ambre gris*: La structure chimique de l'ambréine, premier triterpène tricyclique naturel, a été éclaircie avec précision par les travaux de RUZICKA, LARDON et JEGER, ainsi que par ceux de LEDERER et coll. Il a été démontré que la partie bicyclique de l'ambréine est identique à celle de quelques diterpènes des Plantes et à celle des noyaux *A* et *B* des triterpènes dérivés des amyrines α et β. Nous savons maintenant qu'une partie des substances odorantes de l'ambre gris est dérivée de l'ambréine, par oxydation.

4° *Urine*: PRELOG et coll. ont isolé de l'urine de Jument gravide une série de substances en C_{13}, dérivées de l'ionone; la constitution de tous ces corps a été éclaircie et leur synthèse effectuée. Des recherches de LEDERER concernent les phénols et les acides de l'urine de Jument gravide.

5° *Stéroïdes odorants*: PRELOG et RUZICKA ont isolé, pour la première fois des stéroïdes présentant une odeur musquée. La structure de ces substances a été élucidée et leur synthèse effectuée. Une série de steroïdes a été préparée par synthèse pour étudier le rapport entre constitution et odeur. Trois notes principales se rencontrent parmis ces corps: des odeurs de musc, de bois de cèdre et d'urine.

En considérant les quatre drogues principales utilisée en Parfumerie: musc, civette, castoréum et ambre gris, on constate qu'elles doivent leur parfum à *trois mécanismes biochimiques* différents:

a) transformation d'acides gras banaux en cétones macrocycliques (musc et civette),

b) accumulation d'une multitude de substances hétérogènes provenant de la nourriture végétale et normalement excrétées par d'autres Vertébrés (castoréum),

c) oxydation d'un triterpène inodore (ambre gris).

Nous ne savons rien des substances odorantes si variées et spécifiques qui jouent un rôle important dans l'attraction sexuelle chez les *Insectes*. C'est là un vaste champ ouvert aux chercheurs à venir.

Bibliographie.

1. ALDRICH, T. B.: A Chemical Study of the Secretion of the Anal Glands of *Mephitis Mephitica* (common Skunk) with Remarks on the Physiological Properties of this Secretion. J. exp. Med. **1**, 323 (1896).

2. BATTY, J. W., I. M. HEILBRON and W. E. JONES: Studies in the Polyene Series. III. J. chem. Soc. (London) **1939**, 1556.

3. BAUMANN, E.: Über die Bildung von Hydroparacumarsäure aus Tyrosin. Ber. dtsch. chem. Ges. **12**, 1450 (1879).

4. BECKMANN, E.: Über das Drüsensekret des Stinkdachses. Pharmaz. Zentralhalle Deutschland **37**, 557 (1896); Chem. Zbl. **1896** II, 673.

5. BIELIG, H. J. u. A. HAYASIDA: Über das Verhalten des β-Jonons im Tierkörper. (Biochemische Hydrierungen VIII.) Hoppe-Seyler's Z. physiol. Chem. **266**, 99 (1940).

6. BLOCH, K. and D. RITTENBERG: On the Utilization of Acetic Acid for Cholesterol Formation. J. biol. Chemistry **145**, 625 (1942).

7. BLOMQUIST, A. T. and R. W. HOLLEY: Many-membered Carbon Rings. III. Carboxylic Acid Derivatives of Cycloheptadecane. J. Amer. chem. Soc. **70**, 36 (1948).

8. BLOMQUIST, A. T., R. W. HOLLEY and R. D. SPENCER: Many-membered Carbon Rings. II. A New Synthesis of Civetone and *d,l*-Muscone. J. Amer. chem. Soc. **70**, 34 (1948).

9. BLOMQUIST, A. T. and R. D. SPENCER: Many-membered Carbon Rings. I. Cyclization of Some Bifunctional Ketenes. J. Amer. chem. Soc. **70**, 30 (1948).

10. BOURGOGNE, J.: Lépidoptères; dans «Traité de Zoologie» de P. P. GRASSET. Volume des Insectes Supérieurs. Paris: Masson & Cie. 1950 (sous presse).

11. BOUVEAULT, L. et R. LOCQUIN: Action du sodium sur les éthers-sels des acides gras. Préparation des acyloïnes du type R—CO—CH(OH)—R. Bull. Soc. chim. France (3) **35**, 629 (1926).

12. Büchi, G., O. Jeger u. L. Ruzicka: Zur Kenntnis der Triterpene. (124. Mitt.) Synthese des $\Delta^{5,10}$-1,1-Dimethyl-octalons-(6), eines Abbauproduktes des Ambreins. Helv. chim. Acta **31**, 241 (1948).

13. Büchi, G., K. Seitz u. O. Jeger: Veilchenriechstoffe. (29. Mitt.) Über die Dehydrierung von β- und α-Jonon. Helv. chim. Acta **32**, 39 (1949).

13a. Butenandt, A.: Zur Kenntnis der Sexual-Lockstoffe bei Insekten. Jahrbuch d. Preuß. Akad. Wiss. 23. Nov. 1939.

13b. Caliezi, A. und H. Schinz: Zur Kenntnis der Sesquiterpene und Azulene. (90, Mitteilung.) Die Cyclisation der Farnesylsäure. Helv. chim. Acta. **32**, 2556 (1949).

14. Challenger, F.: Biological Methylations. Annu. Rep. chem. Soc. London **43**, 262 (1947).

15. Chuit, P.: Préparation d'acides polyméthylène-dicarboniques de 11 à 19 atomes de carbone et de quelques-une de leurs dérivés. Helv. chim. Acta **9**, 264 (1926).

16. Collin-Asselineau, C., E. Lederer, D. Mercier et J. Polonsky: Sur quelques produits d'oxydation et de dégradation de l'ambréine. Bull. Soc. chim. France (1950) (sous presse).

16a. Collin-Asselineau, C., E. Lederer et J. Polonsky: Synthèse de quelques acides sesquiterpéniques. Bull. Soc. Chim. France 1950 (sous press).

17. Colonge, J.: Préparation des cétols au moyen des composés amino-magnésiens mixtes. Bull. Soc. chim. France (5) **1**, 1101 (1934).

18. Dakin, H. D.: Further Studies on the Fate of Phenylpropionic Acid and Some of its Derivatives. J. biol. Chemistry **6**, 217 (1909).

19. Delepine, M. et A. Sosa: Dédoublement du *d,l-p*-méthoxy-phényl-1-butanol-3 en ses composants optiquement actifs. Identification du stéréo-isomère lévogyre avec le méthyl-bétuligénol. Bull. Soc. chim. France (5) **9**, 771 (1942).

20. Dürst, O., O. Jeger u. L. Ruzicka: Zur Kenntnis der Trierpene. (137. Mitt.) Über eine Partialsynthese des Ambratriens. Helv. chim. Acta **32**, 46 (1949).

21. Euler, U. S. von: Piperidine as a Normal Pressor Constituent of Human Urine. Acta physiol. scand. **8**, 380 (1944).

22. — Piperidine as a Normal Constituent of Urine. Comm. First Intern. Congress Biochem. Cambridge (England), 1949.

23. Fabre, J. H.: Souvenirs d'entomologie. Études sur l'instinct et les moeurs des Insectes. Paris: Ch. Delagrave. 1882—1907.

23a. Fester, G. A.: Materias oloriferas del reino animal. Rev. Fac. Quim. Ind. Agric. Santa Fé, Argentina **13**, No. 22, 48 pp. (1944).

24. Fester, G. A. u. F. A. Bertuzzi: Drüsen-Sekrete der Alligatoren (Yacarol). Ber. dtsch. chem. Ges. **67**, 365 (1934).

25. — — Über das Sekret des Zorrino, einer südamerikanischen Marderart. Rev. Fac. Quím. ind. agric., Santa Fe, Argentina **5**, 85 (1936); Chem. Zbl. **1937** II, 3336.

26. Fester, G. A., F. A. Bertuzzi u. D. Pucci: Drüsen-Sekret der Alligatoren (Yacarol). (II. Mitt.) Ber. dtsch. chem. Ges. **70**, 37 (1937).

27. Fester, G. A. u. D. Pucci: Vereinfachte Kondensation mit Natriumäthylat. [2,6-Dimethyl-hepten(2)-ol-(7)]. Ber. dtsch. chem. Ges. **69**, 2017 (1936).

28. Fischer, F. G. u. H. J. Bielig: Über die Hydrierung ungesättigter Stoffe im Tierkörper. (Biochemische Hydrierungen VII.) Hoppe-Seyler's Z. physiol. Chem. **266**, 73 (1940).

28a. Flaschentraeger, B. and E. S. Amin: Chemical Attractants for Insects: Sex and Food Odours of the Cotton Leaf Worm and the Cut Worm. Nature (London) **165**, 394 (1950).

29. Ford, E. B.: Butterflies. London: Collins. 1945.

30. Girard, A. et G. Sandulesco: Sur une nouvelle série de réactifs du groupe carbonyle, leur utilisation à l'extraction des substances cétoniques et à la caractérisation microchimique des aldéhydes et cétones. Helv. chim. Acta **19**, 1095 (1936).
31. Goetz, B.: Über weitere Versuche zur Bekämpfung der Traubenwickler mit Hilfe des Sexualduftstoffes. VIII. Anz. Schädlingskunde **15**, 109 (1939); Chem. Zbl. **1940** I, 1409.
32. Grant, J. K.: *p*-Ethylphenylsulphuric Acid in Goat Urine. Biochemic. J. **43**, 523 (1948).
33. Griessmayer, V.: Über das Verhalten von Stärke und Dextrin gegen Jod und Gerbsäure. Liebigs Ann. Chem. **160**, 51 (1871).
34. Grignard, V. et J. Colonge: Nouvelle méthode générale de condensation des cétones. C. R. hebd. Séances Acad. Sci. **194**, 929 (1932).
35. Guibourt, J. N. et G. Planchon: Histoire naturelle des drogues simples. Ou cours d'histoire naturelle professé à l'Ecole de Pharmacie de Paris. 7e éd. Paris: Baillère et Fils. 1876.
36. Haller, H. L., F. Acree and S. F. Potts: The Nature of the Sex Attractant of the Female Gypsy Moth. J. Amer. chem. Soc. **66**, 1659 (1944).
37. Hansley, V. L.: The Preparation of High Molecular Weight Acyloins. J. Amer. chem. Soc. **57**, 2303 (1935).
38. Hardy, E.: The Hyrax, Source of Hyraceum. Perfum. essent. Oil Rec. **37**, 352 (1946).
39. — Animals in Perfumery. I. The Civet. Perfum. essent. Oil Rec. **38**, 276 (1947).
40. — Animals in Perfumery. II. Musk-Rat and Musk-Deer. Perfum. essent. Oil Rec. **38**, 335 (1947).
41. — Scents from Insects; Butterfly and Moth Perfumes and their Extraction. Perfum. essent. Oil Rec. **38**, 403 (1947).
42. — Animals in Perfumery. IV. The Beaver and Castoreum. Perfum. essent. Oil Rec. **39**, 315 (1948).
43. — Animals in Perfumery. V. Lesser Known Sources of Musk. Perfum. essent. Oil Rec. **40**, 93 (1949).
44. Hill, J. W. and W. H. Carothers: Studies of Polymerization and Ring Formation. XX. Many-membered Cyclic Esters. J. Amer. chem. Soc. **55**, 5031 (1933).
45. — — Studies of Polymerization and Ring Formation. XXI. Physical Properties of Macrocyclic Esters and Anhydrides. New Types of Synthetic Musks. J. Amer. chem. Soc. **55**, 5039 (1933).
46. — — Cyclic and Polymeric Formals. J. Amer. chem. Soc. **57**, 925 (1935).
47. Hosking, J. R. u. C. W. Brandt: Über den Diterpenalkohol aus dem Holz von *Dacrydium biforme*. I. Ber. dtsch. chem. Ges. **68**, 1311 (1935).
48. Hunsdiecker, H.: Beiträge zur Kenntnis makrozyklischer Ringsysteme. I. Herstellung und Cyclisierung von (ω-Halogen-acyl)-essigestern. Ber. dtsch. chem. Ges. **75**, 1190 (1942).
49. — Beiträge zur Kenntnis makrozyklischer Ringsysteme. III. Die Synthese des Zibetons (Teil I). Ber. dtsch. chem. Ges. **76**, 142 (1943).
50. — Beiträge zur Kenntnis makrozyklischer Ringsysteme. IV. Synthese des Zibetons (Teil II). Herstellung der *trans*- und *cis*-9-Octadecendicarbonsäure. Ber. dtsch. chem. Ges. **77**, 185 (1944).
51. Hurtley, W. R. H.: Replacement of Halogen in *ortho*-Bromobenzoic acid. J. chem. Soc. (London) **1929**, 1870.
52. Isiguro, T. u. M. Watanabe: Bestandteile der Ambra I. J. pharmac. Soc. Japan **58**, 790 (1938); Chem. Abstr. **33**, 2529 (1939).

53. Janistyn, H.: Über Ambra I. Die Sterine der Ambra. Fette u. Seifen **48**, 501 (1941).
54. Jansen, E. F.: The Isolation and Identification of 2,2'-Dithiolisobutyric Acid from Asparagus. J. biol. Chemistry **176**, 657 (1948).
55. Jeger, O., O. Dürst u. G. Büchi: Zur Kenntnis der Diterpene. (56. Mitt.) Überführung des Manools in ein Umwandlungsprodukt der Abietinsäure. Helv. chim. Acta **30**, 1853 (1947).
56. Jeger, O., O. Dürst u. L. Ruzicka: Zur Kenntnis der Triterpene. (117. Mitt.) Die Konstitution des Ambreins. Helv. chim. Acta **30**, 1859 (1947).
57. Jullien, A., M. D. Garabedian et R. Guibault: Observations relatives aux propriétés pharmacologiques des constituants de la Pourpre chez *Murex trunculus*. C. R. Séances Soc. Biol. Filiales Associées **135**, 1636 (1941).
58. Kapp, E. M. and A. F. Coburn: Urinary Metabolites of Sodium Salicylate. J. biol. Chemistry **145**, 549 (1942).
59. Kerschbaum, M.: Über Lactone mit großen Ringen — die Träger des vegetabilischen Moschus-Duftes. Ber. dtsch. chem. Ges. **60**, 902 (1927).
60. Kettlewell, H. B. D.: Female Assembling Scents, with Reference to an Important Paper on the Subject. Entomologist (London) **79**, 8 (1946).
61. Kharasch, M. S. and F. L. Lambert: Factors Determining the Course and Mechanism of Grignard Reactions. III. The Effect of Metallic Halides on the Reaction between Benzophenone and Methylmagnesium Bromide. J. Amer. chem. Soc. **63**, 2315 (1941).
62. Lederer, E.: Sur les constituants du castoréum. I. Les acides. Bull. Soc. Chim. biol. **23**, 1457 (1941).
63. — Sur les constituants du castoréum. II. Les pigments; isolement et constitution chimique. Bull. Soc. Chim. biol. **24**, 1155 (1942).
64. — Sur les constituants du castoréum. IV. Aldéhydes et cétones. Bull. Soc. Chim. biol. **25**, 1073 (1943).
65. — Isolement de *p*-éthyl-phénol à partir de l'urine de Jument gravide. Bull. Soc. Chim. biol. **25**, 1237 (1943).
66. — Sur les constituants de l'ambre gris. I. Isolement d'une paraffine liquide. Bull. Soc. Chim. biol. **25**, 1239 (1943).
67. — Sur les constituants du castoréum. V. Alcools, ethers, amines. Bull. Soc. Chim. biol. **25**, 1381 (1943).
68. — Chemistry and Biochemistry of the Scent Glands of the Beaver (*Castor fiber*). Nature (London) **157**, 231 (1946).
69. — Travaux récents sur la chimie des parfums d'origine animale. Industrie de la Parfumerie **1**, 110, 148 (1946).
70. — Chemistry and Biochemistry of Some Mammalian Secretions and Excretions. J. chem. Soc. (London) **1949**, 2115.
71. Lederer, E., F. Marx, D. Mercier et G. Pérot: Sur les constituants de l'ambre gris. II. Ambréine et coprostanone. Helv. chim. Acta **29**, 1354 (1946).
72. Lederer, E. et D. Mercier: Obtention d'un dérivé du sclaréol à partir de l'ambréine. Experientia **3**, 188 (1947).
73. — — Sur les constituants de la graisse de laine. IV. La peau de mouton comme organe de la biosynthèse des alcools triterpéniques. Biochim. Biophys. Acta **2**, 91 (1948).
74. Lederer, E., D. Mercier et G. Pérot: Sur les constituants de l'ambre gris. III. Structure chimique de l'ambréine. Bull. Soc. chim. France **1947**, 345.
75. Lederer, E. et J. Polonsky: Sur les constituants du castoréum. III. Quelques phénols. Bull. Soc. Chim. biol. **24**, 1386 (1942).

76. Lederer, E. et J. Polonsky: Synthèse de la dihydroxy-4,4'-dibenzo-α-pyrone, pigment des glandes à parfum du Castor *(Castor fiber)*. Bull. Soc. chim. France **1948**, 831.
77. — — Sur les acides d'urine de Jument gravide. Biochim. Biophys. Acta **2**, 431 (1948).
78. Lederer, E., V. Prelog et R. Schneider: Recherches sur des extraits d'organes. (14ème comm.) Sur la présence de dérivés d'ionone dans le castoréum. Helv. chim. Acta **31**, 2133 (1948).
79. Lederer, E. et R. Tixier: Sur les porphyrines de l'ambre gris. C. R. hebd. Séances Acad. Sci. **224**, 531 (1947).
80. Lederer, E. et L. Velluz: Chimie de la graisse de laine. Industrie de la Parfumerie **2**, 282 (1947).
81. Le Magnen, J.: Un cas de sensibilité olfactive se présentant comme un caractère sexuel secondaire féminin. C. R. hebd. Séances Acad. Sci. **226**, 694 (1948).
82. — Odeurs et Parfums. Collection «Que sais-je?» No. 344. Paris: Presses Universitaires de France. 1949.
82 a. — Étude d'un phénomène de sensibilisation olfactive. C. R. hebd. Séances Acad. Sci. **228**, 122 (1949).
82 b. — Variations spécifiques des seuils olfactifs chez l'homme sous action androgène et oestrogène. C. R. hebd. Séances Acad. Sci. **228**, 947 (1949).
82 c. — Nouvelles données sur le phénomène de l'exaltolide. C. R. hebd. Séances Acad. Sci. **230**, 1103 (1950).
82 d. — L'odeur des hormones sexuelles. C. R. hebd. Séances Acad. Sci. **230**, 1367 (1950).
83. Lutwak-Mann, C.: The Excretion of a Metabolic Product of Salicylic Acid. Biochemic. J. **37**, 246 (1943).
84. McCord, C. P. and W. N. Witheridge: Odors. Physiology and Control. London: McGraw-Hill Book Co. 1949.
85. Marker, R. E., E. L. Wittbecker, R. B. Wagner and D. L. Turner: Sterols CXXXIX. Sapogenins LIX. The Bio-reduction of 4-Dehydrotigogenone. J. Amer. chem. Soc. **64**, 818 (1942).
86. Marshall, P. G.: *Para*-cresol from Urine of Pregnant Mares. Nature (London) **140**, 362 (1937).
87. Meisels, A., O. Jeger u. L. Ruzicka: Zur Kenntnis der Triterpene. (139. Mitt.) Über die Konstitution des α-Amyrins und seine Beziehungen zu β-Amyrin. Helv. chim. Acta **32**, 1075 (1949).
87a. Milas, A. N., W. M. Postman and R. Heggie: Evidence for the Presence of Vitamin A and Carotenoids in the Olfactory Area of the Steer. J. Amer. chem. Soc. **61**, 1929 (1939).
88. Moncrieff, R. W.: The Chemical Senses. New York: J. Wiley & Sons. 1946.
89. — Chemistry of Perfumery Materials. London: United Trade Press Ltd. 1949.
90. Naef & Co.: Verfahren zur Erhöhung der Ausbeute bei der Herstellung von Zibeton aus Zibeth. D. R. P. Nr. 482762 (1927).
91. Naves, Y. R.: Études sur les matières végétales volatiles. XCII. Sur la farnésylidène-acétone. Helv. chim. Acta **32**, 1802 (1949).
92. Naves, Y. R. et G. Mazuyer: Les Parfums Naturels. Paris: Gauthier-Villars. 1939.
93. Okahara, Y.: Untersuchungen über Cetacea. XXXIX. Über einen roten Farbstoff aus Ambra, das Ambraporphyrin. Ein Beitrag zur spektroskopischen Porphyrinuntersuchung. Jap. J. med. Sci., Sect. II **1**, 247 (1927).
94. Pelletier, P. J. et J. Caventou: Sur la nature de la substance adipocireuse de l'ambre gris et sur l'origine de ce produit. J. Pharm. **6**, 49 (1820).

95. PFAU, A. S.: Laboratory Notes. A New Constituent of Lavender Oil. The Composition of Castoreum. Perfum. essent. Oil Rec. **18**, 205 (1927).
96. — Über ein natürliches Vorkommen von Methyl-*n*-octyl-keton. Helv. chim. Acta **15**, 1267 (1932).
97. PRELOG, V., L. FRENKIEL, M. KOBELT u. P. BARMAN: Ein Herstellungsverfahren für vielgliedrige Cyclanone. Helv. chim. Acta **30**, 1741 (1947).
98. PRELOG, V. u. H. FRICK: Veilchenriechstoffe. (25. Mitt.) Über die beiden diastereomeren Tetrahydro-jonane. Helv. chim. Acta **31**, 417 (1948).
99. PRELOG, V. u. J. FÜHRER: Untersuchungen über Organextrakte. (8. Mitt.) Über die Isolierung von 3-Desoxy-equilenin aus dem Harn trächtiger Stuten. Helv. chim. Acta **28**, 583 (1945).
100. PRELOG, V., J. FÜHRER, R. HAGENBACH u. H. FRICK: Untersuchungen über Organextrakte. (11. Mitt.) Über die Konstitution des Alkohols $C_{13}H_{20}O$ aus dem Harn trächtiger Stuten. Helv. chim. Acta **30**, 113 (1947).
101. PRELOG, V., J. FÜHRER, R. HAGENBACH u. R. SCHNEIDER: Untersuchungen über Organextrakte. (13. Mitt.) Über die Isolierung von Jonon-Derivaten aus dem Harn trächtiger Stuten. Helv. chim. Acta **31**, 1799 (1948).
102. PRELOG, V. u. L. RUZICKA: Untersuchungen über Organextrakte. (5. Mitt.) Über zwei moschusartig riechende Steroïde aus Schweinetestes-Extrakten. Helv. chim. Acta **27**, 61 (1944).
103. PRELOG, V., L. RUZICKA, P. MEISTER u. P. WIELAND: Steroïde und Sexualhormone. (113. Mitt.) Untersuchungen über den Zusammenhang zwischen Konstitution und Geruch bei Steroïden. Helv. chim. Acta **28**, 618 (1945).
104. PRELOG, V., L. RUZICKA u. P. WIELAND: Steroïde und Sexualhormone. (90. Mitt.) Über die Herstellung der beiden moschusartig riechenden Δ^{16}-Androstenole-(3) und verwandter Verbindungen. Helv. chim. Acta **27**, 66 (1944).
105. PRELOG, V. u. R. SCHNEIDER: Untersuchungen über Organextrakte. (15. Mitt.) Über die Lage der zweiten Carbonyl-Gruppe im Diketon D (Oxo-*cis*-tetrahydro-jonon) aus dem Harn trächtiger Stuten. Helv. chim. Acta **32**, 1632 (1949).
106. PRELOG, V. u. B. VATERLAUS: Untersuchungen über Organextrakte und Harn. (16. Mitt.) Über die Konstitution des ungesättigten Ketons $C_{13}H_{18}O$ aus dem Harn trächtiger Stuten. Helv. chim. Acta **32**, 2082 (1949).
107. RIBAN, J.: Sur l'ambréine (à partir de l'ambre gris). Bull. Soc. chim. France (4) **11**, 754 (1912).
108. ROSENHEIM, O. and W. W. STARLING: Oxycholesterol. Chem. and Ind. **52**, 1056 (1933).
109. ROSENHEIM, O. and T. A. WEBSTER: The Mechanism of Coprosterol Formation *in vivo*. I. Cholestenone as an Intermediate. Biochemic. J. **37**, 513 (1943).
109 a. — — Precursors of Colprosterol and the Bile Acids in the Animal Organism. Nature (London) **136**, 474 (1935).
110. RUZICKA, L.: Zur Kenntnis des Kohlenstoffringes. I. Über die Konstitution des Zibetons. Helv. chim. Acta **9**, 230 (1926).
111. — Zur Kenntnis des Kohlenstoffringes. VII. Über die Konstitution des Muscons. Helv. chim. Acta **9**, 715 (1926).
112. — Zur Kenntnis des Kohlenstoffringes. VIII. Weitere Beiträge zur Konstitution des Muscons. Helv. chim. Acta **9**, 1008 (1926).
113. — Vielgliedrige Ringe, höhere Terpenverbindungen und männliche Sexualhormone. Les Prix Nobel en 1945. Stockholm. 1947.
114. RUZICKA, L., G. BÜCHI u. O. JEGER: Veilchenriechstoffe. (24. Mitt.) Synthese des Dihydro-γ-jonones, eines Abbauproduktes des Ambreïns. Helv. chim. Acta **31**, 293 (1948).

115. Ruzicka, L., O. Dürst u. O. Jeger: Zur Kenntnis der Triterpene. (113. Mitt.) Überführung des Triterpens Ambreïn in ein Abbauprodukt des Diterpens Manool. Helv. chim. Acta **30**, 353 (1947).
116. Ruzicka, L., H. Gutmann, O. Jeger u. E. Lederer: Zur Kenntnis der Triterpene. (132. Mitt.) Über die Zusammenhänge der Oleanolsäure mit dem Triterpen Ambrein und den Diterpenen Abietinsäure und Manool. Helv. chim. Acta **31**, 1746 (1948).
117. Ruzicka, L. u. M. M. Janot: Höhere Terpenverbindungen. L. Zur Kenntnis des Sclareols. Helv. chim. Acta **14**, 645 (1931).
118. Ruzicka, L. u. F. Lardon: Zur Kenntnis der Triterpene. (105. Mitt.) Über das Ambrein, einen Bestandteil des grauen Ambra. Helv. chim. Acta **29**, 912 (1946).
119. Ruzicka, L., P. Meister u. V. Prelog: Über Steroïde und Sexualhormone. (139. Mitt.) Untersuchungen über den Zusammenhang zwischen Konstitution und Geruch bei den Steroïden. Methylandrostan und Allopregnanderivate. Helv. chim. Acta **30**, 867 (1947).
120. Ruzicka, L., Pl. A. Plattner u. W. Widmer: Über die Herstellung einiger mit der Synthese des Zibetons zusammenhängender Dicarbonsäuren. (1. Mitt.) Herstellung der *cis*- und *trans*-Eikosen-(10)-disäure. Helv. chim. Acta **25**, 604 (1942).
121. Ruzicka, L., V. Prelog u. P. Meister: Steroïde und Sexualhormone. (120. Mitt.) Weitere Untersuchungen über den Zusammenhang zwischen Konstitution und Geruch bei Steroïden. Helv. chim. Acta **28**, 1651 (1945).
122. Ruzicka, L., E. Rey u. A. C. Muhr: Zur Kenntnis der Triterpene. (87. Mitt.) Über verschiedene Umwandlungsprodukte des Lanosterins. Helv. chim. Acta **27**, 472 (1944).
123. Ruzicka, L., G. Salomon u. K. E. Meyer: Vielgliedrige heterozyklische Verbindungen. III. Gewinnung des Hexamethylen-imins und des Hexadecamethylen-imins aus den aliphatischen Bromaminen. Helv. chim. Acta **17**, 882 (1934).
124. Ruzicka, L., H. Schinz u. C. F. Seidel: Zur Kenntnis des Kohlenstoffringes. IX. Über den Abbau von Zibeton, Zibetol und Zibetan. Helv. chim. Acta **10**, 695 (1927).
125. Ruzicka, L., C. F. Seidel u. L. L. Engel: Zur Kenntnis der Diterpene. (53. Mitt.) Oxydation des Sclareols mit Kaliumpermanganat. Helv. chim. Acta **25**, 621 (1942).
126. Ruzicka, L., C. F. Seidel u. M. Pfeiffer: Über die flüchtigen Bestandteile des grauen Ambra. (1. Mitt.) Isolierung von Dihydro-γ-jonon. Helv. chim. Acta **31**, 827 (1948).
127. Ruzicka, L. u. M. Stoll: Zur Kenntnis des Kohlenstoffringes. XIII. Über die Oxydation der 13- bis 17-gliedrigen monozyklischen Ketone mit Caro'scher Säure zu den 14- bis 18-gliedrigen Lactonen. Helv. chim. Acta **11**, 1159 (1928).
128. — — Zur Kenntnis des Kohlenstoffringes. XXVIII. Über die Herstellung von 2-Methyl-, 3-Methyl- und 7-Methyl-cyclopentadecanon-(1). Beiträge zur Synthese des *d,l*-Muscons. Helv. chim. Acta **17**, 1308 (1934).
129. Ruzicka, L., M. Stoll u. H. Schinz: Zur Kenntnis des Kohlenstoffringes. II. Synthesen der carbocyclischen Ketone vom Zehner- bis zum Achtzehnerring. Helv. chim. Acta **9**, 249 (1926).
130. — — — Zur Kenntnis des Kohlenstoffringes. XI. Über den 10-, 11-, 20- und 22-gliedrigen Kohlenstoffring und über die Bildung aliphatischer Ketone neben den cyclischen bei der Zersetzung von Metallsalzen der Polymethylendicarbonsäuren. Helv. chim. Acta **11**, 670 (1928).

131. SABETAY, H. et S. SABETAY: Les travaux récents d'analyse et de synthèse organique et la Chimie des Parfums. Vol. I, 1930—1934. Paris: Gauthier-Villars. 1936.

132. — — Les travaux récents d'analyse et de synthèse organique et la Chimie des Parfums. Vol. II, 1934—1938. Paris: Gauthier-Villars. 1941.

133. SACK, H.: Zur Kenntnis des Zibeths. Chemiker-Ztg. **39**, 538 (1915).

134. SALGUES, R.: Les substances odorantes des concrétions de l'estomac. Industrie de la Parfumerie **2**, 378 (1947).

135. SANDULESCO, G. u. S. SABETAY: Anwendung des Reagenses von GIRARD und SANDULESCO zur Isolierung von Ketonen aus ätherischen Ölen und tierischen Drogen. Riechstoffind. u. Kosmet. **8**, 162 (1937).

136. SANO, T.: The Male Hormone-like Action of Musk. J. pharmac. Soc. Japan **56**, 913 (1936); Chem. Abstr. **32**, 8519 (1938).

137. SCHINZ, H., L. RUZICKA, U. GEYER u. V. PRELOG: Muscopyridin, eine Base $C_{16}H_{25}N$ aus natürlichem Moschus. Helv. chim. Acta **29**, 1524 (1946).

138. SCHOENHEIMER, R., D. RITTENBERG and M. GRAFF: Deuterium as an Indicator in the Study of Intermediary Metabolism. IV. The Mechanism of Coprosterol Formation. J. biol. Chemistry **111**, 183 (1935).

138 a. SÖRENSEN, J. S. and N. A. SÖRENSEN: Studies related to Pristane, II. The Methyl-Octadecanes. Acta Chem. Scand. **2**, 166 (1948).

138 b. — — Studies related to Pristane, III. The Identity of Norphytane and Pristane. Acta Chem. Scand. **3**, 939, (1949).

138 c. SÖRENSEN, N. A. and J. MEHLUM: Studies related to Pristane, I. The Unsaponifiable Matter of the Liver Oil of the Basking Shark. Acta Chem. Scand. **2**, 140 (1048).

139. SOSA, A.: Recherches sur le *Betula alba* L. et le bétuloside. Ann. Chimie **14**, 1 (1940).

140. SPANAGEL, E. W. and W. H. CAROTHERS: Preparation of Macrocyclic Lactones by Depolymerization. J. Amer. chem. Soc. **58**, 654 (1936).

141. SPARHAWK, C. V.: Extraction of Musklike Material from Muskrat Glands. U. S. 2 368 578. Jan. 30, 1945; Chem. Abstr. **39**, 5415 (1945).

142. STEVENS, P. G.: American Musk. II. A Preliminary Note on the Scent Glands of the Beaver. J. Amer. chem. Soc. **65**, 2471 (1943).

143. — American Musk. III. The Scent of the Common Skunk. J. Amer. chem. Soc. **67**, 407 (1945).

144. — American Musk. IV. On the Biological Origin of Animal Musk. Two More Large Ring Ketones from the Muskrat. J. Amer. chem. Soc. **67**, 907 (1945).

145. STEVENS, P. G. and J. L. E. ERICKSON: American Musk. I. The Chemical Constitution of the Musk of the Louisiana Muskrat. J. Amer. chem. Soc. **64**, 144 (1942).

146. — — Musk. U. S. 2364041. Nov. 28, 1944; Chem. Abstr. **39**, 4200 (1945).

147. STOLL, M.: Die Moschusriechstoffe. Fette u. Seifen **46**, 136 (1939).

148. — Synthèses de produits macrocycliques à odeur musquée. (4e comm.) Sur la transformation des cyclanolones en cyclanones. Helv. chim. Acta **30**, 1837 (1947).

149. — Les produits à odeur musquée et l'évolution de la chimie mégacyclique. Chimia (Zürich) **2**, 217 (1948).

150. — Synthèses de produits macrocycliques à odeur musquée. (7e comm.) Préparation du céto-9-heptadecane-dioïque. Helv. chim. Acta **31**, 143 (1948).

151. — Synthèses de produits macrocycliques à odeur musquée. (11e comm.) Réduction de la méthyl-15-cyclopéntadécanolone-2,1 en α-méthyl-cyclopentadécanone. Helv. chim. Acta **31**, 1082 (1948).

152. Stoll, M. et A. Commarmont: Synthèses de produits macrocycliques à odeur musquée. (9e comm.) Transformation de la cyclo-pentadécène-2-one-1 en muscone. Helv. chim. Acta **31**, 554 (1948).
153. — — Synthèses de produits macrocycliques à odeur musquée. (12e comm.) Réduction de la β-méthyl-cyclopentadécanoïne en muscone et γ-méthyl-cyclopentadécanone. Helv. chim. Acta **31**, 1435 (1948).
154. — — Odeur et constitution. II. α- et β-Bicyclofarnésal et β-bicyclofarnésol. Helv. chim. Acta **32**, 1836 (1949).
155. Stoll, M., M. Hinder et L. Ruzicka: Stéroïdes et hormones sexuelles. (151e comm.) Cycloheptadécane-dione-1,9. Helv. chim. Acta **31**, 1176 (1948).
156. Stoll, M. et J. Hulstkamp: Synthèses de produits macrocycliques à odeur musquée. (2e comm.) Sur une amélioration de la préparation des acyloïnes cycliques. Helv. chim. Acta **30**, 1815 (1947).
157. Stoll, M., J. Hulstkamp et A. Rouve: Synthèses de produits macrocycliques à odeur musquée. (8e comm.) Synthèse de la civettone naturelle. Helv. chim. Acta **31**, 543 (1948).
158. Stoll, M. et A. Rouve: Contribution à l'étude des combinaisons carbocycliques. (36e comm.) Études sur la décomposition de sels d'acides polyméthylène-α,ω-dicarboniques. Helv. chim. Acta **27**, 1570 (1944).
159. — — Synthèses de produits macrocycliques à odeur musquée. (3e comm.) Sur les acyloïnes cycliques. Helv. chim. Acta **30**, 1822 (1947).
160. — — Synthèses de produits macrocycliques à odeur musquée. (5e comm.) Nouvelle synthèse de la muscone. Helv. chim. Acta **30**, 2019 (1947).
161. Stoll, M. u. G. Stoll-Comte: Zur Kenntnis des Kohlenstoffringes. XVI. Über den Zusammenhang zwischen Dichte und Molekelanordnung innerhalb einer Reihe homologer aliphatischer und cyclischer Kohlenwasserstoffe. Helv. chim. Acta **13**, 1185 (1930).
162. Suzuki, M.: Untersuchungen über Cetacea. XIV. Über die Natur des Ambra und ihre Bestandteile. Jap. J. med. Sci., Trans., Sect. II **1**, 31 (1925); Chem. Zbl. **1926 I**, 147.
163. Tsujimoto, M.: Saturated Hydrocarbons in the Liver Oil of the Basking Shark. Ind. Engng. Chem. **9**, 1098 (1917).
164. Turfitt, G. E.: The Microbiological Degradation of Steroids. IV. Fission of the Steroid Molecule. Biochemic. J. **42**, 376 (1948).
165. Ueno, S. u. S. Komori: Zusammensetzung des Ityo-Fischöles. J. Soc. chem. Ind. Japan, suppl. Bind. **38**, 345 B (1935); Chem. Zbl. **1936 II**, 1636.
166. Walbaum, H.: Das natürliche Moschusaroma. J. prakt. Chem. (2) **73**, 48 (1906).
167. Walbaum, H. u. A. Rosenthal: Zur Kenntnis der Riechstoffe des Bibergeils (Castoreum). J. prakt. Chem. (2) **117**, 225 (1927).
168. Weitkamp, A. W.: Acidic Constituents of Degras-new Method of Structure Elucidation. J. Amer. chem. Soc. **67**, 447 (1945).
169. Weitkamp, A. W., A. M. Smiljanic and S. Rothman: The Free Fatty Acids of Human Hair Fat. J. Amer. chem. Soc. **69**, 1936 (1947).
170. Wieland, H., O. Schlichting u. R. Jakobi: Untersuchungen über die Gallensäure. XXVI. Über die Natur der Seitenkette und des vierten Ringes. Hoppe-Seyler's Z. physiol. Chem. **161**, 80 (1926).
171. Wigglesworth, V. B.: The Principles of Insect Physiology. 3rd ed. London: Methuen. 1947.
172. Williams, R. T.: Detoxication Mechanisms. The Metabolism of Drugs and Allied Organic Compounds. London: Chapman & Hall Ltd. 1947.
173. Windaus, A.: Überführung des Cholesterins in Koprosterin. XXIV. Mitteilung zur Kenntnis des Cholesterins. Ber. dtsch. chem. Ges. **49**, 1732 (1916).

174. ZIEGLER, K. u. R. AURNHAMMER: Über vielgliedrige Ringsysteme. V. Die Bildungstendenz cyclischer Verbindungen. Liebigs Ann. Chem. **513**, 43 (1934).
175. ZIEGLER, K., H. EBERLE u. H. OHLINGER: Über vielgliedrige Ringsysteme. I. Die präparativ ergiebige Synthese der Polymethylketone mit mehr als 6 Ringgliedern. Liebigs Ann. Chem. **504**, 94 (1933).
176. ZIEGLER, K. u. K. WEBER: Über vielgliedrige Ringsysteme. IV. Die Synthese des *rac.* Muskons. Liebigs Ann. Chem. **512**, 164 (1934).
177. ZOBRIST, F. u. H. SCHINZ: Zur Kenntnis der Sesquiterpene. (86. Mitt.) Die Cyclisation von Farnesyliden-aceton. Helv. chim. Acta **32**, 1192 (1949).

(Reçu le 7 novembre 1949.)

Vorkommen und biochemisches Verhalten der Chinone.

Von O. HOFFMANN-OSTENHOF, Wien.

Mit 1 Textabbildung.

Inhaltsübersicht.

Einleitung.

Die vorliegende Übersicht stellt zum Teil die Erweiterung eines vor nur wenigen Jahren erschienenen Referates über dasselbe Gebiet dar (*129*). Seit damals ist aber eine erfreulich große Anzahl wertvoller Arbeiten über verschiedene Aspekte der Biochemie der Chinone erschienen, so daß auf manchen Teilgebieten eine völlige Neugruppierung des Stoffes erforderlich war.

Die Chinone sind eine altbekannte Gruppe organischer Substanzen. Der Grundkörper der Reihe, das *p*-Benzochinon, wurde 1838 von WOSKRESENSKY, einem Schüler von WÖHLER, in dessen Laboratorium erstmalig synthetisch erhalten. Seither konnte nachgewiesen werden, daß einer großen Anzahl von Naturstoffen Chinonstruktur zukommt. Bakterien, Pilze, Flechten, höhere Pflanzen und Tiere sind imstande, Chinone verschiedenster Konstitution zu synthetisieren, und manchen dieser Stoffe kommt eine nicht geringe biologische Bedeutung zu. Selbst im Mineralreich konnten chinoide Substanzen aufgefunden werden.

Im ersten Teil der vorliegenden Zusammenfassung sollen die natürlich vorkommenden Chinone besprochen werden. Der zweite Abschnitt bringt eine Darstellung der biologischen Wirkungen der natürlichen und synthetischen Chinone; es handelt sich hier vielfach um Probleme, welche gerade in der letzten Zeit in verschiedenen Laboratorien eingehend bearbeitet wurden. Dabei soll auch versucht werden, die bisher gewonnenen Erkenntnisse über den Mechanismus der biologischen Wirkungen der Chinone entsprechend abzuhandeln.

Um diese Übersicht nicht allzu ausgedehnt zu gestalten, wird auf eine ausführliche Darstellung der Chemie und der Physiologie der vitamin-K-aktiven Substanzen verzichtet, da ja hier eine Anzahl ausgezeichneter Referate zur Verfügung steht (*67, 68, 77, 159, 281*). Ebenso wird das Gebiet der bei der Oxydation verschiedener Phenole, Phenylalanin, Tyrosin, Dioxyphenylalanin und Adrenalin entstehenden chinoiden Zwischenprodukte vom Typus des Hallachroms, des Adrenochroms und anderer Orthochinone nur kurz besprochen und im übrigen auf vorhandene Übersichtsreferate und Arbeiten verwiesen (*59, 60, 220* bis *225, 243, 276*).

I. Natürlich vorkommende Chinone.

Das Vorkommen von Chinonen im Mineralreich.

Im Mineralreich konnten chinoide Substanzen bisher nur in zwei Fällen nachgewiesen werden; bei beiden handelt es sich um Anthrachinone. 1923 berichteten WERENSKIOLD und OFTEDAL (*319*) über die Auffindung von 9,10-Anthrachinon in der Asche eines seit vielen Jahren brennenden Kohlenflözes in Spitzbergen. Es handelt sich hier also um eine rezente Bildung. Dem Anthrachinon wurde der mineralogische Name *Hoelit* gegeben.

Ein zweites Vorkommen von Chinonderivaten wurde von TREIBS und STEINMETZ (*308*) wahrscheinlich gemacht. Im Tonschiefer kommen in sehr geringer Menge Farbstoffe vor, welchen auf Grund verschiedener physikalischer und analytischer Untersuchungen nach Ansicht der genannten Verfasser die Struktur von alkylierten Polyoxyanthrachinonen zukommt. Es handelt sich hier um mindestens zwei Substanzen, welchen zu Ehren von C. Graebe, dem Bahnbrecher auf dem Gebiete der Anthracen- und Anthrachinonchemie, die Namen *Graebeit a* und *Graebeit b* gegeben wurde. Graebeit *a* hat die Bruttoformel $C_{18}H_{14}O_8$ oder $C_{17}H_{14}O_8$ und dürfte ein substituiertes Hexaoxyanthrachinon sein.

Chinone bakteriellen Ursprungs.

Zahlreichen Bakterienarten kommt die Fähigkeit zu, Chinone zu bilden. Unter diesen Substanzen ist das Vitamin K_2 von überragender physiologischer Bedeutung.

Schon 1899 berichtete Beijerinck (*31*) über die Bildung eines Chinons durch einen *Streptothrix chromogena* oder auch *S. humifica* genannten Mikroorganismus des Bodens. Beijerinck nahm an, daß es sich um Benzochinon handelt; als Nachweis hierfür wurde allerdings einzig die Fähigkeit dieser Substanz verwendet, Gelatine bei Lichtzutritt unlöslich zu machen. Aus dieser Arbeit ist daher keineswegs zu entnehmen, um welches Chinon es sich in Wirklichkeit gehandelt hat. Aus seiner Beobachtung zog Beijerinck weitgehende Schlüsse über den Mechanismus der Humusbildung im Boden, für welche er die chinonbildenden Bodenbakterien verantwortlich machte. Gegen diese Auffassungen wurden aber bald darauf von Furuta (*99*) berechtigte Gegengründe, so insbesondere die Giftigkeit der Chinone gegenüber verschiedenen Lebewesen, geltend gemacht.

Kluyver und Mitarbeiter (*169*) beobachteten 1939 die Bildung von Calcium- und Magnesiumsalzen des Tetraoxy-*p*-benzochinons (I) durch *Pseudomonas beijerinckii* Hof. Das Pigment wird aus *meso*-Inosit oder auch aus *l*-Inosit im schwach alkalischen Medium bei beschränkter Belüftung gebildet.

Als erstes Naphthochinon bakteriellen Ursprungs wurde 1933 das Phthiocol (II), $C_{11}H_8O_2$ (gelbe Prismen, F. 173°), aus menschlichen Tuberkelbazillen isoliert [Andersen und Newman (*3, 4*)].

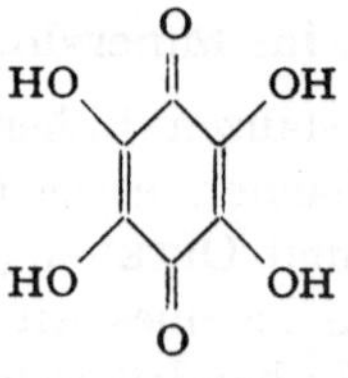

(I.) Tetraoxy-*p*-benzochinon. 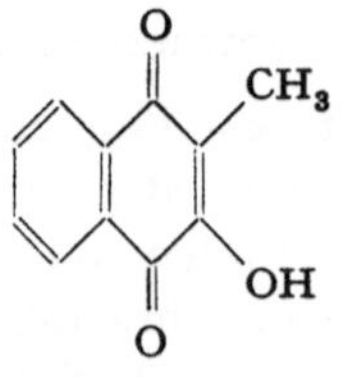 (II.) Phthiocol.

Ebenfalls ein Naphthochinon-Derivat ist das Vitamin K_2 (III), welches in zahlreichen Mikroorganismen (darunter *Bacillus subtilis*, *B. mycoides*, *B. cereus*, *Sarcina lutea*, *Proteus vulgaris*, *Staphylococcus aureus, Mycobacterium tuberculosis, Serratia marcescens, Escherichia coli* und *Eberthella typhosa*) nachgewiesen wurde.

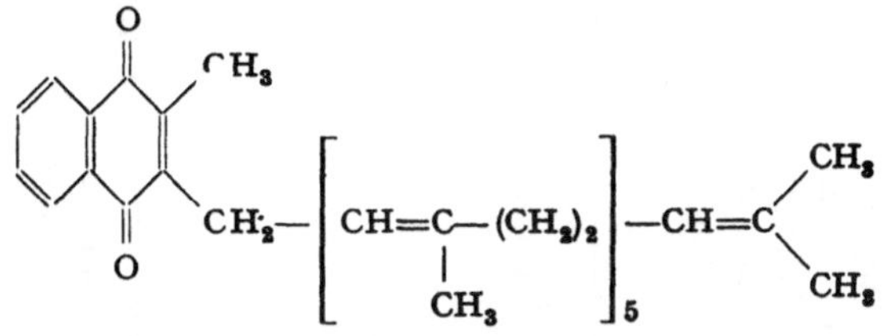

(III.) Vitamin K_2.

In letzter Zeit berichtete EVANS (*85*) über die Bildung von Orthochinonen durch verschiedene phenolspaltende *Bodenbakterien*. So entsteht aus Phenol und aus Brenzcatechin *o*-Benzochinon (IV) als Zwischenprodukt, worauf im weiteren Verlauf der Reaktion der Ring oxydativ aufgespalten wird. Völlig gleichartig verläuft die Oxydation der Benzoesäure durch dieselben Organismen, wobei Protocatechusäure und das entsprechende *o*-Chinon (V) als Zwischenprodukte nachweisbar sind. Diese Vorgänge — oder zumindest deren erste Stufen — werden scheinbar durch Enzyme katalysiert, welche den Phenoloxydasen vom Typus der Tyrosinase nahe verwandt sind.

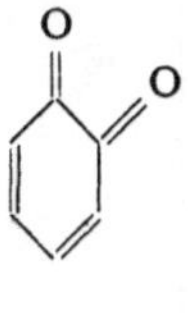

(IV.) *o*-Benzochinon.

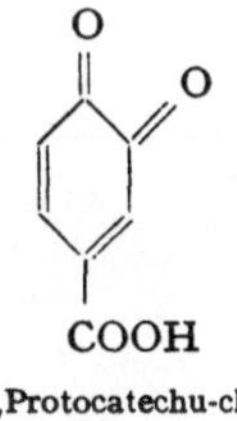

(V.) „Protocatechu-chinon".

Durch Pilze erzeugte Chinone.

Eine erstaunlich große Anzahl von Chinonen verschiedenster Konstitution konnte als End- oder Zwischenprodukte des Stoffwechsels vieler Pilze vorgefunden werden. Die Erforschung dieser Substanzen ist besonders in den Laboratorien von RAISTRICK in London und von KÖGL in München (später Utrecht) durchgeführt worden. Der erstgenannte Autor hat sich vorzüglich mit niedrigeren Pilzen befaßt, wohingegen KÖGL hauptsächlich die Erforschung der Pigmente von höher organisierten Pilzen zu verdanken ist. Über die verschiedenen Produkte des Pilzstoffwechsels, darunter auch die an dieser Stelle zu besprechenden Chinone, sind einige Übersichtsreferate erschienen (*40, 103, 270, 271, 283*).

In Tabelle 1 (S. 158 ff.) sind die von den verschiedenen Pilzen erzeugten Chinone verzeichnet, soweit sie in ihrer Konstitution zumindest teilweise erkannt wurden.

Tabelle 1. Chinoide Pilzpigmente (*A.* = *Aspergillus*, *H.* = *Helminthosporium*, *F.* = *Fusarium* und *P.* = *Penicillium*).

Trivialname und Bruttoformel	Konstitutionsformel	Synthese	F. °	Farbe	Vorkommen	Literaturzitat
Gentisinchinon (VI) $C_7H_6O_3$	O; O; CH_2OH	(*45*)	75—76	gelb	Als chinhydron-artige Molekülverbindung mit Gentisylalkohol (1 Mol Chinon + 3 Mol Alkohol) im Kulturfiltrat von *P. urticae* Bainier	(*84*)
Fumigatin (VII) $C_8H_8O_4$	O; H_3C; OH; OCH_3; O	(*25, 264*)	116	kastanien-braun	*A. fumigatus* Fresenius	(*7*)
Spinulosin (VIII) $C_8H_8O_5$	O; H_3C; OH; HO; OCH_3; O	(*8*)	201	dunkel-purpur	*P. spinulosum* Thom.; *P. cinerascens* Biourge; ein Stamm von *A. fumigatus*	(*7, 9, 46*)

Citrinin (IX) $C_{12}H_{12}O_5$	O HC CH · CH_3 HO CH · CH_3 HOOC O	(*55*)	175 (Zers.)	goldgelb	*P. citrinum* THOM. und verwandte Arten	(*53, 55, 63, 64, 65, 119*)
Phoenicin (X) $C_{14}H_{10}O_6$	O O H_3C OH HO CH_3 O O	(*262, 183*)	230—231 (Zers.)	rot	*P. phoeniceum* VAN BEYMA; *P. rubrum* GRASSBERGER	(*95, 262*)
Oosporein (XI) $C_{14}H_{10}O_8$	O O H_3C OH HO CH_3 HO OH O O	(*183*)	—	mennige-farbig	*Oospora colorans* VAN BEYMA	(*183*)
Polyporsäure (XII) $C_{18}H_{12}O_4$	O HO OH O	(*86*)	280 (Zers.)	braun-violett	*Polyporus nidulans*	(*171*)

Fortsetzung s. S. 160

Fortsetzung der Tabelle 1.

Trivialname und Bruttoformel	Konstitutionsformel	Synthese	F. °	Farbe	Vorkommen	Literaturzitat
Atromentin (XIII) $C_{18}H_{12}O_6$	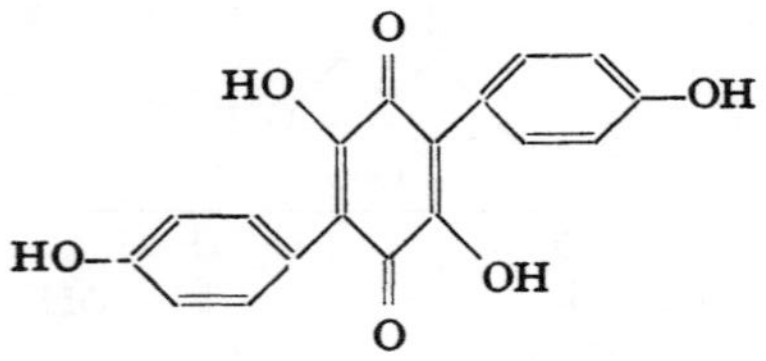 O, HO, OH, HO, OH, O	(*172*)	—	braun, metallisch glänzend	*Paxillus atromentosus*	(*170, 172, 173, 180*)
Muscaruffin (XIV) $C_{25}H_{16}O_9$	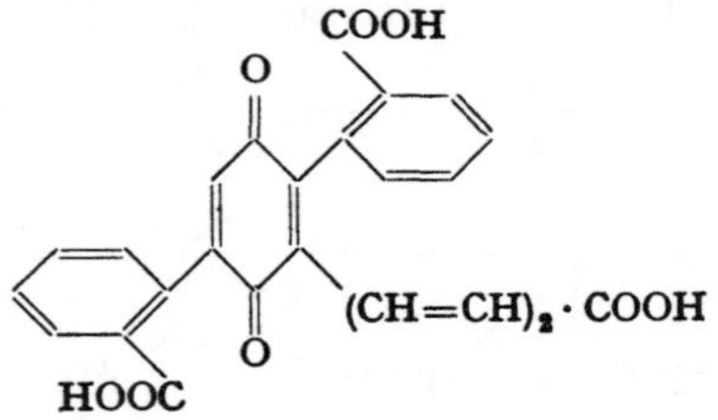 COOH, O, $(CH{=}CH)_2 \cdot COOH$, O, HOOC	—	—	orangerot	*Amanita muscaria* (Fliegenpilz)	(*176*)
Gentisinsäure (XV) $C_7H_6O_4$	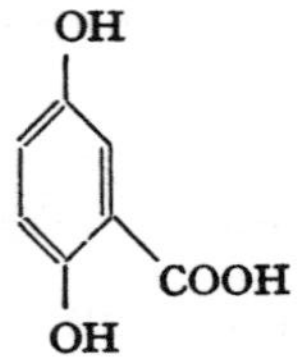 OH, COOH, OH	(*291*)	200	farblos	*P. griseo-fulvum*	(*275*)

Gentisinalkohol (XVI) $C_7H_8O_3$	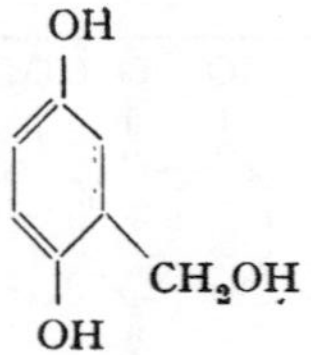 OH; CH_2OH; OH	(*45*)	100	farblos	*P. patulum* Bainier; in *P. urticae* Bainier als Mol.-Verbindung mit Gentisinchinon (siehe VI)	(*41, 45, 84*)
Auroglaucin (XVII) $C_{19}H_{28}O_3$	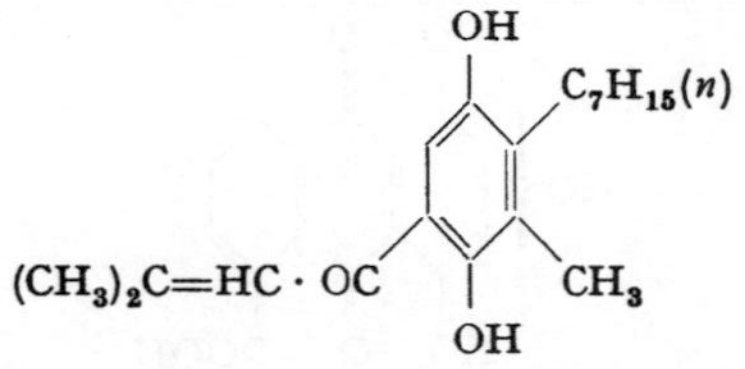 OH; $C_7H_{15}(n)$; $(CH_3)_2C{=}HC \cdot OC$; CH_3; OH	—	153	goldgelb	*A. glaucus* *A. echinulatus*	(*104, 267, 268*)
Flavoglaucin (XVIII) $C_{19}H_{22}O_3$	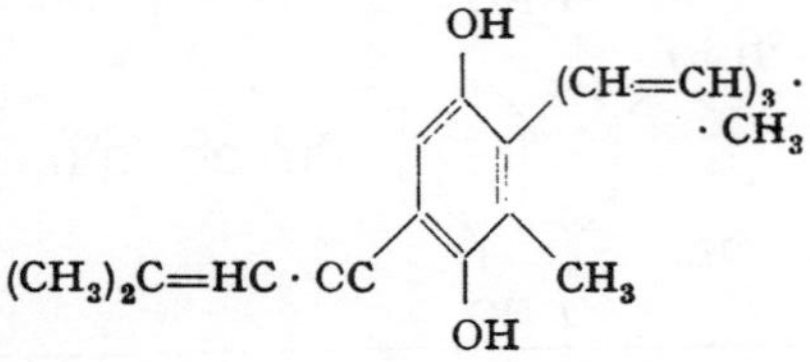 OH; $(CH{=}CH)_3 \cdot \cdot CH_3$; $(CH_3)_2C{=}HC \cdot CC$; CH_3; OH	—	109—110	hellgelb	*A. glaucus*, *A. echinulatus*	(*104, 267, 268*)
Javanicin (XIX) $C_{15}H_{14}O_6$	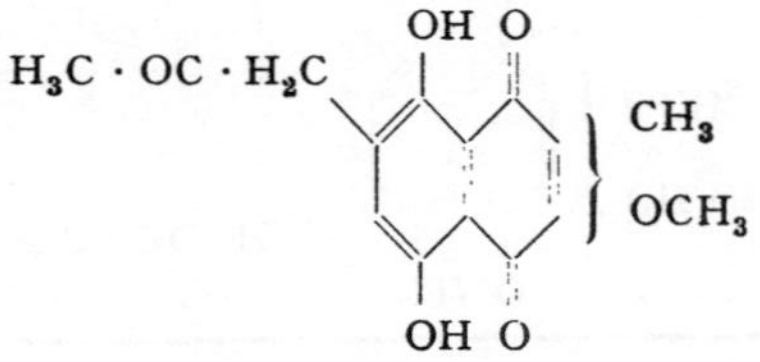 OH O; $H_3C \cdot OC \cdot H_2C$; $\left.\begin{matrix} CH_3 \\ OCH_3 \end{matrix}\right\}$; OH O	—	208	rot	*F. javanicum*	(*15, 16*)

Fortsetzung s. S. 162

Fortsetzung der Tabelle 1.

Trivialname und Bruttoformel	Konstitutionsformel	Synthese	F. °	Farbe	Vorkommen	Literaturzitat
Oxyjavanicin (XX) $C_{15}H_{14}O_7$	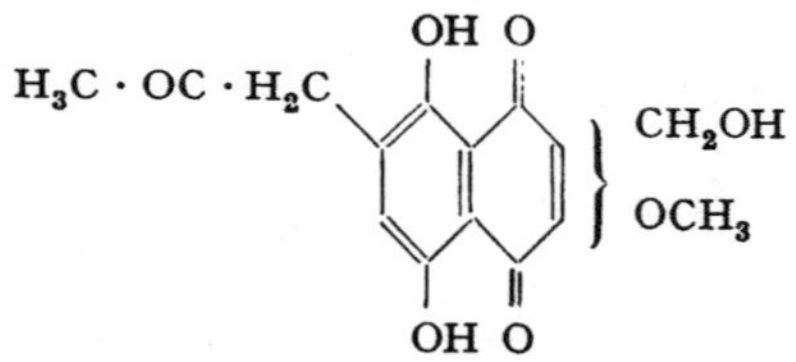 OH O; $H_3C \cdot OC \cdot H_2C$; CH_2OH; OCH_3; OH O	—	178	rot	*F. javanicum*	(*15, 16*)
Solanion (XXI) $C_{15}H_{14}O_6$	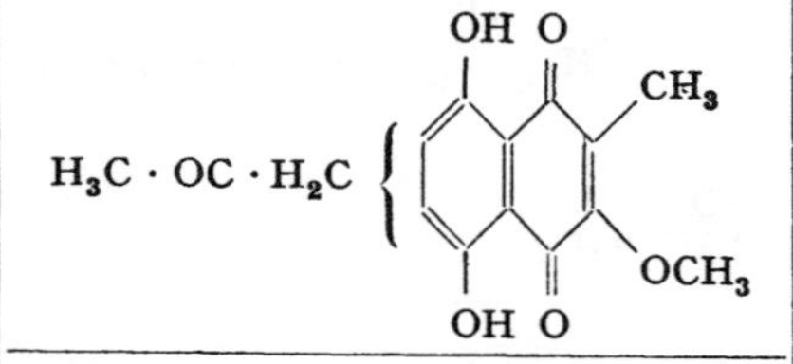 OH O; CH_3; $H_3C \cdot OC \cdot H_2C$; OCH_3; OH O	—	208	orange-gelb	*F. solani* D_2 purple	(*317*)
Boletol (XXII) $C_{15}H_8O_7$	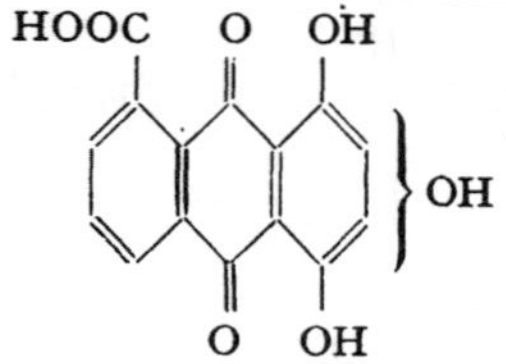 HOOC O OH; OH; O OH	(*175*)	275—280 (Verkohlung)	rot	*Boletus*-Arten	(*174, 175*), bezüglich des entsprechenden Dichinons vgl. S. 168
Emodinsäure (XXIII) $C_{15}H_8O_7$	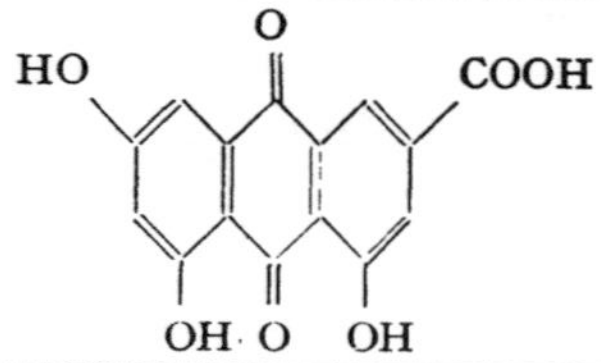 O; HO; COOH; OH O OH	(*6*)	363—365	orange	*P. cyclopium* WESTLING	(*6*)

Chrysophansäure (XXIII a) $C_{15}H_{10}O_4$	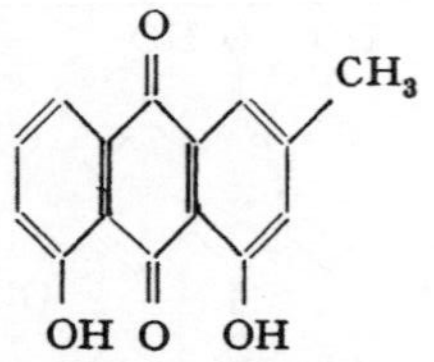	(*81*, *82*)	169	goldgelb	*P. islandicum* SOPP.	(*153 a*)
Helminthosporin (XXIV) $C_{15}H_{10}O_5$	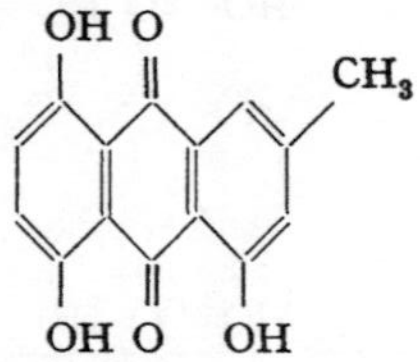	(*273*)	226—227	rotbraun	*H. gramineum* RABENHORST; *H. cynodontis* MARIGNONI; *H. catenarium* DRECHSLER; andere *H.*-Arten	(*56*, *270*, *274*)
Frangula-emodin (XXV) $C_{15}H_{10}O_5$	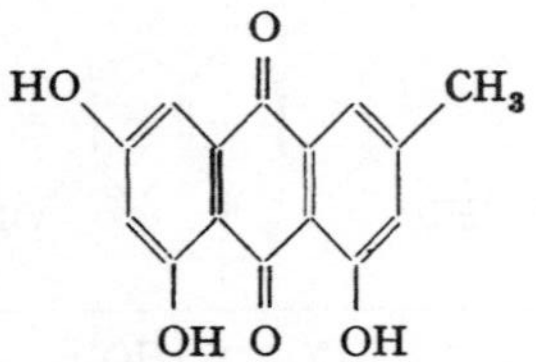	(*83*)	253—254	orange-gelb	*Dermocybe sanguinea*	(*79*)
Catenarin (XXVI) $C_{15}H_{10}O_6$	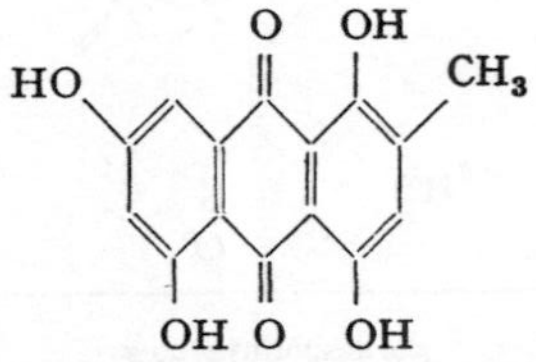	(*12*)	246	rot	*H. catenarium*, *H. gramineum*, *H. velutinum* LINK, *H. tritici-vulgaris* NISIKADO	(*10*, *12*, *56*, *274*)

Fortsetzung s. S. 164

Fortsetzung der Tabelle 1.

Trivialname und Bruttoformel	Konstitutionsformel	Synthese	F. °	Farbe	Vorkommen	Literaturzitat
Islandicin (XXVII) $C_{15}H_{10}O_5$	O OH CH_3 OH O OH	—	218	dunkel-rot	*P. islandicum* SOPP.	(*153 a*)
ω-Oxyemodin (Citro-rosein) (XXVIII) $C_{15}H_{10}O_6$	O HO CH_2OH OH O OH	—	288	orange	*P. cyclopium, P. citro-roseum* DIERCKX	(*6, 264*)
Cynodontin (XXIX) $C_{15}H_{11}O_6$	OH O OH CH_3 OH O OH	(*11*)	224—225	braun	*H. cynodontis, H. euchle-nae* ZIMMERMANN	(*11, 272, 273, 274*)
Physcion (XXX) $C_{16}H_{12}O_5$	O H_3CO CH_3 OH O OH	(*83*)	206—207	orange	*A. glaucus*-Arten	(*22*)

Erythroglaucin (XXXI) $C_{16}H_{12}O_6$	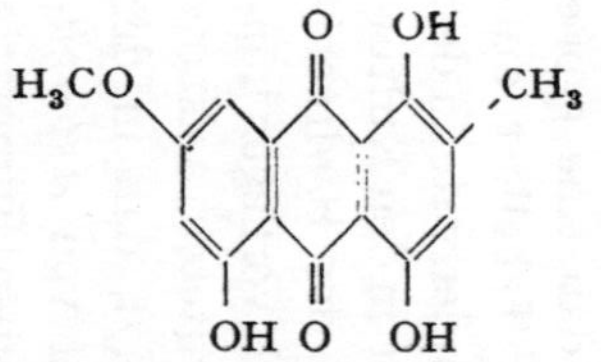 O OH; H_3CO; CH_3; OH O OH	(*10*)	205—206	rot	*A. glaucus*-Arten, *A. ruber*	(*10, 22, 269*)
Tritisporin (XXXII) $C_{15}H_{10}O_7$	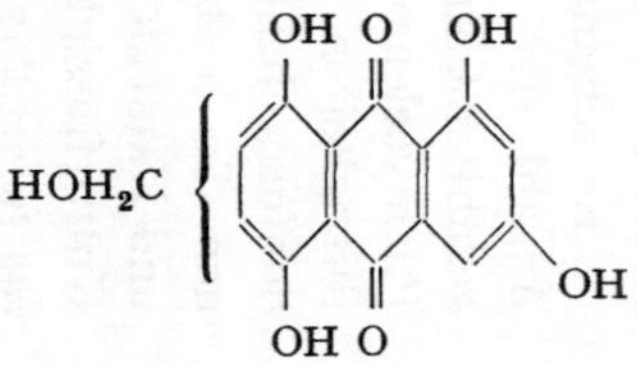 OH O OH; HOH_2C {; OH; OH O	—	260—262	rotbraun	*H. tritici-vulgaris*	(*274*)
Carviolin (XXXIII) $C_{16}H_{12}O_6$	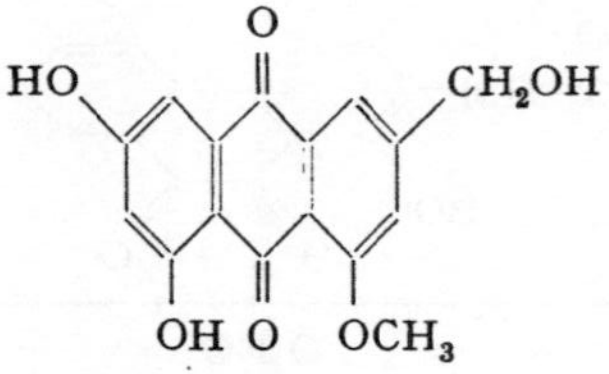 O; HO; CH_2OH; OH O OCH_3	—	286	gelb	*P. carmino-violaceum* BIOURGE	(*124, 125*)
Dermocybin (XXXIV) $C_{16}H_{12}O_7$	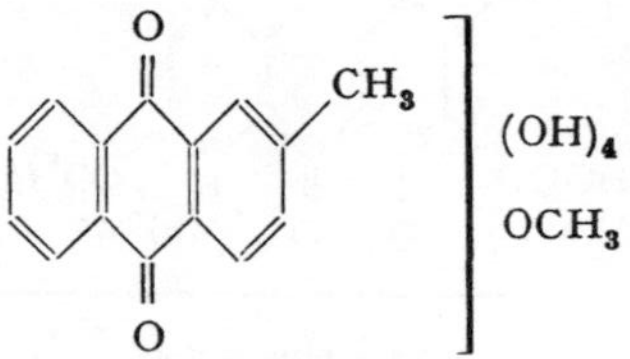 O; CH_3; O; $(OH)_4$ OCH_3	—	228	rot	*Dermocybe sanguinea*	(*179*)

Fortsetzung s. S. 166

Fortsetzung der Tabelle 1.

Trivialname und Bruttoformel	Konstitutionsformel	Synthese	F. °	Farbe	Vorkommen	Literaturzitat
Phomazarin (XXXV) $C_{19}H_{17}O_8N$	H_3CO, $H_9(n)C_4$, O, O, O, COOH, N, OH, OH O	—	179 (Zers.)	orange-gelb	*Phoma terrestris*	(*181, 184, 185*)
Telephorsäure (XXXVI) $C_{20}H_{12}O_9$	HO, HOOC, O, O, OH, OH, $(CH{=}CH)_2 \cdot COOH$	—	> 350	schwarz-violett	*Telephora*-Arten *Hydnum ferrugineum*	(*178*)

Die an zweiter und dritter Stelle in Tabelle 1 stehenden *Toluchinon*-Derivate Fumigatin (VII)* und Spinulosin (VIII), welche verhältnismäßig einfache Verbindungen sind, unterscheiden sich voneinander nur dadurch, daß Spinulosin eine Hydroxylgruppe mehr aufweist als das Fumigatin. Die enge Verwandtschaft dieser Substanzen wird auch biologisch durch die Tatsache nahegebracht, daß unter den gewöhnlich Fumigatin produzierenden Stämmen von *Aspergillus fumigatus* Fresenius es auch solche gibt, welche statt dessen Spinulosin bilden. Den beiden genannten Pigmenten stehen die erst später entdeckten Derivate des 5,5'-Dimethyldibenzochinons, das Phoenicin (X) und das Oosporein (XI), sehr nahe, wie aus den Formeln leicht ersichtlich ist.

Es scheint, daß eine große Anzahl der in Tabelle 1 aufgezählten Chinonderivate in den Pilzmycelen oder in den Kulturflüssigkeiten in der Hydrochinonform ebenfalls vorliegen können. So beobachteten Anslow und Raistrick (*7*), daß in der Kulturflüssigkeit von *Aspergillus fumigatus* neben Fumigatin auch das entsprechende Hydrochinon nachzuweisen ist; ebenso findet sich nach Posternak (*264*) in Kulturen von *Penicillium rubrum* auf bestimmten

* Die Strukturformeln (VI) bis (XXXVI) sind in Tabelle 1, Seite 158 bis 166, enthalten.

Nährböden — besonders Bierwürze — hauptsächlich Tetrahydro-phoenicin neben wenig Phoenicin.

Eine sehr interessante Beobachtung in dieser Richtung machten kürzlich ENGEL und BRZESKI (*84*), welche aus *Penicillium urticae* eine chinhydronartige Verbindung isolieren konnten, die aus einem Molekül Gentisinchinon (VI) und drei Molekülen Gentisinalkohol (XVI) bestand, wogegen BIRKINSHAW und Mitarbeiter (*41*) aus *P. patulum* allein den Gentisinalkohol gewannen.

Es ist jedenfalls anzunehmen, daß die Chinone und die entsprechenden Hydrochinone sich unter biologischen Bedingungen leicht ineinander umwandeln können. Auf Grund dieser Überlegungen sind in Tabelle 1 auch einige Substanzen aufgenommen worden, welche bisher nur in der Hydrochinonform nachgewiesen werden konnten. Die Struktur der aus *Aspergillus*-Arten isolierten substituierten Hydrochinone, Flavoglaucin (XVII) und Auroglaucin (XVIII) wurde erst in allerletzter Zeit aufgeklärt (*268*).

Aus höheren Pilzen konnte KÖGL einige interessante Pigmente isolieren, welche dem Spinulosin insoweit verwandt sind, als sie sich durch die gemeinsame Formel (XXXVII) ausdrücken lassen, wobei im Spinulo-

O
R_1 OH
HO R_2
O

(XXXVII.)

sin R_1 eine Methyl- und R_2 eine Methoxygruppe ist, während in der Polyporsäure (aus *Polyporus nidulans*) R_1 und R_2 Phenylreste, im Atromentin (aus *Paxillus atromentosus*) R_1 und R_2 *p*-Oxyphenylreste sind. In dieselbe gemeinsame Formulierung lassen sich übrigens auch die später zu besprechenden pflanzlichen Farbstoffe Embelin und Maesachinon einordnen (S. 181). Das Muscaruffin, der Farbstoff des Fliegenpilzes, ist hingegen als Diphenylbenzochinon-derivat mit Polyporsäure und Atromentin wohl verwandt, zeigt aber einen komplizierteren Bau.

Derivate der *Naphthochinone* als Stoffwechselprodukte von Pilzen sind bisher in geringer Zahl bekannt geworden. Hier scheint es einzig die Gruppe der *Fusaria* zu sein, welche derartige Substanzen, Javanicin (XIX), Oxyjavanicin (XX) und Solanion (XXI) bildet (S. 161 und 162).

Hingegen sind wiederum Abkömmlinge des *2-Methylanthrachinons* unter den Pilzprodukten besonders häufig. Solche Stoffe finden sich insbesondere bei Getreideparasiten des Genus *Helminthosporium*, weiters bei den Genera *Penicillium* und *Aspergillus*, sowie auch in höheren

Pilzen. Bemerkenswert ist es, daß oftmals dieselben Stoffe oder sehr nahe verwandte Substanzen sowohl in Pilzen als auch in höheren Pflanzen nachgewiesen werden konnten. So findet sich das Frangula-emodin (XXV), ein Produkt verschiedener *Polygonaceae, Leguminosae* und *Rhamnaceae*, nach Kögl (*179*) auch im Pilz *Dermocybe sanguinea*; sein Methyläther, das Physcion (XXX), kommt sowohl frei als auch als Glykosid in *Rumex*-Arten vor, weiters in der Kulturflüssigkeit von *Aspergillus glaucus*-Arten und schließlich als Bestandteil verschiedener Flechten.

Eine interessante Eigenschaft zeigt das bereits 1902 von Bertrand aus *Boletus satanas* isolierte und von Kögl in seiner Struktur weitgehend aufgeklärte Boletol (XXII). Es geht unter dem Einfluß des Luftsauerstoffs und einer im Pilz enthaltenen Oxydase in ein entsprechendes Dichinon (XXXVIII) über. Das letztere ist imstande, blaue Salze zu bilden, wodurch sich die Blaufärbung des Pilzgewebes an der Luft erklären läßt.

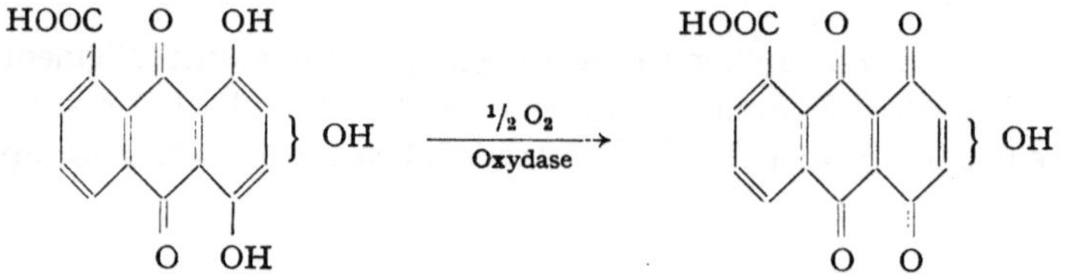

HOOC O OH ... } OH —$^1/_2$ O_2, Oxydase→ HOOC O O ... } OH

(XXII.) Boletol. (XXXVIII.) Dichinon.

In der Literatur findet sich noch eine Reihe von Berichten über aus Pilzen und deren Kulturmedien isolierte Substanzen, welche mit einiger Wahrscheinlichkeit ebenfalls der Anthrachinonreihe angehören, deren Konstitution aber noch nicht mit ausreichender Genauigkeit bestimmt ist, um eine Einordnung in Tabelle 1 (S. 158 ff.) zu rechtfertigen.

So beschreibt Igaraci (*156*) eine Substanz, welcher er den Namen Funiculosin gibt. Funiculosin wurde aus Kulturen von *Penicillium funiculosum* isoliert; es hat die Bruttoformel $C_{15}H_{10}O_5$, kristallisiert in tiefroten Blättchen (F. 218°), enthält keine Methoxygruppen und gibt ein Triacetat sowie ein Tribenzoat. Bei Zinkstaubdestillation entsteht Anthracen, so daß die Annahme des Entdeckers, daß es sich hier entweder um ein Trioxymethylanthrachinon oder um ein Oxymethyldioxyanthrachinon handelt, recht unwahrscheinlich erscheint. In letzter Zeit berichten nun Howard und Raistrick (*153*), daß das von ihnen aus *Penicillium islandicum* Sopp isolierte Anthrachinonderivat, welches unter dem Namen Islandicin in Tabelle 1 aufgenommen wurde, in seinen Eigenschaften weitgehend mit denjenigen, welche beim Funiculosin berichtet wurden, übereinstimmt. Da auch die beiden Stämme *Penicillium funiculosum* und *P. islandicum* nahe verwandt sind, besteht einige Wahrscheinlichkeit, daß die beiden Pigmente identisch sind.

Neben dem in seiner Struktur aufgeklärten Carviolin konnte Hind (*124*) aus Kulturen von *Penicillium carmino-violaceum* noch ein zweites Derivat des 2-Methylanthrachinons isolieren, welchem die Bruttoformel $C_{20}H_{16}O_7$ zukommt und dem der Name Carviolacin gegeben wurde. Das Produkt bildet lichtbraune Nadeln, welche bei 243° unter Zersetzung schmelzen; es gibt ein Triacetat und einen Trimethyläther und geht bei Zinkstaubdestillation in 2-Methylanthracen über. Weitere Erkenntnisse über die Konstitution des Carviolacins scheinen noch nicht vorzuliegen.

Ein Pilzpigment, welches nur mit Einschränkungen unter die Anthrachinonderivate eingereiht werden darf, ist das Phomazarin (XXXV), der Farbstoff von *Phoma terrestris*, das erst in letzter Zeit von KÖGL (*184*, *185*) in seiner Struktur aufgeklärt wurde. Es handelt sich hier um ein Azanthrachinon, also eine Substanz, in welcher ein α-ständiges Kohlenstoffatom in einem benzoiden Kern des Anthrachinonskeletts durch ein Stickstoffatom ersetzt ist. Auf Grund seiner Konstitution könnte das Phomazarin auch als Alkaloid angesprochen werden.

Der Gruppe der *Phenanthrenchinone* kommt, trotz der weiten Verbreitung von Phenanthrenabkömmlingen in der Natur, anscheinend keine besondere Bedeutung zu. Bisher ist nur ein einziges chinoides Pigment aus Pilzen mit Sicherheit als ein Derivat des Phenanthrenchinons erkannt worden. Es ist dies die Telephorsäure (XXXVI), welche übrigens auch in Flechten vorkommt. Von einem zweiten Pigment, welches von KÖGL (*177*, *182*) aus dem Pilz *Peziza aeroginosa* kristallin erhalten wurde, ist ebenfalls anzunehmen, daß es zu den Phenanthrenchinonen gehört, allerdings wahrscheinlich nicht zu den 9,10-Phenanthrenchinonen (wie die Telephorsäure), sondern zu dem in der Natur sonst noch nie festgestellten Typus des 2,7-Phenanthrenchinons.

O= ⟨Strukturformel⟩ =O

(XXXIX.) 2.7-Phenanthrenchinon.

Eine Anzahl weiterer Substanzen, welche als Chinone erkannt wurden, deren Zugehörigkeit zu einem bestimmten Typus aber noch nicht mit Sicherheit festgestellt werden konnte, soll hier anhangsweise besprochen werden.

Aus dem Mycel von *Helminthosporium leersii* konnten ASHLEY und RAISTRICK (*21*) zwei Stoffwechselprodukte isolieren, das gelbe Luteoleersin, $C_{26}H_{38}O_7$ (F. 135°), und das Alboleersin, $C_{26}H_{40}O_7$ (F. 215 bis 216°); farblose Nadeln. Das eine ist wahrscheinlich ein höheres Chinon und das andere das entsprechende Hydrochinon.

Penicilliopsin, das anscheinend mit den Anthrachinonkörpern nahe verwandt ist, wurde von OXFORD und RAISTRICK (*251*) aus *Penicilliopsis clavariaeformis* SOLMS-LAUBACH gewonnen. Es schmilzt bei 330°, hat die Bruttoformel $C_{30}H_{24}O_8$, kristallisiert in orangegelben, winzigen Nadeln, gibt ein Diacetat und ein *bis*-Phenylcarbaminat. Bei Erhitzen wird Frangula-emodin-anthranol gebildet; durch Zinkstaubdestillation in Wasserstoffatmosphäre entsteht 2-Methylanthracen. Bei Oxydation in Piperidinlösung wird ein dunkelviolettes Produkt „Oxypenicilliopsin" ($C_{30}H_{20}O_9$) gebildet. Dessen Farbreaktionen sind von denjenigen der Ausgangssubstanz völlig verschieden. Bei Belichtung des Oxypenicilliopsins scheint eine Isomerisierung einzutreten. Der resultierende Stoff („bestrahltes Oxypenicilliopsin"), dem die gleiche empirische Formel zukommt wie dem Oxypenicilliopsin, zeigt auffallende Ähnlichkeiten mit Hypericin (S. 191), wenngleich eine Identität ausgeschlossen werden muß.

Schließlich konnte vor kurzer Zeit Michael (231) ein kristallines Produkt ($C_{15}H_{16}O_7$) aus der Kulturflüssigkeit von *Oidiodendron fuscum* Rubak isolieren. Die Substanz, Fuscin genannt, kristallisiert in orangegelben Blättchen (F. 230°), und wird in den Kulturfiltraten von seinem Dihydroprodukt, dem Dihydrofuscin, begleitet. Das Verhalten des Fuscins deutet mit einiger Wahrscheinlichkeit auf eine chinoide Struktur, allerdings ohne daß weitere Aussagen gemacht werden können.

Wenn man nun rückblickend die große Zahl der von Pilzen produzierten Chinone betrachtet, so ist es bemerkenswert, daß wir über die chemische Konstitution der meisten dieser Substanzen recht genaue Kenntnis besitzen, aber anderseits noch sehr wenig über ihre Bildungsweise in den Organismen und über ihre biologischen Funktionen wissen.

Die Biosynthese von Pilzchinonen.

Was die *Biosynthese* dieser Produkte betrifft, soll hier vor allem bemerkt werden, daß die vielen von Raistrick und Mitarbeitern aus niedrigen Pilzen isolierten Substanzen durchwegs von Organismen erzeugt werden, welche auf künstlichen Nährböden, deren einzige Kohlenstoffquelle Glucose ist, gewachsen sind. Diese Pilze müssen also Mechanismen besitzen, welche es ihnen ermöglichen, komplizierte und Zuckern gar nicht mehr ähnliche Substanzen aus Glucose aufzubauen.

Vor einigen Jahren hat nun Tatum (302) auf Grund genetischer Überlegungen eine Theorie der Biosynthese der pilzlichen Stoffwechselprodukte aufgestellt. Er stellte fest, daß die meisten, bisher aus Pilzen isolierten Substanzen als Derivate einer ziemlich beschränkten Anzahl von chemischen Grundkörpern erkannt wurden und schließt daraus, daß die verschiedenen vorgefundenen Ringsysteme ihre Herkunft in gemeinsamen strukturellen Einheiten haben. Auf Grund der an dem Brotpilz *Neurospora crassa* erhaltenen Erkenntnisse [vgl. z. B. die ausgezeichnete Zusammenfassung von Beadle (29); dort auch weitere Literatur] setzt er voraus, daß auch die Bildung der oben besprochenen Pilzpigmente unter der Lenkungstätigkeit von spezifischen Genen steht; so ist also die Fähigkeit eines bestimmten Stammes oder einer Art, einen solchen Körper zu synthetisieren, durch den Besitz eines besonderen Gens bedingt.

Tatum stellt eine von einfachen Zuckern ausgehende Reaktionsreihe auf, welche mit der wahrscheinlich in mehreren Stufen erfolgenden Bildung aliphatischer Körper aus den Zuckern beginnt. Aus diesen Substanzen entstehen dann durch mannigfaltige Arten von Ringschlüssen weitere Zwischenprodukte und die Endprodukte, wobei jede dieser Reaktionen durch spezifische Gene kontrolliert wird. Geringere Strukturveränderungen, wie sie etwa durch Oxydation oder Methylierung bewirkt werden, können vor oder nach dem Ringschluß erfolgen, aber auch für sie ist die Mitwirkung spezifischer Gene erforderlich.

Aus den primären (aliphatischen) Zwischenprodukten könnten dann entweder noch recht zuckerähnliche Endprodukte, wie Kojisäure, Patulin und Tetronsäurederivate, anderseits aber auch — nach TATUM über eine zweite Gruppe von Zwischenprodukten — benzoide Verbindungen vom Typus *A* oder *B* entstehen (*R*, R_1 und R_2 stellen C_1-Reste dar, wie CH_3, CH_2OH oder COOH).

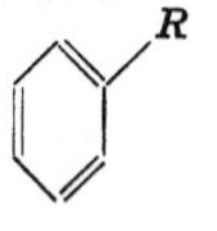

R

R_1

R_2

(XL.) Typus *A*. 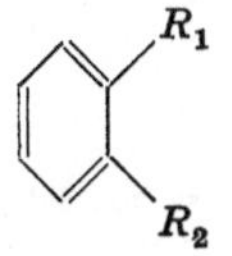 (XLI.) Typus *B*.

Typus *A* wird als Grundstoff der Toluchinonderivate Fumigatin und Spinulosin angenommen; bei Zusammentritt zweier Moleküle vom Typus *A* kann es zur Bildung der Ditoluchinon-Abkömmlinge Phoenicin und Oosporein kommen. Bei einer anderen Art der Vereinigung von zwei Molekülen des Typus *A* mag die Bildung von Benzophenon-Deri-

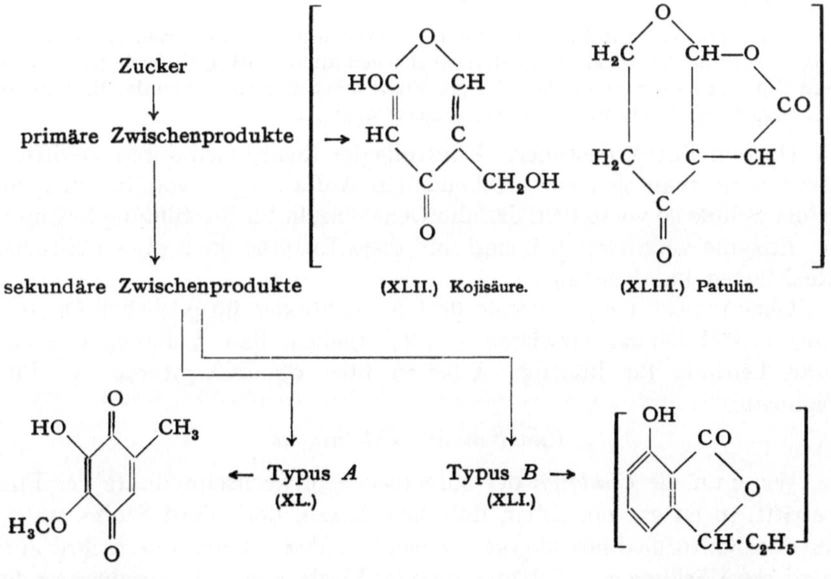

Zucker

↓

primäre Zwischenprodukte → [Kojisäure, Patulin]

↓

sekundäre Zwischenprodukte

O HOC CH HC C C CH_2OH O

(XLII.) Kojisäure.

O H_2C CH—O CO H_2C C—CH C O

(XLIII.) Patulin.

O HO CH_3 H_3CO O

(VII.) Fumigatin.

← Typus *A* (XL.)

Typus *B* → (XLI.)

OH CO O $CH \cdot C_2H_5$

(XLIV.) Mellein.

A + *A* →

O H_3C OH HO O CH_3 O O

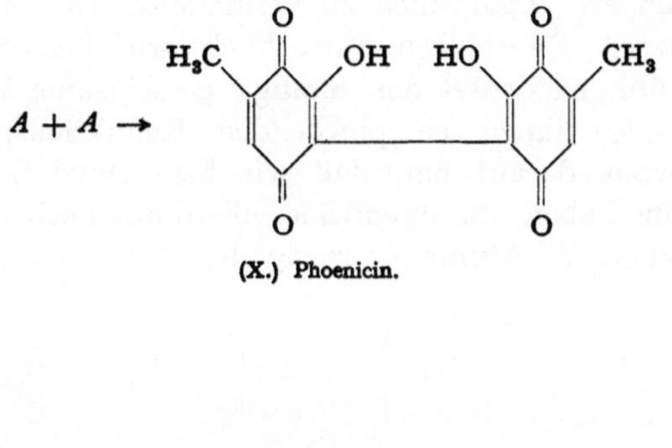

(X.) Phoenicin.

(XLV.) Sulochrin. (XXV.) Frangula-emodin.

↓

(XXVIII.) ω-Oxyemodin, (XXX.) Physcion usw.

Schema der Bildungsweise von Pilzprodukten nach Tatum (*302*) (modifiziert) (s. auch S. 171).

vaten erfolgen, aus denen Abkömmlinge des Xanthons von der Art des Ravenelins in einem weiteren Schritt entstehen könnten.

Bei Zusammentritt von je einem Molekül der Typen *A* und *B* sollten ebenfalls Benzophenon-derivate gebildet werden. In Tatums Schema ist als Zwischenprodukt das von Nishikawa (*243 a*) aus *Oospora sulphurea-ochracea* isolierte Sulochrin eingefügt, aus dem man sich die Bildung von Anthrachinon-derivaten leicht erklären kann.

Es ist möglich, daß die Bildung der Naphthochinonderivate, welche im Schema von Tatum nicht berücksichtigt sind, über einen anderen Weg vor sich geht; Weiss und Nord (*317*) konnten es kürzlich glaubhaft machen, daß eine an dei Biosynthese des Solanions (XXI) beteiligte Stufe Acetaldehyd ist.

Die von Tatum postulierte Kontrolle der Reaktionen durch spezifische Gene kann man sich entsprechend den Auffassungen von Beadle und seiner Schule so vorstellen, daß die Gene einzeln für die Bildung bestimmter Enzyme verantwortlich sind und diese Enzyme dann die spezifischen Reaktionen katalysieren.

Obzwar noch keine der geforderten Reaktionen im pilzlichen Organismus tatsächlich nachgewiesen wurde, ergeben diese Vorstellungen eine gute Leitlinie für künftige Arbeiten über die Biosynthese der Pilzpigmente.

Funktion der Pilzchinone.

Was nun die *Funktion* der chinoiden Stoffwechselprodukte der Pilze betrifft, so ist es vorstellbar, daß eine Anzahl derjenigen Stoffe, welche an das Kulturmedium abgegeben werden, dazu dient, durch ihre antibiotische Wirkung eine Schädigung oder Verdrängung des produzierenden Pilzes durch andere Organismen zu verhindern. In vielen Fällen sind die beschriebenen Substanzen tatsächlich wirkungsvolle Antibiotica (S. 202). Im übrigen deutet das häufige gemeinsame Vorkommen der Chinone mit den ihnen entsprechenden Reduktionsprodukten vom Hydrochinontypus darauf hin, daß wir hier Oxydations-Reduktionssysteme vor uns haben, die irgendeine, allerdings noch keineswegs definierbare Funktion im Atmungsvorgang des Pilzes haben könnten.

Durch Flechten erzeugte Chinone.

Auch aus einer Anzahl von Flechten sind Chinonderivate isoliert worden. Da solche Stoffe in Algen bisher noch niemals nachgewiesen werden konnten, ist es wahrscheinlich die Pilzkomponente der Flechten, welche für die Bildung der Chinone verantwortlich ist. Chinoide Substanzen, die aus Flechten isoliert wurden und deren Konstitution bis zu einem gewissen Grad als aufgeklärt angesehen werden kann, sind in Tab. 2 (S. 174—175) verzeichnet.

Außer den in Tabelle 2 angeführten Substanzen sind noch einige andere Stoffe aus Flechten gewonnen worden, welche mit einiger Wahrscheinlichkeit als Chinone angesprochen werden können. Hierher gehört das Fallacin, welches ASANO und FUZIWARA (*19*) aus *Xanthoria fallax* isolierten. HESSE (*116*, *117*) konnte aus *Neophromium lusitanicum* und aus *Blastenia arenaria* zwei chinonartige Pigmente gewinnen, welche er Neophromin bzw. Blastenin (oder Blasteniasäure) nannte. Neophromin bildet ockerfarbige Nadeln (F. 196° unter Zersetzung); ihm dürfte die Bruttoformel $C_{16}H_{12}O_6$ zukommen. Blastenin schmilzt bei etwa 270° und ist orangerot.

Eine Anzahl weiterer derartiger Stoffe wurde von ZOPF (*335*) beschrieben. Das Rhodophyscin (rote Blättchen, F. 260° unter Zersetzung) und das Endococcin, welches in gelblichgrünen Prismen kristallisiert, lassen sich aus *Physcia endococcina* gewinnen. Aus *Sphaerophorus fragilis* und *Sph. coralloides* isolierte ZOPF das goldgelbe Täfelchen bildende Fragilin. Auch das aus *Haematomma porphyrium* dargestellte Hymenorhodin (*336*) dürfte ein Chinon sein.

Über Bildungsweise und Funktion der chinoiden Flechtenstoffe kann heute noch keine Aussage gemacht werden; die Verhältnisse dürften hier sehr ähnlich wie bei den Pilzen liegen.

Von höheren Pflanzen erzeugte Chinone.

Die größte Anzahl verschiedener Chinonderivate wird durch höhere Pflanzen produziert. Wir finden unter den pflanzlichen Substanzen chinoider Struktur einige geschätzte Farbstoffe, weiters wirksame Bestandteile von Drogen, vitamin-K-aktive Substanzen, und sogar Stoffe, welche Tiere und Menschen gegen Licht sensibilisieren können, wodurch die sogenannten Lichtkrankheiten hervorgerufen werden.

Einige Teile des hier besprochenen Gebietes wurden bereits in verschiedenen Übersichtsreferaten ausführlich behandelt (*80*, *257*, *291a*).

Die in ihrer Konstitution ausreichend charakterisierten Chinone, welche aus höheren Pflanzen isoliert wurden, sind in Tab. 3 (S. 180 bis 191) verzeichnet, die auch die Strukturformeln (XLVIII) bis (LXXXVI) enthält.

Wir finden dort Derivate des Benzochinons, des 1,4-Naphthochinons und des Anthrachinons in großer Anzahl, dagegen konnte bisher nur ein einziges Pigment als Derivat des 1,2-Naphthochinons erkannt werden; ebenso ist die Gruppe der Phenanthrenchinone nur durch einen Ab-

Tabelle 2. Chinoide Flechtenstoffe.

Trivialname und Bruttoformel	Konstitutionsformel	Synthese	F.°	Farbe	Vorkommen	Literaturzitat
Endocrocin (XLVI) $C_{16}H_{10}O_7$	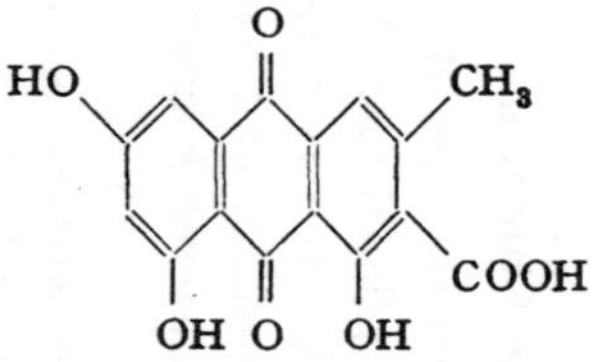 O; HO; CH_3; COOH; OH O OH	—	318 (Zers.)	kupferrot	*Nephromopsis endocrocea* ASAHINA	(17); identisch vielleicht mit Origmaeasäure aus *Sticta origmaea* ACH (334)
Physcion (XXX) (Parietin, Rheochrysidin) $C_{16}H_{12}O_5$	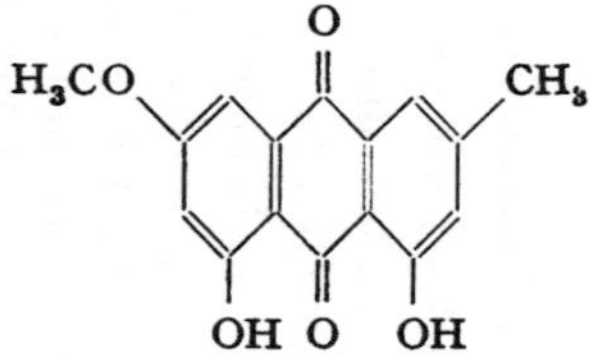 O; H_3CO; CH_3; OH O OH	(83)	206—207	orange	*Xanthonia*-Arten; *Gasparinia elegans*; *Tornatenia*-Arten	(117)
Rhodocladonsäure (XLVII) $C_{17}H_{12}O_9$	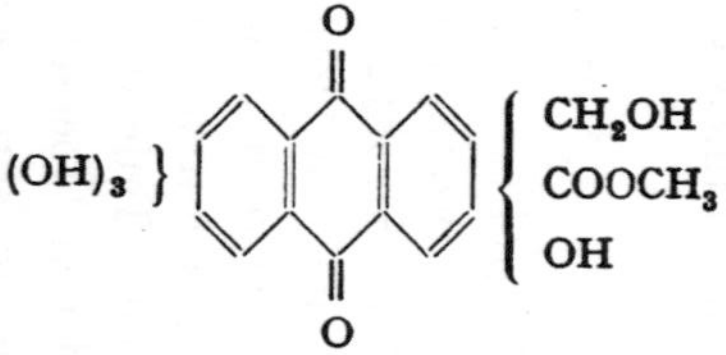 O; $(OH)_3$; CH_2OH, $COOCH_3$, OH; O	—	250—300 (Zers.)	rubinrot	*Cladonia*-Arten	(187)

Solorinsäure (XLVIII) $C_{21}H_{20}O_7$	H_3CO; O; OH O OH; OH; $CO(CH_2)_4CH_3$	—	203	orange-rot	*Solorina crocea*	(*188*, *333*)
Telephorsäure (XXXVI) $C_{20}H_{12}O_9$	HO; HOOC; O; O; OH; OH; $(CH{=}CH)_2 \cdot COOH$	—	> 350	schwarz-violett	*Lobaria vetigera*	(*18*)

kömmling vertreten. Das Hypericin stellt ein höher kondensiertes System dar; es ist anzunehmen, daß Stoffe mit gleicher Wirksamkeit, welche in verschiedenen Pflanzen vorkommen, eine ungefähr gleichartige Struktur aufweisen.

Die *Benzochinon-Derivate* 2,6-Dimethoxybenzochinon (XLVIII) und Thymochinon (IL) waren bereits viele Jahre bekannt und auf synthetischem Wege erhalten worden, bevor noch ihr natürliches Vorkommen nachgewiesen werden konnte. Die Konstitution des Perezons (L) ist auch heute noch nicht mit ausreichender Sicherheit festgestellt. Für Embelin ist mit einiger Wahrscheinlichkeit die in der Tabelle 3 angeführte Struktur anzunehmen, da die Identität des von ASANO und YAMAGUTI (*20*) auf synthetischem Wege erhaltenen Produktes mit natürlichem Embelin nachgewiesen zu sein scheint. Allerdings berichteten schon 1931 HASAN und STEDMAN (*113*) über eine Synthese, bei der die von ihnen erhaltene Substanz eine um ein Kohlenstoffatom längere aliphatische Seitenkette aufweist, als das Embelin nach ASANO und YAMAGUTI.

Die Existenz eines anscheinend weit verbreiteten pflanzlichen Chinons, das auf Grund verschiedener Eigenschaften ebenfalls zu den Derivaten des *p*-Benzochinons gehören soll, dessen chemische Konstitution aber noch nicht feststeht, wurde vor einiger Zeit von KOFLER (*186*) wahrscheinlich gemacht.

Eine Reihe recht interessanter Substanzen gehört zu den Derivaten des 1,4-*Naphthochinons*. Bezüglich des Vitamin K_1 (LXIII) muß, wie erwähnt wurde, auf Übersichtsreferate (*67, 68, 77, 159, 281*) verwiesen werden. Das Lapachol (LVIII) ist mit den in letzter Zeit von Fieser (*89*) beschriebenen Substanzen mit therapeutischer Wirkung gegen Malaria nahe verwandt; im Alkannin (LX) und Shikonin (LXI) liegen Spiegelbildisomere vor, welche aus verschiedenen Pflanzen stammen. Das Dunnion (LXIV) ist der einzige Vertreter der Derivate des 1,2-Naphthochinons.

Anthrachinon-Derivate. Die weitaus zahlreichsten der pflanzlichen Chinon-Derivate gehören zur Gruppe der *Anthrachinone*. Es ist bemerkenswert, daß gegen alle Erwartung oft nahe verwandte Pflanzen verschiedene Anthrachinon-Derivate aufweisen, während die gleichen Substanzen bei einander gar nicht verwandten Pflanzengattungen vorkommen können. Aus der Auffindung bestimmter Anthrachinonabkömmlinge in Pflanzen lassen sich jedenfalls genetische Zusammenhänge nur sehr schwer erschließen.

Mitter und Biswas (*236*) haben darauf hingewiesen, daß die Substitution verschiedener Gruppen in den einzelnen pflanzlichen Anthrachinon-Derivaten bestimmten Regeln gehorcht, welche sie folgendermaßen formulieren:

1. Keine derartige Verbindung enthält mehr als vier Substituenten, nämlich nicht mehr als eine Methyl-, Carbinol- oder Carboxylgruppe (welche sämtlich nur in β-Stellung vorkommen) und nicht mehr als drei Hydroxylgruppen oder deren Methyläther.
2. Die Substanzen mit einer maximalen Anzahl von Substituenten stellen in den einzelnen Pflanzen den Fundamentaltypus dar; die anderen in derselben Pflanzenart vorkommenden Anthrachinone lassen sich vom Fundamentaltypus durch Oxydation, Reduktion oder Veresterung bzw. Ätherbildung ableiten.
3. Von den vier Substituenten stehen zwei (darunter die Methyl- bzw. Carbinol- oder Carboxylgruppe) in β-Stellung, die anderen beiden in α-Stellung.
4. Wenn beide β-ständigen Substituenten in einem der zwei benzoiden Kerne nebeneinander stehen, so befinden sich auch die beiden anderen Substituenten im selben Kern. Zwei β-ständige Hydroxylgruppen können nur in einem Kern nebeneinander vorkommen.
5. Wenn die beiden β-Substituenten sich in verschiedenen Kernen befinden, so sind die α-ständigen Substituenten so verteilt, daß eine symmetrische Anordnung entsteht.

Wie ein Blick auf Tab. 3 zeigt (S. 180 ff.), scheinen diese Regeln weitgehend erfüllt zu sein; Gründe dafür sind allerdings schwer anzugeben. Wahrscheinlich wird erst ein eingehendes Studium der Biosynthese der pflanzlichen Anthrachinon-Derivate eine Erklärung der Zusammenhänge geben.

Wie schon erwähnt wurde, sind einige der angeführten Anthrachinon-Derivate wirksame Bestandteile von abführenden (purgierenden) Drogen

(*Cortex Frangulae, Fructus Rhamnis catharticae, Cascara Sagrada, Folia Sennae, Aloe, Rhizoma Rhei* usw.). Es sind dies das Frangula-emodin (XXV), das Aloe-emodin (LXXVI), die Chrysophansäure (LXXIII) und das Rhein (XXIX). Sie sind frei und glykosidisch gebunden in den Drogen vorhanden. Es scheint aber, daß die glykosidisch gebundenen Anteile vielfach in reduzierter Form als Anthranolglykoside in den Pflanzen vorliegen, wie das z. B. für die Glykoside aus Sennadrogen in einer neueren Arbeit von A. STOLL und Mitarbeitern (*298 a*) gezeigt wurde. Die pharmakologische Wirkung dieser Anthrachinonabkömmlinge läßt sich auch durch synthetische Oxyanthrachinone erzielen. So ist das 1,8-Dioxyanthrachinon ein unter dem Namen „Istizin" bekanntes und viel verwendetes Abführmittel.

Der bekannteste natürliche Anthrachinonfarbstoff ist das Alizarin (LXVI), welches früher in großen Mengen aus der Wurzel des Krapps (*Rubia tinctorium*) gewonnen wurde. Heute wird der Farbstoff nur mehr auf synthetischem Wege hergestellt. Im Krapp kommt Alizarin zum größten Teil glykosidisch gebunden als Ruberythrinsäure (LXXVII) vor; die Konstitution dieses Glykosids wurde von RICHTER (*280*) aufgeklärt. ZEMPLÉN (*332a*) konnte später die von RICHTER postulierte Struktur durch eine elegante Synthese beweisen.

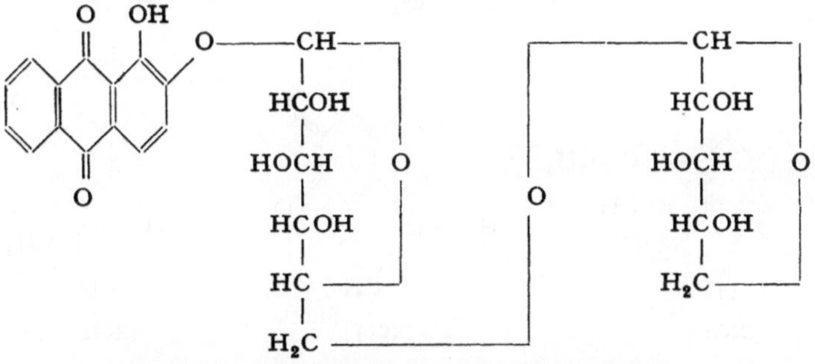

O OH O—CH— HCOH HOCH O HCOH HC H₂C — CH— HCOH HOCH O HCOH H₂C

(LXXXVII.) Ruberythrinsäure. (β-2-Alizarin-primverosid.)

Außer den in Tab. 3 angeführten Anthrachinon-Derivaten ist eine Anzahl weiterer, aus pflanzlichen Organen gewonnener Stoffe mit gleichem Grundskelett beschrieben worden, deren genaue Konstitution aber noch nicht völlig geklärt ist. So konnten OESTERLE und TISZA (*247*) aus der Wurzel von *Morinda citrifolia* zwei Substanzen der Bruttoformel $C_{15}H_{10}O_4$ isolieren, Morindadiol und Soranjidiol.

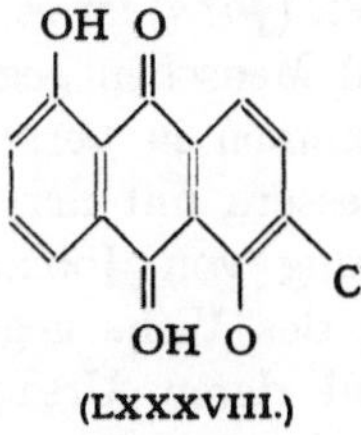

OH O CH₃ OH O

(LXXXVIII.)

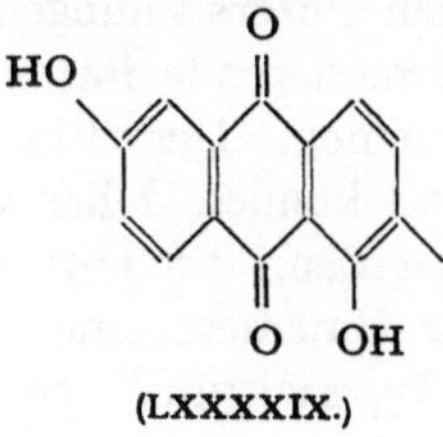

O HO CH₃ O OH

(LXXXXIX.)

Nach MITTER und BISWAS (*236*) dürfte es sich hier um Stoffe der untenstehenden Struktur (LXXXVIII) und (LXXXIX) handeln, wobei die Zuordnung noch nicht ganz festzustehen scheint.

Noch unaufgeklärte Produkte verwandter Konstitution wurden von PERKIN und HUMMEL (*260*) aus *Morinda umbellata* isoliert.

Auch über das sogenannte Natalo-emodin, welches aus den in der Natal-Aloe vorkommenden Glykosiden Nataloin und Homonataloin durch oxydative Spaltung entsteht, besteht bezüglich der Struktur noch keine Klarheit (*202*). Ebenso erwarten die verschiedenen, unter dem Namen Isoemodine bekannten Substanzen, welche aus *Rhamnus-* und *Cassia*-Arten isoliert wurden, die Aufklärung ihrer Konstitution. GORIS und CANAL (*102*) berichteten über die Isolierung eines Anthrachinon-Derivats von der Formel $C_{10}H_{15}O_5$ aus *Populus balsamifera (Salicaceae)*. Die von TSCHIRCH und CRISTOFOLETTI (*309*) aus *Rheum rhaponticum* gewonnenen Anthrachinon-Derivate Chrysorhapontin und Chrysopontin sollen nach HESSE (*118*) mit Chrysophanol (LXXIII) bzw. mit Rhabarberon (LXXVIII) identisch sein.

Phenanthrenchinon-Derivate Als einziger bisher aus höheren Pflanzen isolierter Vertreter der *Phenanthrenchinon-Klasse* wurde von WESSELY und Mitarbeitern (*320, 321*) das Tanshinon A (LXXXV) erkannt. Immerhin ist die in Tabelle 3 angeführte Konstitutionsformel nur eine von den vier möglichen, da zwischen ihr und den Symbolen (XC), (XCI) und (XCII) noch keine Entscheidung getroffen werden konnte.

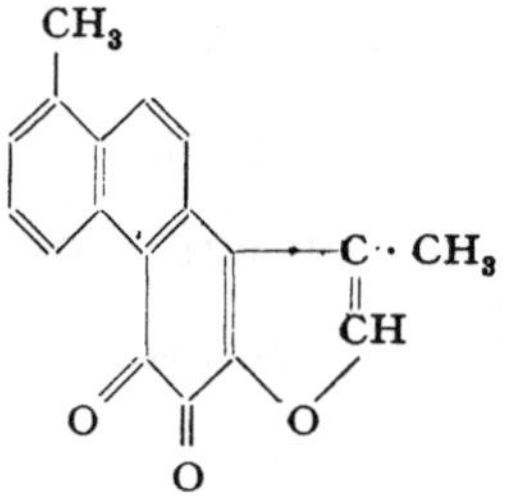

(XC.) 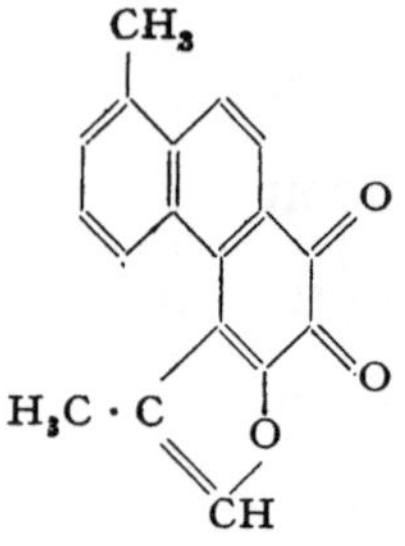 (XCI.) 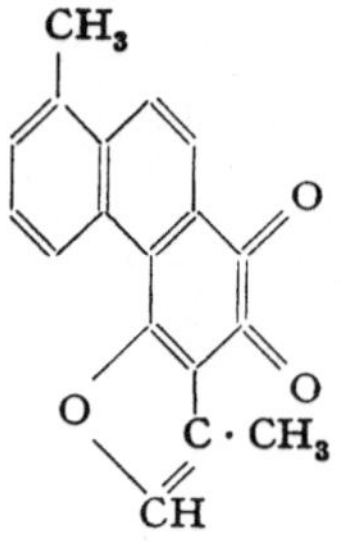 (XCII.)

Photosensibilisierende Chinone.

Das von BROCKMANN und Mitarbeitern (*49, 51*) untersuchte Hypericin (LXXXVI) verdient besonderes Interesse. Die aus dem Johanniskraut *(Hypericum perforatum)* mit Methanol extrahierbare Substanz ist rotviolett und fluoresziert mit roter Farbe.

Schon seit den Untersuchungen von TAPPEINER (*301a*) ist es bekannt, daß gewisse fluoreszierende Farbstoffe Tiere und Menschen gegen Licht sensibilisieren können. Derartige Wirkstoffe kommen in verschiedenen Pflanzen vor und können daher von Pflanzenfressern mit der Nahrung aufgenommen werden. So tritt nach Beimengung von Johanniskraut zum Futter der Haustiere eine Sensibilisierung des Viehs gegen Licht auf, welche „Hyperizismus“ genannt wird und deren Ursache eben

Hypericin ist. Die von BROCKMANN, POHL, MAIER und HASCHAD (*51*) vorgeschlagene Formulierung des Hypericins als Derivat des Naphthodianthrons kann noch nicht als endgültig betrachtet werden, dürfte aber wahrscheinlich den Tatsachen entsprechen.

Verschiedene ähnliche chinoide Wirkstoffe scheinen auch in anderen Pflanzen vorzukommen. Die durch Verfütterung von Buchweizen entstehende Krankheit heißt „Fagopyrismus"; das wirksame Prinzip im Buchweizen wird Fagopyrin genannt. Nach einer Arbeit jüngsten Datums aus dem Laboratorium von BROCKMANN (*52 a*) dürfte das Fagopyrin ebenfalls ein Derivat des 2,2'-Dimethyl-naphthodianthrons sein, welches sich vom Hypericin nur dadurch unterscheidet, daß 1. eine oder beide der nicht α-ständigen OH-Gruppen eine andere Stellung einnehmen als im Hypericin und 2. diese Gruppen mit noch unbekannten Resten ätherartig verknüpft sind. Nach chromatographischen Befunden scheinen im Buchweizen noch andere ähnlich gebaute Farbstoffe vorzukommen. Bei Verfütterung von gewissen Sorten der Platterbse entsteht ein analoger Krankheitszustand („Lathyrismus"); in Südafrika ist „Geeldikop" eine durch *Tribulus terrestris* verursachte Krankheit, die in ihren Erscheinungen den aufgezählten Photosensibilisierungs-phänomenen sehr nahe verwandt ist. Über die Wirkstoffe, welche „Lathyrismus" und „Geeldikop" verursachen, ist noch nichts bekannt.

Biosynthese und Funktion der Chinone höherer Pflanzen.

Eine Anzahl von Chinon-Derivaten, wahrscheinlich zum größten Teil Derivate des *o*-Benzochinons, dürfte durch die Wirkung von Phenoloxydasen entstehen, welche in vielen Pflanzen nachgewiesen wurden. Es handelt sich hier um Fermente, welche imstande sind, ein- und mehrwertige Phenole zu Chinonen zu oxydieren. Verschiedene Autoren nehmen an, daß diese Oxydasen eine maßgebliche Rolle bei Atmungsvorgängen spielen, und zwar trete hier in vielen Fällen ein Phenol-Chinonsystem an die Stelle des bei den Tieren verbreiteten Cytochromsystems.

Dagegen wurde von anderer Seite eingewendet, daß erstens derartige Oxydasen nicht in allen atmenden pflanzlichen Organen nachweisbar sind und zweitens, daß auch in vielen Pflanzen ein Cytochromsystem vorliegt. JAMES (*158*) bezweifelt allerdings die Gültigkeit dieser Gedankengänge und hält es für wahrscheinlich, daß die verschiedenen pflanzlichen Gewebe die Atmungsfunktion mit verschiedenen Mitteln erfüllen können und sowohl das Cytochromsystem, als das Phenol-Chinonsystem und auch das Ascorbinsäuresystem imstande sind, eine solche Rolle zu spielen. Er hält es sogar für möglich, daß unter Umständen ein und dasselbe Gewebe im Verlauf seiner Existenz von einem Mechanismus auf einen anderen umschaltet.

Tabelle 3. Chinone aus höheren Pflanzen.

Trivialname und Bruttoformel	Konstitutionsformel	Synthese	F. °	Farbe	Vorkommen	Literaturzitate und Anmerkungen
2,6-Dimethoxy-*p*-benzochinon (XLVIII) $C_8H_8O_4$	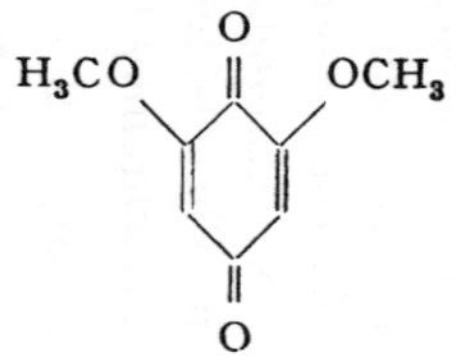 O, H_3CO, OCH_3, O	(*127*)	251	goldgelb	*Adonis vernalis*	(*162*)
Thymochinon (IL) $C_{10}H_{12}O_2$	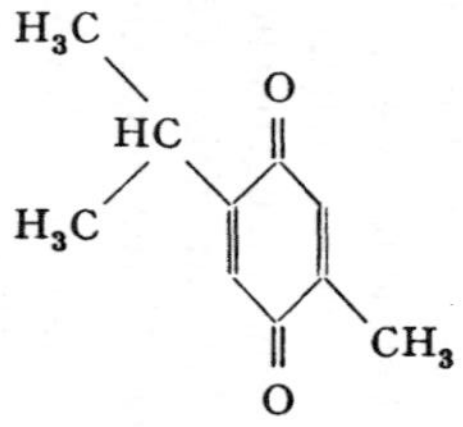 H_3C, HC, H_3C, O, CH_3, O	(*200*)	48	gelb	In ätherischen Ölen aus *Monarda fistula* und *Thujå articulata*	(*108*)
Perezon (Pipitzahoinsäure) (L) $C_{15}H_{20}O_3$	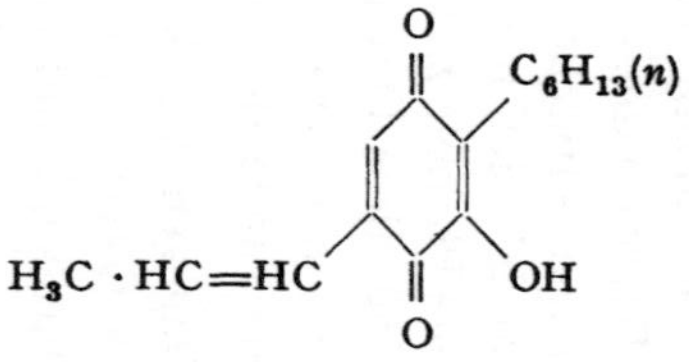 O, $C_6H_{13}(n)$, $H_3C \cdot HC{=}HC$, OH, O	—	104—106	goldgelb	Wurzel von *Trixis pipizahua* (Droge *Radix perziae*)	(*87, 278*) Purgans

Embelin (Embelsäure) (LI) $C_{17}H_{26}O_4$	O, OH, HO, $C_{11}H_{23}(n)$, O	[*20*, *113*(?)]	143	goldgelb	Beeren von *Embelia ribes*	(*20*, *113*, *164*, *165*, *242*) Antihelminthicum
Maesachinon (LII) $C_{26}H_{42}O_4$	O, OH, HO, $C_{20}H_{39}(n)$, O	—	122	orange-rot	Früchte von *Maesa japonica*	(*126*)
Juglon (LIII) $C_{10}H_6O_3$	OH, O, O	(*35*, *328*)	154	gelbrot	In allen grünen Organen des Walnußbaumes und anderer *Juglandaceae*	(*34*, *328*)
Lawson (LIV) $C_{10}H_6O_3$	O, OH, O	(*305*)	zirka 192 (Zers.)	gelb	In den Blättern von Henna *(Lawsonia alba)*	(*307*)

Fortsetzung s. S. 182

Fortsetzung der Tabelle 3.

Trivialname und Bruttoformel	Konstitutionsformel	Synthese	F. °	Farbe	Vorkommen	Literaturzitate und Anmerkungen
2-Methoxy-1,4-naphtho-chinon (LIV a) $C_{11}H_8O_3$	O, O, OCH_3	(305)	183,5	hellgelb	Blüten von *Impatiens balsamina*	(213 a) Fungizid
Plumbagin (LV) $C_{11}H_8O_3$	OH O, O, CH_3	(69, 90)	78—79	orange-gelb	*Plumbago*-Arten, *Drosera rotundifolia*, *Diospyros*-Arten	(71, 163, 217, 254) In der indischen Volksheilkunde verwendet
Droseron (LVI) $C_{11}H_8O_4$	H_3C, OH O, HO, O oder H_3C, O, HO, OH O	—	178	gelb	*Drosera whittackeri*	(215)
Oxydroseron (LVII) $C_{11}H_8O_5$	H_3C, OH O, HO, OH O	(329)	192—193	rot	*Drosera whittackeri*	(215)

Lapachol (LVIII) (Taigusäure, Grönhartin, Tecumin) $C_{15}H_{14}O_3$	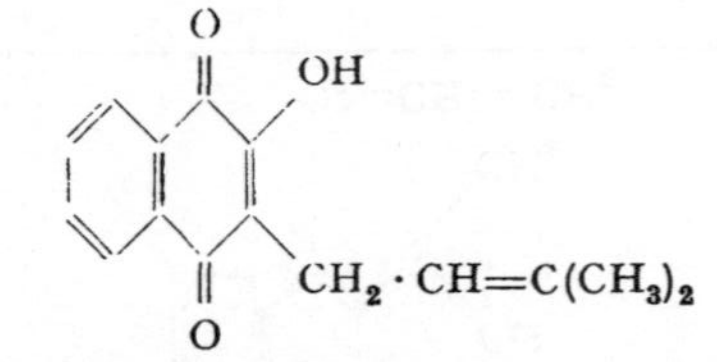 O, OH, O, $CH_2 \cdot CH{=}C(CH_3)_2$	(*88*)	142—143	gelb	Taigu- oder Lapachoholz (verschiedene *Bignoniceae*), Bethabarraholz, Grünherzholz, Holz von *Avicennia tomentosa*	(*150, 151, 255*)
Lomatiol (LIX) $C_{15}H_{14}O_3$	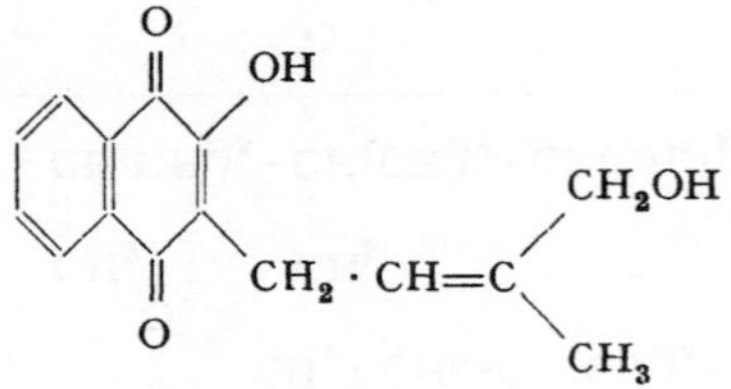 O, OH, O, $CH_2 \cdot CH{=}C(CH_2OH)(CH_3)$	—	127	gelb	Samen von *Lomatia*-Arten	(*152, 279*)
Alkannin (LX) $C_{16}H_{16}O_5$	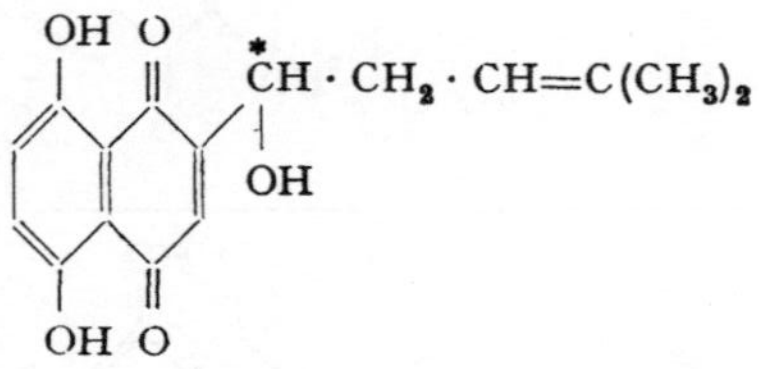 OH O, OH O, $\overset{*}{C}H(OH) \cdot CH_2 \cdot CH{=}C(CH_3)_2$	—	149	braunrot, kupferglänzend	Wurzel von *Alcanna tinctoria*	(*48, 52*)
Shikonin (LXI) $C_{16}H_{16}O_5$	Spiegelbildisomeres des Alkannins	—	149	braunrot	Shikonwurzel (*Lithospermon erythrorhizon*)	(*48. 52*)

Fortsetzung s. S. 184

Fortsetzung der Tabelle 3.

Trivialname und Bruttoformel	Konstitutionsformel	Synthese	F. °	Farbe	Vorkommen	Literaturzitate und Anmerkungen
Alkannan (LXII) $C_{16}H_{18}O_4$	OH O, OH O; $(CH_2)_3 \cdot CH(CH_3)_2$	*(50)*	99	rot	*Alcanna tinctoria* (in sehr geringen Mengen)	*(48, 50)*
Vitamin K_1 (LXIII) (α-Phyllochinon) $C_{31}H_{46}O_2$	O, O; CH_3; $CH_2 \cdot CH{=}C(CH_3) \cdot (CH_2)_3 \cdot CH(CH_3) \cdot (CH_2)_3 \cdot CH(CH_3) \cdot (CH_2)_3 \cdot CH(CH_3)_2$	*(39a, 88a)*	53,5—54,5	hellgelb	In allen grünen Pflanzenteilen	(vgl. S. 176)
Dunnion (LXIV) $C_{15}H_{14}O_3$	O, O; $C(CH_3)_2$; O—CH—CH_3	—	98—99	orange-rot	*Streptocarpus dunnii*	*(265)*

β-Oxyanthrachinon (LXV) $C_{14}H_8O_3$	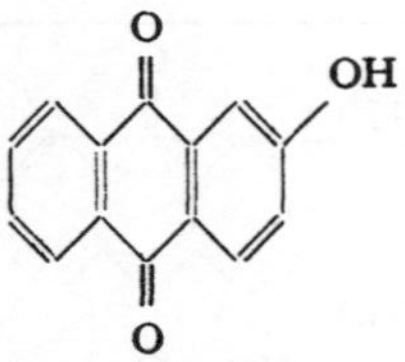	(*106*)	306	gelb	*Oldenlandia umbellata* (*Rubiaceae*)	(*259, 261*)
Alizarin (LXVI) $C_{14}H_8O_4$	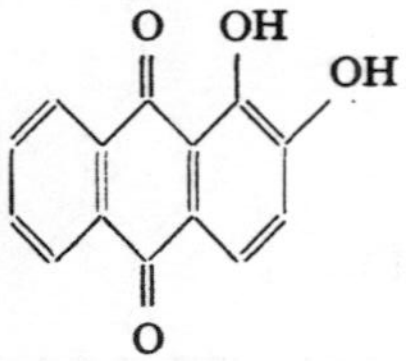	(*105*)	289—290	orange	*Rubia tinctorium*; *Oldenlandia umbellata*; *Morinda citrifolia*; *M. longiflora*; *Concianella maritima* (alle *Rubiaceae*)	Glykosid: Ruberythrinsäure (Alizarinprimverosid) (*280, 332a*) (S. 177)
Purpuroxanthin (LXVII) $C_{14}H_8O_4$	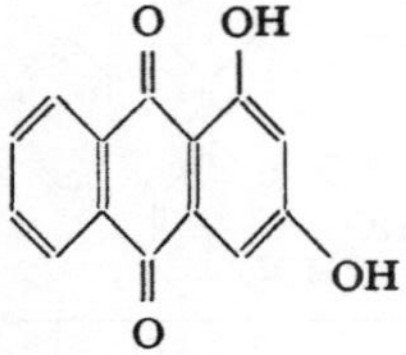	(*213*)	270	gelbrot	*Rubia tinctorium*; *R. munjista*; *R. sikkimensis*	(*258*)
Purpurin (LXVIII) $C_{14}H_8O_5$	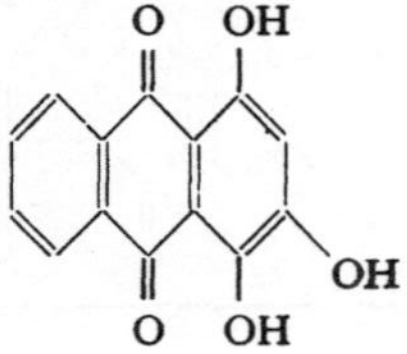	(*212*)	263	orangerot	*R. tinctorium*; *R. munjista*; *R. sikkimensis* und andere *Rubiaceae*	(*258*)

Fortsetzung s. S. 186

Fortsetzung der Tabelle 3.

Trivialname und Bruttoformel	Konstitutionsformel	Synthese	F. °	Farbe	Vorkommen	Literaturzitate und Anmerkungen
Tectochinon (LXIX) $C_{15}H_{10}O_2$	O, O, CH_3	(93)	175—176	gelblich	*Tectona grandis* (*Verbenaceae*): „Teakholz"	(161)
Alizarinmonomethyläther (α) (LXX) $C_{15}H_{10}O_4$	O, OCH_3, OH, O	(255a)	178—179	orange	*R. tinctorium*	(258, 261, 28)
Hystazarinmonomethyläther (LXXI) $C_{15}H_{10}O_4$	O, OCH_3, OH, O	(199, 211, 287)	237	orange	*Oldenlandia umbellata*	(261)
Rubiadin (LXXII) $C_{15}H_{10}O_4$	O, OH, CH_3, OH, O	[234 (?), 198]	290	gelb	*R. tinctorium*; *Morinda longiflora*; *M. citrifolia*; *Galium verum* (alle: *Rubiaceae*)	(299, 28) Glykoside: Rubiadinglucosid; Rubiadinprimverosid

Chrysophansäure (LXXIII) (Chrysophanol) $C_{15}H_{10}O_4$	O; CH_3; OH O OH	(*81, 82*)	196	goldgelb	*Rheum-* und *Rumex*-Arten (*Polygonaceae*); *Rhamnus*-Arten (*Rhamnaceae*); *Cassia*-Arten (*Leguminosae*); *Cluytia similis* (*Euphorbiaceae*)	Glykosid: Chrysophanein (Chrysophan-glucosid)
1,3-Dioxy-6-methyl-anthrachinon (LXXIV) $C_{15}H_{10}O_4$	O; HO; CH_3; OH O	—	269	gelb	*Morinda umbellata*	(*260*)
Morindon (LXXV) $C_{15}H_{10}O_5$	O OH; CH_3; HO; OH O	(*38, 157*)	281—282	orange	*Morinda*-Arten	(*260, 246a*) Glykosid: Morindin (Morindon-2-hexosid)
Aloe-emodin (LXXVI) (Rheum-emodin) $C_{15}H_{10}O_5$	O; CH_2OH; OH O OH	(*235*)	225	orange-gelb	*Aloe*-Arten (*Liliaceae*); *Cassia*-Arten (*Leguminosae*); *Rheum emodi* (*Polygonaceae*)	Glykoside: Barbaloin, Isobarbaloin, β-Barbaloin

Fortsetzung s. S. 188

Fortsetzung der Tabelle 3.

Trivialname und Bruttoformel	Konstitutionsformel	Synthese	F. °	Farbe	Vorkommen	Literaturzitate und Anmerkungen
Frangula-emodin (XXV) $C_{15}H_{10}O_5$	O; HO; CH_3; OH O OH	(*83*, *264*)	259—260	rot-orange	*Rheum-*, *Rumex-* und *Polygonum*-Arten (*Polygonaceae*); *Andira araroba* (*Leguminosae*); *Ventilago maderaspatana* und andere *Rhamnaceae*	Glykoside: Glucofrangulin (Emodinrhamnoglucosid); Frangulin (Emodinrhamnosid); Rhamnoxanthin (Emodinrhamnosid); Rhamnocathartin (Rhamnoxanthinhexosid); Polygonin (Emodinhexosid)
Chrysaron (LXXVII) $C_{15}H_{10}O_5$	O; CH_3; OH; OH; OH O	(*166*)	165—166	goldgelb	*Rheum rhaponticum*	(*166*, *167*) Glykosid: Glucochrysaron (Chrysaron-glucosid)

Rhabarberon (LXXVIII) $C_{15}H_{10}O_5$	OH O; CH_3; OH; OH O	(*166*)	212	gelb	Verschiedene *Rheum*-Arten (*Polygonaceae*); *Rhamnus purshianus* (*Rhamnaceae*)	(*166, 167*)
Rhein LXXIX) $C_{15}H_8O_6$	O; COOH; OH O OH	(*81*)	321	gelb	*Rheum*-Arten (*Polygonaceae*); *Cassia*-Arten (*Leguminosae*)	(*310 a*) Glykosid: Rheinglucosid
Munjistin (LXXX) $C_{15}H_8O_6$	O OH; COOH; OH; O	(*237, 248*)	231	goldgelb	*Rubia munjista, R. tinctorium, R. sikkimensis*	(*289 a*)
Pseudopurpurin (LXXXI) $C_{15}H_8O_7$	O OH; COOH; OH; O OH	(*248*)	222—224	rot	*Rubia*-Arten; *Galium verum, G. mollugo, Crucianella maritima, Sheradia*-Arten und *Asperula*-Arten (alle: *Rubiaceae*)	(*122*) Glykosid: Galiosin (Pseudopurpurinprimverosid)

Fortsetzung s. S. 190

Fortsetzung der Tabelle 3.

Trivialname und Bruttoformel	Konstitutionsformel	Synthese	F. °	Farbe	Vorkommen	Literaturzitate und Anmerkungen
Rubiadinmonomethyläther (α) (LXXXII) $C_{16}H_{12}O_4$	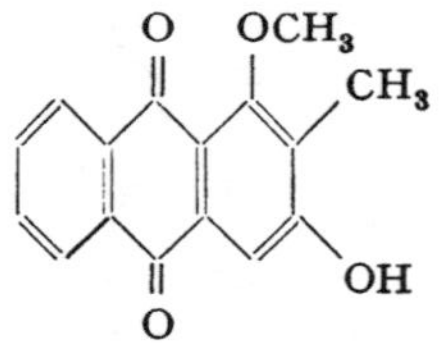	—	291	goldgelb	*Morinda longiflora, M. citrifolia (Rubiaceae)*	*(28, 293)*
2-Dimethyläther des Anthragallols (LXXXIII) und (LXXXIV) $C_{16}H_{12}O_5$	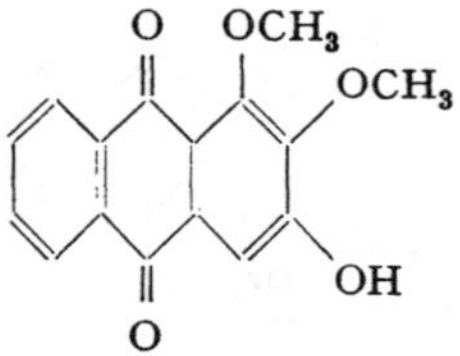 und 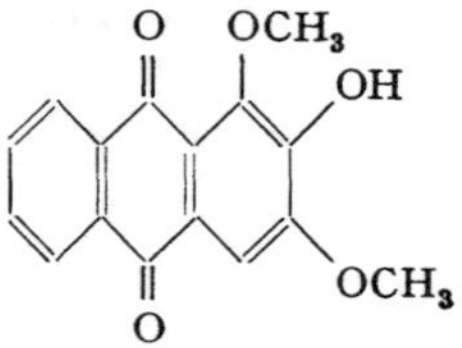	—	209 und 230—232	—	*Oldenlandia umbellata* und *Morinda umbellata (Rubiaceae)*	*(256, 260, 261)*
Physcion (XXX, S. 164 und 174) $C_{16}H_{12}O_5$	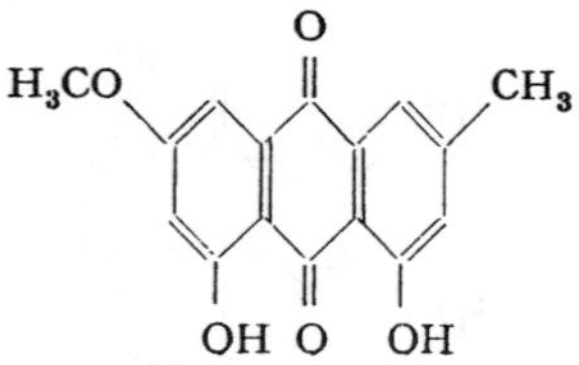	*(83)*	206—207	orange	*Rheum-, Rumex-* und *Polygonum*-Arten *(Polygonaceae)*; *Andira ararota (Leguminosae)*; *Ventilago maderaspatana (Rhamnaceae)*.	Glykosid: Rheochrysin (Physcionglucosid).

Tanshinon I (LXXXV) $C_{18}H_{12}O_3$	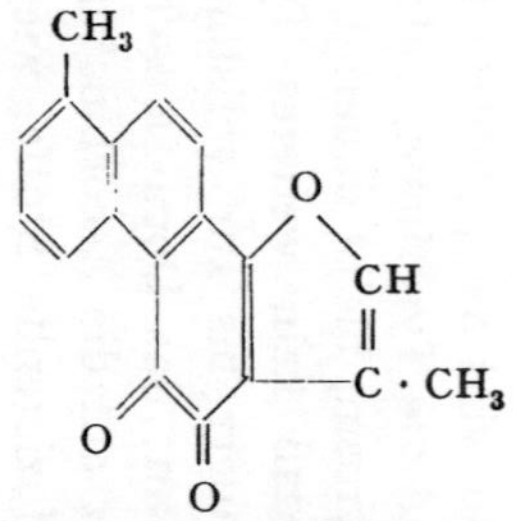 (andere Strukturmöglichkeiten vgl. S. 178)	—	232—234	braun	Droge „Tanshen" (Wurzel von *Salvia miltiorrhiza*)	(*241, 320, 321*)
Hypericin (LXXXVI) $C_{30}H_{16}O_8$	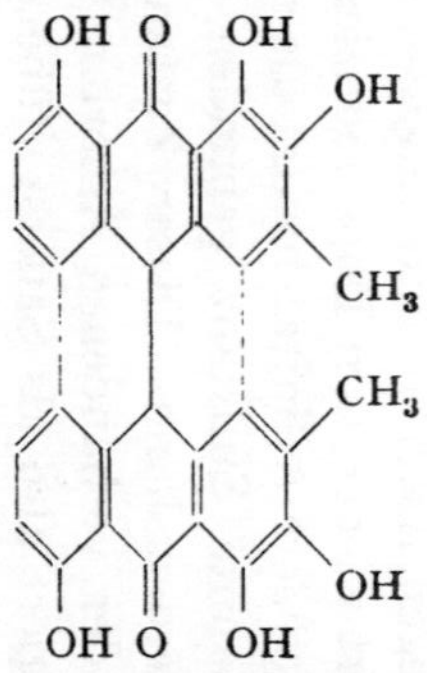 (wahrscheinliche Formel)	—	320 (Zers.)	rotviolett	*Hypericum perforatum*; wahrscheinlich alle *Hypericaceae*	(*51, 285*)

Es scheint allerdings darüber noch keine Klarheit zu bestehen, welche speziellen Phenol-Chinonsysteme tatsächlich als Elektronenüberträger bei der Atmung fungieren sollten. NELSON und Mitarbeiter (*243*, *282*) nehmen an, daß es das System Dioxyphenylalanin (Dopa) (XCIV) und das entsprechende Chinon (Dopachinon) (XCV) ist, welches in den Atmungsmechanismus eingebaut ist. Die Autoren schlagen einen ziemlich komplizierten Mechanismus vor, welcher vom Tyrosin (XCIII) ausgeht; ihre Vorstellungen sind durch Versuche *in vitro* gestützt. Die Tyrosinase pflanzlicher Gewebe, welche sowohl Mono- als auch Diphenole oxydieren kann, soll primär das Tyrosin zum Dopa oxydieren, wobei letztere Substanz eine größere Affinität zum Enzym haben soll als das Tyrosin selbst. Das Enzym übt dann seine diphenol-oxydierende Wirkung am Dopa aus, wodurch das Dopachinon entsteht.

COOH — CH · NH_2 — CH_2 — (Benzolring) — OH

(XCIII.) Tyrosin.

COOH — CH · NH_2 — CH_2 — (Benzolring) — OH, OH

(XCIV.) Dopa.

COOH — CH · NH_2 — CH_2 — (Chinonring) — O, O

(XCV.) Dopachinon.

Die Versuche zeigten, daß in Abwesenheit von Ascorbinsäure das Dopachinon zu braunen, melaninartigen Produkten weiter umgewandelt wird und gleichzeitig noch mehr Tyrosin in gleicher Weise oxydiert wird. Wenn hingegen Ascorbinsäure zugegen ist, so wird das gebildete Dopachinon sofort wieder zum Dopa reduziert und die Tyrosinase, welche zu Dopa eine größere Affinität besitzt als zu Tyrosin, bleibt dauernd durch die erstgenannte Substanz gebunden, so daß kein weiteres Tyrosin oxydiert werden kann. Dieser Zustand dauert bis zur vollständigen Oxydation der vorhandenen Ascorbinsäure an; das Dopa-Dopachinonsystem fungiert also als Sauerstoffüberträger auf die Ascorbinsäure.

In der Pflanze könnten nun andere reduzierende Stoffe, wie etwa das in der Kette der Atmungsvorgänge vor dem Phenol-Chinonsystem stehende Überträgerglied, eine Rolle übernehmen, welche derjenigen der Ascorbinsäure in den beschriebenen Versuchen analog ist. NELSON und DAWSON (*243*) bringen diese Vorstellungen in das folgende Schema.

(H_2O +) Dopachinon → Melaninprodukte.

vorhergehender Teil der Reaktionskette der Atmung } 2 H ⇅ Tyrosinase + O_2

l-Tyrosin + O_2 $\xrightarrow{\text{Tyrosinase}}$ Dioxyphenylalanin

Schema nach NELSON und DAWSON (*243*).

Diesem Schema entsprechend ist also das Dioxyphenylalanin der Elektronenüberträger, welcher unter Mitwirkung der Tyrosinase in manchen pflanzlichen Geweben als terminales System der Atmung mitwirkt. Das *l*-Tyrosin erfüllt hier die Funktion eines Reservoirs, damit die Pflanze stets, falls sich die Notwendigkeit ergeben sollte, neuen Überträger bilden könne. Jede Störung der Reaktionskette, welche die Reduktion des Chinons verhindert, bewirkt, daß zuerst Dopa und dann auch das vorhandene Tyrosin zu melaninartigen Produkten aufoxydiert werden. NELSON und DAWSON erklären so die dunkle Verfärbung von rohen, zerschnittenen Kartoffeln, welche nach einiger Zeit an der Luft eintritt, als eine Störung des normalen Atmungssystems, wodurch eben die melaninartigen Stoffe erscheinen. Es ist natürlich durchaus möglich, daß auch andere derartige Redoxsysteme (wie z. B. Protocatechusäure und das entsprechende Chinon oder Kaffeesäure und ihr Chinon) an der Atmung in ähnlicher Weise teilnehmen, wie es für das System Dopa-Dopachinon wahrscheinlich gemacht wurde.

An dieser Stelle soll auch noch eine Beobachtung von BOSWELL (*44*) erwähnt werden. Kartoffelscheiben wirken in Anwesenheit von Kaffeesäure oxydierend auf Glykokoll, Glutaminsäure und Asparaginsäure; stickstoff-freie Säuren werden hingegen nicht angegriffen. Es ist anzunehmen, daß dieser Effekt durch das Chinon verursacht wird, welches die in den Kartoffeln anwesende Phenoloxydase aus der Kaffeesäure bildet; die oxydierende Wirkung von Chinonen auf manche Aminosäuren ist schon seit längerem bekannt (*168*, *322*). Es ist möglich, wenn auch nicht sehr wahrscheinlich, daß wir auch hier eine Funktion der Chinone im Stoffwechsel der Pflanzen vor uns haben.

Ob aber die vielen in Tabelle 3 genannten Chinone ebenfalls eine Rolle in der Atmung spielen, muß als recht zweifelhaft bezeichnet werden. Worin nun die Funktion dieser Substanzen in der Pflanze besteht, ist noch in völliges Dunkel gehüllt. Ebensowenig wissen wir über die Bildungsweise der verschiedenen pflanzlichen Chinone, wenn wir von ganz wenigen Fällen (z. B. Dopachinon) absehen. Aus Analogiegründen könnte angenommen werden, daß auch hier, wie bei den Pilzen, die Chinone Umwandlungsprodukte von Kohlehydraten sind.

Chinone tierischen Ursprungs.

Einige von Tieren produzierte Substanzen, welche sich durch ihre physiologischen Wirkungen oder durch sonstige bemerkenswerte Eigenschaften auszeichnen, wurden als Chinone erkannt. Darunter finden sich Derivate des *p*-Benzochinons, des *o*-Benzochinons, des 1,4-Naphthochinons und des Anthrachinons.

Der erste Bericht über ein Chinon tierischen Ursprungs stammt von BÉHAL und PHISALIX (*30*). Sie fanden, daß der Tausendfüßler *Julus terrestris*, besonders bei mechanischer oder elektrischer Reizung, ein Hautsekret absondert, welches den Geruch von Chinon zeigt. Ein aus dem Sekret isolierter Stoff verhielt sich in

bezug auf Flüchtigkeit, Löslichkeit und Reaktionen völlig gleichartig wie Benzochinon. Allerdings reichen die in dieser fast fünfzig Jahre zurückliegenden Arbeit verwendeten Methoden nicht aus, um mit Bestimmtheit zu behaupten, daß die Autoren tatsächlich Benzochinon, wie sie annahmen, oder ein anderes Chinon in den Händen hatten. Es wäre wohl eine dringliche Aufgabe, diese Untersuchung mit modernen Mitteln zu überprüfen.

Derivate des *o-Benzochinons* finden sich im Tierreich sehr weit verbreitet. Sie entstehen durch die Wirkung verschiedener Oxydasen auf einzelne Phenole; im weiteren Verlauf der Reaktion werden dann diese Orthochinone in das hochpolymere Melanin, dessen Struktur noch unbekannt ist, verwandelt.

Den Orthochinonen kommt bei verschiedenen Klassen der Invertebraten, auch wenn wir von der Melaninbildung absehen, eine nicht unbedeutende Funktion zu, welcher gerade in letzter Zeit größere Beachtung geschenkt wird. Die verschiedenen Orthochinone, wie Dopachinon oder „Protocatechuchinon", wirken als Gerbstoffe bei der Erzeugung widerstandsfähiger Gerüstproteine mit. Chinongegerbte Eiweißstoffe konnten bei vielen Invertelraten nachgewiesen werden. Das am längsten bekannte Beispiel ist die Cuticula verschiedener Insektenarten (*265 a*); gleichartige chinongegerbte Strukturproteine finden sich aber auch bei Würmern und Mollusken (*52 a, 52 b, 308 a*). Die in Frage kommenden Gewebe erweisen sich als reich an verschiedenen Phenolen und an für diese spezifischen Oxydasen. Folglich kann mit Sicherheit angenommen werden, daß die die Gerbung der Proteine bewirkenden Chinone durch fermentative Oxydation aus den Phenolen entstehen. (Bezüglich des Mechanismus der Chinongerbung vgl. S. 216.)

Der zur Melaninbildung führende Vorgang ist bei zwei biologischen Ausgangsprodukten, dem Tyrosin und dem Adrenalin, schon recht genau untersucht worden; wir geben unten ein Schema über eine mögliche Bildungsweise, wobei bezüglich Einzelheiten auf die in der Einleitung zitierten Übersichtsreferate und Arbeiten verwiesen werden muß.

MAZZA und STOLFI (*226*) konnten das Hallachrom (XCVII) schon 1931 aus dem Polychäten *Halla parthenopea* isolieren. FRIEDHEIM (*95*) hat dann später diese Substanz näher untersucht und ihr Oxydations-Reduktionspotential gemessen. Auf Grund der Tatsache, daß Hallachrom imstande ist, *in vitro* die Atmung von Erythrozyten, Seeigel- und *Ascaris*-Eiern zu steigern, vermutet FRIEDHEIM, daß dieses Pigment eine aktivierende Rolle im Sauerstoffhaushalt von *Halla* ausübt. Er hält es für wahrscheinlich, daß das Hallachrom als Sauerstoffspeicher dienen kann und es dem Tier dadurch ermöglicht, sich längere Zeit im sauerstoffarmen Schlamm des Meeresgrundes aufzuhalten.

Das Adrenochrom (CIII), welches besonders von GREEN und RICHTER (*107*) näher untersucht wurde, hat in letzter Zeit größeres Interesse be-

(*A*.)

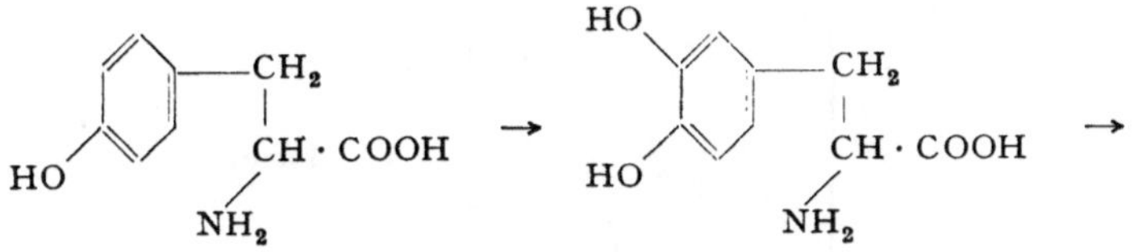

HO—C₆H₄—CH₂—CH(NH₂)·COOH → (HO)₂C₆H₃—CH₂—CH(NH₂)·COOH →

(XCIII.) *l*-Tyrosin. (XCIV.) *l*-3,4-Dioxyphenylalanin (Dopa).

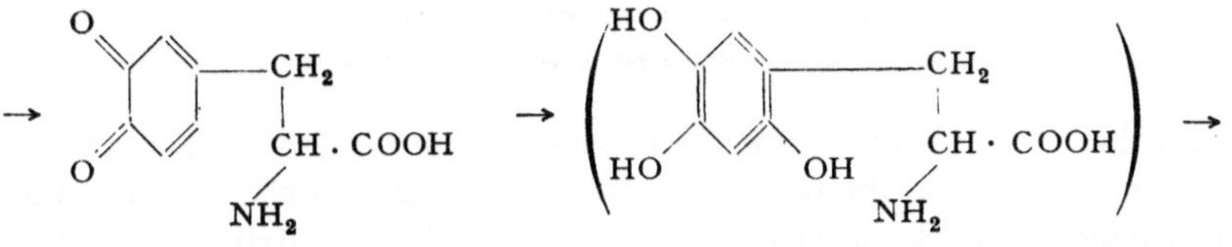

→ O=C₆H₃(=O)—CH₂—CH(NH₂)·COOH → ((HO)₂C₆H₂(OH)—CH₂—CH(NH₂)·COOH) →

(XCV.) Dopachinon. (XCVI.)

→ Hallachrom (CH₂, CH·COOH, NH, O, O) → Decarboxyliertes Hallachrom (CH₂, CH₂, NH, O, O) →

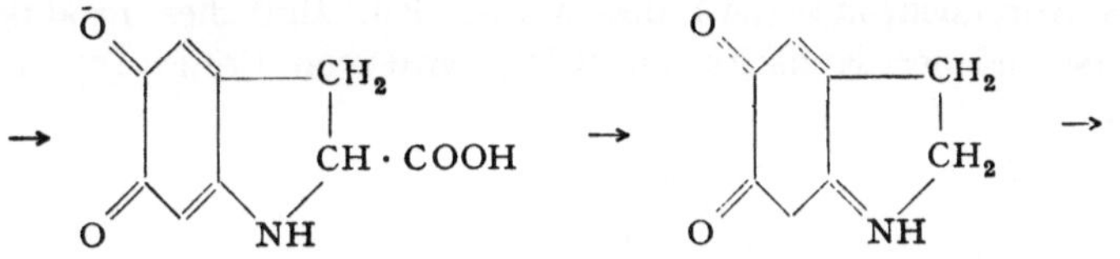

(XCVII.) Hallachrom. (XCVIII.) Decarboxyliertes Hallachrom.

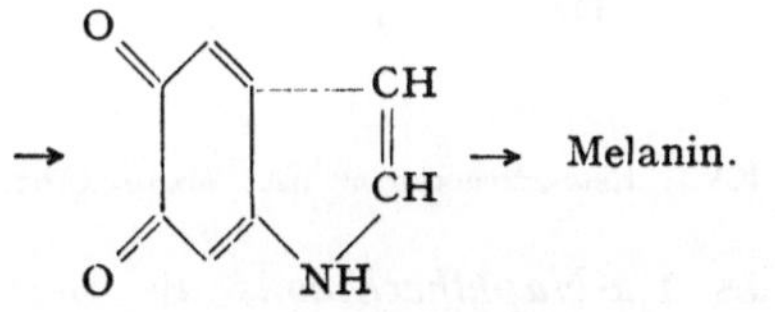

→ Indol-o-chinon (CH, CH, NH, O, O) → Melanin.

(IC.) Indol-*o*-chinon.

(*B*.)

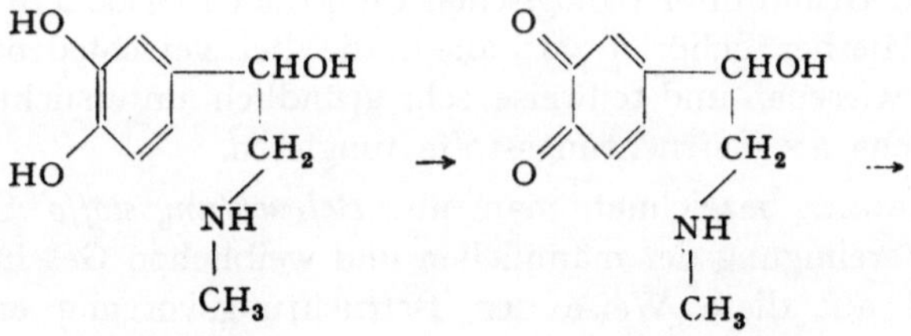

(HO)₂C₆H₃—CHOH—CH₂—NH—CH₃ → O=C₆H₃(=O)—CHOH—CH₂—NH—CH₃ →

(C.) Adrenalin. (CI.) Adrenalinchinon

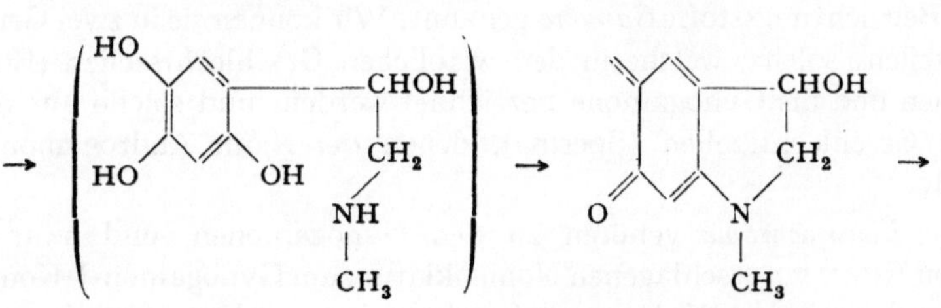

→ ((HO)₂C₆H₂(OH)—CHOH—CH₂—NH—CH₃) → Adrenochrom (CHOH, CH₂, N—CH₃, O, O) →

(CII.) (CIII.) Adrenochrom.

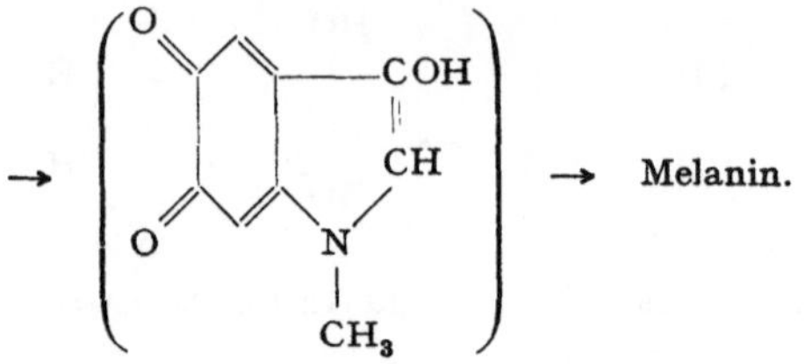

(CIV.) „Oxoadrenochrom" nach Cohen.

Schema der zur Melaninbildung führenden Vorgänge (s. auch S. 195).

anspruchte. Es konnte festgestellt werden, daß dieser Substanz oder vielleicht auch Produkten, welche *in vivo* aus ihr entstehen, eine blutdrucksenkende Wirkung zukommt. Eine ausführliche Arbeit über dieses Gebiet veröffentlichte Marquardt (*220*), der auch eine höhere Oxydationsstufe des Adrenochroms (CV) beschreibt. Ein ähnliches Produkt von etwas verschiedener Konstitution (CIV) wird von Cohen (*57—60*) beschrieben.

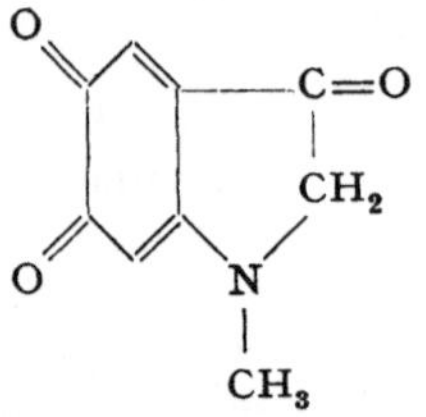

(CV.) „Keto-adrenochrom" nach Marquardt.

An Derivaten des *1,4-Naphthochinons* als tierische Produkte sind bisher nur Pigmente bei den Echinodermen beschrieben worden, denen allerdings auf Grund ihrer biologischen Funktionen besondere Bedeutung zukommt. Hierher gehören vor allem die bei verschiedenen Seeigelarten nachgewiesenen und teilweise sehr gründlich untersuchten Echinochrome, welche als Befruchtungsstoffe fungieren.

Wie bekannt, bezeichnet man als *Befruchtungsstoffe* Substanzen, welche die Vereinigung der männlichen und weiblichen Geschlechtszellen fördern und auf diese Weise den Befruchtungsvorgang ermöglichen. Nach einem Vorschlag von Hartmann und Kuhn (*110, 189*) werden diese Befruchtungsstoffe *Gamone* genannt. Wir können sie in zwei Gruppen unterteilen: solche, welche in den weiblichen Geschlechtszellen (Ei) vorkommen und als Gynogamone bezeichnet werden, und solche aus männlichen Geschlechtszellen (Sperma), denen der Name Androgamone zukommt.

Die *Echinochrome* gehören zu den Gynogamonen, und zwar nach der von Kuhn vorgeschlagenen Nomenklatur zum Gynogamon-I-Komplex, der durch folgende Wirkungen charakterisiert wird: 1. Aktivierung der

Spermienbewegung; 2. chemotaktische Anlockung der Spermatozoen und 3. Antagonismus gegenüber Substanzen des Androgamon-Komplexes.

Echinochrome konnten bisher nur aus einigen Seeigelarten isoliert werden (*Arbacia* und *Strongylocentrotus*). Nach den Untersuchungen von Hartmann, Schartau, Kuhn und Wallenfels (*112, 195*) liegen diese Farbstoffe in den Eiern der Tiere symplex-artig an Proteine gebunden vor. Das von Kuhn und Wallenfels (*194*) aus den ausgereiften, braunroten Ovarien von *Arbacia pustulosa* isolierte Echinochrom A (CVI) (10 mg pro Ovar) kristallisiert in roten Nadeln (F. 220°). Auf Grund der zur Konstitutionsermittlung durchgeführten Versuche (Zinkstaubdestillation, reduzierende Acetylierung, Methylierung mit Diazomethan, Oxydation mit Chromsäure) kommt ihm die Formel eines 2-Äthyl-3,5,6,7,8-pentaoxy-1,4-naphthochinons zu, was auch durch Synthese bestätigt werden konnte (*314*). Es muß allerdings angenommen werden, daß diese Formel die tatsächliche Struktur der Substanz nicht vollständig charakterisiert, da das Echinochrom A ein im Gleichgewichtszustand befindliches Gemisch mehrerer isomerer Chinone darstellen dürfte.

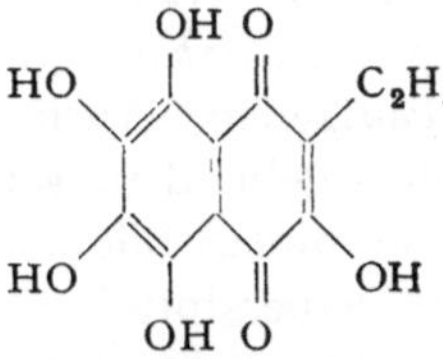

(CVI.) Echinochrom A.

Die hochmolekularen Echinochrom-Symplexe können durch Extraktion mit 4%iger Kochsalzlösung gewonnen werden, und zwar erhält man bei Extraktion des gesamten *Arbacia*-Eies einen ternären Symplex, welcher von Kuhn und Mitarbeitern als *Echinochrom ... Träger ... Hilfsträger* symbolisiert wird; wenn man hingegen das Ei ohne die Gallerthülle extrahiert, erhält man den binären Symplex: *Träger ... Echinochrom.*

Die physiologische Wirkung des Echinochroms A oder seines ternären Symplexes läßt sich durch folgende Versuche charakterisieren: Bringt man geringe Mengen in einen Tropfen Seewasser, in welchem Seeigel-Spermatozoen suspendiert sind, so geraten diese in lebhafte Bewegung („aktivierende Wirkung"). Eine Kapillare mit verdünnter Echinochromlösung, welche in eine Sperma-Suspension gebracht wird, verursacht Anlockung der Spermatozoen und Wanderung in die Kapillare („chemotaktische Wirkung"). Schließlich zeigt sich auch eine auffallende agglutinierende Wirkung des Echinochroms gegenüber den Spermien, welche sich noch in einer Verdünnung von 1 : 2,5 Milliarden nachweisen läßt. Während der ternäre Symplex die geschilderten Effekte in teilweise noch verstärktem Ausmaß entfaltet, wirkt eine Lösung des binären Symplexes nur in Gegenwart des Hilfsträgers aus den Gallerthüllen.

Ovarien, welche sich noch nicht im Stadium der Vollreife befinden, sind violett und enthalten wenig Echinochrom A. Hingegen konnten aus ihnen zwei andere, in ihrer Struktur allerdings noch nicht aufgeklärte Naphthochinon-derivate kristallin erhalten werden, von denen das eine Echinochrom B genannt wird, rot gefärbt ist, und wie Echinochrom in Natronlauge löslich ist, während das zweite, das sogenannte Echinochrom C, violett ist und sich in Natronlauge nicht löst (*195*).

Hartmann und Schartau (*111*) untersuchten auch eine Anzahl anderer Naphthochinone auf ihre Wirkung gegenüber den Spermatozoen von *Arbacia pustulosa*. Juglon (5-Oxy-1,4-naphthochinon) (LIII) ist imstande, die Spermatozoen in ähnlicher Weise wie das Echinochrom zu aktivieren, wogegen ihm die agglutinierenden Effekte nicht zukommen. Naphthazarin (5,8-Dioxy-1,4-naphthochinon) und Naphthopurpurin (5,6,8-Trioxy-1,4-naphthochinon) sind hingegen nicht zur Aktivierung, aber zur Agglutination befähigt. Eine Mischung von Juglon und Naphthazarin kann die Wirkungen von Echinochrom A recht gut nachahmen. Von Interesse ist es auch, daß Naphthazarin unter dem Einfluß von Licht bald die Fähigkeit zur Agglutination verliert und dann eine aktivierende Wirkung besitzt, welcher erst nach langer Zeit schwache Agglutination folgt.

Das Echinochrom A hat nach Wallenfels (*312*) außer seiner geschilderten Rolle als Gynogamon in *Arbacia pustulosa* noch die Funktion eines Farbstoffes der roten Blutzellen. Auch hier ist das Echinochrom an einen hochmolekularen Träger gebunden, welcher aber von demjenigen, mit dem es den Gynogamon-Symplex bildet, verschieden ist. Ob dem Echinochrom die Rolle eines Sauerstoffüberträgers der Blutzellen zukommt, wurde anscheinend noch nicht untersucht; auf Grund seiner Eigenschaften wäre das durchaus denkbar. Das Redoxpotential des Echinochroms A liegt nach Wallenfels und Möhle (*315*) unterhalb von + 100 MV. Eine mit Hilfe einer Additivitätsregel vorgenommene Berechnung ergab ein E_0 von + 76 bis + 81 MV.

Das von Ball (*26*) schon 1936 aus *Arbacia punctulata* kristallin isolierte Pigment, welchem nach Lederer und Glaser (*201*) die Formel $C_{12}H_{10}O_7$ zukommt, dürfte mit dem Echinochrom A identisch sein.

Vom Standpunkt der vergleichenden Biochemie gesehen, erweckt es einiges Interesse, daß unter allen bisher untersuchten Metazoen nur einige wenige Echinodermenarten Naphthochinon-derivate als Bestandteile von Gynogamon-komplexen besitzen. Alle sonstigen bisher in ihrer Struktur bekannten Befruchtungsstoffe gehören zur Klasse der Carotinoide.

Eine ausgezeichnete Übersicht über das Gebiet der Befruchtungsstoffe wurde kürzlich von Bielig und Graf Medem (*39*) veröffentlicht.

Mit den Echinochromen strukturell nahe verwandt sind die *Stachelfarbstoffe* verschiedener Echinodermen. Kuhn und Wallenfels (*196*) gelang es, die Struktur des Stachelfarbstoffes von *Arbacia pustulosa* weitgehend aufzuklären. Es handelt sich um ein Carbinol (CVII), das anscheinend als schwerlösliches Calciumsalz der Dienolform (CVIII) vorliegt. Bei der Abspaltung des Calciums mit Säuren erfolgt gleichzeitig eine Dehydrierung und es entsteht das Keton (CIX). Dem Carbinol wurde von den Autoren der Name Spinochrom A gegeben, während das daraus entstehende Keton die Bezeichnung Spinon A erhielt.

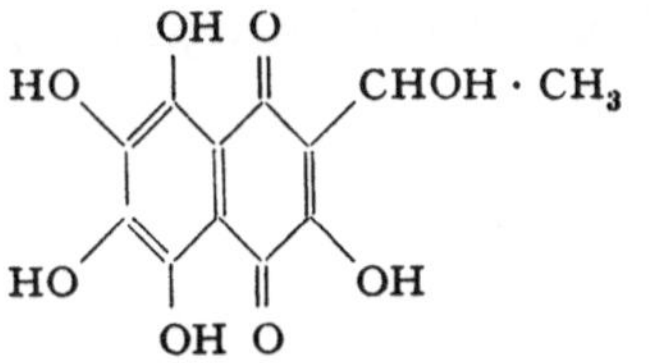

(CVII.) Spinochrom A.

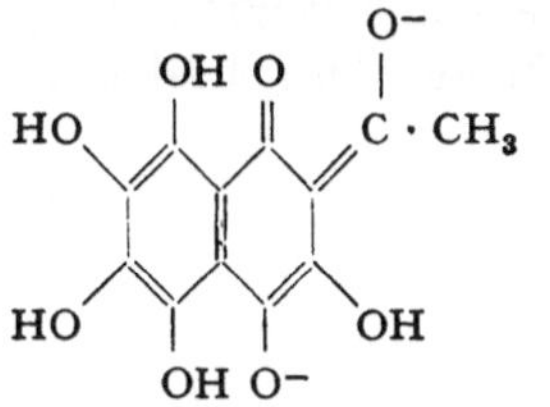

(CVIII.) Dienolform des Spinochrom A.

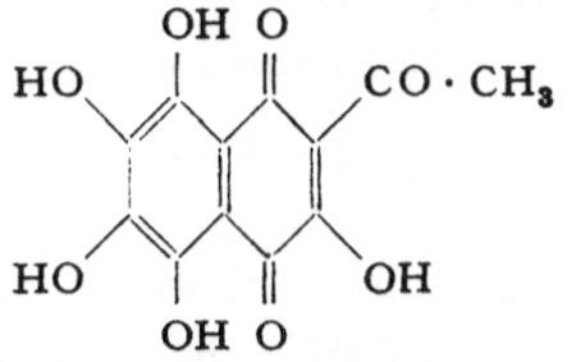

(CIX.) Spinon A.

Ein von Spinon A verschiedenes Pigment ($C_{12}H_{10}O_8$) wurde von LEDERER und GLASER (*201*) aus den Stacheln von *Paracentrotus lividus* isoliert. Im selben Ausgangsmaterial wurde noch ein Farbstoff der Bruttoformel $C_{12}H_{10}O_7$ gefunden, welchen die genannten Forscher für Echinochrom hielten. MUSAJO und MINCHILLI (*240*) konnten aber zeigen, daß das Pigment eine Acetylgruppe enthält, und nehmen an, daß es sich um ein 2-Acetyltetraoxy-naphthochinon handelt.

Drei weitere, kristallisierte Stachelfarbstoffe wurden von KURODA und OSHIMA (*197*) aus Seeigelarten des Stillen Ozeans isoliert, nämlich Spinochrom Aka (aus *Pseudocentrotus depressus*), Spinochrom F (aus *Heterocentrotus mammilatus*) und Spinochrom M (aus *Anthocidaris crassispina*). Es handelt sich hier um Naphthochinon-derivate, welche wahrscheinlich den Pigmenten von *Arbacia pustulosa* in ihrer Konstitution sehr ähnlich sind.

Über das Vorkommen weiterer Naphthochinon-derivate im Tierreich, so vor allem aus der Gruppe der vitamin-K-aktiven Substanzen, ist bis heute noch nichts bekanntgeworden. Es scheint, daß der tierische Körper allgemein nicht imstande ist, Vitamin K zu synthetisieren. Es ist anzunehmen, daß verschiedene Tierarten, welche keinen Nahrungsbedarf an Vitamin K zeigen, dieses deshalb nicht von außen zuführen müssen, weil sie durch die Bakterienflora ihres Darmes in ausreichender Menge versorgt werden.

Unter den Chinonen tierischen Ursprungs finden wir auch eine Anzahl von *Anthrachinon-derivaten*, von denen einige allgemeiner bekannt sind. Hierher gehören die Kermessäure, die Carminsäure, die Laccainsäure und vermutlich auch das Lanigerin.

Die ziegelrote Kermessäure (CX), $C_{18}H_{12}O_9$ (F. 250°; Zers.), ist der Farbstoff der in Südeuropa beheimateten Schildlaus *Lecanium ilicis*. Er wurde von DIMROTH (*73, 76*) in seiner Konstitution aufgeklärt; die untenstehende Formel ist durch Abbaureaktionen weitgehend gesichert.

H_3C O OH
$CO \cdot CH_3$
HO OH
HOOC O OH

(CX.) Kermessäure.

Das färbende Prinzip des Cochenille-Farbstoffs, welcher durch Trocknen und Zerreiben der ungeflügelten Weibchen der hauptsächlich in Zentralamerika vorkommenden Schildlaus *Coccus cacti coccinelliferi* gewonnen wird, ist die Carminsäure. Die Carminsäure (CXI), $C_{22}H_{20}O_{13}$ (Zersetzung zwischen 136 und 205°; $[\alpha]_D = +80{,}5°$), wurde ebenfalls von DIMROTH (*72, 75*) bearbeitet. Seine Abbaureaktionen lassen, analog zur Kermessäure, auf ein Derivat des α-Methylanthrachinons schließen, dem wahrscheinlich die untenstehende Formulierung zukommt.

H_3C O OH
$O . C_6H_{11}O_5$
HO OH
HOOC O OH

(CXI.) Carminsäure.

Über die Natur der aliphatischen C_6-Kette, welche zuckerartigen Charakter besitzt und vier acetylierbare Hydroxyle enthält, ist bisher nichts Näheres bekanntgeworden.

Über die zinnoberrote Laccainsäure, $C_{20}H_{14}O_{10}$, welche aus dem von der Schildlaus *Coccus laccae* erzeugten Stocklack gewonnen wird, welcher früher unter dem Namen „Lac-dye" als Ersatz für Cochenille Verwendung fand, ist noch sehr wenig bekannt. Immerhin ist die Anthrachinonstruktur durch Vergleich der Absorptionsspektra und durch das Ergebnis einer Zinkstaubdestillation weitgehend sichergestellt (*74, 310*). Aus demselben Organismus konnten TSCHIRCH und LÜDY (*310*) noch ein zweites Anthrachinon-derivat, das Erythrolaccin, $C_{15}H_{10}O_6 \cdot H_2O$, gelbe

Kristalle, isolieren; die Autoren schreiben ihm die Konstitution eines Tetraoxy-methyl-anthrachinons zu.

Schließlich berichtete BLOUNT (*43*) über die Isolierung eines Anthrachinon-derivats aus der auf der Weymouthkiefer lebenden Blutlaus *Eriosoma lanigera*. Das Lanigerin ist ein orangefarbenes Pigment, welches der Formel $C_{17}H_{14}O_7$ entspricht; F. 274 bis 276° unter Zers.

II. Biologische Wirkungen der Chinone.

Bald nach der erwähnten ersten Darstellung eines Chinons, des *p*-Benzochinons, durch WOSKRESENSKY (1838) wurde bei dieser Substanz eine physiologische Wirkung vermutet. Schon ihr Entdecker schreibt, daß ihr „... ein durchdringender, die Augen zu Tränen reizender Geruch ..." eigen ist und WÖHLER berichtet 1844: „... sein starker, Nase und Augen reizender Geruch hinterläßt noch lange eine ähnliche Wirkung wie Jod und Chlor ...", Eine etwas spätere Untersuchung von WÖHLER und FRERICHS (*330*) ergab allerdings, daß dem Chinon zumindest nicht die vermutete Giftwirkung zukommt: „... 0,5 g einem Hund gegeben, ließ gar keine Wirkung wahrnehmen; ebensowenig eine größere Gabe von ungefähr 1 g. Es war nicht im Harn zu finden und es war nicht auszumitteln, was aus ihm geworden war ..."

Seit damals sind viele Untersuchungen über die Wirkungen der Chinone auf verschiedene Organismen durchgeführt worden, von denen hier jedoch nur die wichtigsten besprochen werden können.

Wirkungen der Chinone gegenüber Mikroorganismen.

Die *bakteriziden* und *bakteriostatischen Effekte* einiger Chinone wurden erstmalig etwa gleichzeitig von THALHIMER und PALMER (*304*), sowie von COOPER (*62*) beobachtet. MORGAN und COOPER (*239*) untersuchten die Wirkungen verschiedener Chinone auf Bakterienkulturen, wobei sie feststellten, daß — im Gegensatz zu den Verhältnissen bei den Phenolen — bei den Chinonen das bakterizide Vermögen mit dem Aufsteigen in der homologen Reihe geringer wird, so daß Toluchinon oder Thymochinon schwächere antibakterielle Effekte zeigen als das Benzochinon. Thymol und die Kresole sind bekanntlich wirksamere Desinfektionsmittel als das unsubstituierte Phenol. Das antibakterielle Vermögen von halogenierten Chinonen ist höher als dasjenige des Benzochinons; unter den Naphthochinonen erwies sich das 1,2-Isomere wesentlich wirksamer als das 1,4-Naphthochinon. MORGAN und COOPER beobachteten auch, daß Flüssigkeiten biologischen Ursprungs, wie etwa Peptonbrühe, Harn oder Rinderserum, die antibakterielle Wirkung stark hemmen. Aus diesem Grunde schlossen sie, daß die Chinone als Desinfektionsmittel kaum in Frage kämen.

Mit den Arbeiten der letztgenannten Autoren schien das Interesse auf diesem Gebiete zu erlahmen; in den Dreißigerjahren wurde keine einzige erwähnenswerte Untersuchung über die antibakteriellen Eigenschaften der Chinone durchgeführt. Erst seit 1941 belebte sich wieder das Interesse für diese Effekte, was hauptsächlich der Tatsache zuzuschreiben ist, daß um diese Zeit die antibiotischen Wirkungen des Penicillins ihre erste praktische Anwendung erfuhren. Es lag nahe, unter anderen, schon bekannten Stoffwechselprodukten von Pilzen neue, wirksame Antibiotica zu suchen, und so wurden die von der Schule von Raistrick aus Pilzen isolierten Chinone auf ihre antibiotischen Effekte untersucht. Außerdem hatte man etwa zur gleichen Zeit beobachtet, daß das 2-Methyl-1,4-naphthochinon, welches als vitamin-K-aktives Produkt ausgedehnte therapeutische Verwendung findet, auch eine beachtliche antibakterielle Wirkung zeigt.

Oxford, welcher gemeinsam mit Raistrick (*252*) die antibiotischen Effekte von Fumigatin und Spinulosin beschrieb, untersuchte später (*250*) eine größere Anzahl synthetischer Derivate des *p*-Benzochinons auf ihre Wirkungen gegenüber verschiedenen Bakterienarten. Er stellte fest, daß mehrere synthetische Produkte, wie z. B. 3,5-Dimethoxy-toluchinon, Trimethoxy-toluchinon und 2,6-Dimethoxy-benzochinon das Wachstum von *Staphylococcus aureus* weitaus stärker hemmen als das Fumigatin, welches wiederum das Spinulosin um ein Vielfaches übertrifft.

Auf Grund dieser Ergebnisse unternahm es Oxford, bestimmte *Beziehungen zwischen der Konstitution der Chinone und ihrer Wirkung* abzuleiten. So stellt er fest, daß die Einführung einer Hydroxylgruppe in das Molekül die Wirksamkeit vermindert, während Methoxygruppen diese erhöhen. Er beobachtete auch ein beträchtliches Abfallen der Wirkung mit der Zeit und widmete dieser Erscheinung eine besondere Untersuchung (*249*). Er hält es für wahrscheinlich, daß die Chinone eine Bindung mit Eiweißstoffen der bakteriologischen Media eingehen, wodurch die Wirksamkeit gegenüber den Bakterien abgeschwächt wird.

Entsprechende Untersuchungen über Naphthochinone wurden von Armstrong und Mitarbeitern (*13, 14*), später von Calandra und Mitarbeitern (*54*), Moeller (*238*), Alcalay (*1*), Lloyd und Middlebrook (*214*), Vinet (*311*), Atkins und Ward (*23*), sowie Colwell und McCall (*61*) durchgeführt.

Nach Alcalay lassen sich bei den Naphthochinonen, was die Substitution des chinoiden Kerns betrifft, folgende Gesetzmäßigkeiten aufstellen: Die Einführung von Halogenatomen erhöht die Aktivität; die Substitution saurer Gruppen verringert sie hingegen. Aliphatische Aminogruppen, welche in den Kern eingeführt werden, bewirken starke bakteriostatische Wirkung gegenüber allen Bakterien, wogegen aromatische Aminogruppen eine Verstärkung nur gegenüber grampositiven Bakterien verursachen. Eine Substitution im benzoiden Kern bleibt fast ohne Einfluß auf den bakteriostatischen Effekt. Naphthochinon-derivate weisen *in vitro* eine beachtliche Aktivität gegenüber Tuberkelbazillen auf; eine ebensolche Wirkung wird auch von dem schon besprochenen Javanicin (XIX; S. 161), einem Stoffwechselprodukt von *Fusarium javanicum*, ausgeübt.

Eine praktische Anwendung der antibakteriellen Wirkung der Chinone scheint aber nicht in Frage zu kommen. Eine genaue Untersuchung über die Wirksamkeit von 3,5-Dimethoxytoluchinon, welches nach OXFORD (*250*) das *in vitro* aktivste Chinon-derivat gegen *Staphylococcus aureus* ist, wurde von GLOCK und Mitarbeitern (*101*) durchgeführt. Es stellte sich heraus, daß das Präparat Mäusen gegenüber Infektionen mit Staphylokokken, sowie auch mit Streptokokken und mit *Clostridium perfringens* keinerlei Schutz gewährte.

Trotz dieser Ergebnisse wurden die Untersuchungen über die antibakterielle Wirkung der Chinone auch in den letzten Jahren weitergeführt. Das Ziel dieser Arbeiten liegt eher darin, den *Mechanismus der Chinonwirkung* auf die Bakterien aufzuklären und auf diese Art Erkenntnisse über den Zusammenhang zwischen der Konstitution der Substanzen und ihrem Bioeffekt zu erhalten, wozu ja die Chinonderivate durch die leichte Darstellbarkeit geradezu einladen. Andrerseits scheinen die Versuche über die antibiotischen Effekte der Chinone auch geeignet zu sein, Daten über verschiedene Stoffwechselprozesse der Mikroorganismen zu gewinnen.

MARINI-BETTOLO und DEL PIANTO (*218*, *219*) unternahmen es, zwischen den verschiedenen möglichen Wirkungsmechanismen zu unterscheiden; ihre Untersuchung kann aber nicht als quantitativ bezeichnet werden, da sie nur zwischen vollkommener, teilweiser und keiner Hemmung des Wachstums unterscheiden. Immerhin konnten sie einwandfrei feststellen, daß keine direkte Beziehung zwischen den Redoxpotentialen und der antibiotischen Wirkung besteht, welche Annahme allerdings schon vorher von PAGE und ROBINSON (*253*) widerlegt worden war.

Eine ausführliche Untersuchung von GEIGER (*100*) veranlaßte diesen Verfasser zu folgenden Schlüssen: Die Aktivität der Chinone gegenüber gramnegativen Bakterien sei allgemein geringer als gegenüber grampositiven. Für die Wirkung auf gramnegative Bakterien sei die Anwesenheit von freien Orthostellungen neben den chinoiden Carbonylgruppen erforderlich. Sulfhydryl-verbindungen seien imstande, die Wirkung auf gramnegative Bakterien aufzuheben oder zu verhindern, wogegen grampositive Organismen auf diese Weise nicht geschützt werden können.

In letzter Zeit wurden auch im hiesigen Laboratorium derartige Versuche unternommen (*138*), wobei es eine neuentwickelte Methodik gestattete, recht genaue Aussagen über den Grad der partiellen Hemmungen zu machen. Wir konnten die erwähnten Angaben von GEIGER nicht voll bestätigen. In unseren Versuchen erwiesen sich z. B. Verbindungen ohne freie Orthostellung, wie Chloranil und Dichlorthymochinon als besonders starke Hemmstoffe für gramnegative Bakterien; anderseits konnten wir zeigen, daß Sulfhydryl-verbindungen die Wirkungen der Chinone gegenüber allen verwendeten Bakterienarten aufzuheben imstande sind, während nach GEIGER dies nur für die gramnegativen Organismen gelten sollte. Eine bemerkenswerte und anscheinend neue Beobachtung ist, daß eine Anzahl der Wirkstoffe in Konzentrationen, welche gerade unter den noch hemmenden liegen, eine wuchs*anregende* Wirkung zeigen. Es konnten Wachstumsförderungen bis zu fast 50% beobachtet werden.

Während, wie erwähnt, die Arbeiten über die Hemmwirkungen der Chinone auf das Bakterienwachstum bisher nur theoretische Bedeutung erlangt haben, ist es kürzlich FIESER und Mitarbeitern gelungen, eine Anzahl von Substanzen der 2-Oxy-3-alkylnaphthochinon-reihe zu synthetisieren, welche eine bemerkenswerte Aktivität gegenüber *Malaria*-parasiten zeigen. Sie geben eine Übersicht über ihre ausgedehnten Arbeiten (*89*), so daß hier eine kurze Besprechung genügt.

Die genauere Erforschung der Antimalariawirkung von Naphthochinon-derivaten erfolgte, nachdem schon 1943 festgestellt worden war, daß Hydrolapachol (CXII) eine gewisse Aktivität gegenüber *Plasmodium lophurae* in Enten zeigte.

O
CH₂ · CH₂ · CH(CH₃)₂
O

(CXII.) Hydrolapachol.

Im Laufe mehrerer Jahre wurde eine größere Anzahl von verwandten Produkten synthetisch hergestellt, wobei es sich als vorteilhaft erwies, die Alkyl-Seitenkette des Hydrolapachols in 2-Stellung zu verändern. Manche der auf diese Art erhaltenen Stoffe zeigten mehr als hundertfach stärkere Aktivität als die Ausgangssubstanz. Es konnten auch einige bemerkenswerte Regeln über den Einfluß der Konstitutionsveränderungen auf die Wirksamkeit aufgestellt werden:

Bei Verlängerung der Seitenkette des Hydrolapachols durch Einfügung mehrerer Methylengruppen steigt die Wirksamkeit zuerst an, erreicht ein Optimum bei C_9 und sinkt dann wieder ab. Ähnliche Beziehungen und das gleiche Optimum finden sich bei der Reihe der 2-*n*-Alkyl-3-oxy-1,4-naphthochinone und ebenso verhalten sich die Derivate mit verschiedentlich verzweigten Seitenketten. Wenn aber die Seitenkette auch einen alizyklischen Ring enthält, so liegt das Optimum der Aktivität bei C_{10} oder C_{11}. Zwei Ringe oder ein Arylrest verschieben es noch mehr in der Richtung der längeren Kette. Im folgenden sind einige Formelbilder (CXIII bis CXVII) wiedergegeben. Die Nummern mit einem vorgesetzten *M* sind willkürlich zur Bezeichnung gewählt. Der Faktor *Q* gibt die Wirksamkeit der Substanz im Vergleich mit Chinin an.

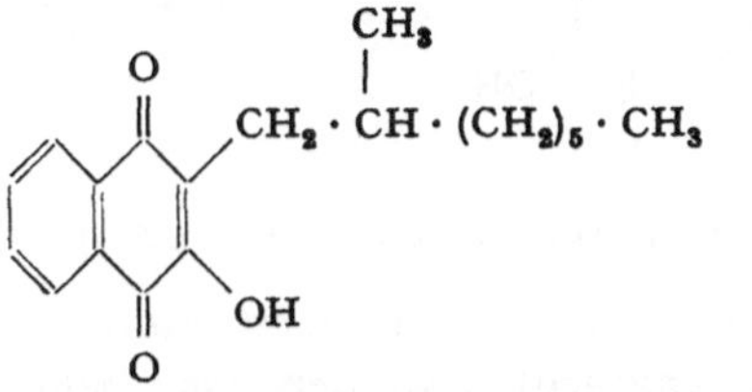

(CXIII.) *M*-285. *Q* = 2,1.

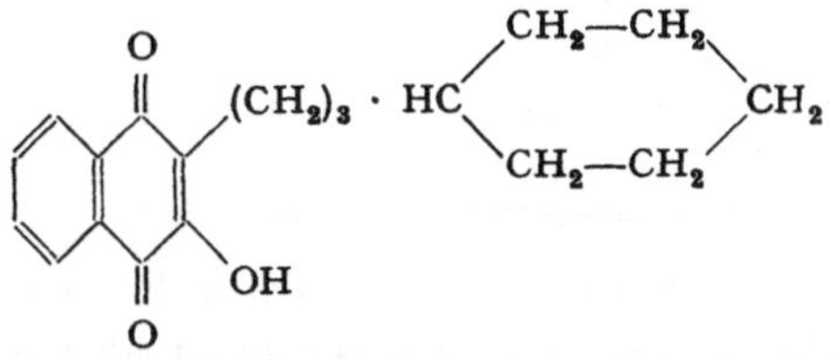

(CXIV.) *M*-1916. *Q* = 0,5.

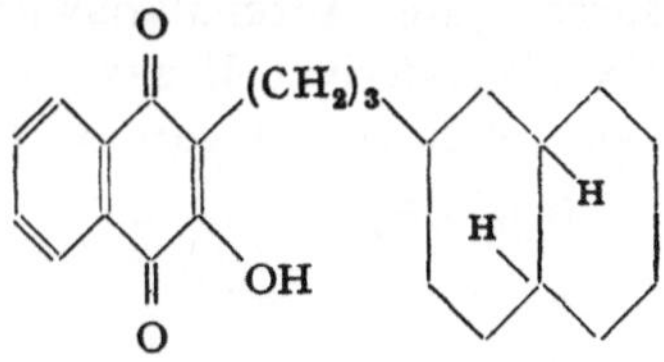

(CXV.) *M*-297. *Q* = 1,2.

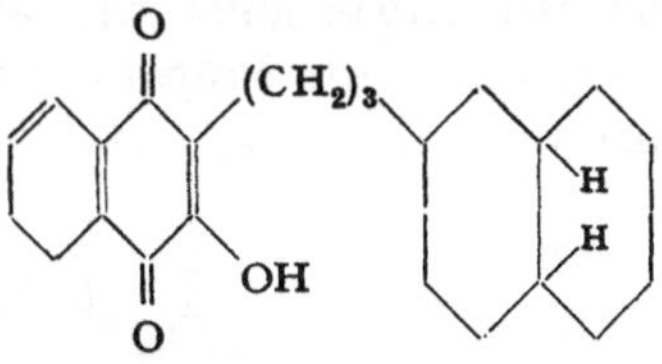

(CXVI.) *M*-2279. *Q* = 2.9.

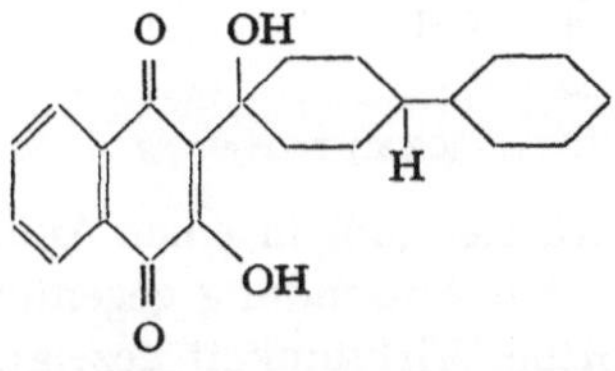

(CXVII.) *M*-2293. *Q* = 15,3.

Anscheinend ist die direkte Kupplung des alizyklischen Ringes an den Naphthochinonkern besonders vorteilhaft, denn *M*-2293 erwies sich als das wirksamste der untersuchten Produkte, soweit es sich um die durch *P. lophurae* verursachte Infektion bei Enten handelte.

Die sterische Anordnung in der Seitenkette ist von großer Bedeutung, was am Beispiel von *M*-297 (CXV) und *M*-2279 (CXVI) ersehen werden kann. Die *trans*-Verbindung ist hier mehr als doppelt so stark wirksam wie die entsprechende *cis*-Form. *M*-2293 (CXVII) ist sogar 13fach wirksamer als sein *cis*-Isomeres.

Klinische Versuche mit derartigen Präparaten, welche an Patienten vorgenommen wurden, die eine Malariatherapie durchgemacht hatten, ergaben allerdings weniger ermutigende Ergebnisse. Der Grund dafür liegt darin, daß die erwähnten Stoffe im menschlichen Organismus sehr schnell chemisch verändert werden. Hier handelt es sich um eine Oxydation der Seitenkette. Die nachstehend gezeigten Abbauprodukte (CXVIII) und (CXIX) sind, obwohl die am Molekül vorgenommene Veränderung als sehr gering erscheint, klinisch völlig inaktiv.

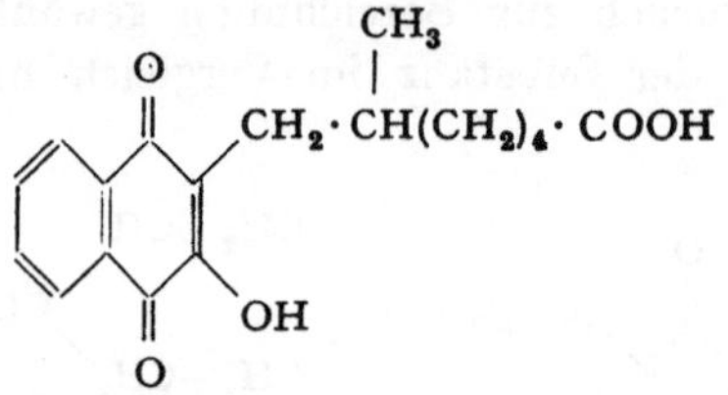

$CH_2 \cdot CH(CH_3)(CH_2)_4 \cdot COOH$

(CXVIII.) Abbauprodukt von *M*-285 (CXIII).

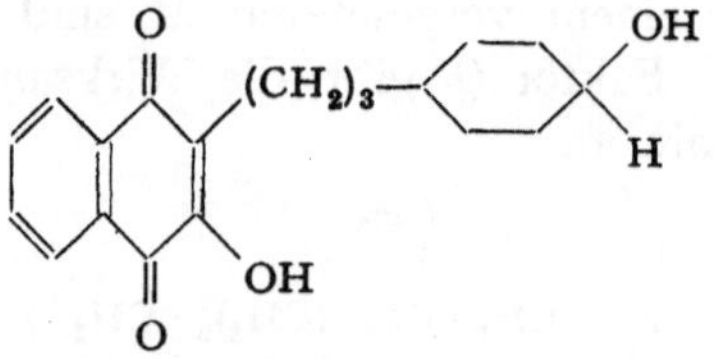

$(CH_2)_3$ — Cyclohexyl-OH

(CXIX.) Abbauprodukt von *M*-1916 (CXIV).

Die Suche nach ähnlich gebauten und ebenso wirksamen Stoffen, welche aber vom Körper nicht so rasch angegriffen werden, war bald erfolgreich. Das durch seine gute Resistenz gegenüber dem Abbau im menschlichen Organismus wie auch durch gute Antimalariawirkung derzeit vorteilhafteste Präparat ist das Oxyalkyl-derivat *M*-2350 (CXX) mit einer verzweigten C_{19}-Kette, welche ein sekundäres Hydroxyl trägt.

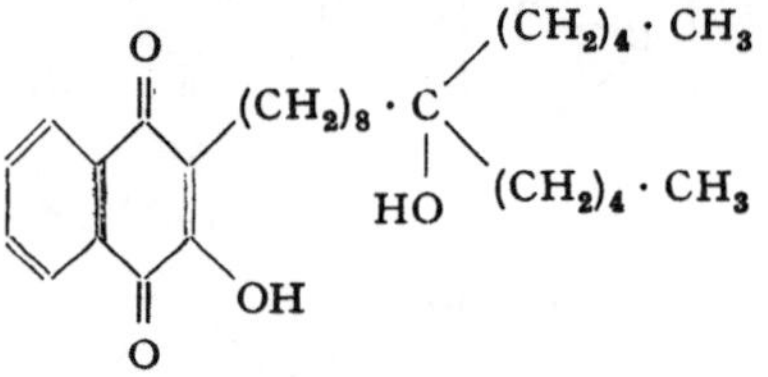

$(CH_2)_8 \cdot C(OH)[(CH_2)_4 \cdot CH_3]_2$

(CXX.) *M*-2350.

Wie FIESER und Mitarbeiter (*89*) in einer Anmerkung berichten, hat dieses Präparat bei der ersten Anwendung gegenüber *Plasmodium vivax*-Infektionen eine erstaunliche Wirksamkeit gezeigt.

Über den Wirkungsmechanismus dieser Substanzen läßt sich noch nichts Bestimmtes aussagen, wenngleich die FIESERsche Schule sich bemüht hat, diese Probleme zu lösen. Der primäre Angriffspunkt der Substanzen dürfte ein an der Atmung der Malariaparasiten beteiligtes Ferment sein. WENDEL (*318*) konnte schon 1946 beobachten, daß die Fähigkeit der Stoffe, die Atmung von Malariaparasiten *in vitro* zu hemmen, mit ihrer Wirksamkeit gegenüber der Vogelmalaria parallel geht. BALL und Mitarbeiter (*27*) untersuchten dann die Wirkung einiger Alkyloxy-naphthochinone auf verschiedene Enzymsysteme und konnten feststellen, daß ein Bernsteinsäure-oxydasesystem, welches aus Bernsteinsäure-dehydrogenase, Cytochrom und Cytochrom-oxydase besteht, bereits durch sehr niedrige Wirkstoffkonzentrationen gehemmt wird, während eine Anzahl anderer Enzyme (Cytochrom-oxydase allein, Xanthin-oxydase, *d*-Aminosäure-oxydase, Katalasen, verschiedene Pyridinnucleotid-spezifische Dehydrogenasen sowie Urease) in ihrer Aktivität nicht beeinträchtigt werden. HEYMANN und FIESER (*121*) fanden allerdings später, daß zwischen der Wirkung auf das Succin-oxydasesystem und der Aktivität der Substanzen gegenüber Malariaparasiten keine befriedigende Übereinstimmung festgestellt werden kann.

Interessanterweise scheint die Wirksamkeit dieser Substanzen weitgehend von ihren lipophilen bzw. hydrophilen Eigenschaften abzuhängen. Die von FIESER (*91*) eingeführte Größe p_E (definiert als das p_H eines wäßrigen Puffers, der imstande ist, den $^1/_{101}$ten Teil des Stoffes aus einem gleichen Volumen Äther zu extrahieren) hat bei den wirksamsten Substanzen die Werte zwischen 10 und 12.

Die malaria-aktiven Naphthochinon-derivate zerstören diejenigen Formen der Malariaparasiten, welche sich außerhalb der Erythrozyten bzw. in den Zellen des reticulo-endothelialen Systems befinden, weshalb sie wahrscheinlich besonders in Kombination mit den bereits bewährten Chemotherapeutica gegen Malaria (Chinin, Atebrin, Plasmochin) eine besonders wertvolle Wirkung entfalten werden. Wie bekannt, unterdrücken Chinin und Atebrin die Trophozoiten innerhalb der roten Blutzellen, sie haben aber wenig Einfluß auf die Formen, welche außerhalb der Erythrozyten existieren. Jedenfalls spricht alles dafür, daß wir in den Oxyalkyl-naphthochinonen neue, wertvolle Heilmittel gegen Malariainfektionen vor uns haben, denen vielleicht noch eine große Bedeutung bei der Bekämpfung dieser Volksseuche mancher Länder zukommen wird.

Auch gegen *niedere Pilze* erweisen sich einige Chinone als sehr wirksam. Wie wir feststellen konnten, wird das Wachstum von *Hefe* durch die verschiedenen Chinone stark gehemmt; durch stärkere Konzentrationen wird die Hefe abgetötet (*137, 149*). Wie bei den Bakterien, findet sich aber auch hier ein bestimmter Konzentrationsbereich, der knapp unter dem noch hemmenden liegt, in welchem die Chinone eine starke wachstumanregende Wirkung entfalten. Um eine möglichst exakte Beschreibung der durch diese Substanzen auf Hefe verursachten Effekte zu erhalten, wurde im hiesigen Laboratorium eine Untersuchungsreihe begonnen, wobei der Einfluß einer größeren Anzahl von Chinonen auf verschiedene Stoffwechselfunktionen der Hefe bei unterschiedlichen Konzentrationen der Wirkstoffe geprüft werden soll. Die ersten Mitteilungen dieser Reihe (*140, 142*) beschäftigen sich mit dem Einfluß auf die Gärung, die Atmung, die Glykogenbildung, die Gesamtlipoid- und Sterinbildung, sowie die Resorption anorganischen Phosphats aus der Nährlösung. Es konnte gezeigt werden, daß in subletalen Konzentrationen die Gärung, die Phosphatresorption und die Bildung von Lipoiden — und hier besonders des Sterinanteils — merklich gefördert werden, während die Atmung und Glykogenbildung der Hefe unter denselben Umständen eine Hemmung erfahren. Unsere Schlußfolgerungen sollen weiter unten besprochen werden.

Schopfer und Boss (*288*) haben vor kurzem die Wirkung einer Anzahl von Naphthochinon-derivaten (darunter einige vitamin-K-aktive Substanzen) auf 44 verschiedene, meist zu den niedrigen Pilzen gehörende Mikroorganismen untersucht. Sie beobachteten, daß die Hemmung des Wachstums durch Nicotinsäure rückgängig gemacht werden kann. Die fungistatische bzw. fungizide Wirkung der Naphthochinone findet übrigens schon seit längerer Zeit praktische Anwendung. Verwendet wird insbesondere das 2,3-Dichlor-1,4-naphthochinon welches sich zur Bekämpfung von gewissen Getreidesamen-schädlingen *(Ustilago)*, von verschiedenen Pilzkrankheiten, welche Hülsenfrüchte befallen, weiters auch gegen die durch *Rhizoctomia*- und *Glomerella*-Arten verursachten Schädigungen von Baumwollsamen, sowie gegen die für das Verschimmeln von Baumwollgeweben verantwortlichen *Stachybotrys*-Arten bewährt hat (*303*).

Eine starke fungistatische Wirkung verschiedener Naphthochinonderivate gegenüber *Monolinia fructicola* wird von Little, Sproston und Foote (*94 a, 213 a, 213 b*) berichtet.

An dieser Stelle soll auch erwähnt werden, daß Naphthochinone einen hemmenden Einfluß auf die Biolumineszenz, sowie auf die Atmung von *Leuchtbakterien* haben (*227, 298*). Nun hält es Spruit (*296, 297*) auf Grund seiner ausführlichen Untersuchungen für wahrscheinlich, daß das Luziferin, also der an dem Phänomen der Biolumineszenz primär beteiligte Wirkstoff, ein 1,4-Naphthohydrochinon-derivat ist, welches in 2-Stellung eine Carbonylgruppe mit einem noch unbekannten weiteren Rest tragen soll.

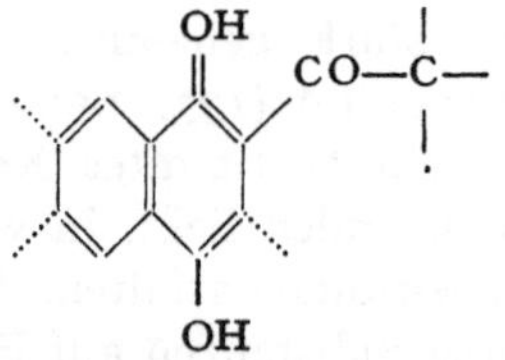

(CXXI.) Formulierung des Luziferins nach Spruit.

Die letztgenannten Hemmeffekte lassen sich auf drei verschiedene Arten erklären: 1. Die Naphthochinone hemmen allgemein die Lebensfunktionen der Organismen; demnach wären sie als unspezifische Zellgifte zu bezeichnen. 2. Die Naphthochinone verdrängen das Luziferin von den Wirkgruppen des bei der Biolumineszenz beteiligten Fermentsystems Luziferase. 3. Die Naphthochinone sind imstande, das Luziferin zu dehydrieren.

Die erste Möglichkeit konnte von Spruit und Schuiling (*298*) ausgeschaltet werden; in Wachstumsversuchen zeigte es sich nämlich, daß Naphthochinonmengen, welche Atmung und Luminiszenz vollständig hemmen, die Zahl der sich entwickelnden Bakterienkolonien überhaupt nicht beeinflussen. Wahrscheinlich ist die dehydrierende Wirkung der Naphthochinone für die Hemmung der Lichtbildung verantwortlich. Wir müssen uns vorstellen, daß der vermutlich die Lumineszenz verursachende Oxydoreduktionsprozeß am Luziferin durch die Chinone derartig gehemmt wird, daß das Luziferin irreversibel zu einer Chinonstufe oxydiert wird. Die Möglichkeit, daß es sich hier vielleicht doch um eine kompetitive Hemmung handelt, ist noch nicht völlig auszuschließen.

Der Einfluß der Chinone auf Vorgänge bei der Zellteilung.

Antimitotische Effekte.

Die Fähigkeit, die Zelle während des Teilungsvorgangs hemmend zu beeinflussen, kommt verhältnismäßig vielen organischen und auch anorganischen Substanzen zu. Wir können hier zwischen Verbindungen unterscheiden, welche eine Hemmung der Zellteilung in irgendeinem Stadium bewirken, wie es z. B. das Colchicin tut (derartige Substanzen werden „Antimitotica" oder „mitosehemmende" Stoffe genannt), und solchen Verbindungen, welche einen Einfluß auf den Zustand, die Gestalt und die Erbeigenschaften der Chromosomen haben und welche „mutagene" Substanzen genannt werden. Ausführliche Berichte über mitosehemmende Stoffe verdanken wir DUSTIN (*78*), LEHMANN (*205*), sowie MEIER und SCHÄR (*229*), während eine ausgezeichnete Übersicht über mutagene Stoffe kürzlich von AUERBACH (24) veröffentlicht wurde. Eine interessante Arbeit über die mögliche Wirkungsweise antimitotischer Substanzen wurde jüngst von LOVELESS und REVELL *(214a)* publiziert.

LEHMANN konnte bereits 1942 beobachten, daß verschiedene Chinone imstande sind, bei einem von ihm selbst gefundenen Testobjekt, beim Ei des Süßwasser-oligochäten *Tubifex*, die Fortentwicklung der Eier vom Ein- oder Zweizellenstadium zur vielzelligen Stufe zu blockieren, ohne dabei die Kerne abzutöten. An einem anderen Versuchsmaterial, nämlich an einer Gewebekultur von Hühnerherz-Fibroblasten, beobachteten MEIER und ALLGÖWER (*228*) eine gleichartige Wirkung der Chinone.

Die teilungshemmende Wirkung der Chinone gegenüber diesen Objekten ist insoweit beachtlich, als diese Substanzen besonders hohe Wirksamkeit im Vergleich zu anderen antimitotisch aktiven Verbindungen zeigen. Während die für die Mitosehemmung erforderliche Minimalkonzentration bei Phenylurethan 1 : 5000, bei Colchicin 1 : 30000 und bei Stilböstrol 1 : 300000 ist, wirkt das 1,2-Benzanthren-3,4-chinon noch bei einer Verdünnung von 1 : 500 Millionen.

In der *Art der Wirkung* unterscheiden sich die einzelnen Chinonklassen wesentlich voneinander. Hierüber liegen ausführliche Berichte von LEHMANN und seinen Mitarbeitern vor (*154. 155, 204, 207, 208*). Das Benzochinon und Derivate, wie Toluchinon, Methoxytoluchinon, Chloranil und auch das Hydrochinon bewirken durchwegs Blockierung der

Teilung, wobei aber der Kernapparat nur wenig geschädigt wird. Doch erfolgt bei den blockierten Keimen meist schon nach einigen Stunden Zellauflösung, während Keime, welche durch Colchicin in ihrer Teilung gehemmt wurden, tagelang weiterleben können. 1,4-Naphthochinon, 1,4-Naphthohydrochinon, sowie 2-Methyl-1,4-naphthochinon blockieren die Furchung des Tubifex-eies in einem kleinen Konzentrationsbereich zwischen 1 : 3 und 1 : 15 Millionen; Cytolyse wird aber nicht bewirkt. 1,2-Naphthochinon verursacht Zellauflösung ohne blockierend zu wirken. 9,10-Anthrachinon erweist sich als völlig inaktiv, hingegen ist das 1,2-Anthrachinon sowohl antimitotisch als auch cytolytisch wirksam, wobei ein antimitotischer Konzentrationsbereich ohne Cytolyse nicht aufgefunden werden konnte.

Der Wirkung des Colchicins am ähnlichsten sind diejenigen Effekte, welche durch Derivate des 9,10-Phenanthrenchinons, sowie höher kondensierte Chinone hervorgerufen werden. Hier sind insbesondere zu erwähnen: das 9,10-Phenanthrenchinon selbst, weiters das Pimanthrenchinon (1,6-Dimethyl-9,10-phenanthrenchinon), das Retenchinon (1-Methyl-7-isopropyl-9,10-phenanthrenchinon), sowie das oben genannte Tetraphenchinon, (1,2-Benzanthren-3,4-chinon), welches die höchste Wirksamkeit der Reihe zeigt.

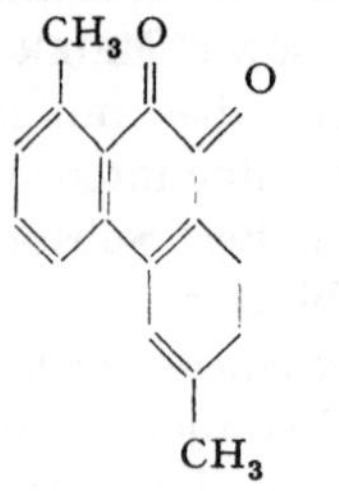

(CXXII.) Pimanthrenchinon. 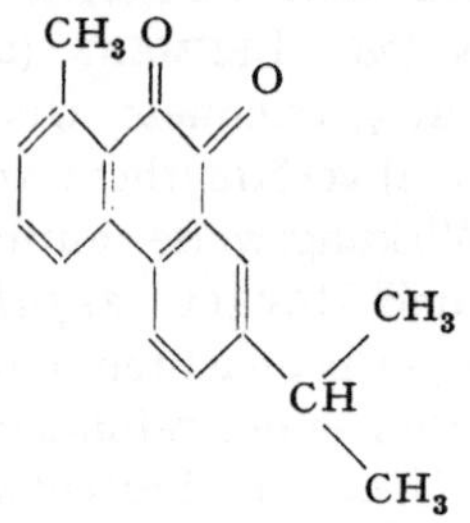 (CXXIII.) Retenchinon. 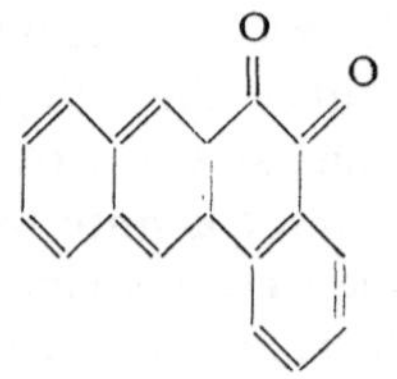 (CXXIV.) Tetraphenchinon.

Es ist von einigem Interesse, daß das dem Tetraphenchinon sehr nahe verwandte Chrysenchinon vollständig wirkungslos ist.

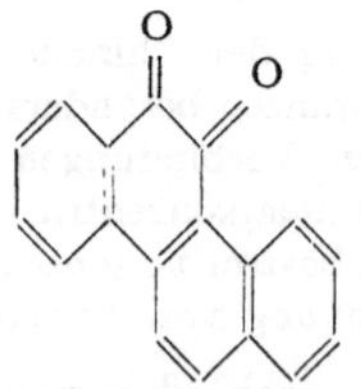

(CXXV.) Chrysenchinon.

Die Wirksamkeit der einzelnen antimitotischen Chinone wird nach LEHMANN (*205*, *206*) in Abb. 1 veranschaulicht.

In einer späteren Arbeit berichten LEHMANN und ANDRES (*206 a*) über die

Wirkung von Acenaphthochinon auf den Kern embryonaler Keime von *Tubifex* und Amphibien. Es wird das Auftreten von Polyploidie festgestellt. Benzanthrachinon bewirkt ebenso wie Colchicin und Podophyllin auch Verschwinden der embryonalen Zellkerne und Bildung sogenannter Pseudokerne, welche im FEULGEN-Test positiv sind. Zwei weitere neue Arbeiten aus demselben Laboratorium (*282 a*, *282 b*) behandeln die Beeinflussung des Meiosevorgangs im *Tubifex*-Ei durch 1,4-Naphthochinon und Phenanthrenchinon.

Nach Untersuchungen von MITCHELL und SIMON-REUSS (*233*) zeigt das synthetische Vitamin-K-Präparat, 2-Methyl-1,4-naphthohydrochinon-diphosphat sowohl gegenüber Hühnerherz-Fibroblasten als auch gegenüber Gewebekulturen von gewissen menschlichen Carcinomen eine mitosehemmende Wirkung. Nach einer späteren Mitteilung aus demselben Laboratorium (*98*) ist das unsubstituierte 1,4-Naphthohydrochinon-diphosphat 1000fach wirksamer als das 2-Methylprodukt. Die Autoren zeigen auch, daß wahrscheinlich nicht die Chinonnatur der Verbindungen — es ist anzunehmen, daß die phosphorylierten Hydrochinone *in vivo* in Chinone umgewandelt werden — für die Wirkung verantwortlich ist, sondern anscheinend das konjugierte System —CO · CH : CH · CO— in *cis*-Konfiguration, da auch Maleinsäure, nicht aber Fumarsäure starke antimitotische Effekte hervorrufen.

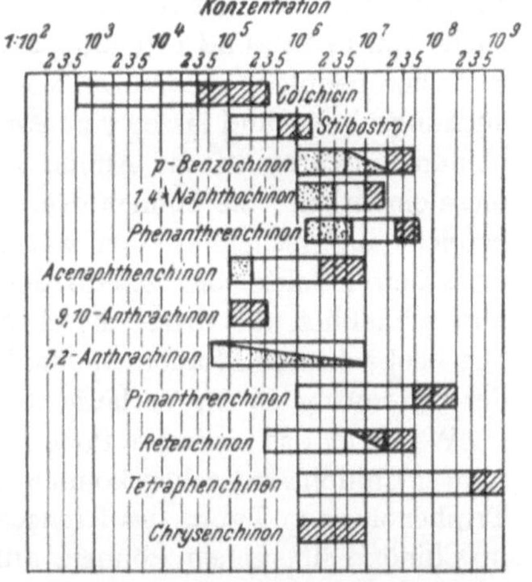

Abb. 1. Der Wirkungsmechanismus antimitotischer Stoffe beim *Tubifex*-Ei. Der antimitotische Bereich wird nach links vom cytolytischen Bereich (punktiert), nach rechts vom entwicklungsstörenden Bereich (schraffiert) begrenzt. [Nach LEHMANN (*205*), (*206*).]

In letzter Zeit berichtete MITCHELL (*232*) über eine therapeutische Anwendung der antimitotischen Wirkung der Chinone gegenüber Carcinomzellen in Kombination mit Röntgenstrahlen.

Über antimitotische Wirkungen gegenüber *Pflanzen*zellen ist bisher noch nicht viel bekannt geworden. NYBOM und KNUTSON (*246*) untersuchten die durch 2-Methylnaphthochinon, weiters durch das Diacetat und das Diphosphat des entsprechenden Hydrochinons verursachte Hemmung der Mitose bei *Allium cepa* (Speisezwiebel). Die Autoren berichten, daß sie insbesondere mit 2-Methyl-1,4-naphthochinon ungewöhnliche Formen der Mitosehemmung beobachten konnten. Die Chromosomen verteilen sich in der Anaphase beliebig in zwei oder manchmal mehreren Gruppen an den Polen. Für diese Erscheinung wird der Name „*distributed c-mitosis*“ (verteilte c-Mitose) vorgeschlagen. (Unter c-Mitose wird allgemein eine colchicinartige Hemmung der Mitose verstanden.)

In einer neuen Arbeit macht Reed (*277*) Chinone, welche aus Polyphenolen entstehen sollten, für die Verkümmerung von durch Zinkmangel geschädigten Pflanzen verantwortlich. Er zeigt, daß eine Mitosehemmung vorliegt, welche durch Cystein rückgängig gemacht werden kann, und nimmt an, daß das Auftreten von Chinonen durch das Fehlen eines Enzymsystems bedingt wird, welches Zink für seine Funktion benötigt.

Mutagene Effekte.

Während in den bisher berichteten Untersuchungen hauptsächlich die entwicklungshemmende Wirkung der Substanzen erforscht und auf mutagene Effekte weniger Wert gelegt wurde, so liegt das Hauptgewicht der Versuche von Levan und Tijo (*209, 210*) auf den *mutagenen Wirkungen*. Diese Forscher benützen die Wurzel von *Allium cepa* (Speisezwiebel) als Untersuchungsobjekt und lassen eine Anzahl von Phenolen, aber auch *p*-Benzochinon und aromatische Amine vom Typus des *p*-Phenylendiamins auf diese einwirken. Sie können drei verschiedene wirksame Konzentrationsbereiche unterscheiden, von welchen der erste derjenige der allgemein toxischen Effekte, der zweite derjenige der c-mitotischen Wirkung ist und schließlich der dritte die chromosomen-verändernden Effekte zeigt. Die beobachteten Phänomene sind Fragmentationen sowie Verklebungen der Chromosomen im Verlaufe der Mitose.

Während das einfache Phenol und andere Substanzen, aus welchen schwer Chinone entstehen können, geringe Effekte bewirken, sind Di- und Triphenole vom Typus des Brenzcatechins und Pyrogallols, welche leicht in Chinone übergehen können, auffallend aktiv. Der Verfasser ist nun der Auffassung, daß die tatsächlich wirksamen Agentien in diesem Falle die Chinone sind, welche aus den Diphenolen entstehen, was wohl dadurch bekräftigt wird, daß die Wurzeln von *Allium cepa* eine stark wirksame Oxydase vom Typus der Tyrosinase aufweisen. Die Tatsache, daß *p*-Benzochinon etwas geringere Aktivität zeigt als die genannten Polyphenole, ist noch kein Beweis gegen diese Annahme, denn *p*-Benzochinon zersetzt sich unter den von Levan und Tijo (*209, 210*) geschilderten Versuchsbedingungen sehr rasch, und es mag wohl sein, daß gerade die dauernde fermentative Neubildung von Chinonen aus den Polyphenolen die hohe Wirksamkeit dieser Substanzen bedingt.

Es wäre übrigens denkbar, daß auch die mutagene Wirkung von Phenolen, welche Hadorn (*109*) bei Behandlung explantierter Ovarien von *Drosophila melanogaster* beobachten konnte, auf die Umwandlung des Phenols in chinoide Substanzen zurückzuführen sei. Hadorn selbst hält es auf Grund gewisser Unregelmäßigkeiten in seinen Versuchsergebnissen für möglich, daß nicht das Phenol selbst, sondern eine in kleinen Mengen vorhandene Begleitsubstanz für die Effekte verantwortlich sei.

Es ist nicht unwahrscheinlich, daß viele der in der Literatur berichteten Wirkungen der Chinone auf das Pflanzenwachstum und vielleicht auch auf Tiere durch die antimitotischen und mutagenen Effekte erklärt werden können.

Dies gilt z. B. von einer älteren Untersuchung über den Einfluß von *p*-Benzochinon auf die Keimung verschiedener Samen (*292*), weiters auch für eine quantitative Versuchsreihe über die Wirkungen verschiedener Chinone auf das Keimlingswachstum von *Lepidium sativum*, welche im hiesigen Laboratorium durchgeführt wurde (*147*). Hier ist es bemerkenswert, daß die Chinone eine stärkere Hemmwirkung entfalten als die von KUHN und Mitarbeitern (*193*) beschriebenen Blastokoline und sogar in manchen Fällen das in dieser Richtung besonders aktive Cumarin übertreffen.

Wirkungen der Chinone auf tierische Organismen.

Wirkung von Chinonen auf Invertebraten.

Über den Einfluß von Chinonen auf Invertebraten scheint, wenn wir von den schon abgehandelten Wirkungen der Echinochrome absehen, bisher noch nicht viel bekannt zu sein.

ANFINSEN (*5*) untersuchte den Einfluß von 2-Oxy-3-(2'-methyloctyl)-1,4-naphthochinon auf die Zellteilung und die Atmung von Eiern von *Arbacia punctulata*. In Konzentrationen von 10^{-7} bis 10^{-8} molar wurde etwa 50%ige Hemmung der Atmung und vollständige Blockierung der Teilung bewirkt.

MEYERHOF und RANDALL (*230*) berichten, daß Adrenochrom und 1,2-Naphthochinon-4-sulfonsäure einen hemmenden Einfluß auf die Atmung, die Glykolyse und die Beweglichkeit bei *Trypanosoma equiperdum* ausüben, wobei Adrenochrom die aktivere Substanz ist.

Versuche über die Einwirkung von Phenanthrenchinon auf Kulturen von *Glaucoma pyriformis* KAHL wurden von JONES und BAKER (*160*) ausgeführt. Wenige Tropfen einer Phenanthrenchinonlösung, welche in eine auf einem Uhrglas befindliche Kultur der Organismen gebracht wird, bewirken, daß sich die einzelnen Individuen in Aggregationen zusammenballen. Die wenigen außerhalb dieser Aggregationen verbliebenen Organismen unterlagen bald der Giftwirkung des Phenanthrenchinons, während die Aggregationen auf noch nicht geklärte Weise imstande sind, den Einzelindividuen Schutz zu bieten, die nach Aufhören der schädlichen Einflüsse die Zusammenballung verließen und weiterleben konnten. Auch eine Anzahl anderer Chinone kann dieselben Effekte hervorrufen.

Eine Untersuchung aus dem Laboratorium des Verfassers (*37, 130*) befaßt sich mit der Giftwirkung verschiedener Chinone auf den Strudelwurm *Planaria gonocephala*. Dieser Organismus hat die Eigenschaft, unter schädigenden Einflüssen, wie höherer Temperatur, ultravioletter Strahlung und toxischer Stoffe, vor dem Todeseintritt graduell zu zerfallen. Es handelt sich hier um eine Histolyse, welche anfangs begrenzt, an bevorzugten Körperstellen auftritt und sich dann mehr oder minder schnell auf den ganzen Organismus fortpflanzt. Bei rechtzeitigem Entzug der schädlichen Einflüsse ist der Wurm imstande, die bereits zerfallenen Körperteile zu regenerieren.

Unter den untersuchten Chinonen erwiesen sich Naphthazarin und Methylnaphthazarin am giftigsten, die Oxynaphthochinone Isonaphthazarin und Lawson hatten die geringste Wirkung. *p*-Benzochinon, Toluchinon, 1,2-Naphthochinon, *p*-Xylochinon sowie die Methoxybenzochinone zeigten alle beachtliche Giftwirkung, welche aber hinter derjenigen der Naphthazarine zurückblieb. In fast allen untersuchten Konzentrationen der Chinone war Zerfall zu beobachten; eine Ausnahme fand sich nur bei den höchsten herstellbaren Konzentrationen von Benzochinon und Toluchinon, welche Todeseintritt bei gleichzeitiger Fixierung verursachten. Es dürfte sich hier um eine unspezifische Eiweißwirkung (Gerbung) handeln. Der Verlauf der Dosiswirkungs-kurven für die Zerfallseffekte läßt darauf schließen, daß der Primärvorgang eine Adsorption des Wirkstoffes ist.

Weniger typisch für die Einflüsse der Chinone auf niedere Tiere ist eine Untersuchung von Bergstermann (*32*) über die Wirkung der Substanzen auf Blutegel-Bauchmarkpräparate. *p*-Benzochinon verursacht bereits in Verdünnung von 1 : 100000 leichte Erregung des Zentrenapparates, während 1 : 20000 eine sehr große Frequenzsteigerung bewirkt, welcher unter zunehmender Kontraktur schließlich bald irreversible Lähmung folgt.

Wirkung von Chinonen auf Vertebraten.

Über die Wirkungen von Chinonen auf Lebensfunktionen *höherer Tiere* ist besonders von pharmakologischer Seite häufig gearbeitet worden. Hohe Konzentrationen verschiedener Chinone, so insbesondere der Benzochinon-derivate, verursachen lokale Gewebsschädigungen, wie Nekrosen. Ein Bericht von Takizawa (*301*) über eine cancerogene Wirkung von Chinonen scheint bisher keine Bestätigung gefunden zu haben.

Eine wichtige und vielbearbeitete Erscheinung ist die durch Chinone verursachte Methämoglobinbildung, welche sowohl *in vivo* als auch *in vitro* beobachtet werden kann. Heubner und seine Schule haben diesem Effekt eine Anzahl von Experimentalarbeiten gewidmet, welche in einem Übersichtsreferat von Heubner (*120*) ausführlich besprochen sind. Nach diesem Autor läßt sich die Reaktion des Hämoglobins mit dem Chinon auf das Bestehen einer Spannung zwischen zwei Redoxsystemen zurückführen, welche er wie folgt symbolisiert:

Die Bedingungen für die Bildung des rechtsstehenden Paares werden mit zunehmender Wasserstoffionenkonzentration günstiger, was der gleichgerichteten Zunahme des Redoxpotentials des Chinon-Hydrochinon-Systems entsprechen soll.

Eine Arbeit aus dem hiesigen Laboratorium (*148*) ergab allerdings bei Vergleich der Methämoglobinbildung durch Chinone verschiedenen Redoxpotentials, daß dieses für das Vermögen der Substanzen, Methämoglobin zu bilden nicht absolut maßgeblich ist. So ist das 1,2-Naphthochinon, welches ein niedrigeres Redoxpotential besitzt als das *p*-Benzochinon, der stärkere Methämoglobinbildner.

Die durch die Chinone verursachte Methämoglobinbildung erklärt zwanglos die durch diese Stoffe verursachten Atmungsstörungen, welche bis zum Erstickungstod führen können, da ja das Methämoglobin nicht imstande ist, die sauerstoffübertragende Funktion des Hämoglobins zu übernehmen.

Über die Beeinflussung des Zentralnervensystems liegen einige ältere Arbeiten vor. So berichtete BAGLIONI schon 1900 über folgenden Versuch (*24 a*):

Ein Frosch wird unter Ausschaltung des Herzens mit einer Mischung von gleichen Teilen defibriniertem, koliertem Ochsenblut und physiologischer Kochsalzlösung, welcher das Chinon zugesetzt wird, von der *Aorta communis* aus durchspült, wobei ein rhythmisch arbeitender Motor die Rolle des Herzens übernimmt. Bei einer Konzentration von *p*-Benzochinon von 1 : 600 in der Blutmischung gerät der Frosch nach zwei bis drei Pumpenstößen in unaufhörliche fibrilläre Zuckungen, die manchmal sogar tetanischen Charakter annehmen. Der Effekt tritt auch in einem Bein mit durchschnittener *Arteria ischiadica* auf, was den Ursprung aus dem Zentralnervensystem beweist. Auf eine ähnliche Beeinflussung deuten die bereits berichteten Arbeiten von BERGSTERMANN (*32*) über die Wirkung von Chinonen auf das Bauchmark des Blutegels.

Von verschiedenen Seiten wird über eine blutdrucksenkende Wirkung der Chinone an Ratten bei subcutaner oder peroraler Applikation berichtet (*97*, *290*). Dieser Effekt läßt sich vielleicht dadurch erklären, daß die Chinone eine aktivierende Wirkung auf das Hypertensinase-System ausüben sollen (*66*), wobei diese Angabe wohl noch der Überprüfung bedarf. Eine andere Möglichkeit des Mechanismus dieser Wirkung wird von SOLOWAY und OSTER (*295*) besprochen. Sie zeigten, daß die hochdruckerregenden Amine durch verschiedene Chinone *in vitro* inaktiviert werden. Eine ausführliche Behandlung unserer derzeitigen Kenntnisse über die beim Zustandekommen des essentiellen Hochdrucks beteiligten Faktoren findet sich bei SCHALES (*284*).

Allgemein toxikologische Arbeiten über einzelne Chinone, welche in größerer Anzahl — so insbesondere bei den vitamin-K-aktiven Stoffen — vorgenommen wurden, bieten in diesem Zusammenhang nur geringes Interesse. Es mag immerhin erwähnt werden, daß anscheinend Derivate des 1,4-Naphthochinons allgemein eine geringere Giftigkeit aufweisen, als sie *p*-Benzochinon-derivaten zukommt.

Zugeführte Chinone scheinen im Körper zu den entsprechenden Hydrochinonen reduziert zu werden, wobei der Organismus sie dann meist in konjugierter Form, z. B. als Glucuronide oder als Ester von Aminosäuren, ausscheidet. Das erstere gilt z. B. vom 3,5-Dimethoxytoluchinon, während das einfache Toluchinon in der Form seines Hydrochinons mit Aminosäuren konjugiert wird (*101*).

Der mögliche Mechanismus von Chinonwirkungen.

Es ist verständlich, daß gerade dadurch, daß die Chinone einfache, leicht variierbare Substanzen sind, ein gewisser Anreiz für die Untersucher besteht, Aufklärung über den Mechanismus einiger der vielen beschriebenen Wirkungen zu suchen.

Eine Anzahl von Effekten, und zwar alle diejenigen, welche durch verhältnismäßig hohe Konzentrationen verursacht werden, lassen sich dadurch erklären, daß die Chinone eine unspezifische „eiweißgerbende" Wirkung ausüben. In der Industrie macht man schon seit vielen Jahren von dieser Eigenschaft Gebrauch; *p*-Benzochinon wird bei der Herstellung bestimmter Ledersorten verwendet, wenn auch die Chinongerberei nur in einem geringen Umfang betrieben wird, da der hohe Preis einen ausgiebigeren Gebrauch verbietet. Meist wird die Chinongerbung in Kombination mit anderen Gerbstoffen vorgenommen.

Es sind anscheinend vor allem, wenn auch nicht ausschließlich, die freien Aminogruppen der Proteine, welche mit den Chinonen reagieren. Dies dürfte in ähnlicher Weise vor sich gehen, wie es FISCHER und SCHRADER (*92*) bei der Reaktion des Glykokolläthylesters mit Benzochinon beobachtet hatten: in diesem Falle können zwei Aminogruppen in den Chinonkern treten, wofür Wasserstoff freigesetzt wird, welcher das Chinon teilweise reduziert.

$$C_6H_4O_2 + 2\,H_2N \cdot H_2C \cdot COOC_2H_5 \xrightarrow{-4H} H_5C_2OOC \cdot H_2C \cdot HN{-}C_6H_2O_2{-}NH \cdot CH_2 \cdot COOC_2H_5$$

(CXXVI.) Reaktion von *p*-Benzochinon mit Glykokollester.

HILPERT und BRAUNS führten Modellversuche mit Benzochinon und Hautpulver durch (*123*). Aus der Menge des verbrauchten Chinons und der Menge des gleichzeitig gebildeten Hydrochinons läßt sich auf den Typus der Reaktion zwischen Chinon und Protein schließen. Tritt nur eine Aminogruppe in den Chinonkern, so muß sich für zwei angewandte Chinonmoleküle ein Molekül Hydrochinon in der Lösung bilden; wenn hingegen Disubstitutionsprodukte entstehen sollten, so müßten auf drei Moleküle Chinon zwei Moleküle Hydrochinon entfallen:

$$(A) \quad 2\,C_6H_4O_2 + NH_2 \cdot R = C_6H_3O_2(NH \cdot R) + C_6H_6O_2,$$

$$(B) \quad 3\,C_6H_4O_2 + NH_2 \cdot R = C_6H_2O_2(NH \cdot R)_2 + 2\,C_6H_6O_2.$$

In saurem, neutralem und alkoholischem Medium verläuft die Reaktion eindeutig nach Typus *A*. Es dürfte also nur eine Aminogruppe mit einem Chinonmolekül reagieren. In alkalischem Medium findet ein Sekundärprozeß statt; es bilden sich polymere Oxychinone, welche vom Protein vermutlich adsorptiv aufgenommen werden. Aber auch dann dürfte die Reaktion zum größten Teil nach *A* verlaufen.

Ein Befund von THOMAS und FOSTER (*306*) macht es wahrscheinlich, daß außer den Aminogruppen auch noch andere Gruppen des Proteinmoleküls mit Chinonen reagieren können. Die Autoren beobachteten, daß desaminiertes Hautpulver auch im sauren Bereich — also ohne Bildung polymerer Oxychinone — imstande ist, sich mit Chinonen umzusetzen. Dieser Befund läßt sich durch eine Reaktion der Chinone mit den SH-Gruppen der Cysteinreste im Eiweißmolekül erklären. KUHN und BEINERT (*191*) haben vor einiger Zeit Modellversuche über die Einwirkung von Chinon auf Cysteinester-chlorhydrat durchgeführt und konnten die Bildung wohldefinierter heterozyklischer Kondensationsprodukte beobachten.

In einer neuen, ausführlichen Arbeit über die chemischen Vorgänge bei der Gerbung (*108 a*) wird auch die Chinongerbung entsprechend abgehandelt.

Die biologische Bedeutung der Eiweißgerbung durch Chinone bei verschiedenen Klassen der Invertebraten wurde bereits auf S. 194 besprochen.

Eine weitere, recht unspezifische Wirkung mancher Chinone ist die Fähigkeit, Aminosäuren oxydativ zu desaminieren. Die Reaktion verläuft nach WIELAND und BERGEL (*322*) über die Bildung einer Iminosäure, welche dann CO_2 abspaltet. Das so gebildete Produkt wird zu Aldehyd und Ammoniak dehydriert.

$$\underset{\displaystyle NH_2}{R \cdot \overset{}{CH} \cdot COOH} \xrightarrow[\text{(Chinon)}]{-2\,H} \underset{\displaystyle NH}{R \cdot \overset{}{C} \cdot COOH} \xrightarrow{-CO_2} \underset{\displaystyle NH}{R \cdot \overset{}{CH}} \xrightarrow{H_2O} R \cdot CHO + NH_3$$

(CXXVII.) Aminosäureabbau durch Chinone.

Dieser Vorgang ist von KISCH und Mitarbeitern (*168*) quantitativ untersucht worden. Eine große Anzahl von Chinonen ist in obigem Sinne wirksam. Es werden durchaus nicht alle Aminosäuren desaminiert; als geeignete Substrate erwiesen sich bisher bei Versuchen *in vitro* Glykokoll, Serin, Leucin, weiters auch Glykokolläthylester, Glycyl-glycin und Glycyl-tyrosin, hingegen werden Isoleucin, Alanin, Valin, α-Aminobuttersäure sowie β-Alanin nicht desaminiert. Inwieweit dieser Reaktion eine Bedeutung beim Zustandekommen biologischer Wirkungen der Chinone zukommt, muß heute noch offen bleiben.

Wirkung der Chinone auf Enzyme.

Verhältnismäßig spezifischer sind die Effekte, welche die Chinone gegenüber einer großen Anzahl von *Fermenten* ausüben, wobei heute die Meinung der meisten Autoren dahin geht, daß wir bei diesen Wirkungen den Schlüssel für viele durch Chinone verursachte biologische Effekte zu suchen haben. Dieses Gebiet ist deshalb in den letzten Jahren von zahlreichen Forschern bearbeitet worden; es muß allerdings zugegeben werden, daß die in diese Arbeitsrichtung gesetzten Hoffnungen sich bisher nur teilweise erfüllt haben.

Die Chinone können anscheinend die einzelnen Fermente auf verschiedene Arten beeinflussen. So reagieren sie vielfach mit Sulfhydrylgruppen, welche für die Enzymaktivität notwendig zu sein scheinen; dieser Mechanismus dürfte der häufigste sein, wenngleich nicht bei allen Fermenten, in welchen für die Wirkung unentbehrliche SH-Gruppen vorliegen, tatsächlich durch Chinone Hemmung eintritt. Bei Enzymen, welche Redoxvorgänge katalysieren, können die Chinone dadurch, daß sie selbst Redoxsysteme darstellen, in diese Vorgänge eingreifen, wobei ihr Effekt als eine Art Redox-pufferwirkung anzusehen ist. In vielen Fällen ist es allerdings zweifelhaft, auf welche Weise die Beeinflussung eines Fermentsystems durch ein Chinon zustande kommt.

Unter den esterspaltenden Enzymen sind es insbesondere die *Phosphatasen*, über deren Hemmbarkeit durch Chinone einiges bekannt ist. Schon 1942 konnte SIZER (*294*) eine Hemmung der alkalischen Phosphomonoesterase (aus Nierengewebe) durch *p*-Benzochinon beobachten. Unsere Untersuchungen (*145*, *146*) beschäftigten sich mit Phosphomonoesterasen verschiedenen Ursprungs, darunter sowohl solche mit saurem als auch mit alkalischem Wirkungsoptimum. Alle diese Fermente waren durch eine größere Anzahl von Chinonen in gleicher Weise hemmbar. Die Wirkungsreihen für die einzelnen Chinone stimmten gut überein, so daß auf einen gleichartigen Hemmungsmechanismus sowohl für die drei untersuchten Phosphomonoesterasen aus Hefe als auch für die entsprechenden Fermente der Niere, der Prostata und des Serums geschlossen werden muß.

Die *Pankreaslipase* und die *Serumcholinesterase* (Pferd) erwiesen sich hingegen gegenüber dem Einfluß von Chinonen als unempfindlich (*135*), obwohl beide Enzyme als Sulfhydryl-fermente angesehen werden.

Versuche über die Wirkung von Chinonen auf die *Amylasen* des menschlichen Speichels und der tierischen Pankreasdrüse (*133*) ergaben, daß diese beiden Fermente durch Chinone überhaupt nicht beeinflußt werden, wenn sie durch die Anwesenheit von Salzen aktiviert sind; in Abwesenheit solcher Aktivatoren üben hingegen Chinone eine merkbare aktivierende Wirkung aus.

Eine Hemmwirkung des *p*-Benzochinons auf die *Urease* wurde 1933 etwa gleichzeitig von QUASTEL (*266*) sowie HELLERMANN und Mitarbeitern (*115*) festgestellt. Die Annahme, daß das Redoxvermögen des Chinons für diesen Effekt verantwortlich ist, wurde jedoch bald darauf durch FISCHGOLD (*94*) widerlegt, welcher zeigen konnte, daß Substanzen vergleichbaren Redoxpotentials keinerlei Wirkung auf die Urease haben, so daß es sich hier um eine spezifische Fähigkeit des Chinons handelt. Eine spätere Arbeit aus dem Laboratorium des Verfassers stellte eine Wirkungsreihe verschiedener Chinone gegenüber der Urease auf (*143*). Dabei konnte festgestellt werden, daß durchaus nicht diejenigen Chinone, welche die stärkste antibiotische Aktivität entfalten, auch auf die Urease die größte Hemmwirkung ausüben. In letzter Zeit berichten SCHOPFER und GROB (*289*) über den Einfluß verschiedener Naphthochinon-derivate auf die Ureasewirksamkeit. Ihre wichtigste Beobachtung ist, daß Substanzen mit Vitamin-K-Aktivität, z. B. 2-Methyl-1,4-naphthochinon, imstande sind, die durch andere Naphthochinone bewirkte Hemmung zumindest teilweise rückgängig zu machen.

Die Wirkung der Chinone auf die Urease dürfte allgemein darauf beruhen, daß diese Substanzen mit Sulfhydrylgruppen des Enzyms reagieren, welche für dessen Aktivität wesentlich sind. Durch Zugabe von verschiedenen Thiolverbindungen (Natriumsulfid, Cystein, Glutathion) läßt sich die Urease in ihrer Wirksamkeit reaktivieren, was sehr für die Richtigkeit der obigen Annahme spricht.

Unter den proteolytischen Fermenten ist insbesondere der Einfluß von Chinonen auf das *Papain* und die *Hefeproteinase* näher untersucht worden. BERSIN und LOGEMANN (*36*) konnten schon 1933 die Hemmbarkeit des Papains durch *p*-Benzochinon nachweisen. HELLERMAN und PERKINS (*114*) haben diese Erscheinung eingehend geprüft und schlagen den folgenden Mechanismus für die Wirkung von Oxydationsmitteln auf das Papain vor:

$$\underset{\text{(aktiv)}}{2\,\text{Enzym—SH}} \underset{\text{Reduktion}}{\overset{\text{Oxydation}}{\rightleftharpoons}} \underset{\text{(inaktiv)}}{\text{Enzym—S—S—Enzym}} + 2\,\text{H}$$

Nach Versuchen aus dem Laboratorium des Verfassers (*132*) ist es allerdings unwahrscheinlich, daß diese Verhältnisse so einfach liegen, da sonst anzunehmen wäre, daß verschiedene Chinone entsprechend ihrem Redoxpotential imstande wären, die Aktivität des Papains zu hemmen, was aber nicht der Fall ist. Das Papain verhält sich gegenüber den einzelnen Chinonderivaten in sehr ähnlicher Weise wie die Urease, was auf einen gleichartigen Hemmungsmechanismus hindeutet.

Eine Arbeit über die Wirkungen verschiedener Chinone auf ein Hefeproteinase-präparat (*144*) ergab, daß dieses Ferment in seinem Verhalten gegenüber den Chinonen weitgehend dem Papain analog ist. Bei diesen Versuchen beobachteten wir gewisse Übereinstimmungen zwischen der Hemmwirkung der einzelnen Chinone gegenüber diesem Hefeferment und der durch dieselben Chinone verursachten Wachstumshemmungen bei der Hefe. Trotzdem ist es meines Erachtens unwahrscheinlich, daß hier ein kausaler Zusammenhang besteht, da heute wohl die allgemeine Auffassung dahingeht, daß die eiweißsynthetisierenden Fermente nicht mit den eiweißspaltenden identisch sind.

Ein Bericht über die Hemmbarkeit einer Proteinase aus *Mycoderma*-Arten durch *p*-Benzochinon war dem Verfasser nur im Auszug zugänglich (*42*).

Noch unveröffentlichte Versuche aus dem hiesigen Laboratorium (*136*) mit dem κ-Toxin von *Clostridium welchii*, welches starke abbauende Wirkung gegenüber Kollagen hat, zeigen, daß auch dieses Toxinferment durch Chinone gehemmt wird.

Der Befund, daß die Hypertensinase durch Chinone aktiviert wird, ist schon an anderer Stelle erwähnt worden (*66*).

KUHN und BEINERT (*190*, *192*) berichteten, daß die *Brenztraubensäure-carboxylase* aus Hefe sehr chinonempfindlich ist, was auch durch unsere Versuche bestätigt werden konnte (*140*). Die Wirkungsreihe der einzelnen Chinone gegenüber der Carboxylase ist von denjenigen gegenüber anderen Fermenten sehr verschieden, was auf einen ungleichen Mechanismus der Hemmwirkung schließen läßt.

Bei ihren schon berichteten Versuchen über die Wirkung von Chinonen auf *Monolinia fructicola* konnten FOOTE und Mitarbeiter (*94a*, S. 208) einen bemerkenswerten Parallelismus zwischen Carboxylase-hemmung und fungistatischer Wirksamkeit feststellen.

Auch über eine Hemmung der *Transaminasen* ist berichtet worden (*47*).

Schon seit längerer Zeit ist der hemmende Einfluß von Chinon auf die Wasserstoffperoxydzersetzung durch *Katalase* bekannt. ALEXEJEW und RUSSINOWA (*2*) untersuchten bereits 1928 diesen Effekt, wobei sie *p*-Benzochinon auf ein Blutkatalase-präparat einwirken ließen. Versuche im Laboratorium des Verfassers zeigten, daß fast allen Chinonen der Benzochinon- und Naphthochinonreihe eine Hemmwirkung gegenüber der Aktivität von Katalasen aus Blut und aus Rinderleber zukommt (*131*, *139*). Ein ähnlicher Effekt wird auch durch die chinoide Struktur besitzenden Indophenole verursacht (*134*).

Über die Beeinflussung der Wasserstoff und Elektronen übertragenden Fermente durch Chinone ist noch nicht viel bekannt, obwohl nach Ansicht des Verfassers gerade die Erforschung dieser Effekte Aussicht bietet, Mechanismen mancher Chinonwirkungen aufzuklären.

Die *Aldehyd-dehydrogenase* der Milch wird nach WIELAND und MITCHELL (*326*) durch *p*-Benzochinon in ihrer Wirksamkeit gehemmt. Ebenso verhalten sich die *Alkohol-dehydrogenasen* der Hefe und des *Acetobacter peroxydans* (*323*, *327*).

Genauere Untersuchungen liegen über die Wirkung von Chinonen auf das *Bernsteinsäure-oxydase*-System vor. Dieses wird sowohl durch *p*-Benzochinon (*33*, *324*, *325*) als auch durch verschiedene Naphthochinon-derivate — darunter die malariawirksamen Verbindungen von FIESER (S. 204) — schon durch geringe Konzentrationen in seiner Aktivität gehemmt (*27*, *121*). Von den beiden bekannten Enzymbestandteilen des Systems ist es die primär an der Bernsteinsäure angreifende Bernsteinsäuredehydrogenase, welche beeinflußt wird, während die Cytochromoxydase gegenüber Chinonen unempfindlich zu sein scheint. Interessanterweise ist die Bernsteinsäure, also das Substrat des Ferments, imstande, dieses gegen die Chinonwirkung teilweise zu schützen, da Fermentlösungen, welche in Abwesenheit des Substrats dem Einfluß der Chinone ausgesetzt werden, viel stärkeren Aktivitätsverlust aufweisen als solche, welchen die Chinone in Anwesenheit von Bernsteinsäure zugesetzt wurden.

Eine von SÜLLMANN (*300*) berichtete Hemmwirkung von *p*-Benzochinon gegenüber dem ungesättigte Fettsäuren oxydierenden Ferment Lipoxydase wird von diesem Forscher mit der allgemeinen antioxygenen Wirkung des Chinons erklärt, worunter man wohl eine Redoxpufferung verstehen soll.

MEYERHOF und RANDALL (*230*) untersuchten den Einfluß von Adrenochrom und 1,2-Naphthochinon-4-sulfonsäure auf die *Glykolyse* in Gehirnextrakten. Die beiden Substanzen verursachen eine beträchtliche Hemmung dieses Vorganges. Die Autoren halten es für wahrscheinlich, daß hiefür die Aktivitätshemmung von Hexokinase und Phosphohexokinase verantwortlich ist. Demnach handelt es sich in diesem Falle um eine Beeinflussung derjenigen Fermente, welche einen Phosphatrest von Adenosintriphosphat auf Glucose bzw. 6-Phosphofructose übertragen. Die Wirkung von 1,2-Naphthochinon-4-sulfonsäure auf die Glykolyse des Gehirnextrakts ist etwa fünfmal so stark wie diejenige des Adrenochroms.

Die oben berichteten Versuche von MEYERHOF und RANDALL über den Einfluß der gleichen Chinone auf die Atmung, die Glykolyse und die Beweglichkeit von *Trypanosoma equiperdum* ergaben, daß auch diese Vorgänge in ähnlichen Konzentrationen wie im Gehirnextrakt von Adrenochrom gehemmt werden, während größere Mengen des Naphthochinons für diese Effekte erforderlich sind.

Die Entdeckung des Naphthochinon-derivats Solanion (XXI; S. 162) als Stoffwechselprodukt von *Fusarium solani* D_2 violett (*317*) veranlaßte NORD und Mitarbeiter (*70, 220a, 244, 245, 316*), den Einfluß des Pigmentes sowie verwandter Naphthochinon-Derivate auf die *Lipoidbildung* verschiedener *Fusarium*-Arten zu prüfen. Es zeigte sich, daß ein grundlegender Unterschied im Verhalten von selbst Pigmente produzierenden Fusarien und solchen, welche keine Farbstoffe bilden, besteht. Solanion hat keinen Einfluß auf die Fettproduktion von pigmentbildenden Fusarien, hingegen werden z. B. bei *Fusarium lini* BOLLEY, welches keinen Farbstoff erzeugt, sowohl das Mycelgewicht, also das Wachstum, wie auch die Fettbildung sehr stark verringert; es scheint dabei eine Veränderung des Fettstoffwechsels stattzufinden. Das unter dem Einfluß von Solanion entstehende Fett enthält weitaus größere Mengen an ungesättigten Fettsäuren (Jodzahl 138), während der unbeeinflußte Organismus ein Fett mit der Jodzahl 84 erzeugt.

In ähnlicher Art scheinen dem Solanion verwandte Naphthochinone auf den Fettstoffwechsel von *Fusarium lini* BOLLEY einzuwirken. Auch hier wird weniger Kohlenhydrat zu Fett umgesetzt, was beweist, daß die Chinone einen Einfluß auf diejenigen Fermentsysteme ausüben, welche für die Umwandlung von Kohlenhydrat in Fett verantwortlich sind. Als wirksamstes Chinon erwies sich das Naphthazarin, welchem das 2-Methyl-1,4-naphthochinon und das Solanion folgen; diesen reihen sich Phthiocol,

Methylnaphthazarin, 1,4-Naphthochinon und Oxydroseron an. In allen Fällen zeigt sich gleichzeitig eine Wachstumshemmung; so verursachen bereits 4 mg Naphthazarin in einem Liter Kulturmedium vollständige Aufhebung des Wachstums.

In einer Arbeit aus neuester Zeit berichten MASELLI und NORD (*220 a*), daß der Einfluß verschiedener Naphthochinone auf die Zusammensetzung der Gesamtlipoide bei *Fusarium lini* hauptsächlich darin besteht, daß größere Mengen von ungesättigten Säuren (Linolsäure, Linolensäure, wahrscheinlich auch Ölsäure) gebildet werden. Die Sterinmenge ist hingegen gegenüber den Kontrollversuchen vermindert.

Die *Fettbildung der Hefe* wird durch Chinone in einer ganz anderen Art beeinflußt (*141*). Durch die meisten der Substanzen wird die Gesamtlipoidmenge nicht verändert; Toluchinon, 5-Methoxytoluchinon und Naphthazarin steigern sogar die Fettproduktion. Die Chinone sind aber bei der Hefe imstande, die Zusammensetzung der Lipoidfraktion grundlegend zu verändern, dies erfolgt durch eine Erhöhung des Sterinanteiles im Gesamtfett. Jedenfalls steht es eindeutig fest, daß die Fermentsysteme, welche für die Lipoidbildung in der Hefe verantwortlich sind, mit denjenigen, welche die gleiche Funktion bei *Fusarium lini* ausüben, in ihren Eigenschaften und in ihrer Beeinflußbarkeit nicht übereinstimmen.

In einer kürzlich erschienenen Mitteilung berichten BERNHAUER und Mitarbeiter (*33 a*) über eigenartige Effekte verschiedener Chinone auf den Glucoseabbau durch *Rhizopus oryzae*. Bei bestimmten Konzentrationen wird die Milchsäurebildung gefördert, die Alkoholbildung hingegen gehemmt.

Die vorhergehende Aufzählung der verschiedenen Wirkungen von Chinonen auf einzelne Fermente und Fermentsysteme ist, wie sich der Verfasser wohl bewußt ist, nicht dazu geeignet, dem Leser Klarheit über die Mechanismen der biologischen Chinonwirkungen zu verschaffen. Tatsächlich sind bisher nur wenige Fälle der Übereinstimmung von Bioeffekt und Enzymbeeinflussung *in vitro* aufgefunden worden. Meistens wird die Chinonbeeinflussung eines biologischen Systems ihr Zustandekommen der gleichzeitigen Wirkung des Chinons auf verschiedene Fermente verdanken, wobei in vielen Fällen schwer zu bestimmen ist, welche Komponente entscheidend ist (*128*). Außer den Fermentbeeinflussungen sind auch noch andere Faktoren zu berücksichtigen; so können Parallelen zwischen Enzymbeeinflussung und Wirkung durch das verschiedene Eindringungsvermögen der einzelnen Substanzen in die Zelle oder das Gewebe verwischt werden.

Für die Annahme, daß in den meisten Fällen die biologische Wirkung der Chinone eine Anzahl von Enzymbeeinflussungen und anderen Faktoren umfaßt, sprechen z. B. die auf S. 207 zitierten Versuche über die Wirkung der Chinone auf verschiedene Lebensfunktionen der Hefe.

Es wurde unter gleichartigen Bedingungen der Einfluß auf Gärung, Atmung, Glykolyse, Fettbildung, Sterinproduktion sowie verschiedene isolierte Enzyme der Hefe geprüft, und es konnte in keinem Falle ein Parallelismus zwischen den Wirkungen auf die einzelnen Systeme festgestellt werden. Ebensowenig ergaben sich Parallelen zwischen einem der besprochenen Enzymsysteme und dem Wachstum. Wenn man den Mechanismus einer Chinonwirkung erforschen will, wird es einer sehr genauen Analyse der Komponenten bedürfen, aus welchen die Gesamtwirkung zusammengesetzt ist.

Schließlich soll hier noch eine Erscheinung geschildert werden, welche von den bisher beschriebenen Effekten (zumindest äußerlich) verschieden ist und der vielleicht in manchen Fällen beim Zustandekommen biologischer Chinonwirkungen eine gewisse Bedeutung zukommt. Bei einigen Naphthochinonen wird ein biochemischer Antagonismus gegenüber vitamin-K-aktiven Substanzen vermutet (*332*). Wie erwähnt, ist Vitamin K_2 in manchen Bakterienarten nachgewiesen worden. Die Organismen müssen also die Fähigkeit haben, das Vitamin zu synthetisieren, und es ist anzunehmen, daß dieses eine noch unbekannte Funktion im Stoffwechsel jener Bakterien ausübt. In einem Falle, bei *Mycobacterium phlei* (JOHNES Bazillus), konnte gezeigt werden, daß dieser Organismus, welcher in Abwesenheit eines nicht identifizierten Faktors im Medium nicht gedeiht, bei Zusatz von Phthiocol oder 2-Methyl-1,4-naphthochinon, also zwei vitamin-K-aktiven Substanzen, zum Nährboden auch ohne den unbekannten Faktor wachsen kann, obzwar dieser nicht mit einer der beiden Substanzen identisch ist.

Bei Annahme einer wesentlichen Rolle der vitamin-K-aktiven Substanzen im Stoffwechsel mancher Bakterien ist es nun vorstellbar, daß strukturell ähnliche Verbindungen, welche aber diese Funktion nicht übernehmen können, den Stoffwechsel dadurch stören, daß sie (gerade vermöge ihrer strukturellen Ähnlichkeit) jene Stellen der Zelle besetzen, an welchen das Vitamin K seine Wirksamkeit entfaltet; das letztere wird dadurch an der Erfüllung seiner Funktion gehindert.

Tatsächlich konnte WOOLLEY (*331*) beobachten, daß die Wirkung des 2,3-Dichlor-1,4-naphthochinons gegenüber verschiedenen Mikroorganismen durch Zugabe von 2-Methyl-1,4-naphthochinon oder Vitamin K_1 teilweise aufgehoben wird, was der Autor als einen Beweis dafür wertet, daß hier ein Antagonismus des 2,3-Dichlor-1,4-naphthochinons gegenüber vitamin-K-aktiven Substanzen vorliegt. Seine Ergebnisse konnten von verschiedenen Forschern bestätigt werden (*137*, *288*).

Trotzdem spricht manches dagegen, daß man diese Erscheinung den klassischen Fällen von biochemischem Antagonismus gleichsetzt. So ist es bisher noch nie gelungen, gerade in den Mikroorganismen, bei welchen die erwähnten Effekte am besten nachweisbar sind, nämlich bei Hefen und anderen niedrigen Pilzen, vitamin-

K-artige Stoffe nachzuweisen. Weiters ist, im Gegensatz zu anderen Fällen von biochemischem Antagonismus, ein Überschuß der vitamin-aktiven Substanz zur Aufhebung der Wirkung des Dichlornaphthochinons erforderlich. Der Verfasser hält es deshalb für wahrscheinlich, daß es sich hier um einen weniger spezifischen Vorgang handelt als in den anderen Fällen, welche als „biochemischer Antagonismus" bezeichnet werden. Der Mechanismus könnte darauf beruhen, daß das weit ungiftigere vitamin-K-aktive Produkt imstande ist, das Dichlornaphthochinon von seinem Angriffspunkt im Organismus zu verdrängen und dadurch dessen Wirkung aufzuheben. Dafür spricht auch besonders, daß SCHOPFER (*289*) bei seinen auf S. 219 erwähnten Versuchen über die Urease eine Abschwächung des mit verschiedenen hochwirksamen Substanzen erzielten Hemmungseffektes durch 2-Methylnaphthochinon beobachten konnte. Da der Urease im Stoffwechsel der von WOOLLEY untersuchten Mikroorganismen sicher nicht eine derartig bedeutende Rolle zukommt, daß ihre Hemmung gleichzeitig eine starke Wachstumsverminderung zur Folge hätte, das Vitamin K aber anderseits auch an der Wirkung der Urease bestimmt unbeteiligt ist, kann dieser Verdrängungsmechanismus nur wenig spezifisch sein.

Schlußbemerkung.

Die Chinone sind seit mehr als einem Jahrhundert bekannt. In dieser Zeit sind mehr als hundert zu dieser Klasse gehörende Substanzen als Naturstoffe entdeckt und größtenteils strukturell aufgeklärt worden. Trotzdem ist es erst in den letzten fünfzehn Jahren gelungen, über die Funktionen einiger weniger dieser Stoffe (Vitamine K, Echinochrome) Klarheit zu erlangen, während sonst die biologische Rolle dieser Substanzen noch so gut wie unerforscht ist. Ebenso wissen wir heute kaum etwas über die Biosynthese der Chinone. Es stehen auf diesem Gebiete noch viele Probleme offen, für welche man interessante Lösungen erhoffen darf.

Unsere Kenntnisse über die biologischen Wirkungen der chinoiden Substanzen sind ebenfalls durch die Arbeit der letzten Jahre sehr erweitert worden, doch auch hier sind noch wesentliche Lücken zu füllen. Der praktische Nutzen der Erforschung dieser Probleme beginnt sich allerdings bereits zu zeigen: Vitamin-K-Präparate haben in der Therapie allgemein Eingang gefunden und dasselbe kann man von den von FIESER und Mitarbeitern entdeckten Naphthochinon-derivaten mit Antimalariawirkung erhoffen. Halogenierte Naphthochinone finden in der Schädlingsbekämpfung Verwendung. Es ist mit Bestimmtheit anzunehmen, daß auch noch manche andere Substanzen der Chinongruppe praktisch verwertbare Wirkungen zeigen werden.

Literaturverzeichnis.

1. ALCALAY, W.: Action de composés naphthoquinoniques sur la croissance des bactéries acido-résistantes et autres. Schweiz. Z. Path. Bakt. **10**, 229 (1947).

2. ALEXEJEW, A. u. K. RUSSINOWA: Wasserstoffacceptoren und die Katalase. I. Über die Einwirkung der Wasserstoffacceptoren auf die Blutkatalase des Menschen *in vitro*. Bull. Inst. Rech. biol. Perm. **6**, 425 (1928); Chem. Zbl. **1929** II, 2055.

3. Anderson, R. J. and M. S. Newman: Lipins of Tubercle Bacilli. XXXIV. Isolation of a Pigment and Anisic Acid. J. biol. Chemistry **101**, 499 (1933).
4. — — Lipins of Tubercle Bacilli. XXXV. Constitution of Phthiocol, the Pigment Isolated from Human Tubercle Bacillus. J. biol. Chemistry **103**, 197 (1933). Vgl. R. J. Anderson: The Chemistry of the Lipoids of the Tubercle Bacillus and Certain Other Microorganisms. Fortschr. Chem. organ. Naturstoffe **3**, 145 (1939).
5. Anfinsen, C. B.: Inhibitory Action of Naphthoquinones on Respiration Process—Inhibition of Cleavage and Respiration in the Eggs of *Arbacia punctulata*. J. cellular comparat. Physiol. **29**, 333 (1947).
6. Anslow, W. K., J. Breen and H. Raistrick: Studies in the Biochemistry of Micro-organisms. LXIV. Emodic Acid and ω-Hydroxy-emodin, Metabolic Products of a Strain of *Penicillium cyclopium* Westling. Biochemic. J. **34**, 159 (1940).
7. Anslow, W. K. and H. Raistrick: Studies in the Biochemistry of Microorganisms. LVII. Fumigatin and Spinulosin, Metabolic Products of *Aspergillus fumigatus* Fresenius resp. *Penicillium spinulosum* Thom. Biochemic. J. **32**, 687 (1938).
8. — — Studies in the Biochemistry of Micro-organisms. LVIII. Synthesis of Spinulosin, a Metabolic Product of *Penicillium spinulosum* Thom. Biochemic. J. **32**, 803 (1938).
9. — — Studies in the Biochemistry of Micro-organisms. LIX. Spinulosin, a Metabolic Product of a Strain of *Aspergillus fumigatus* Fresenius. Biochemic. J. **32**, 2288 (1938).
10. — — Studies in the Biochemistry of Micro-organisms. LXVII. The Molecular Constitution of Catenarin and Erythroglaucin. Biochemic. J. **34**, 1124 (1940).
11. — — Studies in the Biochemistry of Micro-organisms. LXVIII. Synthesis of Cynodontin, a Metabolic Product of *Helminthosporium cynodontis*. Biochemic. J. **34**, 1546 (1940).
12. — — Studies in the Biochemistry of Micro-organisms. LXIX. Synthesis of Catenarin, a Metabolic Product of Species of *Helminthosporium*. Biochemic. J. **35**, 1006 (1941).
13. Armstrong, W. D. and J. W. Knutson: Effects of Quinones on Acid Formation in Saliva. Proc. Soc. exp. Biol. Med. **52**, 307 (1943).
14. Armstrong, W. D., W. Spink and J. Kahnke: Antibacterial Effects of Quinones. Proc. Soc. exp. Biol. Med. **53**, 230 (1942).
15. Arnstein, H. R. V. and A. H. Cook: Production of Antibiotics by Fungi. III. J. chem. Soc. (London) **1947**, 1021.
16. Arnstein, H. R. V., A. H. Cook and M. S. Lacey: An Antibacterial Pigment from *Fusarium javanicum*. Nature (London) **157**, 133 (1946).
17. Asahina, Y. u. F. Fuzikawa: Untersuchungen über Flechtenstoffe. LV. Über Endocrocin, ein neues Oxyanthrachinonderivat. Ber. dtsch. chem. Ges. **68**, 1558 (1935).
18. Asahina, Y. u. S. Shibata: Untersuchungen über Flechtenstoffe. XCIV. Über das Vorkommen der Telephorsäure in Flechten. Ber. dtsch. chem. Ges. **72**, 1531 (1939).
19. Asano, M. u. S. Fuziwara: Über den Farbstoff von *Xanthoria fallax* (Hepp.) Arn. (Vorl. Mitt.). J. pharmac. Soc. Japan **56**, 185 (1936).
20. Asano, M. u. K. Yamaguti: Über die Konstitution des Embelins. J. pharmac. Soc. Japan **60**, 34 (1940).
21. Ashley, J. N. and H. Raistrick: Studies in the Biochemistry of Microorganisms. LVI. Luteoleersin and Alboleersin, Metabolic Products of *Hel-*

minthosporium leersii Atkinson. A Possible Oxidation-Reduction System. Biochemic. J. **32**, 449 (1938).

22. Ashley, J. N., H. Raistrick and T. Richards: Studies in the Biochemistry of Micro-organisms. LXII. The Crystalline Colouring Matters of Species in the *Aspergillus glaucus*-Series. Biochemic. J. **33**, 1291 (1939).

23. Atkins, P. and J. L. Ward: Antibacterial Effects of Vitamin-K Analogues. Brit. J. exp. Pathol. **26**, 120 (1945).

24. Auerbach, C.: Chemical Mutagenesis. Biol. Rev. Cambridge philos. Soc. **24**, 355 (1949).

24a. Baglioni, S.: Zitiert nach Handbuch der experimentellen Pharmakologie, herausgegeben von A. Heffter, Bd. 1, S. 871. Berlin: Springer-Verlag. 1923.

25. Baker, W. and H. Raistrick: Derivatives of 1,2,3,4-Tetrahydroxy-benzenes. VII. The Synthesis of Fumigatin. J. chem. Soc. (London) **1941**, 670.

26. Ball, E. G.: Oxydation-Reduction. 22. Lapachol, Lomatiol and Related Compounds. J. biol. Chemistry **114**, 649 (1935).

27. Ball, E. G., C. B. Anfinsen and O. Cooper: The Inhibitory Action of Naphthoquinones on Respiratory Processes. J. biol. Chemistry **168**, 257 (1947).

28. Barrowcliff, M. and F. Tutin: Chemical Investigation of the Root and the Leaves of *Morinda longiflora*. J. chem. Soc. (London) **91**, 1907 (1907).

29. Beadle, G. W.: Some Recent Developments in Chemical Genetics. Fortschr. Chem. organ. Naturstoffe **5**, 300 (1948).

30. Béhal, A. et P. Phisalix: La quinone, principe actif du venin de *Julus terrestris*. C. R. hebd. Séances Acad. Sci. **131**, 1004 (1900).

31. Beijerinck, M. W.: Über Chinonbildung durch *Streptothrix chromogena* und Lebensweise dieses Mikroben. Zbl. Bakteriol., Parasitenkunde Infektionskrankh., Abt. II **6**, 1 (1899).

32. Bergstermann, H.: Die Wirkung von Oxyphenolen und aromatischen Aminen und deren Oxydationsprodukten auf Blutegelbauchmarkpräparate. Biochem. Z. **317**, 228 (1944).

33. Bergstermann, H. u. W. Stein: Über die Wirkung von Chinonen und anderen thiolgruppenbindenden Giften auf die Bernsteinsäuredehydrase. Biochem. Z. **317**, 217 (1944).

33a. Bernhauer, K., J. Rauch und J. N. Miksch: Über die Säurebildung durch *Rhizopus*-Arten. II. Zur Milchsäurebildung in der Submerskultur. Biochem. Z. **320**, 178 (1950).

34. Bernthsen, A.: Über das Juglon. Ber. dtsch. chem. Ges. **17**, 1945 (1884).

35. Bernthsen, A. u. A. Semper: Über die Konstitution des Juglons und seine Synthese aus Naphthalin. Ber. dtsch. chem. Ges. **20**, 934 (1887).

36. Bersin, T. u. W. Logemann: Über den Einfluß von Oxydations- und Reduktionsmitteln auf die Aktivität von Papain. Hoppe-Seyler's Z. physiol. Chem. **220**, 209 (1933).

37. Bertalanffy, L. v., O. Hoffmann-Ostenhof and O. Schreier: A Quantitative Study of the Toxic Action of Quinones on *Planaria gonocephala*. Nature (London) **158**, 948 (1946).

38. Bhattacharya, R. and J. L. Simonsen: A Synthesis of Morindone. J. Indian Inst. Sci., Ser. A **10**, 6 (1927).

39. Bielig, H.-J. u. Graf F. Medem: Wirkstoffe der tierischen Befruchtung. Experientia **5**, 11 (1949).

39a. Binkley, S. B., L. C. Cheney, W. F. Holcomb, R. W. McKee, S. A. Thayer, D. W. MacCorquodale and E. A. Doisy: The Constitution and Synthesis of Vitamin K_1. J. Amer. chem. Soc. **61**, 2558 (1939).

40. BIRKINSHAW, J. H.: Some Aspects of Fungal Metabolism; with Particular Reference to the Production of Antibiotics. Trans. Brit. Mycol. Soc. **30**, 50 (1948).
41. BIRKINSHAW, J. H., A. BRACKEN and H. RAISTRICK: Studies in the Biochemistry of Micro-organisms. 72. Gentisyl Alcool, a Metabolic Product of *Penicillium patulum* BAINIER. Biochemic. J. **37**, 726 (1943).
42. BLAGOVESCHENSKY, A. V. u. I. A. SOROKINA: Wirkung von Oxydantien auf Hefeproteinase. Bull. Biol. Méd. exp. URSS **4**, 176 (1937); Chem. Zbl. **1938** II, 869.
43. BLOUNT, B. K.: The Chemistry of Insects. II. Examination of the Woolly Aphis and of the White Pine Chermes. J. chem. Soc. (London) **1936**, 1034.
44. BOSWELL, J. G.: Oxidation Systems in the Potato Tuber. Ann. of Botany **9**, 55 (1945).
45. BRACK, A.: Isolierung von Gentisinalkohol neben Patulin aus dem Kulturfiltrat eines Penicilliumstammes und über einige Derivate des Gentisylalkohols. Helv. chim. Acta **30**, 1 (1947).
46. BRACKEN, A. and H. RAISTRICK: Studies in the Biochemistry of Micro-organisms. 75. Dehydrocarolic Acid, a Metabolic Product of *Penicillium cinerascens* BIOURGE. Biochemic. J. **41**, 569 (1947).
47. BRAUNSTEIN, A. E.: Transamination and the Integrative Functions of the Dicarboxylic Acids in Nitrogen Metabolism. Adv. Protein Chem. **3**, 1 (1947).
48. BROCKMANN, H.: Die Konstitution des Alkannins, Shikonins und Alkannans. Liebigs Ann. Chem. **521**, 1 (1935).
49. — Lichtkrankheiten durch fluoreszierende Pflanzenfarbstoffe. Forsch. u. Fortschr. **19**, 299 (1943).
50. BROCKMANN, H. u. K. MÜLLER: Über die Synthese des Alkannans und anderer Alkylnaphthochinone. Liebigs Ann. Chem. **540**, 51 (1939).
51. BROCKMANN, H., F. POHL, K. MAIER u. M. N. HASCHAD: Über das Hypericin, den photodynamischen Farbstoff des Johanniskrautes *(Hypericum perforatum)*. Liebigs Ann. Chem. **553**, 1 (1942).
52. BROCKMANN, H. u. H. ROTH: Über spiegelbildliche Naturfarbstoffe. Naturwiss. **23**, 246 (1935).
52 a. BROCKMANN, H., E. WEBER u. E. SANDLER: Fagopyrin, ein photodynamischer Farbstoff aus Buchweizen *(Fagopyrum esculentum)*. Naturwiss. **37**, 43 (1950)
52 b. BROWN, C. A.: Quinone Tanning in the Animal Kingdom. Nature (London) **165**, 275 (1950).
52 c. — Protein Skeletal Materials in the Invertebrates. Exp. Cell Res., Suppl **1**, 351 (1949).
53. BROWN, J. P., N. J. CARTWRIGHT, A. ROBERTSON and W. B. WHALLEY: Structure of Citrinin. Nature (London) **162**, 72 (1948).
54. CALANDRA, J. C., O. E. FANCHER and L. S. FOSDICK: Effect of Synthetic Vitamin K and Related Compounds on the Rate of Acid Formation in Saliva. J. dental. Res. **23**, 31 (1944).
55. CARTWRIGHT, N. J., A. ROBERTSON and W. B. WHALLEY: A Synthesis of Citrinin. Nature (London) **163**, 94 (1949).
56. CHARLES, J. H. V., H. RAISTRICK, R. ROBINSON and A. R. TODD: Studies in the Biochemistry of Micro-organisms. 28. Helminthosporin and Hydroxyhelminthosporin, Metabolic Products of *Helminthosporium gramineum* RABENHORST. Biochemic. J. **27**, 499 (1933).
57. COHEN, G. N.: Sur le premier produit de transformation de l'adrénochrome au cours de la mélanisation de l'adrénaline. C. R. hebd. Séances Acad. Sci. **220**, 796 (1945).

58. COHEN, G. N.: Sur le sort de l'oxoadrénochrome au cours de la mélanisation de l'adrénaline. C. R. hebd. Séances Acad. Sci. **220**, 927 (1945).
59. — Études sur la mélanisation. I. Sur le premier produit de transformation de l'adrénochrome au cours de la mélanisation de l'adrénaline. Bull. Soc. Chim. biol. **28**, 104 (1946).
60. — Études sur la mélanisation. IV. Structures des précurseurs de mélanines. Règles et exceptions. Bull. Soc. Chim. biol. **29**, 265 (1947).
61. COLWELL, C. A. and M. MCCALL: The Mechanism of Bacterial and Fungus Growth Inhibition by 2-Methyl-1,4-naphthoquinone. J. Bacteriol. **51**, 659 (1946).
62. COOPER, E. A.: On the Relations of Phenol and *meta*-Cresol to Proteins; a Contribution to Our Knowledge of the Mechanism of Disinfection. Biochemic. J. **6**, 362 (1912).
63. CORE, T. S., T. B. PANSE and K. VENKATARAMAN: Citrinin. Nature (London) **157**, 333 (1946).
64. COYNE, F. P., H. RAISTRICK and R. ROBINSON: Studies in the Biochemistry of Micro-organisms. 15. Molecular Structure of Citrinin. Philos. Trans. Roy. Soc. London, Ser. B **220**, 297 (1931).
65. CRAM, D. J.: The Structure of Citrinin. J. Amer. chem. Soc. **70**, 440 (1948).
66. CROXATTO, H., R. CROXATTO and H. GUTIERREZ: Quinones and Hypertensinase Activity. Rev. med. y alimentación **6**, 68 (1943/44); Chem. Abstr. **38**, 5009 (1944).
67. DAM, H.: Vitamin K, Its Chemistry and Physiology. Adv. Enzymol. **2**, 286 (1942).
68. — Vitamin K. Vitamins and Hormones **6**, 28 (1948).
69. DE BURUAGA u. F. VERDÚ: Synthese des Plumbagins. An. Soc. españ. Fisica Quim. **32**, 830 (1934); Chem. Zbl. **1935** I, 3146.
70. DESCHAMPS, I.: On the Mechanism of Enzyme Action. XXXIV. The Influence of a Pigment from *Fusarium solani* D_2 Purple on the Composition of Fats Formed in *Fusaria*. Arch. Biochemistry **20**, 457 (1949).
71. DIETERLE, H. u. E. KRUTA: Über einen Inhaltsstoff von *Drosera rotundifolia*. Arch. Pharmaz. Ber. dtsch. pharmaz. Ges. **274**, 457 (1936).
72. DIMROTH, O.: Über die Carminsäure. Liebigs Ann. Chem. **339**, 1 (1913).
73. DIMROTH, O. u. H. FICK: Über den Farbstoff des Kermes. III. Liebigs Ann. Chem. **411**, 315 (1916).
74. DIMROTH, O. u. S. GOLDSCHMIDT: Über den Farbstoff des Stocklacks. Liebigs Ann. Chem. **399**, 62 (1913).
75. DIMROTH, O. u. H. KAMMERER: Über die Carminsäure. Ber. dtsch. chem. Ges. **53**, 471 (1920).
76. DIMROTH, O. u. W. SCHEURER: Über den Farbstoff des Kermes. II. Liebigs Ann. Chem. **399**, 43 (1913).
77. DOISY, E. A., S. B. BINKLEY and S. A. THAYER: Vitamin K. Chem. Reviews **28**, 477 (1941).
78. DUSTIN, P., Jr.: Some New Aspects of Mitotic Poisoning. Nature (London) **159**, 794 (1947).
79. EDER, R. u. F. HAUSER: Untersuchungen über Derivate des β-Methylanthrachinons. V. Über Frangula-Emodin, Emodinsäure und Derivate derselben. Helv. chim. Acta **8**, 126 (1925).
80. EDER, R. u. B. SIEGFRIED: Über natürliche Oxy- und Oxymethylanthrachinone. Pharmac. Acta Helvetiae **14**, 1 (1939).
81. EDER, R. u. C. WIDMER: Untersuchungen über Derivate des β-Methylanthrachinons. I. Synthese der Chrysophansäure und des 1,5-Dioxy-2-methylanthrachinons. Helv. chim. Acta **5**, 3 (1922).

82. EDER R. u. C. WIDMER: Untersuchungen über Derivate des β-Methylanthrachinons. II. Weitere Beiträge zur Synthese der Chrysophansäure. Helv. chim. Acta **6**, 419 (1923).
83. — — Untersuchungen über Derivate des β-Methylanthrachinons. III. Synthese des Frangulaemodins. Helv. chim. Acta **6**, 966 (1923).
84. ENGEL, B. G. u. W. BRZESKI: Über die Isolierung eines Chinhydrons von Gentisinalkohol und Oxymethyl-*p*-benzochinon (Gentisinchinon) aus dem Kulturfiltrat von *Penicillium urticae* BAINIER. Helv. chim. Acta **30**, 1472 (1947).
85. EVANS, W. C.: Oxidation of Phenol and Benzoic Acid by Some Soil Bacteria. Biochemic. J. **41**, 373 (1947).
86. FICHTER, F.: Über synthetische *p*-dialkylierte Dioxychinone. Liebigs Ann. Chem. **361**, 363 (1908).
87. FICHTER, F., M. JETZER u. R. LEEPIN: Über synthetische *p*-dialkylierte Dioxychinone und über Oxyperezon. Liebigs Ann. Chem. **395**, 1 (1912).
88. FIESER, L. F.: The Alkylation of Hydroxynaphthoquinone. III. A Synthesis of Lapachol. J. Amer. chem. Soc. **49**, 857 (1927).
88a. — Synthesis of 2-Methyl-3-phytyl-1,4-naphthoquinone. J. Amer. chem. Soc. **61**, 2559 (1939).
89. FIESER, L. F., E. BERLINER, F. J. BONDHUS, F. C. CHANG, W. G. DAUBEN, M. G. ETTLINGER, G. FAWAZ, M. FIELDS, M. FIESER, C. HEIDELBERGER, H. HEYMANN, A. M. SELIGMAN, W. R. VAUGHAN, A. G. WILSON, E. WILSON, M. WU, M. T. LEFFLER, K. E. HAMLIN, R. J. HATHAWAY, E. J. MATSON, E. E. MOORE, R. T. RAPALA and H. E. ZAUGG: Naphthoquinone Antimalarials. I. General Survey. J. Amer. chem. Soc. **70**, 3151 (1948).
90. FIESER, L. F. and J. T. DUNN: Synthesis of Plumbagin. J. Amer. chem. Soc. **58**, 572 (1936).
91. FIESER, L. F., M. G. ETTLINGER and G. FAWAZ: Naphthoquinone Antimalarials. XV. Distribution between Organic Solvents and Aqueous Buffers. J. Amer. chem. Soc. **79**, 3228 (1948).
92. FISCHER, E. u. H. SCHRADER: Verbindungen von Chinon mit Aminosäureestern. Ber. dtsch. chem. Ges. **43**, 525 (1910).
93. FISCHER, O.: Über Methylanthracen und einige Verbindungen desselben. Ber. dtsch. chem. Ges. **8**, 675 (1875).
94. FISCHGOLD, H.: Relation between Activity of Urease and Oxidation-Reduction Potential. Biochemic. J. **28**, 406 (1934).
94a. FOOTE, M. W., J. E. LITTLE and T. J. SPROSTON: On Naphthoquinones as Inhibitors of Spore Germination of Fungi. J. biol. Chemistry **181**, 481 (1949).
95. FRIEDHEIM, E. A. H.: Das Pigment von *Halla parthenopea*, ein akzessorischer Atmungs-Katalysator. Biochem. Z. **259**, 257 (1933).
96. — Recherches sur la biochimie des champignons inférieurs. I. Isolement du pigment rouge de *Penicillium phoeniceum* (Phoenicine). Helv. chim. Acta **21**, 1464 (1938).
97. FRIEDMAN, B., S. SOLOWAY, J. MARCUS and B. S. OPPENHEIMER: Quinones as Blood Pressure-Reducing Agents in Hypertensive Rats. Proc. Soc. exp. Biol. Med. **51**, 195 (1942).
98. FRIEDMANN, E., D. H. MARRIAN and I. SIMON-REUSS: Analysis of Antimitotic Action of Certain Quinones. Brit. J. Pharmacol. Chemotherapy **3**, 263 (1948).
99. FURUTA, T.: Giftwirkung von Chinon. Bull. Coll. Agric. Tokyo **4**, 407 (1901); Chem. Zbl. **1902** II, 385.
100. GEIGER, W. B.: The Mechanism of the Antibacterial Action of Quinones and Hydroquinones. Arch. Biochemistry **11**, 23 (1946).

101. GLOCK, G. E., R. H. THORP and E. WIEN: Antibacterial Action of 4,6-Dimethoxytoluquinone and Its Fate in the Animal Body. Biochemic. J. **39**, 308 (1945).

102. GORIS, A. et H. CANAL: Étude sur la composition chimique des bourgeons de *Populus balsamifera*. Bull. Soc. chim. France (5) **3**, 1982 (1936).

103. GOULD, B. S.: Chemical Compounds Formed from Sugars by Molds. Scientific Rep. Ser. Nr. 7. Sugar Research Foundation Inc. New York. 1947.

104. GOULD, B. S. and H. RAISTRICK: Studies in the Biochemistry of Microorganisms. 40. Crystalline Pigments of Species in the *Aspergillus glaucus* Series. Biochemic. J. **28**, 1640 (1934).

105. GRAEBE, C. u. C. LIEBERMANN: Über künstliches Alizarin. Ber. dtsch. chem. Ges. **2**, 332 (1869).

106. — — Über Anthracenderivate. Liebigs Ann. Chem. **160**, 121 (1871).

107. GREEN, D. E. and D. RICHTER: Adrenaline and Adrenochrome. Biochemic. J. **31**, 596 (1937).

108. GRIMAL, E.: Sur l'essence de bois de *Thuya articulata* d'Algérie. C. R. hebd. Séances Acad. Sci. **139**, 927 (1904).

108 a. GUSTAVSON, K. H.: Some Protein-Chemical Aspects of Tanning Processes. Adv. Protein Chem. **5**, 353 (1949).

109. HADORN, E., S. ROSIN u. G. BERTANI: Ergebnisse der Mutationsversuche mit chemischer Behandlung von *Drosophila*-Ovarien in vitro. Proc. 8th Int. Congr. Genetics (Suppl. Hereditas), p. 256 (1949).

110. HARTMANN, M.: Geschlecht und Geschlechtsbestimmung im Tier- und Pflanzenreich. Berlin: W. de Gruyter & Co. 1939.

111. HARTMANN, M. u. O. SCHARTAU: Untersuchungen über die Befruchtungsstoffe der Seeigel. I. Biol. Zbl. **59**, 571 (1939).

112. HARTMANN, M., O. SCHARTAU u. K. WALLENFELS: Untersuchungen über die Befruchtungsvorgänge des Seeigels. II. Biol. Zbl. **60**, 398 (1940).

113. HASAN, K. H. and E. STEDMAN: The Constitution and the Synthesis of Embelic Acid (Embelin), the Active Principle of *Embelia ribes*. J. chem. Soc. (London) **1931**, 2112.

114. HELLERMAN, L. and M. E. PERKINS: Activation of Enzymes. II. Papain Activity as Influenced by Oxidation-Reduction and by Action of Metal Compounds. J. biol. Chemistry **107**, 241 (1934).

115. HELLERMAN, L., M. E. PERKINS and W. M. CLARK: The Urease Activity Influenced by Oxidation and Reduction. Proc. nat. Acad. Sci. U. S. A. **19**, 855 (1933).

116. HESSE, O.: Beitrag zur Kenntnis der Flechten und ihrer Bestandteile. II. J. prakt. Chem. (2) **57**, 409 (1898).

117. — Beitrag zur Kenntnis der Flechten und ihrer Bestandteile. III. J. prakt. Chem. (2) **58**, 465 (1898).

118. — Über die Rhapontikwurzel und die österreichische Rhabarber. J. prakt. Chem. (2) **77**, 321 (1908).

119. HETHERINGTON, A. C. and H. RAISTRICK: Studies in the Biochemistry of Micro-organisms. 14. A New Yellow Colouring Matter Produced from Dextrose by *Penicillium citrinum* THOM. Philos. Trans. Roy. Soc. London, Ser. B **220**, 269 (1931).

120. HEUBNER, W.: Methämoglobinbildende Gifte. Ergebn. Physiol., biol. Chem. exp. Pharmakol. **43**, 9 (1940).

121. HEYMANN, H. and L. F. FIESER: Naphthoquinone Antimalarials. XXI. Antisuccinate Oxidase Activity. J. biol. Chemistry **176**, 1359 (1948).

122. Hill, R. and D. Richter: Anthraquinone Colouring Matters. Galiosin, Rubiadin Primveroside. J. chem. Soc. (London) **1936**, 1714.
123. Hilpert, S. u. F. Brauns: Über Chinongerbung. Collegium (Darmstadt) **1925**, 64.
124. Hind, H. G.: The Colouring Matters of *Penicillium carmino-violaceum* Biourge. Biochemic. J. **34**, 67 (1940).
125. — The Constitution of Carviolin, a Colouring Matter of *Penicillium carmino-violaceum* Biourge. Biochemic. J. **34**, 577 (1940).
126. Hiramoto, M. Über das Maesachinon, einen Farbstoff aus den Früchten von *Maesa japonica*. Proc. Imp. Acad. (Tokyo) **15**, 220 (1939).
127. Hofmann, A. W.: Über dreisäurige Phenole im Buchenholzteeröl und über den Ursprung des Cedrirets. Ber. dtsch. chem. Ges. **11**, 329 (1878).
128. Hoffmann-Ostenhof, O.: Mechanism of the Antibiotic Action of Quinones. Science (New York) **105**, 549 (1947).
129. — Die Biochemie der Chinone. Experientia **3**, 137, 176 (1947).
130. Hoffmann-Ostenhof, O., L. v. Bertalanffy u. O. Schreier: Untersuchungen über bakteriostatische Chinone und andere Antibiotica. VII. Quantitative Versuche über die Giftwirkung von Chinonen auf *Planaria gonocephala*, einem Strudelwurm aus der Gattung der *Tricladida*. Mh. Chem. **79**, 61 (1948).
131. Hoffmann-Ostenhof, O. u. E. Biach: Untersuchungen über bakteriostatische Chinone und andere Antibiotica. II. Hemmungswirkungen verschiedener Antibiotica auf die Wasserstoffperoxydzersetzung durch Blutkatalase. Mh. Chem. **76**, 319 (1947).
132. — — Untersuchungen über bakteriostatische Chinone und andere Antibiotica. IV. Hemmungswirkungen verschiedener Antibiotica auf die Eiweißspaltung durch Papain. Mh. Chem. **78**, 53 (1948).
133. — — Untersuchungen über bakteriostatische Chinone und andere Antibiotica. VIII. Wirkungen verschiedener Chinone auf die Stärkespaltung durch Amylasen tierischen Ursprungs. Mh. Chem. **79**, 248 (1948).
134. Hoffmann-Ostenhof, O., E. Biach u. S. Gierer: Hemmungswirkungen verschiedener Indophenole auf die Wasserstoffperoxydzersetzung durch Blutkatalase. Experientia **3**, 108 (1947).
135. Hoffmann-Ostenhof, O., E. Biach, S. Gierer u. O. Kraupp: Untersuchungen über bakteriostatische Chinone und andere Antibiotica. XIII. Durch Chinone nicht hemmbare Esterasen. Mh. Chem. **79**, 576 (1948).
136. Hoffmann-Ostenhof, O., A. Fasler u. W. Frisch: Unveröffentlichte Versuche.
137. Hoffmann-Ostenhof, O. u. H. Fellner-Feldegg: Die Hemmung des Hefewachstums durch verschiedene Chinone. Mh. Chem. **80**, 648 (1949).
138. — — Die Hemmung des Wachstums verschiedener Bakterienarten durch Chinone. Mh. Chem. **80**, 720 (1949).
139. Hoffmann-Ostenhof, O. u. S. Gierer: Untersuchungen über bakteriostatische Chinone und andere Antibiotica. IX. Hemmung der Aktivität kristallisierter Katalase aus Rinderleber durch Chinone und Indophenole. Mh. Chem. **79**, 379 (1948).
140. Hoffmann-Ostenhof, O. u. E. Kriz: Der Einfluß von Chinonen auf verschiedene Stoffwechselfunktionen der Hefe. Mh. Chem. **80**, 678 (1949).
141. — — The Influence of Some Quinone Derivatives on the Lipide Formation in Yeast. Arch. Biochemistry **24**, 459 (1949).
142. — — Der Einfluß von Chinonen und anderen Wirkstoffen auf die Resorption von radioaktivem Phosphat durch Hefe. Mh. Chem. **81**, 90 (1950).

143. HOFFMANN-OSTENHOF, O. u. W. H. LEE: Untersuchungen über bakteriostatische Chinone und andere Antibiotica. I. Hemmungswirkungen verschiedener Antibiotica auf die Harnstoffzersetzung durch Urease. Mh. Chem. **76**, 180 (1946).
144. HOFFMANN-OSTENHOF, O. u. H. MOSER: Untersuchungen über bakteriostatische Chinone und andere Antibiotica. XII. Hemmungswirkungen verschiedener Chinone auf die Eiweißspaltung durch Hefeproteinasen. Mh. Chem. **79**, 570 (1948).
145. HOFFMANN-OSTENHOF, O. u. E. PUTZ: Untersuchungen über bakteriostatische Chinone und andere Antibiotica. XI. Hemmwirkungen verschiedener Chinone auf einige Phosphomonesterasen der Hefe. Mh. Chem. **79**, 421 (1948).
146. — — Hemmwirkungen verschiedener Chinone auf Phosphomonoesterasen tierischen Ursprungs. Mh. Chem. (im Druck).
147. HOFFMANN-OSTENHOF, O. u. G. REITMAIER: Untersuchungen über bakteriostatische Chinone und andere Antibiotica. VI. Über die Hemmung des Keimlingswachstums von *Lepidium sativum* durch verschiedene Chinonderivate. Mh. Chem. **78**, 277 (1948).
148. HOFFMANN-OSTENHOF, O., W. WEIS u. O. KRAUPP: Untersuchungen über bakteriostatische Chinone und andere Antibiotica. III. Versuche über die durch verschiedene Benzochinon- und Naphthochinonderivate verursachte Methämoglobinbildung *in vitro*. Mh. Chem. **77**, 86 (1947).
149. HOFFMANN-OSTENHOF, O., P. WERTHEIMER u. K. GRATZL: Die Wirkung von Chinonen auf das Hefewachstum. Experientia **3**, 327 (1947).
150. HOOKER, S. C.: The Constitution of "Lapachic Acid" (Lapachol) and Its Derivatives. J. chem. Soc. (London) **61**, 611 (1892).
151. — The Constitution of Lapachol and Its Derivatives. J. chem. Soc. (London) **69**, 1355 (1896).
152. — Lomatiol (Hydroxyisolapachol). J. chem. Soc. (London) **69**, 1381 (1896).
153. HOWARD, B. H. and H. RAISTRICK: Studies in the Biochemistry of Microorganisms. 80. The Colouring Matter of *Penicillium islandicum* SOPP. I. 1:4:5-Trihydroxymethyl-anthraquinone. Biochemic. J. **44**, 227 (1949).
153a. — — Studies in the Biochemistry of Micro-organisms. 81. The Colouring Matters of *Penicillium islandicum* SOPP. II. Chrysophanic Acid, 4:5-Dihydroxy--2-methylanthraquinone. Biochemic. J. **46**, 49 (1950).
154. HUBER, W.: Über die antimitotische Wirkung von Naphthochinon und Phenanthrenchinon auf die Furchung von *Tubifex*. Rev. Suisse Zool. **54**, 61 (1947).
155. — Der Mitoseablauf bei *Tubifex* unter dem Einfluß von Naphthochinon und Phenanthrenchinon. Rev. Suisse Zool. **52**, 349 (1945).
156. IGARACI, H.: J. agric. chem. Soc. Japan **15**, 225 (1939); zitiert nach *153*.
157. JACOBSON, R. A. and R. ADAMS: Trihydroxy-methylanthraquinones. V. J. Amer. chem. Soc. **47**, 283 (1925).
158. JAMES, W. O.: The Respiration of Plants. Annu. Rev. Biochem. **15**, 417 (1946).
159. JOHN, W.: Über das antihämorrhagische Vitamin K. Angew. Chem. **54**, 209 (1941).
160. JONES, R. F. and H. G. BAKER: Formation of Aggregations of *Glaucoma pyriformis* KAHL by Means of Penanthraquinone and Other Substances. Nature (London) **157**, 554 (1946).
161. KAFUKU, K. and K. SEBE: On Tectoquinone, the Volatile Principle of the Teak Wood. Bull. chem. Soc. Japan **7**, 114 (1932).
162. KARRER, W.: Über das Vorkommen von 2,6-Dimethoxychinon in *Adonis vernalis*. Helv. chim. Acta **13**, 1424 (1930).

163. KATTI, M. C. T. u. V. N. PATWARDHAN: Chemische Untersuchung der Wurzelrinde von *Plumbago rosea* LINN. J. Indian. Inst. Sci., Ser. A **15**, 9 (1932); Chem. Zbl. **1932** II, 1458.

164. KAUL, R., A. C. RAY and S. DUTT: The Constitution of the Active Principle of *Embelia ribes*. I. J. Indian chem. Soc. **6**, 577 (1929).

165. — — — The Constitution of the Active Principle of *Embelia ribes*. II. J. Indian chem. Soc. **8**, 231 (1931).

166. KEIMATSU, S. u. J. HIRANO: Studien zur Synthese von Trioxymethylanthrachinonen. J. pharmac. Soc. Japan **49**, 20 (1929).

167. — — Studien zur Synthese von Trioxymethylanthrachinonderivaten. VI. J. pharmac. Soc. Japan **51**, 19 (1931).

168. KISCH, B.: Nichtenzymatische Zwischenkatalysatoren (Stoffüberträger und Aktivatoren). In: C. OPPENHEIMER: Handbuch der Biochemie. Erg.-Werk I A. Jena: Fischer. 1933.

169. KLUYVER, A. J., T. HOF and A. G. J. BOEZARDT: On the Pigment of *Pseudomonas beijerinckii* HOF. Enzymologia (Den Haag) **7**, 257 (1939).

170. KÖGL, F.: Untersuchungen über Pilzfarbstoffe. I. Über das Atromentin. Liebigs Ann. Chem. **440**, 19 (1924).

171. — Untersuchungen über Pilzfarbstoffe. V. Konstitution der Polyporsäure. Liebigs Ann. Chem. **447**, 78 (1925).

172. — Untersuchungen über Pilzfarbstoffe. VII. Die Synthese des Atromentins. Liebigs Ann. Chem. **465**, 243 (1928).

173. KÖGL, F. u. H. BECKER: Untersuchungen über Pilzfarbstoffe. VI. Die Konstitution des Atromentins. Liebigs Ann. Chem. **465**, 211 (1928).

174. KÖGL, F. u. W. B. DEIJS: Untersuchungen über Pilzfarbstoffe. XI. Über Boletol, den Farbstoff des blau anlaufenden Boleten. Liebigs Ann. Chem. **515**, 10 (1934).

175. — — Untersuchungen über Pilzfarbstoffe. XII. Die Synthese von Boletol und Isoboletol. Liebigs Ann. Chem. **515**, 23 (1934).

176. KÖGL, F. u. H. ERXLEBEN: Untersuchungen über Pilzfarbstoffe. VIII. Über den roten Farbstoff des Fliegenpilzes. Liebigs Ann. Chem. **479**, 11 (1930).

177. — — Untersuchungen über Pilzfarbstoffe. X. Über das Xylendein, den Farbstoff des grünfaulen Holzes (II). Liebigs Ann. Chem. **484**, 65 (1930).

178. KÖGL, F., H. ERXLEBEN u. L. JÄNECKE: Untersuchungen über Pilzfarbstoffe. IX. Die Konstitution der Telephorsäure. Liebigs Ann. Chem. **482**, 105 (1930).

179. KÖGL, F. u. J. J. POSTOWSKY: Untersuchungen über Pilzfarbstoffe. II. Über den Farbstoff des blutroten Hautkopfes (*Dermocybe sanguinea* WOLFF). Liebigs Ann. Chem. **444**, 1 (1925).

180. — — Untersuchungen über Pilzfarbstoffe. III. Über das Atromentin (II). Liebigs Ann. Chem. **445**, 159 (1925).

181. KÖGL, F. u. J. SPARENBURG: Untersuchungen über Pilzfarbstoffe. XIII. Über Phomazarin, den Farbstoff von *Phoma terrestris* HANSEN. Recueil Trav. chim. Pays-Bas **59**, 1180 (1940).

182. KÖGL, F. u. G. v. TAEUFFENBACH: Untersuchungen über Pilzfarbstoffe. IV. Über das Xylendein, den Farbstoff des grünfaulen Holzes (I). Liebigs Ann. Chem. **445**, 170 (1925).

183. KÖGL, F. u. G. C. VAN WESSEM: Untersuchungen über Pilzfarbstoffe. XIV. Über Oosporein, den Farbstoff von *Oospora colorans* VAM BEYMA. Recueil Trav. chim. Pays-Bas **63**, 5 (1944).

184. — — Untersuchungen über Pilzfarbstoffe. XV. Über Phomazarin, den Farbstoff von *Phoma terrestris* HANSEN (II). Recueil Trav. chim. Pays-Bas **63**, 251 (1944).

185. KATTI, M. C. T. u. V. N. PATWARDHAN: Untersuchungen über Pilzfarbstoffe. XVI. Über Phomazarin, den Farbstoff von *Phoma terrestris* HANSEN (III). Recueil Trav. chim. Pays-Bas **64**, 23 (1945).
186. KOFLER, M.: Über ein pflanzliches Chinon. Festschrift für E. C. BARELL, S. 199. Basel. 1946.
187. KOLLER, G. u. H. HAMBURG: Über die Rhodocladonsäure. Mh. Chem. **68**, 202 (1936).
188. KOLLER, G. u. H. RUSS: Über die Konstitution der Solorinsäure. Mh. Chem. **70**, 54 (1937).
189. KUHN, R.: Über die Befruchtungsstoffe und geschlechtsbestimmenden Stoffe bei Pflanzen und Tieren. Angew. Chem. **53**, 1 (1940).
190. KUHN, R. u. H. BEINERT: Über das aus krebserregenden Azofarbstoffen entstehende Fermentgift. Ber. dtsch. chem. Ges. **76**, 904 (1943).
191. — — Über die Umsetzung von Cystein mit Chinon. Ber. dtsch. chem. Ges. **77**, 606 (1944).
192. — — Hemmstoffe der Carboxylase. Ber. dtsch. chem. Ges. **80**, 101 (1947).
193. KUHN, R., D. JERCHEL, F. MOEWUS, E. F. MÖLLER u. H. LETTRÉ: Über die chemische Natur der Blastokoline und ihre Einwirkung auf keimende Samen, Pollenkörner, Hefen, Bakterien, Epithelgewebe und Fibroblasten. Naturwiss. **31**, 468 (1943).
194. KUHN, R. u. K. WALLENFELS: Über die chemische Natur des Stoffes, den die Eier des Seeigels *(Arbacia pustulata)* absondern, um die Spermatozoen anzulocken. Ber. dtsch. chem. Ges. **72**, 1407 (1939).
195. — — Echinochrome als prosthetische Gruppen hochmolekularer Symplexe in den Eiern von *Arbacia pustulosa*. Ber. dtsch. chem. Ges. **73**, 458 (1940).
196. — — Über den Stachelfarbstoff von *Arbacia*. Ber. dtsch. chem. Ges. **74**, 1594 (1941).
197. KURODA, C. u. H. OSHIMA: Die Pigmente der Seeigel und die Synthese verwandter Verbindungen. Proc. Imp. Acad. (Tokyo) **16**, 214 (1940).
198. KUSAKA, T.: Über die Synthese des Rubiadins. J. pharmac. Soc. Japan **55**, 110 (1935).
199. LAGODZINSKI, K.: Über 2,3-Dioxyanthracen. Liebigs Ann. Chem. **341**, 90 (1905).
200. LALLEMAND, A.: Über das Thymianöl und das Thymol. Liebigs Ann. Chem. **101**, 119 (1857).
201. LEDERER, E. et R. GLASER: Sur l'échinochrome et le spinochrome. C. R. hebd. Séances Acad. Sci. **207**, 454 (1938).
202. LÉGER, E.: Les aloins. II. Ann. Chimie (9) **8**, 265 (1917).
203. LEHMANN, F. E.: Prüfung zellteilungshemmender Substanzen an einem neuen Testobjekt. Verh. Ver. Schweiz. Physiol. 1942.
204. — Der Auf- und Abbau des Mitoseapparats beim *Tubifex*-Ei und seine stoffliche Beeinflußbarkeit. Rev. Suisse Zool. **52**, 342 (1945).
205. — Chemische Beeinflussung der Zellteilung. Experientia **3**, 223 (1947).
206. — Die Wirkung antimitotischer Stoffe auf das Ei von *Tubifex* und den regenerierenden Schwanz der *Xenopus*-Larve. Exp. Cell Res., Suppl. **1**, 156 (1949).
206 a. LEHMANN, F. E. u. G. ANDRES: Chemisch induzierte Kernabnormitäten. Rev. Suisse Zool. **55**, 280 (1948).
207. LEHMANN, F. E., W. BERNHARD, H. HADORN u. M. LÜSCHER: Zur entwicklungsphysiologischen Wirkungsanalyse von antimitotischen Stoffen. Experientia **1**, 232 (1945).

208. LEHMANN, F. E. u. H. HADORN: Vergleichende Wirkungsanalyse von zwei antimitotischen Stoffen, Colchicin und Benzochinon, am *Tubifex*-Ei. Helv. physiol. pharmac. Acta **4**, 11 (1946).

209. LEVAN, A. and J. HIN TIJO: Chromosome Fragmentation Induced by Phenols. Hereditas **34**, 250 (1948).

210. — — Induction of Chromosome Fragmentation by Phenols. Hereditas **34**, 453 (1948).

211. LIEBERMANN, C.: Über ein neues Dioxyanthrachinon, das Hystazarin. Ber. dtsch. chem. Ges. **21**, 250 (1888).

212. LIEBERMANN, C. u. F. GIESEL: Über die Reductionsprodukte des Chinizarins. Ber. dtsch. chem. Ges. **10**, 606 (1877).

213. LIEBERMANN, C. u. S. VON KOSTANECKI: Über die Färbereieigenschaften und die Synthesen der Oxyanthrachinone. Liebigs Ann. Chem. **240**, 255 (1887).

213 a. LITTLE, J. E., T. J. SPROSTON and M. W. FOOTE: Isolation and Antifungal Action of Naturally Occurring 2-Methoxy-1,4-naphthoquinone. J biol. Chemistry **174**, 335 (1948).

213 b. — — — Synthesis and Antifungal Action of 2-Methyl-mercapto-1,4-naphthoquinone. J. Amer. chem. Soc. **71**, 1124 (1949).

214. LLOYD, J. B. and G. MIDDLEBROOK: Bacteriostatic Activity of Some New Derivatives of Diaminodiphenyl Sulfone and Naphthoquinones against the Tubercle Bacillus. Amer. Rev. Tubercul. **49**, 539 (1942).

214a. LOVELESS, A. and S. REVEIL: New Evidence on the Mode of Action of "Mitotic Poisons". Nature (London) **164**, 938 (1949).

215. MACBETH, A. K., J. R. PRICE and F. L. WINZOR: The Colouring Matters of *Drosera Whittakeri*. I. J. chem. Soc. (London) **1935**, 325.

216. MADINAVEITIA, J.: Untersuchung des 2-Methyl-1,4-naphthochinons. An. Soc. españ. Fisica Quím. **31**, 750 (1933); Chem. Zbl. **1934** II, 940.

217. MADINAVEITIA, J. u. M. GALLEGO: Untersuchung des Plumbagins. An. Soc. españ. Fisica Quím. **26**, 263 (1928); Chem. Zbl. **1929** I, 662.

218. MARINI-BETTOLO, G. B. e E. DEL PIANTO: Contributo alla conoscenza del meccanismo d'azione dell'antibiosi tra microorganismi. Pontif. Acad. Sci. Comment. **10**, 87 (1946).

219. — — Contributo alla conoscenza del meccanismo d'azione dell'antibiosi tra microorganismi. II. Pontif. Acad. Sci. Comment. **11**, 33 (1947).

220. MARQUARDT, P.: Die Auf- und Abbaustufen des Adrenalins. Enzymologia (Den Haag) **12**, 167 (1946).

220 a. MASELLI, J. A. and F. F. NORD: Influence of Certain Naphthaquinones on the Composition of the Lipides Formed by *Fusarium lini* BOLLEY. Arch. Biochemistry **24**, 235 (1949).

221. MASON, H. S.: The Chemistry of Melanin. II. The Oxidation of Dihydroxyphenylalanine by Mammalian Dopa Oxidase. J. biol. Chemistry **168**, 433 (1947).

222. — Classification of Melanins. N. Y. Acad. Sci., Spec. Publs. **4**, 399 (1948).

223. — The Chemistry of Melanin. III. Mechanism of the Oxidation of Dihydroxyphenylalanine by Tyrosinase. J. biol. Chemistry **172**, 83 (1948).

224. MASON, H. S., H. KAHLER, R. C. MACCARDLE and A. J. DALTON: The Chemistry of Melanin. IV. Electron Micrography of Natural Melanin. Proc. Soc. exp. Biol. Med. **66**, 421 (1947).

225. MASON, H. S. and C. I. WRIGHT: The Chemistry of Melanin. V. Oxidation of Dihydroxyphenylalanine by Tyrosinase. J. biol. Chemistry **180**, 235 (1949).

226. MAZZA, F. P. u. G. STOLFI: Untersuchungen über einen Farbstoff von *Halla parthenopea* COSTA. Arch. Scienze biol. **16**, 183 (1931); Chem. Zbl. **1933** I, 1462.

227. MCELROY, W. D. and D. M. KIPNIS: The Mechanism of Inhibition of Bioluminescence by Naphthoquinones. J. cellular comparat. Physiol. **30**, 359 (1947).

228. MEIER, R. u. M. ALLGÖWER: Zur Charakterisierung zellteilungswirksamer Substanzen an der Gewebekultur. Experientia **1**, 57 (1945).

229. MEIER, R. et B. SCHÄR: Différentiation de l'action antimitotique sur la cellule animale normale in vitro. Experientia **3**, 358 (1947).

230. MEYERHOF, O. and L. O. RANDALL: Inhibitory Effects of Adrenochrome. Arch. Biochemistry **17**, 171 (1948).

231. MICHAEL, S. E.: Studies in the Biochemistry of Micro-organisms. 79. Fuscin, a Metabolic Product of *Oidodendron fuscum* RUBAK. I. Preparation, Properties and Antibacterial Activity. Biochemic. J. **43**, 528 (1948).

232. MITCHELL, J. S.: Histological Changes Produced by Large Doses of Tetrasodium-2-methyl-1 : 4-naphthohydroquinone Diphosphate in Some Human Tumours. Experientia **5**, 293 (1949).

233. MITCHELL, J. S. and I. SIMON-REUSS: Combinations of Some Effects of X-Radiation and a Synthetic Vitamin K Substitute. Nature (London) **160**, 98 (1947).

234. MITTER, P. C.: Synthesis of Rubiadin. Nature (London) **120**, 729 (1927).

235. MITTER, P. C. and D. BANERJEE: Synthesis of Aloe-emodin. J. Indian chem. Soc. **9**, 375 (1932).

236. MITTER, P. C. and H. BISWAS: On an Inductive Method for the Study of Natural Products. I. J. Indian chem. Soc. **5**, 769 (1928).

237. — — Über die Synthese des Munjistins. Ber. dtsch. chem. Ges. **65**, 622 (1932).

238. MÖLLER, E. F.: unveröffentlichte Versuche; zitiert nach *313*.

239. MORGAN, G. T. and E. A. COOPER: Bactericidal Action of the Quinones and Allied Compounds. J. Soc. chem. Ind. **43**, 352 (1924).

240. MUSAJO, L. u. M. MINCHILLI: Über ein zweites Pigment aus den Stacheln von *Paracentrotus*. Boll. sci. Fac. Chim. ind. Bologna **3**, 113 (1942); Chem. Zbl. **1943** I, 1275.

241. NAKAO, M. and T. FUKUSHINA: The Chemical Constitution of *Salvia miltiorrhiza*. J. pharmac. Soc. Japan **54**, 154 (1934).

242. NARGUND, K. S. and B. W. BHIDE: The Constitution of Embelin. J. Indian chem. Soc. **8**, 237 (1931).

243. NELSON, J. M. and C. R. DAWSON: Tyrosinase. Adv. Enzymol. **4**, 99 (1944).

243 a. NISHIKAWA, H.: Über Sulochrin, einen Bestandteil des Myceliums von *Oospora sulphurea-ochracea*. Acta Phytochim. (Japan) **11**, 167 (1939).

244. NORD, F. F., J. V. FIORE, G. KREITMAN and S. WEISS: On the Mechanism of Enzyme Action. XL. The Interaction of Solanione, Riboflavin and Nicotinic Acid in the Carbohydrate—Fat Conversion by Certain *Fusaria*. Arch. Biochemistry **23**, 480 (1949).

245. NORD, F. F., J. V. FIORE and S. WEISS: On the Mechanism of Enzyme Action. XXXIII. Fat Formation in *Fusaria* in the Presence of a Pigment Obtained from *Fusarium solani* D_2 Purple. Arch. Biochemistry **17**, 345 (1948).

246. NYBOM, N. and B. KNUTSON: Studies on *c*-Mitosis in *Allium cepa*. Hereditas **33**, 220 (1947).

246 a. OESTERLE, O. A. u. E. TISZA: Zur Kenntnis des Morindins. Arch. Pharmaz. Ber. dtsch. pharmaz. Ges. **245**, 534 (1907).

247. — — Über die Bestandteile der Wurzelrinde von *Morinda citrifolia* L. Arch. Pharmaz. Ber. dtsch. pharmaz. Ges. **246**, 150 (1908).

248. OSCHMANN, M.: Die Pigmentfarben tierischen und pflanzlichen Ursprungs und ihre Synthese. Peintures-Pigments-Vernis **18**, 34 (1943); Chem. Zbl. **1943** II, 1097.

249. OXFORD, A. E.: On the Chemical Reaction Occuring Between Certain Substances which Inhibit Bacterial Growth and the Constituents of Bacteriological Media. Biochemic. J. **36**, 438 (1942).
250. — Anti-Bacterial Substances from Moulds. V. The Bacteriostatic Powers of the Methyl Ethers of Fumigatin and Spinulosin and Other Hydroxy-, Methoxy- and Hydroxymethoxy-Derivatives of Toluquinone and Benzoquinone. Chem. and Ind. **61**, 189 (1942).
251. OXFORD, A. E. and H. RAISTRICK: Studies in the Biochemistry of Microorganisms. 66. Penicilliopsin, the Colouring Matter of *Pennicilliopsis clavariaeformis* SOLMS-LAUBACH. Biochemic. J. **34**, 790 (1940).
252. — — Anti-Bacterial Substances from Moulds. IV. Spinulosin and Fumigatin, Metabolic Products of *Penicillium spinulosum* THOM and *Aspergillus fumigatus* FRESENIUS. Chem. and Ind. **61**, 128 (1942).
253. PAGE, J. E. and F. A. ROBINSON: An Examination of the Relationship between the Bacteriostatic Activity and the Normal Reduction Potentials of Substituted Quinones. Brit. J. exp. Pathol. **24**, 89 (1943).
254. PARIS, R. et H. MOYSE-MIGNON: Pouvoir antimicrobien et présence de plumbagol chez deux *Diospyros* africains. C. R. hebd. Séances Acad. Sci. **228**, 2063 (1949).
255. PATERNO, E.: Ricerche sull'acido lapacico. Gazz. chim. ital. **12**, 337 (1882).
255a. PERKIN, A. G. and R. C. STOREY: The Migration of the Acyl Group in Partly Acylated Phenolic Compounds. J. chem. Soc. (London) **1928**, 229.
256. PERKIN, A. J.: Methylethers of Some Hydroxyanthraquinones. J. chem. Soc. (London) **91**, 2066 (1907).
257. PERKIN, A. J. and A. E. EVEREST: Natural Organic Colouring Matters. London: Longsman, Green & Co. 1918.
258. PERKIN, A. J. and J. J. HUMMEL: The Colouring Principle of *Rubia sikkimensis*. J. chem. Soc. (London) **63**, 1157 (1893).
259. — — The Colouring and Other Principles Contained in Chay Root. J. chem. Soc. (London) **63**, 1160 (1893).
260. — — The Colouring and Other Principles Contained in Mang-Kondu. J. chem. Soc. (London) **65**, 851 (1894).
261. — — The Colouring Matter and Other Principles Contained in Chay Root. III. J. chem. Soc. (London) **67**, 817 (1895).
262. POSTERNAK, T.: Recherches sur la biochimie des champignons inférieurs. II. Sur la constitution et la synthese de la phoenicine et sur quelques nouveaux dérivés de la 4,4'-di-toluquinone. Helv. chim. Acta **21**, 1326 (1938).
263. POSTERNAK, T. et J.-P. JACOB: Recherches sur la biochimie des champignons inférieurs. III. Sur le pigment de *Penicillium citro-roseum* DIERCKX. Helv. chim. Acta **23**, 237 (1940).
264. POSTERNAK, T., J.-P. JACOB et H. RUELIUS: Une synthèse nouvelle de l'émodine et la synthèse de la fumigatine. C. R. hebd. Séances Soc. Physique Hist. natur. Genève **58**, 223 (1941).
265. PRICE, J. R. and R. ROBINSON: A New Natural Pigment of the Naphthalene Series. Nature (London) **142**, 147 (1938).
265a. PRYOR, M. G. M., P. B. RUSSELL and A. R. TODD: Protocatechuic Acid, the Substance Responsible for the Hardening of the Cockroach Ootheca. Biochemic. J. **40**, 627 (1946).
266. QUASTEL, J. H.: Action of Polyhydric Phenols on Urease; Influence of Thiol Compounds. Biochemic J. **27**, 1116 (1933).
267. QUILICO, A. u. L. PANIZZI: Chemische Untersuchungen über *Aspergillus echinulatus*. Ber. dtsch. chem. Ges. **76**, 348 (1943).

268. QUILICO, A., L. PANIZZI and E. MUGNAINI: Structure of Flavoglaucin and Autoglaucin. Nature (London) **164**, 26 (1949).

269. RAISTRICK, H.: Production of Polyhydroxyanthraquinones by Moulds. Enzymologia **4**, 76 (1937).

270. — Certain Aspects of the Biochemistry of Lower Fungi. Ergebn. Enzymforsch. **7**, 316 (1938).

271. — Biochemistry of the Lower Fungi. Annu. Rev. Biochem. **9**, 571 (1940).

272. RAISTRICK, H., R. ROBINSON and A. R. TODD: Studies in the Biochemistry of Micro-organisms. 32. Cynodontin (1:4:5:8-Tetrahydroxy-2-methylanthraquinone), a Metabolic Product of *Helminthosporium cynodontis* MARIGNONI and *H. euchlenae* ZIMMERMANN. Biochemic. J. **27**, 1170 (1933).

273. — — — Synthesis of Helminthosporin. J. chem. Soc. (London) **1933**, 488.

274. — — — Studies in the Biochemistry of Micro-organisms. 37. Production of Hydroxyanthraquinones by Species of *Helminthosporium;* Isolation of Tritisporin, a Metabolic Product of *H. tritici-vulgaris* NISIKADO; Molecular Constitution of Catenarin. Biochemic. J. **28**, 559 (1934).

275. RAISTRICK, H. and P. SIMONART: Studies in the Biochemistry of Micro-organisms. 29. 2:5-Dihydroxybenzoic Acid (Gentisic Acid), a New Product of the Metabolism of Glucose by *Penicillium griseo-fulvum* DIERCKX. Biochemic. J. **27**, 628 (1933).

276. RAPER, H. S.: Tyrosinase. Ergebn. Enzymforsch. **1**, 270 (1932).

277. REED, H. S.: The Action of Quinones on Mitosis. Experientia **5**, 237 (1949).

278. REMFRY, F. G. P.: Perezone. J. chem. Soc. (London) **103**, 1076 (1913).

279. RENNIE, E. H.: A Colouring Matter from *Lomatia ilicifolia* and *Lomatia longifolia*. J. chem. Soc. (London) **67**, 784 (1895).

280. RICHTER, D.: Anthraquinone Colouring Matters: Ruberythric Acid. J. chem. Soc. (London) **1936**, 1701.

281. RIEGEL, B.: Vitamin K. Ergebn. Physiol., biol. Chem. exp. Pharmakol. **43**, 133 (1940).

282. ROBINSON, E. S. and J. M. NELSON: Tyrosine-Tyrosinase Reaction and Aerobic Plant Respiration. Arch. Biochemistry **4**, 111 (1944).

282a. RÖTHELI, A.: Auflösung und Neubildung der Meiosespindel von *Tubifex* nach chemischer Behandlung. Rev. Suisse Zool. **56**, 322 (1949).

282b. — Chemische Beeinflussung plasmatischer Vorgänge bei der Meiose des *Tubifex*-Eies. Z. Zellforsch. **35**, 62 (1950).

283. SANNIÉ, C.: Pigments et substances antibiotiques des champignons et des bactéries. Exp. ann. Biochimie méd. **6**, 225 (1946).

284. SCHALES, O.: Kidney Enzymes and Essential Hypertension. Adv. Enzymol. **7**, 513 (1947).

285. SCHEIBE, G. u. A. SCHÖNTAG: Lichtabsorption und Fluoreszenz des Hypericins. Ber. dtsch. chem. Ges. **75**, 2019 (1943).

286. SCHLEMMER, F. u. O. GENTNER: Pharmazeutisch-chemische Untersuchungen über den Mönchsrhabarber. Arch. Pharmaz. Ber. dtsch. pharmaz. Ges. **278**, 252 (1940).

287. SCHOELLER, A.: Über das Hystazarin. Ber. dtsch. chem. Ges. **21**, 2503 (1888).

288. SCHOPFER, W.-H. et M. L. BOSS: Recherches sur la rôle de la vitamine K et de diverses quinones chez les plantes. Un méchanisme possible de l'effet antibiotique de la vitamine K. Arch. Science (Genève) **1**, 521 (1948).

289. SCHOPFER, W.-H. u. E. C. GROB: Über den Einfluß von 1,4-Naphthochinonderivaten mit Vitamin-K- oder Antivitamin-K-charakter auf die Urease. Helv. chim. Acta **32**. 829 (1949).

289a. SCHUNK, E. u. H. RÖMER: Über Munjistin, ε-Purpurin und Purpurocarbonsäure. Ber. dtsch. chem. Ges. **10**, 790 (1877).

290. SCHWARZ, H. and W. M. ZIEGLER: Influence of Vitamin K Preparations on Blood Pressure in Hypertensive Rats. Proc. Soc. exp. Biol. Med. **55**, 160 (1944).

291. SENHOFER, G. u. F. SARLAY: Über direkte Einführung von Carboxylgruppen in Phenole und aromatische Säuren. IV. Mh. Chem. **2**, 448 (1881).

291a. SIEGFRIED, B.: Über natürliche und synthetische Oxyanthrachinone und Oxymethylanthrachinone. Dissert. Eidg. Techn. Hochsch. Zürich 1938.

292. SIGMUND, W.: Über die Einwirkung von Stoffwechselendprodukten auf die Pflanzen. II. Biochem. Z. **62**, 299 (1914).

293. SIMONSEN, J. L.: Note on the Constituents of *Morinda citrifolia*. J. chem. Soc. (London) **117**, 561 (1920).

294. SIZER, I. W.: The Action of Certain Oxidants and Reductants upon the Activity of Bovine Phosphatase. J. biol. Chemistry **145**, 405 (1942).

295. SOLOWAY, S. and K. A. OSTER: Inactivation of Pressure Amines by Quinones and Related Diketones. Proc. Soc. exp. Biol. Med. **50**, 108 (1942); zitiert nach *284*.

296. SPRUIT, C. J. P.: Naphthochinonen en Bioluminescentie. Dissert. Utrecht 1946.

297. — The Chemical Nature of Luciferins. Enzymologia **13**, 191 (1949).

298. SPRUIT, C. J. P. and A. L. SCHUILING: On the Influence of Naphthoquinones on the Respiration and Light Emission of *Photobacterium phosphoreum*. Recueil Trav. chim. Pays-Bas **64**, 219 (1945).

298a. STOLL, A., B. BECKER u. W. KUSSMAUL: Die Isolierung der Anthraglykoside aus Sennadrogen. Helv. chim. Acta **32**, 1892 (1949).

299. STOUDER, F. D. and R. ADAMS: Polyhydroxy-Methylanthraquinones. IX. J. Amer. chem. Soc. **49**, 2043 (1927).

300. SÜLLMANN, H.: Inhibitoren der enzymatischen Oxydation ungesättigter Fettsäuren. Helv. chim. Acta **26**, 1114 (1943).

301. TAKIZAWA, N.: Über die cancerogene Wirkung von bestimmten Chinonverbindungen. Proc. Imp. Acad. (Tokyo) **16**, 309 (1940); Chem. Zbl. **1941** I, 653.

301a. TAPPEINER, H. VON: Die photodynamische Erscheinung (Sensibilisierung durch fluoreszierende Stoffe). Ergebn. Physiol., biol. Chem. exp. Pharmakol. **8**, 698 (1909).

302. TATUM, E. L.: Biochemistry of Fungi. Annu. Rev. Biochem. **13**, 667 (1944).

303. TER HORST, W. P. and E. L. FELIX: 2,3-Dichloro-1,4-naphthoquinone – Potent Organic Fungicide. Ind. Engng. Chem. **35**, 1255 (1943).

304. THALHIMER, W. and B. PALMER: A Comparison of the Bactericidal Action of Quinone with that of Some of the Commoner Disinfectants. J. infect. Diseases **9**, 181 (1911).

305. THIELE, J. u. E. WINTER: Über die Einwirkung von Essigsäureanhydrid und Schwefelsäure auf Chinone. Liebigs Ann. Chem. **311**, 341 (1900).

306. THOMAS, A. W. and S. B. FOSTER: The Behavior of Deaminized Collagen. Further Evidence in Favor of the Chemical Nature of Tanning. J. Amer. chem. Soc. **48**, 489 (1926).

307. TOMMASI, G.: Studii sull'Henna (*Lawsonia inermis* L.). Sulla costituzione del Lawson. II. Gazz. chim. ital. **50** I, 263 (1920).

308. TREIBS, A. u. H. STEINMETZ: Über das Vorkommen von Anthrachinonfarbstoffen im Mineralreich (Graebeit). Liebigs Ann. Chem. **506**, 171 (1933).

308a. TRUEMAN, E. R.: Quinone-tanning in the Mollusca. Nature (London) **165**, 397 (1950).

309. TSCHIRCH, A. u. U. CHRISTOFOLETTI: Über die Rhaponticwurzel. Arch. Pharmaz. Ber. dtsch. pharmaz. Ges. **243**, 443 (1905).

310. Tschirch, A. u. Jr. Lüdy: Über den Stocklack. Helv. chim. Acta **6**, 994 (1923).
310a. Tutin, F. and H. W. B. Clewer: The Constituents of Rhubarb. J. chem. Soc. (London) **99**, 946 (1911).
311. Vinet, A.: L'activité bactériostatique dans la série du méthylnaphthoquinone et sa relation à l'isomorphie. C. R. Séances Soc. Biol. Filiales Associées **139**, 155 (1945).
312. Wallenfels, K.: Der Farbstoff der roten Blutzellen des Seeigels *Arbacia pustulata*. Ber. dtsch. chem. Ges. **76**, 323 (1943).
313. — Symbiose und Antibiose. Die Chemie **58**, 1 (1945).
314. Wallenfels, K. u. A. Gauhe: Synthese von Echinochrom A. Ber. dtsch. chem. Ges. **76**, 325 (1943).
315. Wallenfels, K. u. W. Möhle: Reduktions-Oxydations-Potentiale der Naphthochinone. Ber. dtsch. chem. Ges. **76**, 924 (1943).
316. Weiss, S., J. V. Fiore and F. F. Nord: On the Mechanism of Enzyme Action. XXXVIII. Effect of Different Naphthoquinones on the Fat Formation in *Fusarium lini* Bolley. Arch. Biochemistry **22**, 314 (1949).
317. Weiss, S. and F. F. Nord: On the Mechanism of Enzyme Action. XXXVII. Solanione, a Pigment from *Fusarium solani* D_2 Purple. Arch. Biochemistry **22**, 288 (1949).
318. Wendel, W. B.: Influence of Naphthaquinones on the Respiratory and Carbohydrate Metabolism of Malarial Parasites. Federat. Proc. **5**, 406 (1946).
319. Werenskiold, W. and J. Oftedal: A Burning Coal Seam at Mt. Pyramide Spitsbergen. Resultater av de Norske Statsunderstottede Spitsbergenekspeditioner I. Oslo, 1923; Chem. Zbl. **1923** I, 1390.
320. Wessely, F. von u. A. Bauer: Über die Konstitution des Tanshinons. I. Ber. dtsch. chem. Ges. **75**, 617 (1942).
321. Wessely, F. von u. S. Wang: Über einen neuartigen natürlichen Chinonfarbstoff aus der Klasse eines Phenanthrofurans. Ber. dtsch. chem. Ges. **73**, 19 (1940).
322. Wieland, H. u. F. Bergel: Über den Mechanismus der Oxydationsvorgänge. VIII. Zum oxydativen Abbau der Aminosäuren. Liebigs Ann. Chem. **439**, 196 (1924).
323. Wieland, H. u. O. B. Claren: Über den Mechanismus der Oxydationsvorgänge. XXXIX. Die Dehydrierung der Hefe bei Gegenwart von Methylenblau und Chinon. Liebigs Ann. Chem. **509**, 182 (1934).
324. Wieland, H. u. K. Frage: Über den Mechanismus der Oxydationsvorgänge. XX. Beiträge zur Kenntnis der Bernsteinsäuredehydrase. Liebigs Ann. Chem. **477**, 1 (1929).
325. Wieland, H. u. A. Lawson: Über den Mechanismus der Oxydationsvorgänge. XXVII. Weitere Beiträge zur Kenntnis der Dehydrase des Muskelgewebes. Liebigs Ann. Chem. **485**, 193 (1932).
326. Wieland, H. u. W. Mitchell: Über den Mechanismus der Oxydationsvorgänge. XXIX. Über die dehydrierenden Enzyme der Milch. IV. Liebigs Ann. Chem. **492**, 156 (1932).
327. Wieland, H. u. J. J. Pistor: Über den Mechanismus der Oxydationsvorgänge. IL. Über das dehydrierende Enzymsystem von *Acetobacter peroxydans*. II. Liebigs Ann. Chem. **535**, 205 (1938).
328. Willstätter, R. u. A. S. Wheeler: Über die Isomerie der Hydrojuglone. Ber. dtsch. chem. Ges. **47**, 2796 (1914).
329. Winzor, F. L.: The Colouring Matters of *Drosera wittackeri*. III. J. chem. Soc. (London) **1935**, 336.

330. WÖHLER, F. u. A. FRERICHS: Über die Veränderungen, welche namentlich organische Stoffe bei ihrem Übergang in den Harn erleiden. Liebigs Ann. Chem. **65**, 335 (1848).
331. WOOLLEY, D. W.: Observations on Antimicrobial Action of 2,3-Dichloro-1,4-naphthoquinone and Its Reversal by Vitamin K. Proc. Soc. exp. Biol. Med. **60**, 225 (1945).
332. — Biological Antagonism between Structurally Related Compounds. Adv. Enzymol. **6**, 129 (1946).
332a. ZEMPLÉN, G. u. R. BOGNÁR: Synthese der Ruberythrinsäure. Ber. dtsch. chem. Ges. **72**, 913 (1939).
333. ZOPF, W.: Zur Kenntnis der Flechtenstoffe. Liebigs Ann. Chem. **284**, 107 (1894).
334. — Zur Kenntnis der Flechtenstoffe. VIII. Liebigs Ann. Chem. **317**, 110 (1901).
335. — Zur Kenntnis der Flechtenstoffe. XIV. Liebigs Ann. Chem. **340**, 276 (1905).
336. — Zur Kenntnis der Flechtenstoffe. XV. Liebigs Ann. Chem. **346**, 82 (1907).

(Eingelaufen am 14. Dezember 1949.)

Cactus Alkaloids and Some Related Compounds.

By L. Reti, Buenos Aires.

Contents.

I. Introduction.

In their monumental work "The Cactaceae", BRITTON and ROSE (*19*) record 1235 species belonging to the three tribes which constitute the family of the Cacti. The actual number of the species must be considerably higher.

Cacti occur frequently in the more arid and less accessible regions of the American Continent, nearly always within very narrow and definite borderlines. The habitat of a species is in many instances a single valley located in a remote, uninhabited region of the Cordillera. Thus the collection of flowering specimens fit for botanical identification is sometimes extremely difficult.

On the other hand, cacti are apt to develop individual variations in their characteristic morphological features, rendering the definition of a species difficult and often illusory. Specimens taken from their normal habitat to botanical gardens or arboreta often die, degenerate or stop flowering.

Taking into account all these difficulties, it is not surprising to find considerable differences of opinion among botanists on the taxonomy of the cactaceae. A considerable number of species have not been well defined and in many cases different names have been given to the same species. The index of BRITTON and ROSE records not less than 7000 binomials.

The study of the chemical composition of the cacti is even more arduous and, for the reasons given above, trustworthy material is not easily obtained owing to taxonomic difficulties, the geographical distribution and the very limited areas of dispersion. In some instances costly expeditions must be organized; the collected material must be transported for long distances and the preservation of this material, containing up to 90% water, is difficult. Finally, existing phytosanitary regulations should not be overlooked.

The data found in the literature on the chemical composition of cacti are meager and incomplete. Although the occurrence of alkaloids has been denoted in some 30 species, in many instances the material examined has been scanty and not well defined. The chemical constitution of the bases is known only with reference to nine species.

Nevertheless, we do know that cacti are alkaloidipherous; future studies undoubtedly will show the presence of many new alkaloids and the occurrence of some known alkaloids will be demonstrated in further species.

Cactus alkaloids are of simple chemical constitution. They are either substituted β-phenylethylamines, evidently related to the naturally occurring aromatic amino acids (tyrosine, dihydroxyphenylalanine, etc.), or they are tetrahydroisoquinolines that could originate from the bases just mentioned by condensation and cyclization through the action of simple organic compounds containing only one or two carbon atoms in their molecules.

Obviously, under these conditions, it will be quite possible to find bases of the same type in plants of widely separated botanical families or even in products of animal metabolism. It has, therefore, been considered convenient to include in the present survey not only alkaloids isolated from the cacti but also a number of naturally occurring bases of similar constitution. Amino acids, epinephrine, ephedrine and related compounds will only be mentioned without going into details. We will also point out certain relations and implications within the wider field of general biochemistry. We have tried to compile a rather complete bibliography and to include all essential experimental data known on this subject, since a great number of pertinent papers have been published a number of years ago or in journals with a rather limited circulation.

General surveys on cactus alkaloids have been published previously by RETI (*162*, *164*, *166*) as well as by SPÄTH and BECKE (*201*); see also the book of HOBSCHETTE (*77*) and the chapter on alkaloids of the cactaceae in HENRY's monograph (*65*); for peyote cf. EWELL (*35*), SAFFORD (*181*), REUTTER (*169*) and AMOROSA (*1*).

II. Historical.

Centuries before the white man arrived, the Indian tribes of the Southern part of North America, from Oklahoma and Arkansas through almost the whole of Mexico, used and revered the *peyote*.

The chroniclers of the Spanish conquest of Mexico are the first who mention this drug and describe its use and properties. BERNARDINO DE SAHAGUN, a Franciscan monk and missionary, writes as follows in his "Historia General de las Cosas de Nueva España" (the manuscript, dated 1560, was printed only in 1829): "There is another plant like earth nopal, called peyote; it is white, grows in the Northern parts, and produces in those who eat or drink it terrible or ludicrous visions; the inebriation lasts for two or three days and then disappears. The Chichimecas eat it commonly, it gives them strength and incites them to battle, alleviates fear, and they feel neither hunger nor thirst, and they say it protects them from every kind of danger."

This first reference to peyote summarizes in a remarkable manner all our knowledge on the nature of this astonishing drug. SAHAGUN also points out accurately the botanical nature of the plant which is a "tuna de la tierra", i.e., a cactus. The word "tuna", of Mexican origin, is used in Spanish America only to indicate cacti.

In BRITTON and ROSE's monograph (*19*), however, it is claimed that SAHAGUN supposed the plant to be a fungus and termed it teonânactl or "sacred mushroom". In 1591 Doctor CARDENAS in his book, "Primera parte de los problemas y secretos maravillosos de las Indias", mentions the "satanic" plant called peyote, used by the Mexican indians to evoke the devil and predict the future. The naturalist FRANCISCO HERNANDEZ, who lived during the reign of Philip II, described the same plant in his book "Quatro libros de la Naturaleza y virtudes de las plantas y animales que están recividos en el uso de Medicina en la Nueva España", Mexico 1615 (published in 1790). He says that those who ate its root could predict the attacks of enemies, their fortunes in the future or reveal the hiding place of stolen goods. The Catholic church attributed diabolic properties to the magic effects of peyote and urged the priests to make inquiries about it in the confessional. NICOLÁS DE LEÓN (1611), JACINTO DE LA SERNA (1626), BARTOLOMEO DE ALUA (1634) and Father JOSÉ ORTEGA (1754) mentioned the peyote and called it "raíz diabólica".

As the Spaniards conquered Nayarit, they were astonished by the extraordinary resistance of the defenders of the Sierra de Alica. Eating peyote, they could walk several days without water, food or sleep (MATHIAS DE LA MOTTA PADILLAS: Historia de la Conquista de la Nueva Galicia; written in 1742, printed in Mexico 1870).

The use and cult of the peyote remained alive among the Indian tribes of the area mentioned in spite of church and state prohibitions. Merchants in Indian territories call it mescal or mescal buttons, the Mexicans on the Rio Grande, pellote, peyote or peyotl.

PETRULLO reported (*132*) that in 1918 a numerous sect, restricted to Indians, was founded in Oklahoma. It was called the "Peyote-Church" and joined in a strange synthesis old Mexican, Christian and local religious rites.

Further details and bibliography on the uses and religious cult of the peyote will be found in LEWIN (*104–107*), MOONEY (*122*), DIGUET (*31*), LUMHOLTZ (*117*), NEWBERNE and BURKE (*124*), and especially in the books by ROUHIER (*180*) and BERINGER (*11*).

Up to 1886 nothing was known of the nature and character of the "mescal buttons". At that time LOUIS LEWIN, travelling in America came to know the plant and obtained specimens which he examined. HENNINGS recognized it as a new species of *Anhalonium* and named the plant *Anhalonium lewinii*. LEWIN's examination showed that the drug contained alkaloids, and a crystallized base, anhalonine, was isolated. However, anhalonine is not responsible for the sensory excitation caused by peyote. LEWIN's discovery raised considerable interest in the pharmacological effect of the drug and in the chemical compounds which account for it. This interest was extended to the whole family of the Cactaceae, which, until then, had been considered as being free of alkaloids.

In 1894 HEFFTER (*58–63*) started to study the cacti from chemical as well as from pharmacological points of view. He isolated anhaline from *Anhalonium fissuratum* and pellotine from *A. williamsii*. In 1896 HEFFTER (*60*) was able to isolate and identify the active hallucinatory principle of *A. lewinii*. This new alkaloid was called *mezcaline*. At the same time two other bases were separated, viz. anhalonidine and lophophorine. In 1899 KAUDER (*87*) found in the same plant a new alkaloid, anhalamine, and the already known pellotine. In 1901 HEYL examined different species of cacti and discovered pectenine in *Cereus pecten aboriginum* (*75*). The same author isolated carnegine from *Carnegiea gigantea* (*76*) in 1928.

Up to 1919 little was known about the chemical structure of the cactus bases. In this year SPÄTH published the first of a series of important papers on "*Anhalonium* alkaloids". This work was continued up to 1939 and includes also the bases contained in a few other cactaceae (*192–215*).

The clarification of the structure and the syntheses of all of the members of this class of compounds of such biochemical interest, must be credited to SPÄTH and his collaborators. The accomplishment is all the more remarkable since SPÄTH had to contend with a scarcity of material; several fundamental structures were determined on very small samples, left over from HEFFTER's and HEYL's experiments.

SPÄTH and his co-workers isolated from *Anhalonium lewinii*, in addition to the already mentioned bases, anhalinine and anhalidine in 1935 (*199, 200*); N-methyl-mezcaline in 1937 (*203*); N-acetyl-mezcaline in 1938 (*204*), and O-methyl-*d*-anhalonidine (*205*) in 1939. SPÄTH established also the structure of carnegine and proved that HEYL's "pectenine" and carnegine were identical (*196, 211*). In collaboration with OREKHOV (*212, 206*) he established conclusively the constitution of the alkaloids of *Salsola richteri* (Chenopodiaceae). The bases of this plant show remarkable structural analogies with the cactus alkaloid carnegine.

In 1933 RETI (*161*) isolated hordenine and a natural quaternary base, related to the former, viz. candicine, from the Argentine cactus *Trichocereus candicans*. The same bases were found later in other *Trichocereus* species (*167, 49*). A new quaternary base, coryneine, was isolated from *Stetsonia coryne* in 1934 by RETI, ARNOLT and LUDUEÑA (*168*). Trichocereine, a new cactus alkaloid, together with mezcaline, were found in the Argentine giant cactus *Trichocereus terscheckii* [RETI (*163*)].

III. Occurrence of Alkaloids in Cacti.

As mentioned above, only a few species belonging to the family of the Cactaceae have been examined from a chemical point of view. However, the early investigations of LEWIN (*104–107*), HEFFTER (*58–64*) and HEYL (*75, 76*), and the more recent ones of HERRERO-DUCLOUX (*68–72*), RETI (*161–168*) and others, seem to indicate that the faculty of producing and storing alkaloidal substances should be considered one of the characteristics of this botanical family. This ability is more noticeable in the tribe of the Cereae, while in the few species of the Opuntiae investigated, only minor quantities of basic substances have been found. No observation of the existence of alkaloids in members of the third tribe,—the Pereskieae—has been recorded so far.

Alkaloidal substances have been isolated from the species given in the following list, in which the botanical denomination of the original publications has been adhered to, adding, when necessary, the synonyms, in the first place those mentioned in the monograph by BRITTON and ROSE (*19*).

List of Cactaceae from Which Alkaloids of Known Structure Have Been Isolated.

1. *Anhalonium lewinii* (HENNINGS).
   Lophophora williamsii (LEMAIRE) COULTER; BRITTON and ROSE.
   Echinocactus lewinii (HENNINGS).
   Anhalonium williamsii (LEMAIRE) etc.

In the chemical literature *Anhalonium lewinii*, *A. williamsii* and *A. jourdanianum* are mentioned as different species. However, botanists definitely recognize only one species (BRITTON and ROSE; SCHUMANN). It would be worth while to investigate, using fresh and well identified material, whether only pellotine is present in *A. williamsii* as stated by HEFFTER. Such findings may give support to a revision of the taxonomy of these cacti.

The plant grows from Central Mexico to Southern Texas, and is an object of commerce, carried out by some of the Indian tribes, although this is forbidden by law. The globular plants are sliced into 3 or 4 sections and then dried in the sun; these dried pieces are the "mescal buttons" of the trade.

The plant is also known as pellote, peyote, peyotl; it is called challote in Starr County, Texas. Interest in the cactus alkaloids arose when the remarkable use by the Indian tribes and the strange pharmacological properties of this little plant were known.

Eleven bases have been isolated from *Anhalonium lewinii*: mezcaline, N-methylmezcaline, N-acetylmezcaline, anhalamine, anhalidine, anhalinine, anhalonidine, pellotine, O-methylanhalonidine, anhalonine, and lophophorine (*192–215, 87, 104–107*). An extensive study on various varieties of peyote and their alkaloidal contents was published by BECCARI (*10*).

2. *Anhalonium fissuratum* (ENGELMANN).
   Ariocarpus fissuratus (ENGELMANN), SCHUMANN.

Known as "living rock". Occurrence: Western Texas; northern Coahuila, and Zacatecas, Mexico.

According to HEFFTER (*58*) it contains anhaline (hordenine).

3. *Cereus pecten-aboriginum* (ENGELMANN).
   Pachycereus pecten-aboriginum (ENG.), BRITTON and ROSE.

Grows in Chihuahua, Sonora, Colina, Lower California, Mexico.

Examined by HEYL (*75*), it was found to contain pectenine (identical with carnegine).

4. *Carnegiea gigantea* (ENGELMANN), BRITTON and ROSE.
   Cereus giganteus (ENGELMANN).

This is the "saguaro", the giant cactus of Arizona. The Saguaro blossom is the official state flower of the state of Arizona. Occurrence: Arizona, Southeastern California, and Sonora, Mexico. According to HEYL (*76*) it contains carnegine.

5. *Trichocereus candicans* (GILLIES), BRITTON and ROSE.

Occurrence: Northwestern Argentina, Cordoba, San Luis, Mendoza, etc.

NIEDFELD (*125*) first observed the presence of alkaloids in this plant. LEWIS and LUDUEÑA (*108*) studied the pharmacological properties of extracts; later RETI (*161*) isolated from the same species hordenine (0,5–5%) as well as the quaternary base, candicine (0,5–5%). The natural ratio of the two alkaloids varies widely.

6. *Trichocereus lamprochlorus* (LEMAIRE), BRITTON and ROSE.

Occurrence: same as that of the former.

It contains the same alkaloids as *T. candicans*, but in smaller quantities (*167*).

7. *Trichocereus terscheckii* (PARMENTIER), BRITTON and ROSE.

Giant cactus of Argentina, called "cardón grande". Occurrence: Northwestern Argentina, La Rioja, Catamarca, Tucumán, etc.

In contains, according to RETI (*163*); trichocereine and mezcaline.

8. *Trichocereus spachianus* (LEMAIRE), RICCOBONO.

Occurrence: Western Argentina (BRITTON and ROSE). Argentine authors question this statement.

It contains candicine (*49*).

9. *Stetsonia coryne* (SALM-DYCK), BRITTON and ROSE.

It is one of the most striking tree-like cacti in South America and often forms the dominant feature of the landscape on the high plains of Northern Argentina. Occurrence: Cordoba, La Rioja, Santiago del Estero, etc.

RETI, ARNOLT and LUDUEÑA (*168*) found in this cactus coryneine (1%), an interesting quaternary catechol compound.

List of Cactaceae Which Contain Alkaloids of Undetermined Structure.

1. *Opuntia vulgaris* (MILLER).

Grows in Argentina, Uruguay, Brazil, Cuba, etc. FAIVELEY (*36*) reports having found alkaloids in the flowers. FALCO and HILBURG (*37*) found basic substances in the stems, which precipitate with alkaloid reagents.

2. *Cereus grandiflorus* (MILLER).

Selenicereus grandiflorus (LINNAEUS), BRITTON and ROSE.

Grows in Jamaica and Cuba. The drug was recommended as a heart tonic. SULTAN (*217*) isolated 2% of an alkaloid called cactine. The presence of alkaloids was confirmed by BOINET and BOY-TEISSIER (*15*).

3. *Mamillaria centricirrha* (LEMAIRE).

Neomamillaria magnimamma (HAWORTH), BRITTON and ROSE.

Occurrence: Central Mexico; HEFFTER (*61*).

4. *Phyllocactus ackermannii* (SALM-DYCK).

Epiphyllum ackermannii (HAWORTH). BRITTON and ROSE list it as a hybrid. Grows in Mexico; HEFFTER (*61*).

5. *Phyllocactus russelianus* (SALM-DYCK).

Schlummbergera russeliana (GARDNER), BRITTON and ROSE.

Found in the Organ Mountains, Brazil; HEFFTER (*61*).

6. *Echinocactus myriostigma* (SALM-DYCK).

Astrophytum myriostigma (LEMAIRE).

It grows in Northern Central Mexico; HEFFTER (*61*)

7. *Cereus peruvianus* (LINNAEUS) MILLER.

Occurrence: Southeastern coast of South America; HEFFTER (*61*).

8. *Echinocereus mamillosus* (RÜMPLER).

Listed as a hybrid by BRITTON and ROSE; HEFFTER (*61*).

9. *Echinocactus visnaga* (HOOKER).

It grows on the highlands of San Louis Potosí, Mexico. BRITTON and ROSE record a single plant 3 meters high, 1,3 meters in diameter, weighing 2000 kilograms. Cf. HEFFTER (*61*).

10. *Anhalonium prismaticum* (LEMAIRE).

Ariocarpus retusus (SCHEIDWEILER).

Occurrence: States of Coahuila, Zacatecas and San Louis Potosí, Mexico; HEFFTER (*58*).

11. *Mammillaria uberiformis* (ZUCCARINI).

Dolichotele uberiformis (ZUCCARINI), BRITTON and ROSE.

It grows in Central Mexico; LEWIN (*106*).

12. *Rhipsalis conferta* (SALM-DYCK).

Rhipsalis teres (VELLOZO) STEUDEL.

It grows in the States of Minas Geraes, Rio de Janeiro and Sao Paulo, Brazil; LEWIN (*106*).

13. *Pilocereus argentianus* (Orcutt).
Lophocereus schottii (Engelman), Britton and Rose.
Occurrence: Southern Arizona, Lower California, Sonora, Mexico. Heyl (*75*) found in this plant considerable amounts (5,8%) of an amorphous base, m.p. 82–86°, called pilocereine, $C_{30}H_{44}N_2O_4$. Amorphous salts; 13,48% methoxyl.

14. *Pachycereus marginatus* (De Candolle), Britton and Rose.
Occurrence: Hidalgo, Querétaro, Guanajuato, Mexico. Roca (*175*) found in this cactus some unidentified alkaloids, furthermore, tyrosine and tyrosinase.

15. *Gymnocalycium gibbosum* (Haworth) Pfeiffer.
Occurs in Argentina. Herrero-Ducloux (*69*) isolated from this plant small amounts of alkaloids, with reactions similar to those of mezcaline, anhalonine and lophophorine.

16. *Gymnocalycium multiflorum* (Hook), Britton and Rose.
Occurrence: Argentina (Córdoba, Catamarca), Brazil, Uruguay, Paraguay. Herrero-Ducloux (*71*) found very small quantities of an alkaloid, similar to the β-alkaloid of *Gymnocalycium gibbosum* (mezcaline-like).

17. *Echinopsis eyriesii* (Turpin) Zuccarini.
Occurs in Southern Brazil, Uruguay and Entre Ríos, Argentina. Herrero-Ducloux (*68*) states the presence of very small quantities of alkaloids.

18. *Trichocereus sp. aff. T. Terscheckii.*
Herrero-Ducloux (*72*) found small amounts of a non-phenolic base.

19. *Trichocereus thelegonoides* (Spegazzini), Britton and Rose.
Occurs in Northern Argentina. It contains alkaloids (*49*).

20. *Trichocereus thelegonus* (Weber), Britton and Rose.
Occurs in Northwestern Argentina (Tucumán): It contains alkaloids (*49*).

21. *Trichocereus huascha* (Weber), Britton and Rose.
Occurs in Northern Argentina (Catamarca). It contains alkaloids (*49*).

IV. Location of the Alkaloids in Tissues of the Cacti.

Only very few observations are available on this subject. Janot and Bernier (*80*) found that in pellote, the alkaloids are almost exclusively located in the internal cells of the cortical parenchyma at the top of the plant. In *Trichocereus candicans*, Niedfeld (*125*), utilizing microchemical methods, observed that the alkaloids are mostly situated in the chlorophyllaceous cortical parenchyma. In *Trichocereus terscheckii* (Reti, unpublished), the following comparative estimations have been made: green epidermis (dry), 0,29%; and central parts, including cortical parenchyma (dry), 0,45% of total alkaloids.

V. Extraction and Isolation of Cactus Alkaloids.

Fresh cacti, with a water content of 90–95% are difficult to handle; the juice is not easily separated from the fibers and the presence of mucilaginous substances causes trouble during extraction. Thus, it is preferable to dry the plants immediately after they are collected, in order to avoid losses and deterioration. The plants should be cleaned, the spines extracted with nippers, and the remaining material cut in thin slices and dried in the sun or better in a low temperature dryer (40–60°). The dried substance is easy to grind and the powder can be stored of necessary.

The powdered material is extracted following the usual procedure. Extractions of the drug and isolation of the alkaloids from *Anhalonium lewinii* have been described by HEFFTER (*60*), KAUDER (*87*), TOMASO (*219*), SPÄTH and BECKE (*201*), as well as by STEINER-BERNIER (*216*); and the extraction of alkaloids from *Pilocereus sargentianus* and *Cereus pecten-aboriginum*, by HEYL (*75*, *76*).

The following is a practical procedure, with general applicability.

The powdered material is extracted with alcohol (85–95%) in a Soxhlet or in a percolator. A small amount (1%) of acetic or formic acid may be added. The extract is filtered and concentrated in vacuo to a small volume. Water is added and the last traces of alcohol are eliminated by evaporation under reduced pressure. An intensely colored solution is obtained in which resins, chlorophyll, etc., are often suspended. The decanted and still acid solution is extracted several times with ether, in order to eliminate impurities; it is then made alkaline with ammonia or sodium carbonate and the alkaloids are extracted with ether, chloroform or another suitable solvent. Thus, a crude solution of the free bases is obtained, which can be purified by shaking the solution with dilute mineral acids: some non-basic substances remain in the solvent and the alkaloids transfer to the aqueous layer. Alkalinization, followed by further extraction with a solvent, gives a relatively pure solution of the free bases.

From *Anhalonium lewinii* as many as eleven bases have been isolated. SPÄTH and BECKE's process (201) runs as follows: The drug is extracted with cold alcohol and the aqueous solution of the syrup obtained by evaporating the extract in vacuo is treated with dilute hydrochloric acid. The filtered solution is alkalinized with sodium hydroxide and extracted with ether. The ether solution (*a*) contains the non-phenolic bases; the aqueous solution (*b*), the phenolic bases.

(*a*) After evaporation of the solvent the free bases are distilled in vacuo. On treatment with dilute sulfuric acid, mezcaline sulfate crystallizes. The filtrate from the same is alkalinized, extracted with ether and the bases treated with HCl (1:6): Anhalonine hydrochloride crystallizes. By evaporating the mother liquors crystals of anhalinine hydrochloride are separated. After a complicated treatment of the filtrate, a further quantity of mezcaline and a little lophophorine can be obtained.

(*b*) The solution is neutralized with HCl and, after adding potassium carbonate, extracted with ether. The residue of the evaporated solvent is dissolved in HCl and anhalamine hydrochloride crystallizes. The mother liquors are concentrated and alcohol is added: anhalonidine hydrochloride crystallizes. From the filtrate pellotine is obtained as picrate.

In some cases it may be advantageous to extract the alkaloids from the alkalinized dried plant with low boiling solvents, such as ether, chloroform, etc. OREKHOV and PROSKURNINA (*128*) doubled the yield of salsoline by using dichloroethane instead of alcohol to extract the powdered herb of *Salsola richteri*. In view of RETI's discovery of *quaternary ammonium bases* in several cacti, special care should be taken not to overlook their presence. Isolation procedures have been described by RETI and his co-workers for *Trichocereus candicans* (*161*), *Trichocereus lamprochlorus* (*167*), and *Stetsonia coryne* (*168*). The alkaline aqueous solution, from which all the soluble bases have been extracted, is acidified and the quaternary bases are precipitated with suitable reagents, such as picric acid, picrolonic acid, MAYER's solution, etc. From the precipitate the quaternary ammonium salts can be recovered; for example, candicine iodide has been prepared by treating the suspension of the precipitate (obtained with MAYER's reagent) with hydrogen sulfide.

VI. The Chemistry of the Cactus Alkaloids and Some Related Natural Bases.

A. β-Phenylethylamines.

1. *β-Phenylethylamine*, $C_8H_{11}N$, [benzene ring]–CH_2–CH_2–NH_2

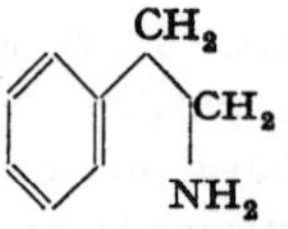

Phenylethylamine has been isolated in many instances from the putrefactive decomposition products of proteins [cf. GUGGENHEIM (*46*)]; but is also occurs as a plant constituent. LEPRINCE (*103*) found in mistletoe a base which should be regarded as phenylethylamine; and WHITE (*228*) isolated the compound from many species of acacia. It does not occur in cacti.

β-Phenylethylamine is a colorless oil, with strong alkaline reaction, $D^{24°} = 0{,}9580$, b.p. 198° (760 mm.). It is slightly soluble in water, readily soluble in alcohol and ether. Hydrochloride, m.p. 217°; oxalate, m.p. 218°; picrate, m.p. 171–174°; chloraurate, m.p. 98–100°; the chloroplatinate crystallizes from alcohol containing HCl in golden leaflets, $(C_8H_{11}N \cdot HCl)_2PtCl_4$; it is insoluble in water, m.p. 253–254°.

2. *N-Methyl-β-phenylethylamine*, $C_9H_{13}N$, [benzene ring]–CH_2–CH_2–$NH \cdot CH_3$, was isolated by YURASHEVSKII (*231*, *232*) from the Chenopodiacea *Arthrophytum leptocladum* M. POP, where it occurs together with dipterine (N-methyltryptamine) and leptocladine (3:4-dimethyl-3:4:5:6-tetrahydro-4-carboline).

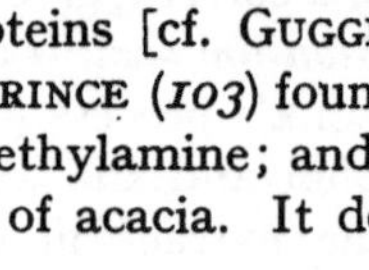

Colorless oil, b.p. 73–75° (4 mm.). Hydrochloride, m.p. 161–162°, picrate 141–142° (from alcohol); picrolonate, m.p. 217–218° (from alcohol); chloroplatinate, m.p. 220–221°; methiodide of the methyl compound, m.p. 227–228°.

3. *Tyramine* (*p*-Hydroxy-*β*-phenylethylamine), $C_8H_{11}ON$, HO–C_6H_4–CH_2–CH_2–NH_2

Tyramine, as a secondary degradation product of tyrosine can be expected to appear in several metabolic and fermentation processes involving proteins. Many observations of this nature are surveyed by GUGGENHEIM (*46*). It occurs, however, also as an unquestionable primary constituent both in the animal and vegetable kingdom. BARGER (*3, 4*) found small quantities in ergot, accompanied by other bases [see also FREUDWEILER (*40*); FUNCK and FINK (*42*)]. It occurs together with tyrosine and histamine in the salivary or venom glands of cephalopods [HENZE (*66, 67*); BOTTAZZI (*17*)]. CRAWFORD and WATANABE (*28, 29*) found the same base in several species of American mistletoes, viz. *Phoradendron flavescens, Ph. villosum, Ph. californicum.* OSTEMBERG (*130*) isolated tyramine from the amines of the European mistletoe, *Viscum album.* ULLMANN (*222*) considers tyramine as the active principle of the thistle, *Silybum marianum.* According to SCHMALFUSS and HEIDER (*182*) the common broom, *Sarothamnus scoparius*, contains tyramine and 3:4-dihydroxy-phenylethylamine. Tyramine has not yet been found in cacti.

Tyramine crystallizes from alcohol in white, hexagonal leaflets, m.p. 161°; b.p. 175–178° (8 mm.); it is soluble at 15° in 95 parts of water and in 10 parts of hot alcohol; slightly soluble in amyl alcohol, much less in ether or chloroform. Well crystallized salts: Hydrochloride, from conc. HCl, very soluble in water, m.p. 268–269°. Picrate, m.p. 206°. Dibenzoate, m.p. 174°. Dicarbomethoxy derivative, m.p. 100,5°; oxalate, m.p. 203–204°.

BARGER (*4*) synthesized the base by reduction of *p*-hydroxyphenylacetonitrile with sodium. Two other syntheses have been described by BARGER and WALPOLE (*9*): A *p*-hydroxy group was introduced into phenylethylamine by nitration, reduction, diazotization, etc.; or anisaldehyde was converted into *p*-methoxyphenylpropionamide, from which, by HOFMANN degradation, tyramine was obtained. ROSENMUND (*176*) condensed anisaldehyde with nitromethane, reduced first the *ω*-nitrostyrene compound to the oxime, which, in a second operation stage was reduced to the amine. Elimination of the *o*-methyl group gave tyramine. KONDO and SHINOZAKI (*92*) simplified the ROSENMUND synthesis by reducing the nitrostyrene compound electrolytically to *p*-methoxyphenylethylamine. SLOTTA and ALTNER (*187*) prepared tyramine starting from phenylethylbromide by nitration, reduction etc. Further syntheses have been described by Koessler and Hanke (*91*), Kindler and Peschke (*89*), and Buck (*21, 22*).

A well known method of preparation is the thermal decarboxylation of tyrosine. Waser (*224*) has improved the yield by heating the amino acid suspended in a high boiling solvent.

4. *Hordenine* (Anhaline, *p*-hydroxy-*β*-phenylethyl-dimethylamine), $C_{10}H_{15}ON$,

$$HO-C_6H_4-CH_2-CH_2-N(CH_3)_2$$

Hordenine.

This compound was found first in the cactacea *Anhalonium fissuratum* by Heffter (*58*) in 1894 and named anhaline. In 1906 Leger (*94–102*) isolated from barley malt germs a base, which he called hordenine. Gaebel (*43*) made the same discovery almost simultaneously. Subsequently, it was shown by Späth that anhaline is identical with hordenine (*192*). Hashitani (*55, 56*) found hordenine in other cereal seedlings, e.g. barley (*Hordeum sativum*) contains 0,17%, *Panicum miliaceum* 0,24%, and *Andropogon sorghum* 0,07%. Some other cereal seedlings contained only traces. Reti (*161*) found hordenine and the corresponding quaternary ammonium-base called candicine in the Argentine cacti *Trichocereus candicans* (0,5–5%) and *Trichocereus lamprochlorus* (0,3%) (*167*). Torquati (*221*), Reilhes (*160*) and Raoul (*141–150*) have studied the formation of hordenine during the germination of barley. Unsprouted barley contains no hordenine; but as soon as four days after germination the alkaloid content reaches a peak (0,45%) and disappears again after about a month.

Hordenine gives a positive Millon test.

Methods of estimation have been described by Janot and Faudemay (*81*), Hashitani (*56*), Arnolt (*2*), Raoul (*146*), Gonnard (*44*), and Pedinelli (*131*).

The base crystallizes well in colorless prisms, m.p. 117–118°; b.p. 173–174° (11 mm.); it sublimes at 140–150°; is readily soluble in water, alcohol, ether, and chloroform. It shows an alkaline reaction and liberates ammonia from its salts. Hordenine forms the following well-crystallized salts: Hydrochloride, m.p. 176,5 to 177,5°; sulfate, m.p. 209–211°; picrate, m.p. 139–140°; picrolonate, m.p. 219–220°; methiodide, m.p. 230–231°; acetyl-hordenine hydriodide, m.p. 176–177°; reineckate, m.p. 176–178° (*44*); benzoylhordenine, m.p. 47–48°.

Léger (*100*) showed that O-methyl-hordenine methiodide yields trimethylamine and *p*-vinylanisole. O-acetylhordenine is oxidized to *p*-acetoxybenzoic acid. Hordenine is therefore *p*-hydroxy-*β*-phenylethyl-dimethylamine which has been confirmed by several syntheses.

BARGER (*5*) starting from phenylethyl alcohol obtained phenylethyl chloride, which, when heated with dimethylamine, yielded dimethyl-phenylethylamine. By nitration, reduction, diazotization, etc., a base identical with natural hordenine was obtained. ROSENMUND (*177*) condensed *p*-methoxybenzaldehyde with nitromethane and reduced the obtained nitrostyrene to *p*-methoxy-phenylethylamine. With methyl iodide a mixture of bases was formed from which, after demethylation with boiling HI, hordenine was obtained in a low yield. VOSWINCKEL (*223*) treated *p*-$CH_3O \cdot C_6H_4CO \cdot CH_2Cl$ with dimethylamine, eliminated the O-methyl with HI and reduced to hordenine. EHRLICH and PISTSCHIMUKA (*34*), converted tyrosol, (*p*)-$HO \cdot C_6H_4 \cdot CH_2 \cdot CH_2OH$, into *p*-hydroxy-phenylethylchloride, $HO \cdot C_6H_4 \cdot CH_2 \cdot CH_2Cl$, which, with dimethylamine, yielded hordenine.

SPÄTH and SOBEL (*215*) as well as KINDLER (*89*) followed more complicated routes for the synthesis. RAOUL (*142, 144*) obtained hordenine (in 50% yield) by methylating tyramine with formaldehyd and formic acid.

5. *Candicine* (*p*-hydroxy-*β*-phenylethyl-trimethyl-ammonium hydroxide), $C_{11}H_{19}O_2N$,

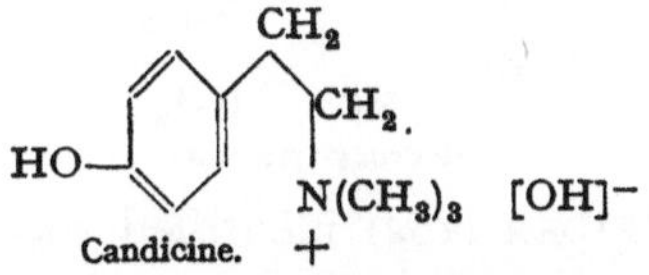

HO— [benzene ring] —CH_2—CH_2—$\overset{+}{N}(CH_3)_3$ $[OH]^-$

Candicine.

was found by RETI (*161*) in the Argentine cactus, *Trichocereus candicans* and by RETI and ARNOLT (*167*) in *T. lamprochlorus*. In both plants it is accompanied by approximately the same amounts of hordenine; however, the ratio of the two bases is variable. *T. candicans* may contain as much as 5% candicine, while *T. lamprochlorus* has only 0,3–0,5%. Candicine occurs also in *T. spachianus* (*49*).

Candicine, like other quaternary compounds, cannot be extracted with some usual solvents. It was isolated by precipitating the purified plant extract with MAYER's reagent; by treating the precipitate with hydrogen sulfide the base was then recovered in the form of its iodide.

Candicine forms well crystallized salts. The slightly soluble iodide is typical; it has been known for a long time as hordenine methiodide, m.p. 230–231°; picrate, m.p. 162 to 163°; chloroplatinate, m.p. 208–209°; chloroaurate 127–128°; iodomercurate (precipitated with MAYER's reagent), m.p. 187°.

Candicin gives a red color reaction with MILLON's reagent. Candicine iodide when heated with silver hydroxide gives the free base which upon heating with alkali yields trimethylamine. O-Methylcandicine is converted by permanganate oxidation into anisic acid. The candicine structure has been established by comparing the derivatives of the natural base with those obtained from synthetic hordenine methiodide.

6. *O-Methyltyramine-N-methylcinnamide*, N-(2-*p*-anisylethyl)-N-methylcinnamide, $C_{19}H_{21}O_2N$,

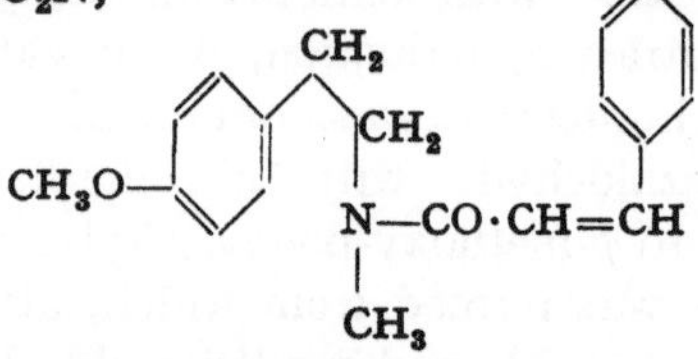

CH_2
CH_2
CH_3O—
N—CO·CH=CH
CH_3

LA FORGE and BARTHEL (*93*) isolated this compound from the bark of Southern prickly ash, ***Zanthoxylum Clava Herculis*** LAM., where it occurs together with berberine and asarinin. It has been obtained synthetically by the same authors who treated O,N-dimethyltyramine with cinnamyl chloride.

The amide crystallizes from petroleum ether; m.p. 76°. Hydrolysis gives cinnamic acid and O,N-dimethyltyramine. Hydrochloride of the amine, m.p. 181–182°; picrate, m.p. 112°. Oxidation gives anisic acid.

7. *3-Hydroxy-tyramine* (3:4-dihydroxy-β-phenylethylamine), $C_8H_{11}O_2N$,

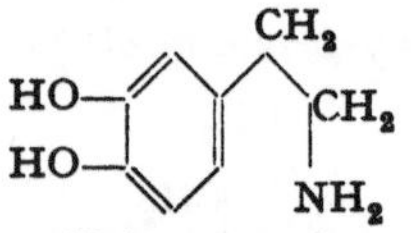

CH_2
HO—
CH_2
HO—
NH_2

3-Hydroxy-tyramine.

SCHMALFUSS and HEIDER (*182*) identified the blood pressure-raising substances contained in the pod of the common broom, ***Sarothamnus scoparius*** (***Cytisus scoparius*** L.), as tyramine and hydroxy-tyramine. Both may be considered as melanin precursors. The amines were isolated with the help of their carbomethoxy derivatives. BUELOW and GISVOLD (*23*) found hydroxy-tyramine in the roots of ***Hermidium alipes*** S. WATSON, a Nyctaginacea. This compound has not yet been isolated from cacti, but the corresponding quaternary base is the cactus alkaloid coryneine. Another such alkaloid, carnegine, may be considered as a bio-product of the interaction of hydroxy-tyramine and acetaldehyde, followed by methylation. As a matter of fact, SCHÖPF and BAYERLE (*185*) obtained a surprisingly high yield of nor-carnegine, by carrying out this condensation under mild ("physiological") conditions, i.e. at p_H 5 and 25°.

Hydrochloride, white crystals, m.p. 241° (uncor.); hydrobromide, m.p. 212°; tricarbomethoxy derivative, m.p. 92–93°; tribenzoyl derivative, m.p. 141°; picrate, m.p. 189°; styphnate, m.p. 206°.

The base can be obtained by heating 3:4-dihydroxy-phenylalanine above the melting point. A synthesis starting from tyramine was described by WASER and SOMMER (*225*). According to HOLZ and CREDNER (*78*), tyramine, in aqueous solution, when irradiated with ultraviolet light in the presence of air, is partly converted into 3:4-dihydroxyphenylethylamine. The same substance appears when oxygen is bubbled through a solution of tyramine and ascorbic acid. In both cases hydrogen peroxide seems to be an intermediate product.

8. *Coryneine* (hydroxy-candicine; 3:4-dihydroxy-β-phenylethyl-trimethyl-ammonium hydroxide), $C_{11}H_{19}O_3N$,

$$\text{(HO)}_2C_6H_3\text{—}CH_2\text{—}CH_2\text{—}\overset{+}{N}(CH_3)_3\ [OH]^-$$

Coryneine.

This interesting quaternary base was isolated by RETI, ARNÓLT and LUDUEÑA (*168*) from the Argentine Cactus *Stetsonia coryne*, following the same technique as applied in the case of candicine. Special precautions had to be taken to avoid degradation of this sensitive catechol compound. The plant contains approximately 1% of the base.

The composition of the chloride, m.p. 200°, obtained by treating the solution of the iodide with silver chloride, was found to be $C_{11}H_{18}O_2NCl$. The molecule is methoxyl-free, however, three N-methyl groups are present. By O-methylation and subsequent permanganate oxidation, veratric acid is formed. Coryneine salts give the characteristic color reactions of catechol derivatives.

Coryneine and candicine, are interesting examples of substances found in nature long after they had been synthesized in the laboratory. BARGER and EWINS (*8*) prepared synthetic 3:4-dihydroxy-β-phenylethyltrimethyl-ammonium chloride. BARGER and DALE (*7*) studied its sympathomimetic action (1910). The natural and the synthetic substance showed identical pharmacodynamic effects.

9. *Mezcaline* (3:4:5-trimethoxy-β-phenylethylamine), $C_{11}H_{17}O_3N$,

$$(CH_3O)_3C_6H_2\text{—}CH_2\text{—}CH_2\text{—}NH_2$$

Mezcaline.

Mezcaline, the active hallucinatory principle of the "mescal buttons" or "pellote", was isolated by HEFFTER (*60*) in 1896. Pellote (*Anhalonium lewinii*) contains up to 6% mezcaline. RETI (*163*) observed the presence of mezcaline in the Argentine cactus *Trichocereus terschecki*, which contains 0,2% trichocereine (dimethyl-mezcaline) and 0,05% mezcaline. HERRERO-DUCLOUX (*69*) had supposed that one of the bases extracted from the cactus *Gymnocalycium gibbosum* was mezcaline.

Microchemical reactions of mezcaline have been described by ROSENTHALER (*179*), HERRERO-DUCLOUX (*70*), and BOLLAND (*16*).

The free base is a colorless oil [Kindler and Peschke (*90*); crystals, m.p. 35–36°], with strong alkaline reaction, b.p. 180° (12 mm.). Soluble in water, alcohol, chloroform, only slightly soluble in ether. It absorbs carbon dioxide from the air and solidifies to carbonate. The sulfate, $(C_{11}H_{17}O_3N)_2 \cdot H_2SO_4 \cdot 2\,H_2O$, is particularly suited for isolation since it is insoluble in alcohol, only slightly soluble in cold water but very soluble in hot water; it forms brilliant prisms, m.p. 183–186°. The hydrochloride forms colorless crystals, m.p. 181° (Kindler and Peschke found m.p. 184°); picrate, m.p. 216–218° (Kindler and Peschke, m.p. 222°); the chloroaurate crystallizes with 1 H_2O, orange needles, m.p. 140–141°; chloroplatinate, straw-yellow needles, m.p. 187–188°; benzoyl-derivative, m.p. 120–121° (Kindler and Peschke, m.p. 123°); *m*-nitrobenzoylderivative, m.p. 161–162°; dimethyl-mezcaline methiodide, m.p. 225°.

Heffter (*59, 62, 63*) determined the empirical formula of mezcaline and found that upon oxidation the base yields trimethylgallic acid. Unfortunately, mezcaline behaves in methyl-imide determinations as though it contained an N-methyl group. Heffter (*64*) synthesized 3:4:5-trimethoxy-benzyl-methylamine and found that it is not identical but isomeric with mezcaline.

The irregularity mentioned was confirmed by Späth (*192*) who, guided by biogenetical considerations, arrived at the correct structure, in spite of the confusing analytical evidence. In Späth's mezcaline synthesis 3:4:5-trimethoxy-benzoyl chloride was reduced by the Rosenmund method (*178*) to the corresponding aldehyde, which when condensed with nitromethane yielded ω-nitro-3:4:5-trimethoxystyrene. This was reduced with zinc dust and acetic acid to the corresponding oxime and, the latter was further reduced, by sodium amalgam, to 3:4:5-trimethoxy-β-phenylethylamine, i.e. mezcaline.

Considerably later than the Späth synthesis, we find in the literature several other successful syntheses of mezcaline.

Slotta and Heller (*186, 188*) prepared their starting material, viz, trimethoxy-phenylpropionic acid by condensation of the substituted benzaldehyde with malonic acid and reduction of the resulting cinnamic acid. Mezcaline was then obtained by Hofmann degradation of the trimethoxy-phenylpropionamide.

Kindler and Peschke (*90*) synthesized very pure, crystallized mezcaline by condensation of 3:4:5-trimethoxy-benzaldehyde with KCN, acetylation, and catalytic reduction to the amine.

Slotta and Szyska (*190, 191*) improved Späth's first synthesis, obtaining mezcaline directly by the electrolytic reduction of ω-nitro-trimethoxystyrene. Hahn and Wassmuth (*50, 54*) started from elemicine and prepared first trimethoxy-phenylacetaldehyde by ozonisation. The oxime was then reduced to mezcaline. Kindler et al. (*88–90*) improved the catalytic reduction of ω-nitrostyrenes to the corresponding phenylethylamines. Hahn and Rumpf (*51*) described the preparation of mezcaline by reduction of ω-nitro-trimethoxystyrene with ADAM's catalyst. See also a review by JENSCH (*85*).

Several isomers of mezcaline as well as mezcaline-like compounds have been synthesized by JANSEN (*82–84*); SLOTTA and HELLER (*188*); SLOTTA and SZYSKA (*190*, *191*); SLOTTA and MÜLLER (*189*); GRACE (*45*); IWAMOTO and HARTUNG (*79*), and HEY (*74*). None of these compounds causes the euphoric state produced by mezcaline.

10. *N-Methylmezcaline*, $C_{12}H_{19}O_3N$,

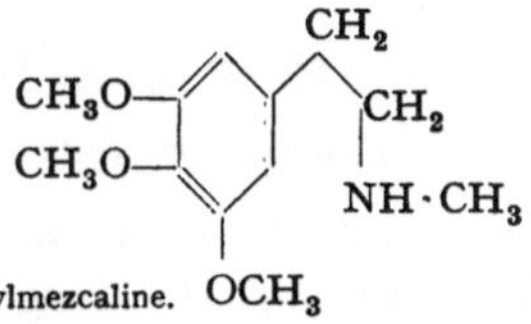

N-Methylmezcaline.

was isolated by SPÄTH and BRUCK (*203*) who worked up the mother liquors originating from the crystallization of the non-phenolic bases of "mescal-buttons". The identity of this base was established by comparison with synthetic N-methylmezcaline which can be obtained by the condensation of mezcaline with benzaldehyde, subsequent methylation of the benzal derivative with methyliodide and hydrolysis of the quaternary base.

Picrate, m.p. 177,5—178°; trinitro-*m*-cresolate, m.p. 189,5–190,5°; *p*-nitrobenzoyl-derivative, m.p. 142–143°.

11. *N-Acetylmezcaline*, $C_{13}H_{19}O_3N$,

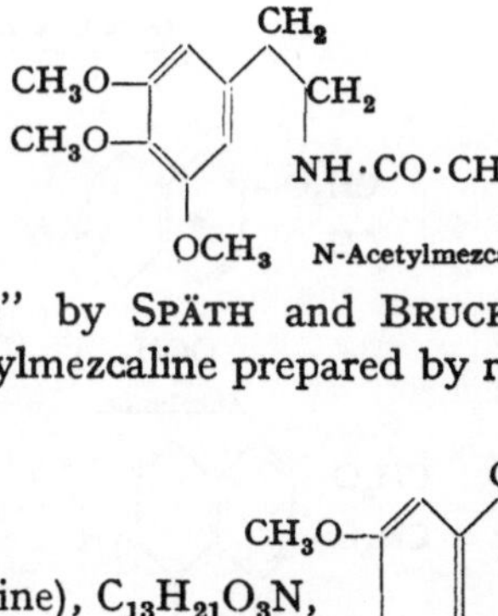

N-Acetylmezcaline.

was detected in the "mescal buttons" by SPÄTH and BRUCK (*204*); m.p. 93–94°. It is identical with N-acetylmezcaline prepared by refluxing mezcaline with acetic acid at 170–175°.

12. *Trichocereine* (N-dimethylmezcaline), $C_{13}H_{21}O_3N$,

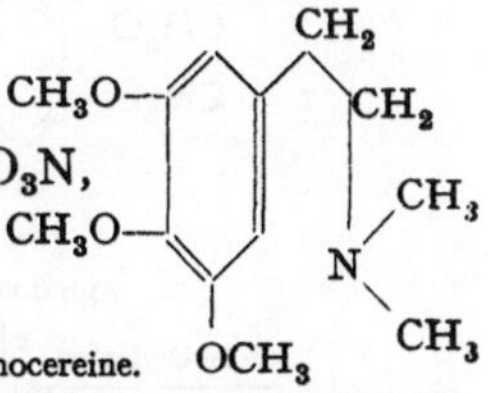

Trichocereine.

was found by RETI (*163*), together with mezcaline, in the cactacea *Trichocereus terscheckii*. The dried plant contains 0,2% trichocereine and 0,05% mezcaline.

Colorless oil, of basic reaction; distills in vacuum without decomposition. Soluble in water, alcohol, ether, chloroform. The salts crystallize well. Hydrochloride, m.p. 205°; picrate, yellow needles, m.p. 169–170°; methiodide, m.p. 225–226°; picrate of the quaternary base, m.p. 165,5°.

Trichocereine contains three methoxyl and two N-methyl groups. Upon oxidation with permanganate trimethylgallic acid is obtained. With dimethylsulfate and potassium iodide the compound yields a methiodide (m.p. 225–226°) which, mixed with dimethylmezcaline methiodide (m.p. 225°), melts without depression. Similar results were obtained by comparing other derivatives of the natural and the synthetic substance. Trichocereine is, therefore, N-dimethyl-mezcaline.

B. Tetrahydro-isoquinolines

found in *Anhalonium lewinii* (pellote) (cf. Table).

1. *Anhalamine*, $C_{11}H_{15}O_3N$, was first isolated by KAUDER (*87*). According to HEFFTER (*63*) the drug contains 0,1% anhalamine.

The base crystallizes in microscopic needles, m.p. 189–191°; hydrochloride, from water with $2\,H_2O$, m.p. 258°, from alcohol with $1\,H_2O$; sulfate, colorless prisms,

Table 1. Tetrahydroisoquinoline Bases Obtained from *Anhalonium lewinii*.

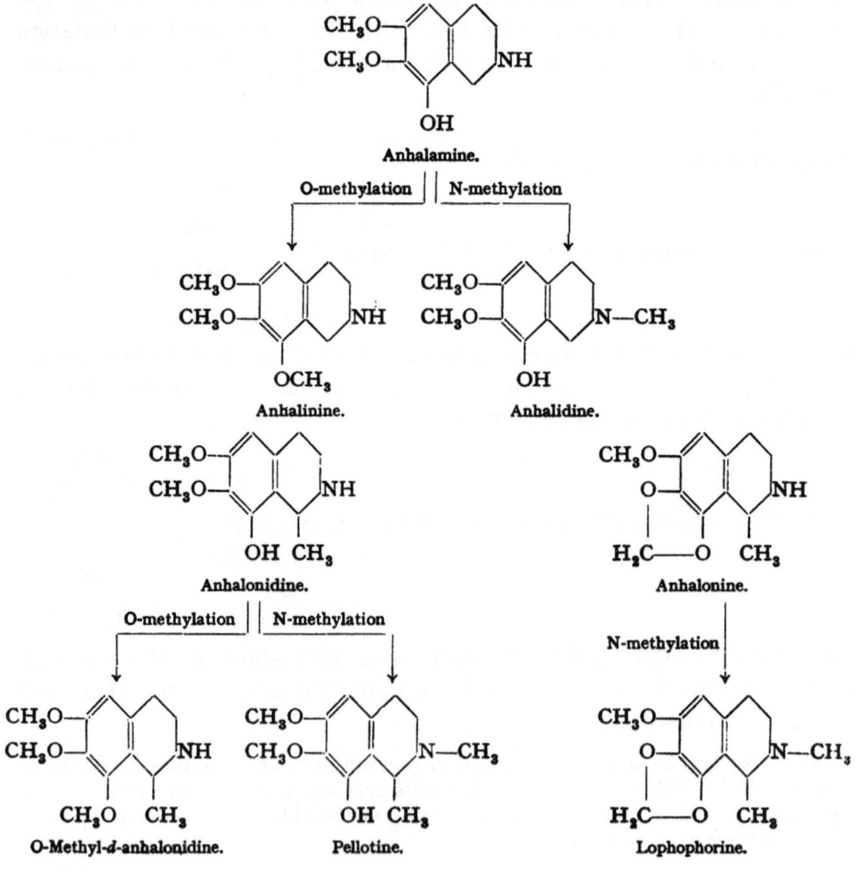

Anhalamine.

O-methylation | N-methylation

Anhalinine. Anhalidine.

Anhalonidine. Anhalonine.

O-methylation | N-methylation N-methylation

O-Methyl-*d*-anhalonidine. Pellotine. Lophophorine.

very soluble in water, less in alcohol. Well crystallized chloroplatinate and chloroaurate. Picrate, m.p. 237–240°; monobenzoyl derivative, m.p. 167,5°; dibenzoyl-derivative, m.p. 128–129°; N-*m*-nitrobenzoyl derivative, m.p. 174–175°; O,N-dimethylanhalamine methiodide, m.p. 211,5–212,5°.

O-Methylanhalamine is termed anhalinine, and N-Methylanhalamine is anhalidine.

2. *Anhalinine*, $C_{12}H_{17}O_3N$, was isolated by SPÄTH and BECKE (*199*) (yield, 0,01%).

Free base, m.p. 61–63°; hydrochloride, white crystals, m.p. 248–250°; picrate, m.p. 184–185°; cloroaurate, m.p. 139–140°; chloroplatinate, m.p. 207–208°; *m*-nitrobenzoyl derivative, m.p. 147–148°; methiodide, m.p. 211,5–212,5°.

3. *Anhalidine*, $C_{12}H_{17}O_3N$ was found by SPÄTH and BECKE (*200*) (yield, 0,001%).

Free base, m.p. 131–133°. Sublimes in high vacuum at 85–95°. O-Methylanhalidine methiodide, m.p. 211,5–212,5°.

4. *Anhalonidine*, $C_{12}H_{17}O_3N$, was discovered by HEFFTER (*60*). According to this author pellote contains as much as 5% of the alkaloid.

The free base crystallizes in small octahedra, m.p. 160–161°; picrate, m.p. 201–208°; N-benzoyl derivative, m.p. 189°; dibenzoyl derivative, m.p. 125–126°; N-*m*-nitrobenzoyl derivative, m.p. 207–208°; N-methylanhalonidine hydroiodide = = pellotine hydriodide, m.p. 125-130°: N-methyl-anhalonidine methiodide = pellotine methiodide, m.p. 199°

5. *Pellotine* (N-methylanhalonidine), $C_{13}H_{19}O_3N$, was isolated by HEFFTER (*59*) from *Anhalonium williamsii* (0,74% of the fresh plant) and was found later by KAUDER (*87*) in *A. lewinii*. In HEFFTER's opinion, however, the pellotine content of the latter drug originated from a contamination by *A. williamsii*. This opinion is shared by LEWIN (*106*). Morphologically, both plants are difficult to differentiate. The botanists are inclined to consider *Anhalonium lewinii*, *A. williamsii* and *A. jourdanianum* as a single species (SCHUMANN: *Anhalonium williamsii*; BRITTON and ROSE: *Lophophora williamsii*).

The base is only slightly soluble in water; crystallizes from alcohol, m.p. 111–112°; the salts have a bitter taste. Hydriodide, m.p. 125–130°; picrate, m.p. 167–169°; chloroaurate, m.p. 147–148°; methiodide, m.p. 199°; O-methylpellotine methiodide, m.p. 226–227°.

SPÄTH and KESZTLER (*210*) prepared the optically active forms of pellotine and studied their racemization, to determine whether the compound is present in the plant in the inactive form or is racemized when manipulated or during the ageing of the drugs. By means of *d*-tartaric acid a fraction with $[\alpha]_D^{17} = -15{,}2°$ was obtained from racemic pellotine. Considering the ease with which this base undergoes racemization, the authors suppose that optically active pellotine is present in the plant.

6. *O-Methyl-d-anhalonidine*, $C_{13}H_{19}O_3N$, Späth and Bruck (*205*) found very small quantities of a new isoquinoline alkaloid in the mother liquors from the crystallization of the non-phenolic bases of ***Anhalonium lewinii***. Its structure was established by analytical and synthetic methods.

It is an oil, b.p. 140° (0,05 mm.). Optically active, $[\alpha]_D^{16} = + 20{,}7°$ (methanol). It yields a characteristic 2:4:6-trinitrobenzoyl derivative, m.p. 259–260°, $[\alpha]_D^{14} = + 39{,}7°$ (methanol). The *d,l*-form had been synthesized by Späth (*193*) as early as 1921.

7. *Anhalonine*, $C_{12}H_{15}O_3N$, was discovered by Lewin (*104–106*). According to Heffter (*60*) the drug contains about 3% anhalonine.

The base crystallizes from light petroleum in needles, m.p. 85,5°, $[\alpha]_D = - 56{,}3°$ (chloroform); hydrochloride, $[\alpha]_D^{17} = - 41{,}9°$; N-methylanhalonine methiodide = lophophorine methiodide, m.p. 223°. Heated to its melting point the quaternary iodide is racemized and then melts at 242–243°.

Späth and Kesztler (*209*) prepared optically active forms of synthetic anhalonine base, with the following properties: *l*-anhalonine, m.p. 85–86°, $[\alpha]_D^{25} = - 56{,}3°$ (chloroform); *d*-anhalonine, m.p. 84,5–85,5°, $[\alpha]_D^{25} = + 56{,}7°$. The synthetic *l*-form, when methylated with formaldehyde and formic acid, gave an N-methyl derivative, $[\alpha]_D^{25} = - 47{,}3°$ (chloroform), identical with natural lophophorine; picrate, m.p. 162–163°.

8. *Lophophorine* (N-methyl-*l*-anhalonine), $C_{13}H_{17}O_3N$, was detected by Heffter (*60*) (yield 0,5%).

It is an oily base, $[\alpha]_D = - 47°$ (chloroform). Hydrochloride, $[\alpha]_D^{17} = - 9{,}47°$; picrate, m.p. 162–163°; methiodide, m.p. 223°; trinitro-*m*-cresolate of the quaternary compound, m.p. 171–172°; picrate of the quaternary compound, m.p. 211–212°.

Structure and Synthesis of the Isoquinoline Bases Obtained from Anhalonium lewinii.

Anhalonidine and Pellotine. Degradation experiments disproved Späth's first assumption that the other bases in "mescal buttons", were structurally similar to the then already clarified mezcaline. Späth, therefore, gathered experimental evidence for their structure, mainly of synthetic nature, working on the hypothesis that their structures were of the isoquinolinic pattern (*193*). Starting from N-acetyl mezcaline the following route was followed:

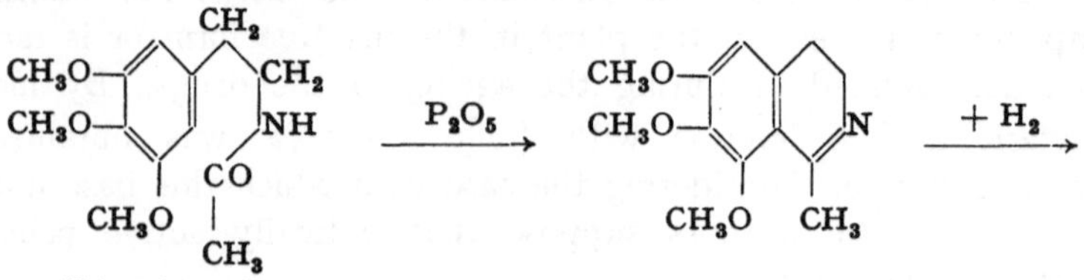

CH_3O, CH_3O, CH_3O — CH_2–CH_2–NH–CO–CH_3 $\xrightarrow{P_2O_5}$ CH_3O, CH_3O, CH_3O — N, CH_3 $\xrightarrow{+ H_2}$

N-Acetyl-mezcaline.

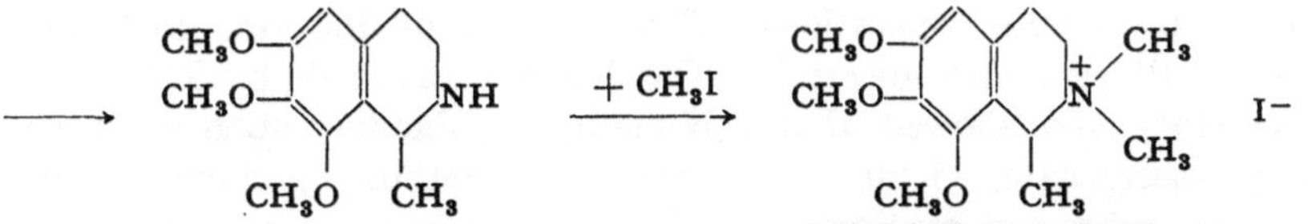

O-Methylpellotine methiodide.

The quaternary iodide obtained was shown to be identical with O-methylpellotine methiodide. Pellotine and anhalonidine yield on complete methylation, the same product, and since anhalonidine is a secondary base, pellotine must be N-methylanhalonidine.

In a further study SPÄTH (*195*) described the synthesis of anhalonidine and pellotine. By treating the O,N-diacetyl derivative of the 5-hydroxy-3:4-dimethoxy-phenylethylamine with phosphorous pentoxide, reducing the dihydroisoquinoline formed and splitting off the O-acetyl group, anhalonidine was obtained. However, the position of the free hydroxyl group in anhalonidine (and pellotine) had still to be determined; two different structures were possible, since the ring closure might have taken place either in *ortho* (I) or *para* position (II) to the hydroxyl group:

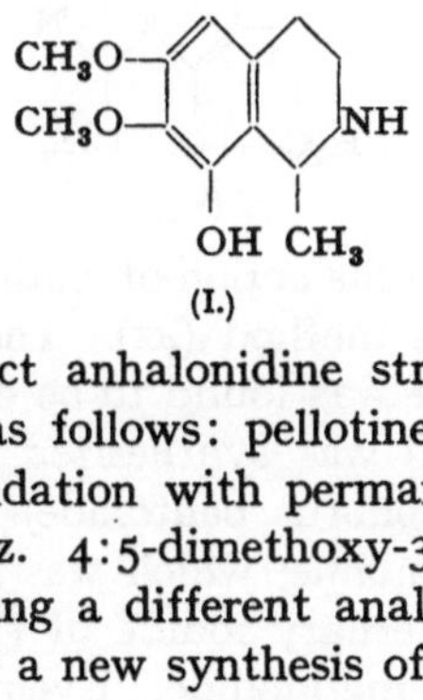

(I.) 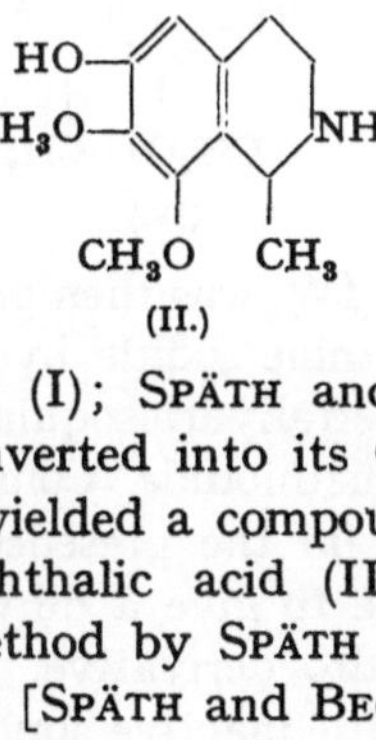 (II.)

The correct anhalonidine structure is (I); SPÄTH and PASSL (*213*) proved this as follows: pellotine was converted into its O-ethyl ether, which on oxidation with permanganate yielded a compound of known structure, viz. 4:5-dimethoxy-3-ethoxyphthalic acid (III). This was confirmed using a different analytical method by SPÄTH and BOSCHAN (*202*), and by a new synthesis of pellotine [SPÄTH and BECKE (*197*)].

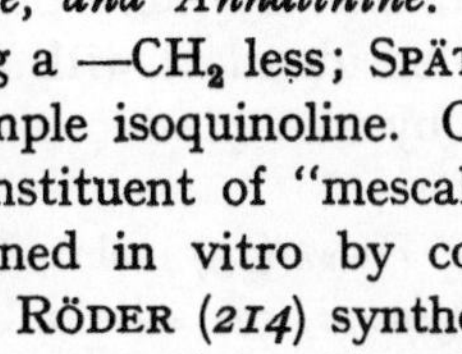

(III.)

Anhalamine, Anhalidine, and Anhalinine. Anhalamine differs from anhalonidine by containing a $—CH_2$ less; SPÄTH (*193*) therefore inferred that the former was a simple isoquinoline. O-Methylanhalamine [later found to be a natural constituent of "mescal buttons" and called anhalinine (*199*)] was obtained in vitro by condensing mezcaline with formaldehyde. SPÄTH and RÖDER (*214*) synthesized anhalamine by condensation of 5-benzyloxy-3:4-methoxy-phenylethylamine with formaldehyde, and further operations. The position of the free hydroxyl, however, still remained uncertain. The decision was made by SPÄTH and BECKE (*198*) who showed that O,N-diethylanhalamine, upon oxidation with permanganate, yielded the same 4:5-dimethoxy-3-ethoxyphthalic acid as was obtained by SPÄTH and PASSL (*213*) in the case of pellotine. All known phenolic tetrahydroisoquinolines which occur in the "mescal buttons" contain, therefore, the free hydroxyl in the 8-position.

Anhalidine is N-methylanhalamine (*200*).

Anhalonine and Lophophorine. It was demonstrated by SPÄTH and GANGL (*207*) that, contrary to the belief of HEFFTER, lophophorine is N-methyl-anhalonine. They found also that both alkaloids contain a methylenedioxy group, besides the earlier known methoxyl group. Starting from the assumption of a methylisoquinoline structure and considering a biochemical relationship with the other *Anhalonium* bases of known constitution, the structures (IV) and (V) were discussed for anhalonine:

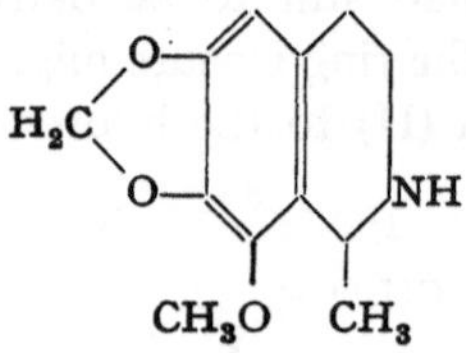

(IV.) 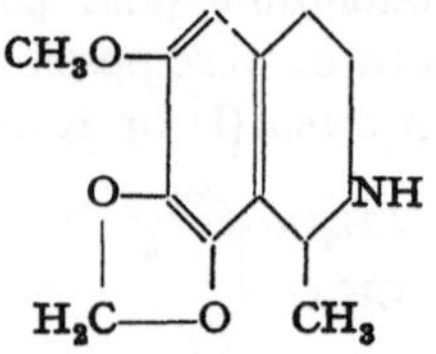 (V.)

Compound (IV), was then prepared by the action of methylmagnesium iodide on cotarnine iodide by FREUND's method (*41*). The quaternary iodide of this tetrahydroisoquinoline base was found to be different from lophophorine methiodide. Compound (V) was synthesized likewise, viz. by condensing (in the presence of phosphorus pentoxide) acetyl-homomyristicylamine to give a dihydroisoquinoline, which was then reduced to the tetrahydro derivative. The quaternary iodide of (V) proved to be identical with inactive lophophorine methiodide. Eventually, structure (V) for lophophorine was confirmed by SPÄTH and BECKE (*199*) by the isolation of isocotarnic acid (3:4-methylenedioxy-5-methoxy-phthalic acid) from the oxidation products of the quaternary base corresponding to anhalonine.

SPÄTH and KESZTLER (*209*) achieved the synthesis of anhalonine and lophophorine. Synthetic *d,l*-anhalonine (*207*), was resolved into the optical antipodes by fractional crystallization of the *l*-tartrate. The *l*-form obtained proved to be identical with natural anhalonidine, and lophophorine was obtained by methylation of this substance with formaldehyde and formic acid.

9. *Carnegine*, $C_{13}H_{19}O_2N$. HEYL (*75*) isolated in 1901 from the Mexican cactus, *Cereus pecten aboriginum*, an alkaloid in form of a crystalline hydrochloride (yield, 0,65%) and termed it pectenine.

The same author (*76*) found (1928), in the American cactus, *Carnegiea gigantea*, a base with the composition $C_{13}H_{19}O_2N$. This alkaloid was named carnegine, and several of its crystalline derivatives were characterized. In 1929 SPÄTH (*196*) established the structure of carnegine and described its synthesis. Finally, a few months later, SPÄTH (*211*) reported that carnegine and pectenine were identical.

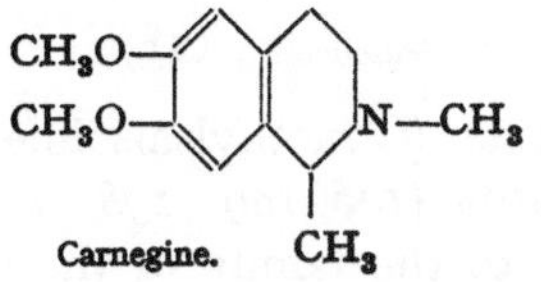

CH_3O, CH_3O, $N—CH_3$, CH_3

Carnegine.

Carnegine is an optically inactive colorless syrup, b.p. 170° (1 mm.). The salts are crystalline: hydrochloride, m.p. 210–211°; hydrobromide, m.p. 228°; picrate, m.p. 212–213°; methiodide, m.p. 210–211°; and trinitro-*m*-cresolate, m.p. 169–170°.

SPÄTH synthesized carnegine, without having carried out cleavage experiments, making use only of the empirical formula as well as the presence of two methoxyl groups, and being guided by the evident structural relationships with the *Anhalonium* bases. Starting from N-acetyl-homoveratrylamine, the following route was followed:

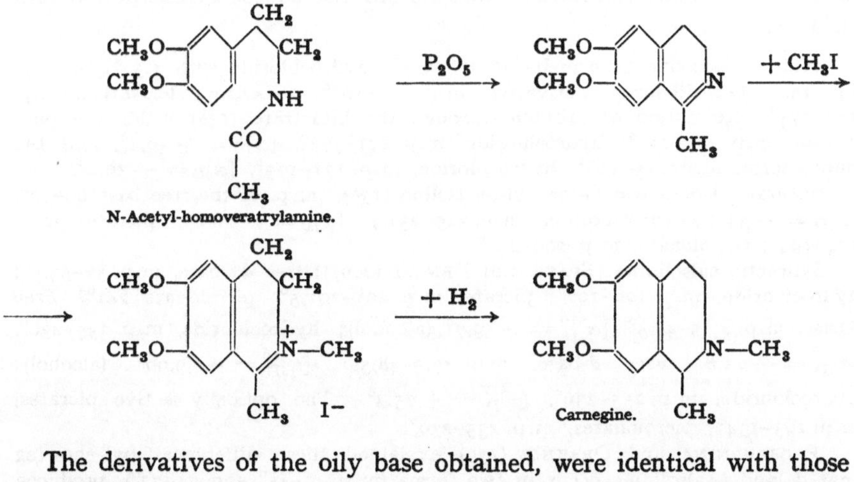

CH_2, CH_3O, CH_3O, CH_2, NH, CO, CH_3 $\xrightarrow{P_2O_5}$ CH_3O, CH_3O, N, CH_3 $\xrightarrow{+ CH_3I}$

N-Acetyl-homoveratrylamine.

$\longrightarrow$ CH_2, CH_3O, CH_3O, CH_2, $\overset{+}{N}—CH_3$, CH_3 I^- $\xrightarrow{+ H_2}$ CH_3O, CH_3O, $N—CH_3$, CH_3

Carnegine.

The derivatives of the oily base obtained, were identical with those of naturally occurring carnegine and pectenine. Furthermore, another a priori possible structure (the product of ring closure in *o*-position to one of the methoxy-groups) could be excluded by the observation that oxidation of the dihydro compound with permanganate gave *m*-hemipinic and not ordinary hemipinic acid.

SCHÖPF and BAYERLE (*185*) obtained "*nor*-carnegine" under mild ("physiological") conditions (p_H 5, at 25°) by condensing hydroxytyramine with acetaldehyde.

10. *Salsoline*, $C_{11}H_{15}O_2N$, *and* 11. *Salsolidine*, $C_{12}H_{17}O_2N$ are not cactus alkaloids but their structure reveals a surprising analogy with carnegine. Actually, carnegine is O,N-dimethyl-salsoline.

HO—, CH_3O—, NH, CH_3

Salsoline.

Salsoline and salsolidine (O-methylsalsoline) have been found by OREKHOV and PROSKURNINA (*126–129, 138, 139*) in the desert plant, *Salsola richteri*, belonging to the family of the *Chenopodiaceae*. A third alkaloid, of unknown constitution (salsamine), occurs in traces in the drug.

The plant yields by extraction with dichloroethane, 0,32% of salsoline. The alkaloid isolated from old plants is optically inactive, but if a recent crop is extracted, a mixture of *d,l*- and *d*-salsoline results. Salsolidine occurs in the plant likewise as a mixture of the *d,l*- and *l*-forms (*139*). The properties of these natural alkaloids are therefore somewhat different from those of the synthetic forms. The optically active bases are stable with respect to racemizing agents and thus racemization must have taken place in the plant tissue itself and not in the course of extraction and isolation.

Natural salsoline base melts at 218–221°; hydrochloride with 1,5 H_2O, m.p. 141–152°; O,N-dibenzoyl derivative, m.p. 166–168°; N-benzoyl derivative, m.p. 172–174°. Resolution of salsoline through the bitartrate (*138*) yields the pure *d*-form, m.p. 215–216°; hydrochloride, m.p. 171–172°, $[\alpha]_D = +40{,}1°$, and the pure *l*-form, m.p. 215–216°; hydrochloride, m.p. 171–173°, $[\alpha]_D = -39{,}2°$.

Natural salsolidine = O-methyl-*l*-salsoline (*138*), m.p. of the free base 69–70°, $[\alpha]_D = -53°$; hydrochloride, m.p. 229–231°, $[\alpha]_D = -26{,}2°$; picrate, m.p. 194–195°; picrolonate, m.p. 220–221°.

Synthetic salsolidine [SPÄTH and DENGEL (*206*)]: free *d,l*-base, m.p. 53–53,5°; hydrochloride, m.p. 196–197°; picrate, m.p. 201–201,5°; picrolonate 241°. Free *l*-base, m.p. 47,5–48,5°, $[\alpha]_D^{16} = -59{,}7°$ (alcohol); hydrochloride, m.p. 235–236°, $[\alpha]_D^{18} = -24{,}8$. Free *d*-base, m.p. 47,5–48,5°; $[\alpha]_D^{16} = +59{,}90°$ (alcohol); hydrochloride, m.p. 235–236°, $[\alpha]_D^{17} = +25{,}3°$. The optically active picrates, m.p. 193–194°; picrolonates, m.p. 235–236°.

PROSKURNINA and OREKHOV (*139*) explained these differences by showing that *d*- and *l*-salsolidine occur in two forms (m.p. 41–45° and 71–73°), produced respectively by distillation in vacuum and crystallization from water; both forms gave identical HCl salts, m.p. 233–235°.

The third alkaloid, salsamine, is not found in all samples. M.p. of the base: 155–157°; picrate, m.p. 213–214°; picrolonate, 220–221°.

Salsoline, when methylated with diazomethane yields O-methyl-salsoline, the *l*-form of which is salsolidine. O,N-methylated salsoline is identical with carnegine. O-methylsalsoline, oxidized with permanganate, yields *m*-hemipinic acid. The position of the hydroxyl group was established by SPÄTH, OREKHOV and KUFFNER (*212*) by synthesis, starting from isovanillin. Salsolidine was synthesized by SPÄTH and DENGEL (*206*).

The natural order *Rhoeadales* (including *Papaveraceae* and *Fumariaceae*) is characterized by a high number of alkaloidiferous species. So far, all the isolated bases possess the benzyl-isoquinoline skeleton. A few species, however, contain small quantities of tetrahydroisoquinolines of simple structure and are related to the cactus alkaloids.

12. *Corypalline*, $C_{11}H_{15}O_2N$, was found by MANSKE (*118*) in *Corydalis pallida* and in the seeds of *C. aurea*. The free base melts at 168°; picrate,

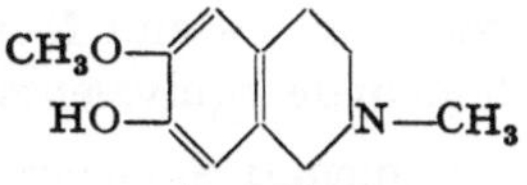

CH_3O—, HO—, N—CH_3

Corypalline.

m.p. 178°. On methylation it yields the well known 2-methyl-6:7-dimethoxy-tetrahydroisoquinoline (m.p. 82°), and upon ethylation, 2-methyl-6-methoxy-7-ethoxytetrahydro-isoquinoline (m.p. 65°). The synthesis of corypalline was accomplished by a route parallel to that used by SPÄTH, OREKHOV and KUFFNER (*212*) in their synthesis of salsoline.

13. *Hydrohydrastinine*, $C_{11}H_{13}O_2N$, is a degradation product of hydrastine or cotarnine, but, according to SPÄTH and JULIAN (*208*) it occurs also in *Corydalis tuberosa*.

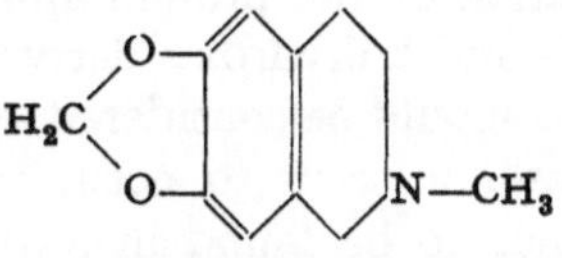

O, H_2C, O, N—CH_3

Hydrohydrastinine.

The base melts at 66°.

14. *Hydrocotarnine*, $C_{12}H_{15}O_3N$, is a well known hydrolytic product of narcotine. HESSE (*73*) found iti n opium, both as a free base and in form of salts.

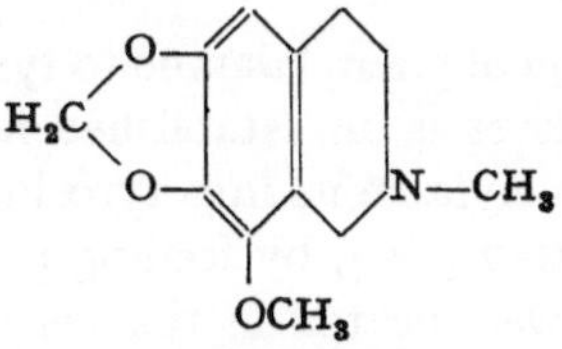

O, H_2C, O, N—CH_3, OCH_3

Hydrocotarnine.

Colorless plates, from light petroleum, m.p. 55,5–56,5°; hydrobromide, m.p. 236–237°, sparingly soluble in water.

The preparation of hydrastinine, hydrohydrastinine and hydrocotarnine, starting from narcotine and cotarnine has been described by PYMAN and REMFRY (*140*), TANAKA, MIDZUNO and OKAMI (*218*), and TOPCHIEV (*220*).

VII. Biogenesis of the Cactus Alkaloids and Their Relationship to Other Natural Products.

There is a general agreement that cactus alkaloids and some other structurally related natural bases are biogenetically linked to the naturally occurring aromatic amino acids, such as phenylalanine, tyrosine, and 3:4-dihydroxyphenylalanine. A scheme of the possible biochemical reactions leading from tyrosine to the best known natural phenylethylamine and tetrahydro-isoquinoline bases is given in Table 2, p. 270. This scheme involves only very simple and biologically plausible reactions, such as decarboxylation, oxidation, O- and N-methylation, ring closure with formaldehyde or acetaldehyde equivalents, etc.

Although so far no experimental evidence is available to support this attractive scheme, many authors are inclined to accept it, at least as a working hypothesis. Thus, since Pictet's (*133*) first suggestions, biogenetical speculation of this kind has proved to be extraordinarily useful in the elucidation of the structure and in the lines of the synthesis of numerous natural products.

The biological formation of simple and more complex natural amines is thoroughly discussed in Guggenheim's fundamental treatise (*46*).

Naturally occurring β-phenylethylamines may have been formed by biological decarboxylation of the protein amino acids, phenylalanine and tyrosine. For the di- and trihydroxy derivatives the corresponding substituted phenylalanines should be considered, although 3:4-dihydroxyphenylalanine (dopa) does not seem to occur in proteins and the trihydroxy derivative still has to be found in nature. However, it is not necessary to suppose that alkaloids are formed directly from the amino acids of proteins. Several workers, especially Guggenheim suggest the possibility that both amino acids and alkaloids originate from common parent substances.

The biological oxidation of phenylalanine to tyrosine and its conversion to 3:4-dihydroxy derivatives is an established fact. In the animal body the direct conversion of phenylalanine into tyrosine has been demonstrated by Moss and Schoenheimer (*123*), by feeding phenylalanine with firmly bound deuterium which was located in the benzene ring. According to Bernheim and Bernheim (*12*), this conversion can be regarded as a step in normal intermediary metabolism and is probably dependent on a specific enzyme system.

The oxidation of tyrosine, tyramine or N-methyltyramine by tyrosinase has been the subject of extended studies which need not be considered exhaustively in the present survey. The reaction sequence follows the route outlined by Raper (*151*, *152*) and Dulière and Raper (*33*):

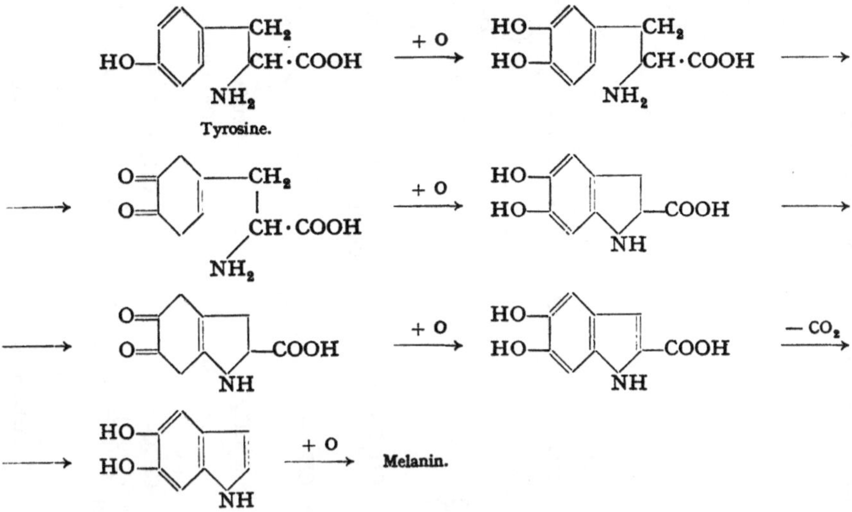

HO— CH_2 / CH·COOH / NH_2 (Tyrosine.) $\xrightarrow{+O}$ HO— HO— CH_2 / CH·COOH / NH_2 ⟶

⟶ O= O= CH_2 / CH·COOH / NH_2 $\xrightarrow{+O}$ HO— HO— —COOH / NH ⟶

⟶ O= O= —COOH / NH $\xrightarrow{+O}$ HO— HO— —COOH / NH $\xrightarrow{-CO_2}$

⟶ HO— HO— NH $\xrightarrow{+O}$ Melanin.

The biological conversion of phenylalanine into epinephrine, in mammalian tissues (involving oxidation, decarboxylation and methylation) is strongly supported by the work of GURIN and DELLUVA (*47*). *d,l*-Phenylalanine, labelled with ^{14}C in the carboxyl group and α-carbon, was converted to epinephrine in which the ^{14}C was located in the terminal carbon of the side chain. Similar results were obtained with tritium-labelled phenylalanine. The result suggests that this biological conversion not only involves decarboxylation of phenylalanine (or one of its derivatives) but that the aminoethyl side chain formed remains attached to the benzene nucleus during the biological synthesis.

It may be assumed that the biosynthesis of β-phenylethylamines in plant tissues follows a similar pattern. The general occurrence of decarboxylases, tyrosinases as well as melanization phenomena gives strong support to this hypothesis, although reliable experimental evidence, such as has been obtained in the case of animal metabolism, is not yet available.

It is interesting that the tertiary and quaternary tyramine derivatives found in the cacti, such as hordenine and candicine, although they take up appreciable amounts of oxygen, do not form pigments when oxidized by tyrosinase [DULIÈRE and RAPER (*33*)]. However, tyrosine or some other derivative must be present in all cacti since; the darkening of cut stems, preceded by a red phase, is characteristic for the whole family. ROCA (*175*) found in the Mexican cactus, ***Pachycereus marginatus***, tyrosine, tyrosinase, and unidentified alkaloidal substances. Tyrosinase has been observed also in ***Trichocereus candicans*** which contains the alkaloids candicine and hordenine.

Table 2. Biochemical Derivation of the Cactus Alkaloids and Some Related Natural Bases.

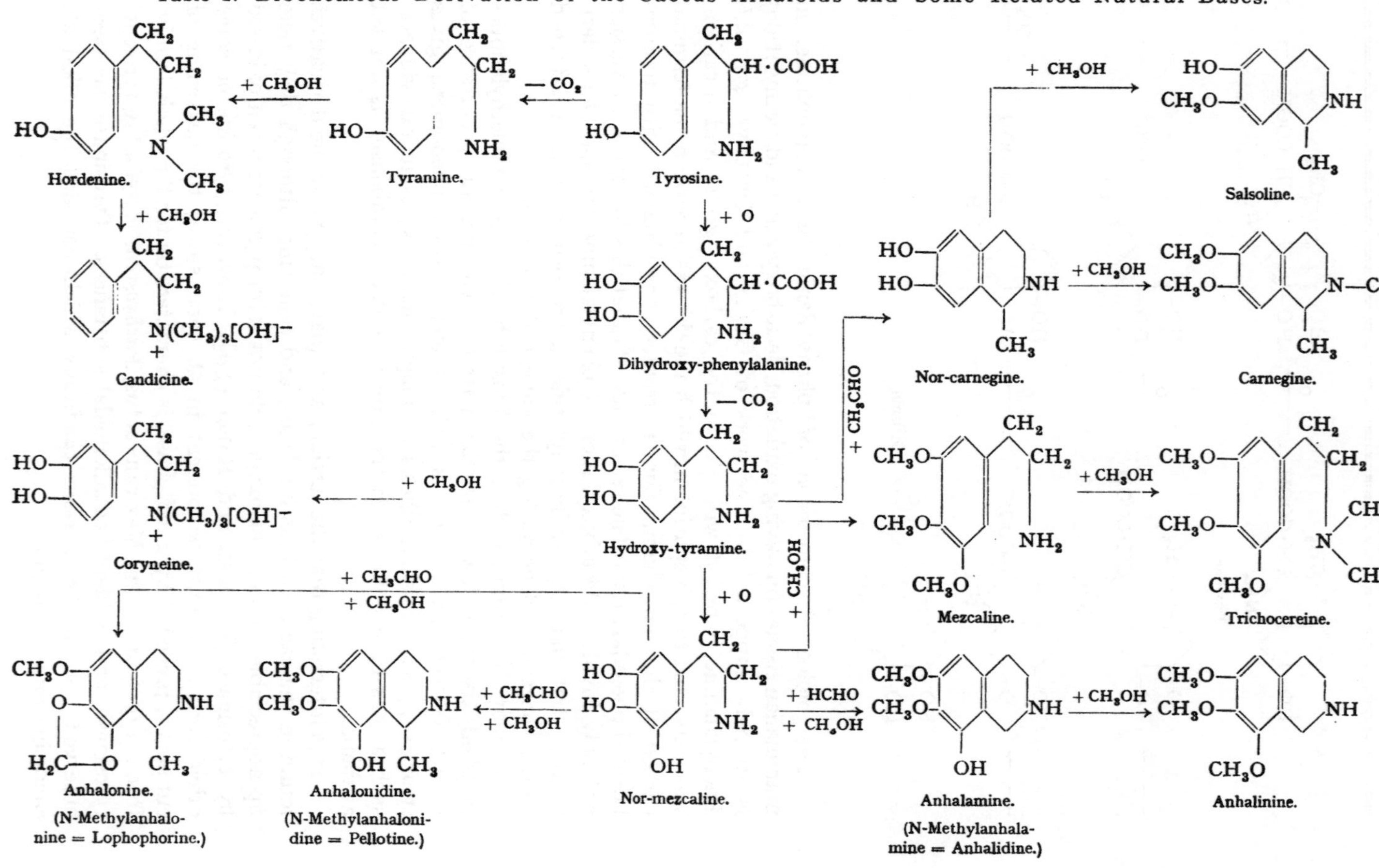

It would be of importance to repeat, using modern isotope techniques, the investigations of RAOUL (*141–150*) who attempted to demonstrate the biological conversion of tyrosine to hordenine in barley.

It is a generally accepted opinion that formaldehyde (or some derivative) accounts for the frequent occurrence of N-methyl, O-methyl and methylenedioxy groups in alkaloids. Methylation may precede decarboxylation as is shown by N-methyltyrosine, which has been repeatedly found in nature and is known under the names of angeline, andirine, geoffroyine, surinamine, and rhatanine (*65*).

WINTERSTEIN and TRIER (*229*) advanced in 1910 a hypothesis on the formation of the benzylisoquinoline alkaloids from aromatic amino acids of the tyrosine type. Two molecules of dihydroxy-phenylalanine were assumed to participate in such a biosynthesis, leading to norlaudanosine:

2 Moles of 3,4-dihydroxyphenylalanine. → → Nor-laudanosine.

Laudanosine can be built up from norlaudanosine by methylation; and papaverine, by methylation and dehydrogenation; hydrastine and berberine, by introduction of another carbon atom (formaldehyde equivalent), etc. According to ROBINSON (*173*) who elaborated this scheme to include practically all the isoquinoline alkaloids of then known structure, the success of the WINTERSTEIN-TRIER hypothesis has been striking and it is a widely accepted system in organic chemistry.

We may suppose that the simple isoquinoline derivatives found in cacti are biosynthesized in a similar way, viz. by the action of formaldehyde or acetaldehyde on substituted phenylethylamines according to a suggestion made by SPÄTH (*193*). For example, mezcaline and formaldehyde would yield anhalinine:

CH_2
CH_3O CH_2
CH_3O NH_2 $\xrightarrow{-H_2O}$ CH_3O CH_3O NH
HCHO
OCH_3 OCH_3

Mezcaline and formaldehyde. Anhalinine.

It is interesting to note that Pictet and Spengler (*134*), many years before Späth's discoveries, synthesized some simple isoquinoline bases starting from phenylethylamines and formaldehyde, and discussed the possibility that alkaloids are produced in plants by a similar condensation process.

Barger (*6*), Robinson (*173*), Polonovski (*135*), and others worked out schemes in which the origin of natural isoquinolines was discussed. Barger considered tyrosine as a precursor of a large group of isoquinoline alkaloids and admitted the possibility of the introduction of additional hydroxyl groups into a preformed benzene ring, similar to the conversion of tyrosine to dopa by tyrosinase. The participation of three molecules of dihydroxy-phenylalanine in the biological formation of emetine was considered recently by Robinson (*174*), following Woodward's suggestions concerning the biogenesis of strychnine (*230*). The Pictet-Winterstein-Trier hypothesis has turned out to be a valuable tool in alkaloid chemistry, first in the restricted field of the benzyl-isoquinolines, later also in other instances, such as that of the harmala alkaloids.

In the absence of experimental proof, the following arguments can be considered as supporting the hypothesis that cactus isoquinolines are related to the phenylethylamine bases:

1. Simultaneous occurrence of substituted phenylethylamines and isoquinoline derivatives not only in the same family but in the same species (*Anhalonium lewinii*).

2. The fact that all the isoquinolines keep the original side chain in the form of a completely hydrogenated bridge. (Actually, the bases are not isoquinolines but tetrahydroisoquinolines.)

3. Natural phenylethylamine bases carry one to three (free or methylated) phenolic groups in the positions 4 or 3:4 or 3:4:5, with reference to the side chain. The same substituents and in the same positions occur in the natural tetrahydroisoquinolines, which points to postulated ring closure.

4. The facility with which similar ring closures are performed *in vitro* under conditions which can be considered as being comparable to physiological ones.

The first experimental demonstration of the simplicity of the manner of alkaloid biosyntheses was Robinson's classical tropinone synthesis Later Schöpf (*183*, *184*) demonstrated that the formation of alkaloids *in vitro* may proceed at room temperature and in nearly neutral media. The SCHÖPF syntheses (begun in 1932) are simple in method and often very efficient in yield. Particularly interesting is the condensation of the naturally occurring 3:4-dihydroxy-phenylethylamine with acetaldehyde at pH 5 and 25°, as carried out by SCHÖPF and BAYERLE (*185*). A substance was obtained in 83% yield which differs from the natural alkaloids carnegine and salsoline only in the degree of methylation (cf. Table 2, p. 270). HAHN and SCHALES (*52, 53*) did not accept SCHÖPF's opinion that in the biosynthesis of isoquinoline alkaloids only phenylethylamines containing free hydroxyl groups, in which the H atom involved in the ring closure is activated, can be considered. They succeeded in preparing some benzyl-isoquinolines at pH 5 and 25° from phenylethylamines containing methoxy and methylenedioxy groups, although this condensation proceeded more slowly and the yields were lower.

We must stress at this point that it would be misleading to consider such biosynthetical methods as true replicas of the processes which are taking place in the living plant. The optical activity of several isoquinoline bases (salsoline, salsolidine, anhalonine, lophophorine, etc.) cannot be reconciled with SCHÖPF and BEYERLE's (*185*) opinion that compounds of this type are formed from phenylethylamines without the participation of enzyme systems.

SPÄTH and KESZTLER (*210*) demonstrated the easy racemization of anhalonidine and they concluded that in the living tissue the compound is probably present in an optically active form. OREKHOV and PROSKURNINA (*129*) extracted optically active salsoline from fresh plants of *Salsola richteri*, while stored crops yielded the racemic alkaloid. Thus, although a connection between natural phenylethylamines and isoquinoline derivatives appears to exist, the biosynthesis of the latter follows a still unknown course, in which optically active intermediaries and enzymatic processes may be involved.

There is an evident biochemical relationship between tyrosine derivatives and some substances of the *indole* type. One very important reaction sequence of this nature is the previously mentioned biological conversion of tyrosine or tyrosine derivatives to oxygenated compounds with indolic nucleus. It is remarkable that hydroxy-phenylalanine (tyrosine) and indole derivatives were repeatedly found in the same genus or species. The following examples may suffice:

Hordenine and gramine (2-dimethylamino-3-methylindole) in *Hordeum* species.

N-Methyl-β-phenylethylamine and N-methyl-tryptamine in *Arthrophytum leptocladum* (*231, 232*).

Adrenaline and bufotenine (N-dimethyl-hydroxytryptamine) in several toad venoms (*Bufo marinus, B. arenarum, B. paracnemis*, etc.) [cf. DEULOFEU (*30*)].

In the biogenesis of some indole alkaloids, tryptophane (or its derivatives) appears to react in a process which is analogous to the formation of isoquinoline bases from phenylalanine derivatives [cf. ROBINSON (*173*)]. The simplest example would be the formation of tetrahydroharman by condensation of tryptamine with acetaldehyde:

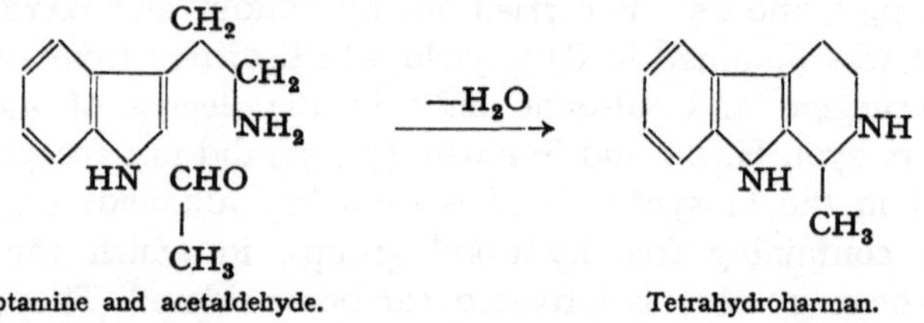

$\xrightarrow{-H_2O}$

Tryptamine and acetaldehyde. Tetrahydroharman.

However, some alkaloids of the indole type may be derived without taking tryptophane into account. Discussing the biogenesis of the Strychnos bases, WOODWARD (*230*) proposes a theory in which 3:4-dihydroxyphenylalanine is considered as precursor.

In the author's opinion, it would be advisable to extend such biogenetical conjectures by embracing the whole field of biochemistry and not limiting them to the comparative evaluation of the chemical structures of amino acids and alkaloids. It is clear that on one hand, the problem is linked with the still unknown origin of the aromatic amino acids themselves; and on the other, nitrogen-free combinations should also be considered because of the facility with which amination and deamination processes take place in living organisms.

The "chemical architecture" of a number of other natural products bears an impressive resemblance to the structure of the phenylethylamine and isoquinoline bases isolated from cacti. Thus, among the constituents of *essential oils*, some compounds are found which must be related to the isoquinoline alkaloids. They contain the same skeleton and the same substituents in analogous positions, as shown by the formulas (VI)–(XI). Indeed, whenever isoquinoline alkaloids were synthesized in the laboratory, some derivatives of these naturally occurring

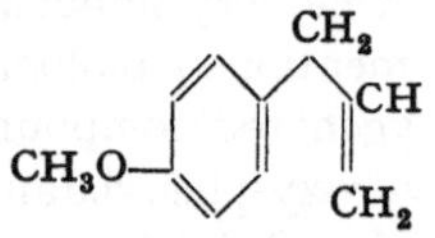

(VI.) Anethol. 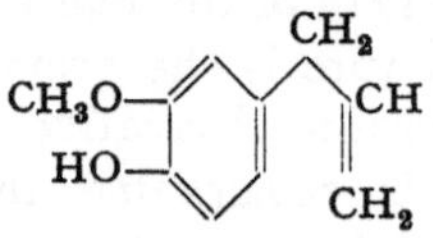 (VII.) Eugenol. 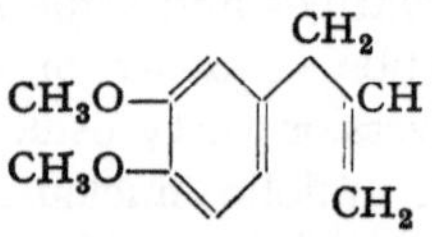 (VIII.) Methyleugenol.

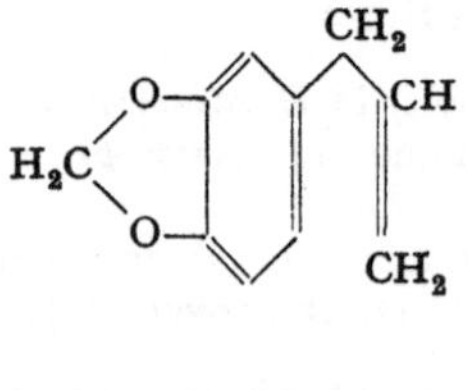

(IX.) Saffrole. 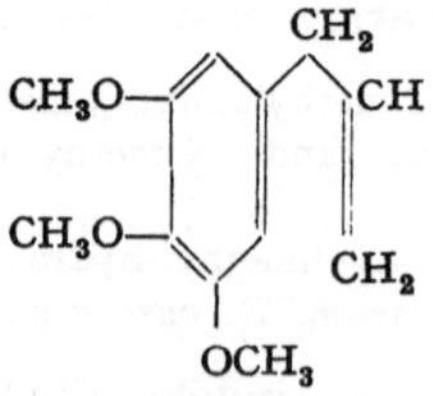 (X.) Elemicine. 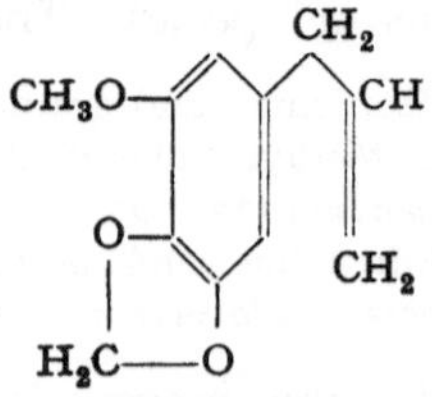 (XI.) Myristicine.

phenols were used as starting materials. This statement is especially valid for the compounds bearing a dioxymethylene group, for the introduction of which no practical method is yet available.

The chemistry of the alkaloids and the chemistry of the essential oils thus appears to be linked by such features which may be far beyond the present manipulation of the laboratory.

However, the essential oils are not the only substances which share a common skeleton with some plant alkaloids. RETI (*165*) considered a number of natural products (constituents of essential oils, resins, flavones, tannins, anthocyanins, alkaloids, etc.) and connected their biogenesis with one of the mass syntheses occurring in nature, viz. that of lignin formation. Without discussing the still open problem of the *lignin* structure [FREUDENBERG (*39*); BRAUNS (*18*)], we may note that there is a general agreement concerning the nature of the building stones. Lignin is built up of substituted phenylpropane units of the following types:

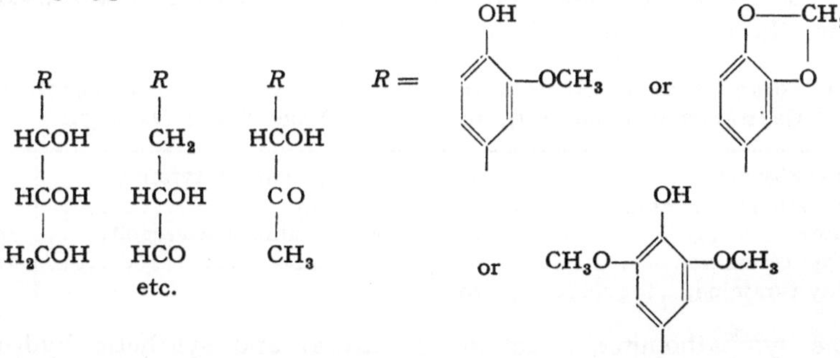

$$R-HCOH-HCOH-H_2COH \quad R-CH_2-HCOH-HCO\ \text{etc.} \quad R-HCOH-CO-CH_3$$

$$R = \text{(4-hydroxy-3-methoxyphenyl, } OH, OCH_3\text{)} \text{ or (3,4-methylenedioxyphenyl, } O-CH_2-O\text{)} \text{ or (4-hydroxy-3,5-dimethoxyphenyl, } CH_3O, OH, OCH_3\text{)}$$

According to FREUDENBERG (*39*) an approach to the understanding of the lignin biosynthesis will be made only when the origin of the phenols, especially those of the C_9 group, becomes known.

We see how the problem of the biogenesis of the cactus alkaloids seems to be related to that of many other natural products which, quantitatively, by far outnumber the relatively scarce alkaloids and the limited amounts of the amino acids. It is a salient problem whether aliphatic amino acids are involved in the biosynthesis of the substances mentioned, or if, rather, aromatic amino acids are built up by the addition of ammonia to nitrogen-free compounds possessing the preformed $C_6 \cdot C_3$ skeletons.

It is possible that alkaloids, amino acids and nitrogen-free substances are formed in the plant from common starting materials by parallel routes as demonstrated by the similarity of their structure.

We may express the hope, that with help of the new isotope techniques and modern chemical genetics, more light will be thrown on these fascinating problems of biochemistry.

VIII. Pharmacological Effects of Cactus Alkaloids and Some Related, Naturally Occurring Bases.

A chapter on the pharmacology of the cactus alkaloids has been written by Joachimoglu and Keeser (*86*), in Heffter's pharmacological Handbook.

Phenylethylamine, Tyramine, Hydroxy-tyramine. After Barger's discovery of tyramine in ergot and the clarification of the adrenaline structure, much attention has been given to the pharmacological properties of compounds of analogous structure, of which those just mentioned are the simplest representatives. A pertinent survey can be found in Guggenheim's book (*46*), and recent advances have been reviewed by Beyer (*14*).

Barger and Dale (7) studied the adrenaline-like effect of a large number of compounds, and they found that approximation to adrenaline in the chemical structure is, on the whole, linked with an increase in the sympathomimetic potency and also with a sharper specificity of the action. The effect on blood pressure in decerebrated cats was used to measure the activity of such compounds (Table 3).

Table 3. Some Relative Effects on the Blood Pressure in Decerebrated Cats (isoamylamine = 1), according to Barger and Dale (7).

Compound	Effect	Compound	Effect
β-Phenylethylamine	2–3	N-Methylhydroxy-tyramine (Epinin)	100
β-Phenylethylmethylamine	2–3	*Nor*-adrenaline (Arterenol)	1000
Tyramine	10	*l*-Adrenaline	1077
N-Methyl-tyramine	10		
Hydroxy-tyramine	20		

The sympathomimetic action of natural and synthetic hydroxy-tyramine was discussed by Raymond-Hamet (*159*). The effects of some cactus alkaloids on the frog heart have been discussed by Mogilewa (*121*).

Hordenine. The pharmacological action of anhaline (hordenine) was first described by Heffter (*58*). In frogs it causes paralysis of the central nervous system without previous excitation. Camus (*24–27*) reported that the compound is slightly antiseptic; it also has an inhibitory effect on some soluble ferments. In mammals it shows relatively low toxicity. Small doses have no effect on the circulation of the blood; larger ones raise the blood pressure and accelerate the pulsation; in very large doses hordenine causes death by arrest of respiration.

Rietschel (*171, 172*) found that the pressure effect is not of central origin and that hordenine stimulates the heart muscle. Although much less active than adrenaline, it is analogous in its action, resembling ephedrine rather than adrenaline. According to Raymond-Hamet (*155, 156, 158*) and Ludueña (*116*), hordenine displays a nicotine-like action.

In large doses it decreases or reverses the hypertensive effect of adrenaline [RAYMOND-HAMET (*157*)].

Candicine was first examined by BARGER and DALE (*7*), and its pharmacological properties were thoroughly studied by LEWIS and LUDUEÑA (*108*, *109*) as well as by LUDUEÑA (*110–114*, *116*).

Candicine displays a nicotine-like action on the visceral nervous system, it first stimulates and then blocks the ganglionic synapse. It has no muscarine-like effect. In the dog, intravenous injection provokes hypertension owing to vasoconstriction due to stimulation of vaso-constrictor nerves and secretion of adrenaline from the adrenal gland. The adrenaline secretory effect is not modified significantly by yohimbine, cocaine or atropine. Sparteine and tetrapropylammonium iodide check this effect completely. Large doses (6 mg./kg.) of candicine have a curare-like action in the dog; this effect has also been observed in the toad, *Bufo arenarum*. The L 50 is 5 mg. per 100 g. for the rat, death occurring owing to respiratory paralysis.

Coryneine had already been examined by BARGER and DALE (*7*); further data were given by RETI, ARNOLT and LUDUEÑA (*168*). The action is very similar to, but more intense than that of candicine.

Mezcaline. In the frog, mezcaline causes narcotic effects (dose, 15 to 30 mg.). In rats the lethal dose is 20 mg./100 g. (LUDUEÑA). Rabbits are extraordinarily resistant to mezcaline [SLOTTA and MÜLLER (*189*)]; injection of even 0,1–0,25 g. of the hydrochloride per kilo body weight causes no visible symptoms. Dogs and particularly cats are more sensitive. A dog after having received a dose of 0,2 g. of mezcaline hydrochloride showed the following strange behavior: the dog started to whine and bark but not at the observer, but towards the opposite side of the cage; when called, it turned and wagged its tail.

RAYMOND-HAMET (*153*, *154*) found that small doses of mezcaline do not have any effect on the blood pressure of dogs; while larger ones (20. mg/kg.) caused hypotension. Mezcaline is antagonistic to the pressor action of adrenaline and the subsequent vagal effect. SLOTTA and MÜLLER (*189*) observed that 40 to 50% of the mezcaline fed to rabbits is excreted in form of trimethoxyphenylacetic acid; the latter is not found in human urine after the administration of mezcaline. According to RICHTER (*170*) mezcaline is excreted by humans unchanged, at least to a high extent (recovery, 58%). BERNHEIM and BERNHEIM (*13*) studied the mechanism of mezcaline oxidation in the rabbit. GRACE (*45*) observed that intravenously injected mezcaline produces respiratory depression and a fall in blood pressure in anesthesized cats and dogs. It stimulates the contractions of the intestine and uterus in situ but not that of the isolated organs.

The most striking physiological effect of the pellote is the production of *visual color hallucinations*. According to Heffter (*62*) amongst the pellote alkaloids only mezcaline is responsible for these symptoms, although Lewin (*105*) attributes a certain hallucinatory power to anhalonidine. Taking into account the high percentage of isoquinoline bases in pellote and their strong toxicity, the effects of intoxication by pellote are evidently not identical to those produced by the administration of pure mezcaline. Nevertheless, in both cases, the visual hallucinations, described by different observers are of a similar nature.

Pellote inebriation has been described by numerous authors such as Prentiss and Morgan (*136, 137*), Heffter (*62*), Weir (*227*), Havelock (*57*), Rouhier (*180*), Hobschette (*77*), and others. The action of pure mezcaline salts was reported by Dixon (*32*), Buchanan (*20*), Foerster (*38*), Beringer (*11*), Marinesco (*119*), etc. In Beringer's book (*11*) numerous cases are reported and a full bibliography is given.

Marinesco (*120*) has also published an interesting study in which the impressions of two artists during mezcaline intoxication are described, and six colored plates, painted during the effect of the drug, are reproduced.

Color visions were provoked by doses of 0,36–0,44 g. of mezcaline sulfate distributed in several subcutaneous injections (Marinesco). The state of intoxication and the visions lasted for 5 to 10 hours. The effects varied widely in different individuals. The most characteristic symptom is that of wonderful visual color hallucinations. Clear consciousness is generally preserved and the subject is fully aware of his condition. Sensory illusions and transposition of sensorial excitation are the interesting factors in this inebriation. Ordinary objects appear to be marvelous. Sounds and music are "seen" in color. In comparison, the impressions of everyday life seem pale and inert. Color symphonies and new, unknown colors of unimaginable beauty and brilliancy are perceived. Euphoria is not always present. Hallucinations of hearing, taste or other senses were reported more rarely. Bradycardia, nausea, a feeling of oppression in the chest, faintness, and headache may also occur.

The interest in these remarkable properties of mezcaline has led to the synthesis of numerous similar compounds (p. 259). However, the slightest structural changes destroy the typical effects of mezcaline. Until now no natural or synthetic compound with the pharmacological properties of mezcaline has been found.

Trichocereine was first examined by Ludueña (*115*). The lethal dose in the rat is approximately 22 mg. of hydrochloride per 100 g. weight. In this animal trichocereine causes excitation, tremor convulsions, paralysis of the extremities, and respiratory paralysis. In dogs large doses provoke a fall in blood pressure. Ingestion of 0,55 g. had no apparent effects in a self experiment of LUDUEÑA; especially, no effects of a sensory nature were noted.

Anhalonine was examined by HEFFTER (*61*); 5–10 mg., when injected in the frog, produced an increase in the reflex excitability after a phase of paresis. In the rabbit similar symptoms are observed but general hyperexcitability predominates.

Anhalonidine [HEFFTER (*61*)]. Doses of 20–25 mg. of the hydrochloride produced narcosis in the frog followed by increased excitability. Larger doses caused complete paralysis. Doses of 30–50 mg. provoked a curarizing effect. No significant symptoms have been observed in mammals.

Pellotine [HEFFTER (*61*)]. In doses of 5–10 mg. pellotine caused temporary convulsions in frogs, and the same effects were observed in dogs or cats. Several authors, cited by JOACHIMOGLU and KEESER (*86*) believe that pellotine could be used in man as a relatively safe narcotic.

Lophophorine [HEFFTER (*61*)] is the most toxic of the bases obtained from *Anhalonium lewinii*; 0,25–1 mg. of injected hydrochloride, provokes a long lasting tetany in the frog. Although the animal recovers, the increased excitability may last for several days. There is no action on the isolated frog's heart. In rabbits 7 mg. of lophophorine per kg. body weight produces hyperexcitability and accelerated respiration; 12,5 mg. per kg. provokes tetany; and 15–20 mg. per kg. is the lethal dose. Intravenous injection of 2,5 mg. causes an increase in blood pressure; larger doses a fall. There is no effect on the heart.

Carnegine. Its pharmacological action is very similar to that of the isoquinoline bases obtained from *Anhalonium lewinii* [HEYL (*75*); MOGILEWA (*121*)]. The lethal dose in the frog is 3–4 mg.; the injection of 2–3 mg. of hydrochloride produces increased reflex excitability and convulsions; larger doses cause paresis. Carnegine provokes convulsions also in warm-blooded animals.

Pilocereine. Doses of 1–2 mg. of the hydrochloride provoke central paralysis in the frog. The heart dilates more in diastole, and the systolic contractions become gradually weaker, without any change in the heart rate. Finally, the heart stops in diastole, apparently due to damaged heart muscle. Death in warm-blooded animals is due to stopping of the heart [HEYL (*75*)].

Salsoline. According to GVISHIANI (*48*), salsoline resembles papaverine in its effects on blood circulation, and hydrastinine in its action on smooth muscles. Its use in the treatment of hypertension has been reported by WASTL (*226*).

References.

1. Amorosa, M.: Alkaloids of the Cactaceae. The Peyotl or Mescal Cactus. Ann. chim. farm. (Suppl. Farm. ital.) Ag., 77—87 (1938) [Chem. Abstr. **32**, 9092 (1938)].

2. Arnolt, R. I.: Dosaje de la anhalina (hordenina). Actas Trab. V. Congr. Nac. Méd. **3**, 376 (1934); An. Asoc. quím. argent. **24**, 19 B (1936).

3. Barger, G.: *p*-Hydroxyphenylethylamine, an Active Principle of Ergot Soluble in Water. Pharmac. J. **83**, 141 (1909).

4. — Isolation and Synthesis of *p*-Hydroxyphenylethylamine, an Active Principle of Ergot Soluble in Water. J. chem. Soc. (London), Proc. **25**, 162; J. chem. Soc. (London) **95**, 1123—1128 (1909).

5. — Synthesis of Hordenine, the Alkaloid from Barley. J. chem. Soc. (London), Proc. **25**, 289; J. chem. Soc. (London) **95**, 2193-2197 (1909).

6. — Isoquinoline and other Alkaloids. IX. Congr. intern. quim. pura aplicada **4**, 97—122 (1934).

7. Barger, G. and H. H. Dale: Chemical Structure and Sympathomimetic Action of Amines. J. Physiology **41**, 19—59 (1910).

8. Barger, G. and A. J. Ewins: Phenolic Derivatives of *p*-Phenylethylamine. J. chem. Soc. (London), Proc. **26**, 248; J. chem. Soc. (London) **97**, 2253—2261 (1911).

9. Barger, G. and G. S. Walpole: Further Syntheses of *p*-Hydroxyphenylethylamine. J. chem. Soc. (London), Proc. **25**, 229; J. chem. Soc. (London) **95**, 1720—1724 (1909).

10. Beccari, E.: Farmacognosia del Peyotl. II. Farmacognosia microscopica e chimica. Arch. Farmacol. sperim. Sci. affini **61**, 161—185 (1936).

11. Beringer, K.: Der Meskalinrausch. Seine Geschichte und Erscheinungsweise. Berlin: J. Springer. 1927.

12. Bernheim, M. L. C. and F. Bernheim: The Production of a Hydroxyphenyl Compound from *l*-Phenylalanine Incubated with Liver Slices. J. biol. Chemistry **152**, 481 (1944).

13. — — The Oxidation of Mescaline and Certain other Amines. J. biol. Chemistry **123**, 317—326 (1938).

14. Beyer, K. H.: Sympathomimetic Amines. The Relation of Structure to their Action and Inactivation. Physiologic. Rev. **26**, 169—197 (1946).

15. Boinet, E. et J. Boy-Teissier: Étude sur l'action cardiaque du «*Cactus grandiflorus*». Bull. de Thérapeut. **121**, 343—349 (1891).

16. Bolland, A.: Mikrochemische Studien. Mh. Chem. **32**, 117—131 (1911).

17. Bottazzi, P.: Untersuchungen über die „hintere Speicheldrüse" von *Octopus macropus*. Arch. int. Physiol. **18**, 313—331 (1921).

18. Brauns, F. E.: Lignin. Fortschr. chem. organ. Naturstoffe **5**, 175—240 (1948).

19. Britton, N. L. and J. N. Rose: The Cactaceae. Washington: The Carnegie Institution of Washington. 1919—1923.

20. Buchanan, D. N.: Med. Ann. **1931**, 12 (cited from Henry).

21. Buck, J. S.: Catalytic Reduction of Mandelonitriles. J. Amer. chem. Soc. **55**, 2593—2597 (1933).

22. — Reduction of Hydroxymandelonitriles. A New Synthesis of Tyramine. J. Amer. chem. Soc. **55**, 3388—3390 (1933).

23. Buelow, W. and O. Gisvold: Phytochemical Investigation of *Hermidium alipes*. J. Amer. pharmac. Assoc. **33**, 270—274 (1944).

24. Camus, L.: L'hordénine, son degré de toxicité, symptômes de l'intoxication. C. R. hebd. Séances Acad. Sci. **142**, 110—113 (1906).

25. CAMUS, L.: Action du sulfate d'hordénine sur la circulation. C. R. hebd. Séances Acad. Sci. **142**, 237—239 (1906).
26. — Action du sulfate d'hordénine sur les ferments solubles et sur les microbes. C. R. hebd. Séances Acad. Sci. **142**, 350—352 (1906).
27. — Étude physiologique du sulfate d'hordénine. Arch. int. Pharmacodynam. Thérap. **16**, 43 (1906).
28. CRAWFORD, A. C. and W. K. WATANABE: *p*-Hydroxyphenylethylamine, a Pressor Compound in American Mistletoe. J. biol. Chemistry **19**, 303—304 (1914).
29. — — The Occurrence of *p*-Hydroxyphenylethylamine in Various Mistletoes. J. biol. Chemistry **24**, 169—172 (1916).
30. DEULOFEU, V.: The Chemistry of the Constituents of Toad Venoms. Fortschr. Chem. organ. Naturstoffe **5**, 241—266 (1948).
31. DIGUET, L.: Le Peyote et son usage rituel chez les Indiens du Nayarit. J. Soc. des Américanistes de Paris [nouv. série] **4**, 21—29 (1907).
32. DIXON, W. E.: Physiological Action of the Alkaloids Derived from *Anhalonium Lewinii*. J. Physiology **25**, 69—86 (1899).
33. DULIÈRE, W. L. and H. S. RAPER: Tyrosinase-Tyrosine Reaction. VII. The Action of Tyrosinase on Certain Substances Related to Tyrosine. Biochemic. J. **24**, 239—249 (1930).
34. EHRLICH, F. u. P. PISTSCHIMUKA: Synthesen des Tyrosols und seine Umwandlung in Hordenin. Ber. dtsch. chem. Ges. **45**, 2428—2437 (1912).
35. EWELL, E. E.: Chemistry of the Cactaceae. J. Amer. chem. Soc. **18**, 624—643 (1896).
36. FAIVELEY, J.: Contribution à l'étude des cactées opuntiées. Thèse Doct. Méd. Paris 1920.
37. FALCO, F. and S. HILBURG: Investigación de alcaloides en las cactáceas del génere Opuntia. Rev. Fac. Quím. ind. agríc., Santa Fe, Argentina **15/16**, N° 26, 71—73 (1946/47).
38. FOERSTER, E.: Selbstversuch mit Meskalin. Z. ges. Neurol. Psychiatr. **127**, Nr. 1—2.
39. FREUDENBERG, K.: Polysaccharides and Lignin. Annu. Rev. Biochem. **8**, 81—112 (1939).
40. FREUDWEILER, R.: Chemische Zusammensetzung des Mutterkorns und Bestimmung der wirksamen Bestandteile. Pharmac. Acta Helvetiae **7**, 116—138, 139—172, 191—197 (1932).
41. FREUND, M. u. H. H. REITZ: Zur Kenntnis des Cotarnins; Verhalten desselben gegen Grignard-Lösungen. Ber. dtsch. chem. Ges. **39**, 2219—2237 (1906).
42. FUNCK, E. u. F. FINK: *Secale cornutum*. Mikrochem. ver. Mikrochim. Acta **29**, 269—272 (1941).
43. GAEBEL, O. G.: Über das Hordenin. Arch. Pharmaz. Ber. dtsch. pharmaz. Ges. **244**, 435—441 (1906).
44. GONNARD, P.: Sur le reineckate d'hordénine. Bull. Soc. Chim. biol. (Paris) **21**, 617—619 (1939).
45. GRACE, G. S.: The Action of Mescaline and Some Related Compounds. J. Pharmacol. exp. Therapeut. **50**, 359—372 (1934).
46. GUGGENHEIM, M.: Die biogenen Amine, 3. Aufl. Basel: S. Karger. 1940.
47. GURIN, S. and A. M. DELLUVA: The Biological Synthesis of Radioactive Adrenaline from Phenylalanine. J. biol. Chemistry **170**, 545—550 (1947).
48. GVISHIANI, G. S.: Über die Pharmakologie des Salsolins. J. Physiol. USSR **24**, 1174—1180 (1938) [Chem. Zbl. **1939**, I, 463].

49. Haagen-Smit, A. J. and M. Olivier: Private communication.

50. Hahn, G.: Synthese des Mezcalins (Entgegnung auf die „Berichtigung" von K. H. Slotta u. G. Szyszka). Ber. dtsch. chem. Ges. **67**, 1210—1211 (1934).

51. Hahn, G. u. F. Rumpf: Über β-(Oxy-phenyl)-äthylamine und ihre Umwandlungen. V. Kondensation von Oxy-phenyl-äthylaminen mit α-Ketonsäuren. Ber. dtsch. chem. Ges. **71**, 2141—2153 (1938).

52. Hahn, G. u. O. Schales: Über β-(Oxy-phenyl)-äthylamine und ihre Umwandlungen. II. Synthese weiterer Amine und der entsprechenden (Oxy-phenyl)-essigsäuren aus natürlichen Allylverbindungen. Ber. dtsch. chem. Ges. **67**, 1486—1493 (1934).

53. — — Über β-(Oxy-phenyl)-äthylamine und ihre Umwandlungen. III. Synthese von Benzyl-isochinolinen unter physiologischen Bedingungen. Ber. dtsch. chem. Ges. **68**, 24—29 (1935).

54. Hahn, G. u. H. Wassmuth: Über β-(Oxy-phenyl)-äthylamine und ihre Umwandlungen. I. Synthese des Mezcalines. Ber. dtsch. chem. Ges. **67**, 696—708 (1934).

55. Hashitani, Y.: Chemical Constituents of Malt Germs, especially Hordenine. J. Tokyo chem. Soc. **40**, 647—667 (1919).

56. — Occurrence of Hordenine in Seedlings of Cereals. J. Tokyo chem. Soc. **41**, 545—556 (1920).

57. Havelock, E.: Mescal a New Artificial Paradise. Annu. Rep. Smithsonian Inst. **1898**, 537—548.

58. Heffter, A.: Über Pellote. Ein Beitrag zur pharmakologischen Kenntnis der Kakteen. Naunyn-Schmiedebergs Arch. exp. Pathol. Pharmakol. **34**, 65—86 (1894).

59. — Über zwei Kakteenalkaloide. Ber. dtsch. chem. Ges. **27**, 2975—2979 (1894).

60. — Über Kakteenalkaloide. II. Ber. dtsch. chem. Ges. **29**, 216—229 (1896).

61. — Über Pellote. Beiträge zur chemischen und pharmakologischen Kenntnis der Kakteen. Naunyn-Schmiedebergs Arch. exp. Pathol. Pharmakol. **40**, 385—429 (1898).

62. — Über Kakteenalkaloide. III. Ber. dtsch. chem. Ges. **31**, 1193—1199 (1898).

63. — Über Kakteenalkaloide. IV. Ber. dtsch. chem. Ges. **34**, 3004—3015 (1901).

64. Heffter, A. u. R. Capellmann: Versuche zur Synthese des Mezcalins. Ber. dtsch. chem. Ges. **38**, 3634—3640 (1905).

65. Henry, T. A.: The Plant Alkaloids, 4th ed. Philadelphia and Toronto: The Blakiston Company. 1949.

66. Henze, M.: *p*-Oxyphenyläthylamin, das Speicheldrüsengift der Cephalopoden. Hoppe-Seyler's Z. physiol. Chem. **87**, 51—58 (1913).

67. — Über den Tyramin- und Tyrosingehalt der Speicheldrüse der Cephalopoden; zugleich Methodisches zur Mikrobestimmung der beiden Substanzen. Hoppe-Seyler's Z. physiol. Chem. **182**, 227—240 (1929).

68. Herrero-Ducloux, E.: Datos químicos sobre el *Echinopsis Eyriesii* (Turpin) Zucc. Rev. Fac. Ci. quím., Univ. nac. La Plata **6**, II, 43—49 (1930).

69. — Datos químicos sobre el *Gymnocalycium Gibbosum* (Haw.) Pfeiff. Rev. Fac. Ci. quím., Univ. nac. La Plata **6**, 75—85 (1930).

70. — Nota sobre algunas reacciones microquímicas de la mezcalina. Rev. farmaceutica **74**, 87—99 (1931).

71. — Datos químicos sobre el *Gymnocalycium multiflorum* (Hook) Britton and Rose. Rev farmaceutica **74**, 251—261 (1932).

72. — Datos químicos sobre el *Trichocereus Sp*. aff. *T. Terschecki*. Rev. farmaceutica **74**, 375—381 (1932).

73. HESSE, O.: Beiträge zur Kenntnis der Opium-Basen. Liebigs Ann. Chem. Suppl. **8**, 261 (1872).

74. HEY, P.: Synthesis of a New Homolog of Mescaline. Quart. J. Pharmac. Pharmacol. **20**, 129—134 (1947).

75. HEYL, G.: Über das Vorkommen von Alkaloiden und Saponinen in Kakteen. Arch. Pharmaz. Ber. dtsch. pharmaz. Ges. **239**, 451—473 (1901).

76. — Über ein Alkaloid aus *Carnegiea gigantea* (ENGELM.) BRITTON und ROSE. Arch. Pharmaz. Ber. dtsch. pharmaz. Ges. **266**, 668—673 (1928).

77. HOBSCHETTE, A.: Les Cactacées Médicinales. Paris: Gaston Doin & Cie. 1929.

78. HOLZ, P. and K. CREDNER: Conversion of Tyramine to Hydroxytyramine by Irradiation. Naunyn-Schmiedebergs Arch. exp. Pathol. Pharmakol. **202**, 150—154 (1943); [Chem. Abstr. **38**, 1755 (1944)].

79. IWAMOTO, H. K. and W. H. HARTUNG: Amino Alcohols. XIV. Methoxyl Derivatives of Phenylpropanolamine and 3,5-Dihydroxyphenylpropanolamine. J. org. Chemistry **9**, 513—517 (1944).

80. JANOT, M. M. et M. BERNIER: Essai de localisation des alcaloïdes dans le peyotl. Bull. Sci. pharmacol. **40**, 145—153 (1933).

81. JANOT, M. M. et P. FAUDEMAY: Silicotungstate cristallisé d'hordénine. Dosage de l'hordénine. Bull. Sci. pharmacol. **39**, 288—293 (1932).

82. JANSEN, M.: Über das β-2,4,5-Trimethoxyphenyläthylamin. Chem. Weekbl. **26**, 421—422 (1929).

83. — β-2,4,5-Trimethoxyphenylethylamine, an Isomer of Mescalin. Recueil Trav. chim. Pays-Bas **50**, 291—312 (1931).

84. — Derivatives of Some Nuclear Methoxylated β-Phenylethylamines. Recueil Trav. chim. Pays-Bas **50**, 617—637 (1931).

85. JENSCH, H.: Zur Synthese des Mescalins. Med. u. Chem. Abhandl. med.-chem. Forschungsstätten. I. G. Farben **3**, 408—411 (1936).

86. JOACHIMOGLU, G. u. E. KEESER: Kakteenalkaloide, in: A. HEFFTER, Handbuch der Experimentellen Pharmakologie, Bd. II, S. 1104—1113. Berlin: Springer-Verlag. 1924.

87. KAUDER, E.: Über Alkaloide aus *Anhalonium Lewinii*. Arch. Pharmaz. Ber. dtsch. pharmaz. Ges. **237**, 190—198 (1899).

88. KINDLER, K.: Über neue und verbesserte Wege zum Aufbau von pharmakologisch wichtigen Aminen. I. Arch. Pharmaz. Ber. dtsch. pharmaz. Ges. **265**, 389—415 (1927).

89. KINDLER, K. u. W. PESCHKE: Über neue und verbesserte Wege zum Aufbau von pharmakologisch wichtigen Aminen. IV. Über Synthesen von Tyramin und Epinin. Arch. Pharmaz. Ber. dtsch. pharmaz. Ges. **270**, 340—353 (1932).

90. — — Über neue und über verbesserte Wege zum Aufbau von pharmakologisch wichtigen Aminen. VI. Über Synthesen des Mescalins. Arch. Pharmaz. Ber. dtsch. pharmaz. Ges. **270**, 410 (1932).

91. KOESSLER, K, and M. T. HANKE: Studies on Proteinogenous Amines. V. The Preparation of *p*-Hydroxyphenylethylamine Hydrochloride (Tyramine Hydrochloride). J. biol. Chemistry **39**, 585—592 (1919).

92. KONDO, T. and Y. SHINOZAKI: Preparation of *p*-Hydroxyphenylethylamine. J. pharmac. Soc. Japan **49**, 267—269 (1929).

93. LA FORGE, F. B. and W. F. BARTHEL: A New Constituent Isolated from Southern Prickly-ash Bark. J. org. Chemistry **9**, 250—253 (1944).

94. LÉGER, E.: Sur l'hordénine: alcaloïde nouveau retiré des germes, dits touraillons, de l'orge. C. R. hebd. Séances Acad. Sci. **142**, 108—110 (1906).

95. — Sur la constitution de l'hordénine. Bull. Soc. chim. France **35**, 868—882 (1906).

96. LÉGER, E.: Sur la constitution de l'hordénine. C. R. hebd. Séances Acad. Sci. **143**, 234—236 (1906).
97. — Sur la constitution de l'hordénine. C. R. hebd. Séances Acad. Sci. **143**, 916 (1906).
98. — Sur quelques dérivés de l'hordénine. C. R. hebd. Séances Acad. Sci. **144**, 208—210 (1907).
99. — Sur la constitution de l'hordénine. J. Pharmac. Chim. **25**, 5—9 (1907).
100. — Sur la constitution de l'hordénine. C. R. hebd. Séances Acad. Sci. **144**, 488—491 (1907).
101. — Sur la constitution de l'hordénine. Bull. Soc. chim. France **1**, 148—151 (1907).
102. — Sur quelques dérivés de l'hordénine. J. Pharmac. Chim. **25**, 273—283 (1907).
103. LEPRINCE, M.: Contribution à l'étude chimique du Gui (*Viscum album*). C. R. hebd. Séances Acad. Sci. **145**, 940—941 (1907).
104. LEWIN, L.: Über *Anhalonium Lewinii*. Naunyn-Schmiedebergs Arch. exp. Pathol. Pharmakol. **24**, 401—411 (1888).
105. — Über *Anhalonium Lewinii* und andere Kakteen. Naunyn-Schmiedebergs Arch. exp. Pathol. Pharmakol. **34**, 374—391 (1894).
106. — Über *Anhalonium Lewinii* und andere giftige Kakteen. Ber. dtsch. bot. Ges. **12**, 283—290 (1894).
107. — Phantastica. New York: E. P. Dutton & Co. 1931.
108. LEWIS, J. T. et F. P. LUDUEÑA: Sur l'action de l'extrait de *Trichocereus candicans* (BRITTON et ROSE) et de ses principes actifs, l'hordénine ou anhaline et les sels de *p*-oxyphényléthyltriméthylammonium, sur la sécrétion d'adrénaline. C. R. Séances Soc. Biol. Filiales Associées **114**, 814—816 (1933); Rev. Soc. Arg. Biol. **9**, 352—364 (1933).
109. — — Action de l'atropine sur la sécrétion d'adrénaline produite par l'excitation du nerf splanchnique, la nicotine et les ammoniums quaternaires. C. R. Séances Soc. Biol. Filiales Associées **116**, 1085 (1934); Rev. Soc. Arg. Biol. **10**, 105—110 (1934).
110. LUDUEÑA, F. P.: Action pharmacodynamique d'un extrait de *Trichocereus candicans* (BRITTON et ROSE). C. R. Séances Soc. Biol. Filiales Associées **114**, 809—811 (1933); Rev. Soc. Arg. Biol. **9**, 335—343 (1933).
111. — Effet de la yohimbine sur l'action de l'extrait de *Trichocereus candicans* et de ses principes actifs sur l'excitabilité du nerf splanchnique. C. R. Séances Soc. Biol. Filiales Associées **114**, 953—954 (1933); Rev. Soc. Arg. Biol. **9**, 449—452 (1933).
112. — Effet du sulfate de spartéine sur l'action de l'extrait de *Trichocereus candicans* et de ses principes actifs sur l'excitabilité du nerf splanchnique. C. R. Séances Soc. Biol. Filiales Associées **114**, 951—953 (1933); Rev. Soc. Arg. Biol. **9**, 453—456 (1933).
113. — Effet de la cocaïne sur l'action de l'extrait de *Trichocereus candicans* et de ses principes actifs sur l'excitabilité du nerf splanchnique. C. R. Séances Soc. Biol. Filiales Associées **114**, 950—951 (1933); Rev. Soc. Arg. Biol. **9**, 457—460 (1933).
114. — Quelques données sur la pharmacologie de la candicine. C. R. Séances Soc. Biol. Filiales Associées **118**, 593 (1935); Rev. Soc. Arg. Biol. **10**, 441—446 (1934).
115. — Pharmacologie de la trichocéréine, alcaloïde du *Trichocereus Terscheki* (PARM.) BRITTON et ROSE. C. R. Séances Soc. Biol. Filiales Associées **121**, 368—369 (1936); Rev. Soc. Arg. Biol. **11**, 604—610 (1935).

116. LUDUEÑA, F. P.: La naturaleza química y la acción farmacodinámica de los alcaloides del *Trichocereus candicans*. Rosario. 1934.

117. LUMHOLTZ, C.: Unknown Mexico. New York: C. Scribner's Sons. 1902.

118. MANSKE, R. H. F.: The Alkaloids of Fumariaceous Plants. XIV. Corypalline, Corlumidine and their Constitution. Canad. J. Res., Sect. B **15**, 159—167 (1937).

119. MARINESCO, G.: Recherches sur l'action de la mescaline. Presse méd. **41**, 1433—1437 (1933).

120. — Visions colorées produites par la mescaline. Presse méd. **41**, 1864—1866 (1933).

121. MOGILEWA, A.: Über die Wirkung einiger Kakteenalkaloide auf das Froschherz. Naunyn-Schmiedebergs Arch. exp. Pathol. Pharmakol. **49**, 137—156 (1903).

122. MOONEY, J.: The Mescal Plant and Ceremony. Therap. Gazette **20**, 7—11 (1896).

123. MOSS, A. R. and R. SCHOENHEIMER: The Conversion of Phenylalanine to Tyrosine in Normal Rats. J. biol. Chemistry **135**, 415—429 (1940).

124. NEWBERNE, R. and C. BURKE: Peyotl. An Abridged Compilation from the Files of the Bureau of Indian Affairs. Chilocco, Ok. 1923.

125. NIEDFELD, H. A.: Primera contribución al estudio microquímico y localización de principios activos en las cactáceas argentinas. Alcaloides del *Trichocereus candicans* (GILLES) BRITTON and ROSE. Rosario. 1931. (Tesis del profesorado suplente; Universidad del Litoral.)

126. OREKHOV, A. u. N. PROSKURNINA: Über die Alkaloide von *Salsola Richteri*. Ber. dtsch. chem. Ges. **66**, 841—843 (1933).

127. — — Über die Alkaloide von *Salsola Richteri*, II. Die Konstitution des Salsolins. Ber. dtsch. chem. Ges. **67**, 878—884 (1934).

128. — — Alkaloids from *Salsola richteri* KAREL. Khim. Farm. Prom. **1934**, Nr. 2, 8—10 [Chem. Abstr. **28**, 5460 (1934)].

129. — — Alkaloids of *Salsola richteri*. Bull. Acad. Sci. URSS, Sér. chim. **1936**, 957—959 [Chem. Abstr. **31**, 5365 (1937)].

130. OSTEMBERG, Z.: Further Studies on the Occurrence of *p*-Hydroxyphenylethylamine in Mistletoes. Proc. Soc. exp. Biol. Med. **12**, 174—175 (1915).

131. PEDINELLI, M.: Nuovo metodo dí determinazione quantitativa dell'ordenina. Ann. Chim. applicata **31**, 410—414 (1941).

132. PETRULLO, V.: The Diabolic Root. Philadelphia: Univ. of Pennsylvania. 1934.

133. PICTET, A.: Über die Bildungsweise der Alkaloide in den Pflanzen. Arch. Pharmaz. Ber. dtsch. pharmaz. Ges. **244**, 389—396 (1906).

134. PICTET, A. u. T. SPENGLER: Über die Bildung von Isochinolin-Derivaten durch Einwirkung von Methylal auf Phenyl-äthylamin, Phenyl-alanin und Tyrosin. Ber. dtsch. chem. Ges. **44**, 2030—2036 (1911).

135. POLONOVSKI, M.: Formation des Alcaloïdes dans la Plante. Cours-Conferénces. Paris: Maison de la Chimie. 1936.

136. PRENTISS D. W. and F. P. MORGAN: *Anhalonium Lewinii* (Mescal Buttons). A Study of the Drug with Special Reference to its Physiological Action upon Man. Therap. Gaz. **19**, 577—585 (1895).

137. — — Therapeutic Uses of Mescal-Buttons (*Anhalonium Lewinii*). Therap. Gaz. **20**, 4—7 (1896).

138. PROSKURNINA, N. et A. OREKHOV: Sur les alcaloïdes de *Salsola Richteri*. III. Sur la salsoline optiquement active, ainsi que sur l'isolement de deux alcaloïdes nouveaux. Bull. Soc. chim. France [5] **4**, 1265—1271 (1937).

139. — — Sur les alcaloïdes de *Salsola Richteri*. IV. Sur quelques propriétés de la salsolidine. Bull. Soc. chim. France **6**, 144—148 (1939).

140. Pyman, F. L. and F. G. P. Remfry: Isoquinoline Derivatives. VII. Preparation of Hydrastinine from Cotarnine. J. chem. Soc. (London) **101**, 1595—1607 (1912).
141. Raoul, Y.: Nouvelle technique de dosage de l'hordénine. C. R. hebd. Séances Acad. Sci. **199**, 425—427 (1934).
142. — Nouvelle synthèse de l'hordénine. C. R. hebd. Séances Acad. Sci. **204**, 74—76 (1937).
143 — Dosage de la tyrosine dans les matières premières végétales. C. R. hebd. Séances Acad. Sci. **204**, 197—200 (1937).
144. — Sur la formation de l'hordénine par un processus d'allure biologique. Bull. Soc. Chim. biol. (Paris) **19**, 675—685 (1937).
145. — A propos du rôle et de l'origine des alcaloïdes. Bull. Sci. pharmacol. **44**, 114—120 (1937).
146. — Dosage de la tyrosine dans les substances végétales. Bull. Soc. Chim. biol. (Paris) **19**, 846—858 (1937).
147. — Evolution de l'hordénine dans l'Orge et relations éventuelles de cet alcaloïde avec la tyrosine. C. R. hebd. Séances Acad. Sci. **205**, 450—452 (1937).
148. — Origine et rôle de l'hordénine. Ann. Fermentat. **3**, 129—148 (1937).
149. — Evolution comparée de l'hordénine dans l'Orge au cours de la germination. Ann. Fermentat. **3**, 385—405 (1937).
150. — The Microchemical Technique of Localizing Alkaloids (with Special Reference to Hordenine). Rev. cytol. cytophysiol. végétales **4**, 92—100 (1939) [Chem. Abstr. **36**, 5957 (1942)].
151. Raper, H. S.: The Tyrosinase-Tyrosine Reaction. V. Production of *l*-3,4-Dihydroxyphenylalanine from Tyrosine. Biochemic. J. **20**, 735—742 (1926).
152. — Tyrosinase-Tyrosine Reaction. VI. Production from Tyrosine of 5,6-Dihydroxyindole and 5,6-Dihydroxyindole-2-carboxylic acid—the Precursors of Melanin. Biochemic. J. **21**, 89—96 (1927).
153. Raymond-Hamet, M.: Sur l'action physiologique de la mezcaline, alkaloïde principal du Peyotl. Bull. Acad. Méd. **105**, 46—54 (1931).
154. — Neue Beobachtungen über die physiologische Wirkung des Mescalins. Naunyn-Schmiedebergs Arch. exp. Pathol. Pharmakol. **169**, 97—113 (1933).
155. — Action de l'hordénine sur le rein à vaisseaux sectionnés puis anastomosés à ceux du cou. C. R. Séances Soc. Biol. Filiales Associées **113**, 386—387 (1933).
156. — Sur les effets cardiaques de l'hordénine. C. R. Séances Soc. Biol. Filiales Associées **113**, 875—877 (1933).
157. — Sur une propriété physiologique encore inconnue de l'hordénine. C. R. Séances Soc. Biol. Filiales Associées **121**, 112—115 (1936).
158. — L'action nicotinique de l'hordénine n'est pas supprimée par l'introduction dans la molécule d'un second oxhydrile phénolique, celui-ci en position méta. C. R. hebd. Séances Acad. Sci. **209**, 67 (1939).
159. — L'oxytyramine doit-elle être tenue pour une amine véritablement sympathomimétique? C. R. Séances Soc. Biol. Filiales Associées **133**, 570—572 (1940).
160. Reilhes, R.: Sur la localisation histochimique de l'hordénine dans les plantules d'Orge. C. R. Séances Soc. Biol. Filiales Associées **122**, 852—854 (1936).
161. Reti, L.: Sur les alcaloïdes de la cactacée *Trichocereus candicans* (Britton et Rose). C. R. Séances Soc. Biol. Filiales Associées **114**, 811—814 (1933); Rev. Soc. Arg. Biol. **9**, 344—351 (1933).
162. — Los Alcaloides de las Cactáceas. An. Asoc. quím. argent. **23**, 26—40 (1935).
163. — Nuovi alcaloidi di cactacee argentine. Atti X. Congr. int. Chim., Roma **5**, 396—405 (1939).

164. RETI, L.: Alcaloides de las cactáceas y substancias naturales relacionadas. Ciencia e Invest. 3, 405—411 (1947).
165. — Le relazioni fra le sostanze aromatiche vegetali e la loro origine. Atti I° Congr. Naz. Chim. Ind. Milano 1924.
166. — Cactus Alkaloids. Proc. Conf. Cultivation Drug and Assoc. Econ. Plants Calif. **1947**, 110—113.
167. RETI, L. and R. I. ARNOLT: Alcaloides del *Trichocereus lamprochlorus* (LEM.) BRITTON and ROSE. Actas y Trabajos del V° Congr. Nac. de Medicina, Rosario 3, 39 (1935).
168. RETI, L., R. I. ARNOLT et F. P. LUDUEÑA: Sur un alcaloïde du *Cereus coryne* SALM. (1850). C. R. Séances Soc. Biol. Filiales Associées **118**, 591—593 (1935); Rev. Soc. Arg. Biol. **10**, 437—440 (1934).
169. REUTTER, L.: Peyotl als sensorielle Droge. Schweiz. Apotheker-Ztg. **62**, 441—443 (1924).
170. RICHTER, D.: Elimination of Amines in Man. Biochemic. J. **32**, 1763—1769 (1938).
171. RIETSCHEL, H. G.: Zur Pharmakologie des Hordenins. Naunyn-Schmiedebergs Arch. exp. Path. Pharmakol. **186**, 387—408 (1937).
172. — Zur Pharmakologie des Hordenins. Klin. Wschr. **16**, 714—715 (1937).
173. ROBINSON, R.: Molecular Architecture of Some Plant Products. IX Congr. intern. quim. pura aplicada (Madrid) **5**, 17—38 (1934) (Publ. 1936).
174. — Structure and Biogenesis of Emetine. Nature (London) **162**, 524 (1948).
175. ROCA, J.: Investigaciones preliminares sobre la presencia de tirosinasa y tirosina en el *Pachycereus marginatus*. An. Inst. Biol. (Mexico) **7**, 97—107 (1936).
176. ROSENMUND, K. W.: Über *p*-Oxyphenyl-äthylamin. Ber. dtsch. chem. Ges. **42**, 4778—4783 (1909).
177. — Die Synthese des Hordenins, eines Alkaloids aus Gerstenkeimen, und über α-*p*-Oxyphenyl-äthylamin. Ber. dtsch. chem. Ges. **43**, 306—313 (1910).
178. — Über eine neue Methode zur Darstellung von Aldehyden. I. Ber. dtsch. chem. Ges. **51**, 585—594 (1918).
179. ROSENTHALER, L.: Beiträge zum Nachweis organischer Verbindungen. IV. Mikrochemische Reaktionen des Mezcalins. Pharmaz. Ztg. **76**, 653—654 (1931).
180. ROUHIER, A.: La plante qui fait les yeux émerveillés. Paris: Gaston Doin & Cie. 1927.
181. SAFFORD, W. E.: Peyote, the Narcotic Mescal Button of the Indians. J. Amer. med. Assoc. **77**, 1278 (1921).
182. SCHMALFUSS, H. u. A. HEIDER: Tyramin und Oxytyramin, blutdrucksteigernde Schwarzvorstufen des Besenginsters *Sarothamnus scoparium* WIMM. Biochem. Z. **236**, 226—230 (1931).
183. SCHÖPF, C.: Zur Frage der Biogenese der Naturstoffe: Alkaloidsynthesen unter physiologischen Bedingungen. IX Congr. intern. quim. pura aplicada **4**, 189—198 (1934).
184. — Die Synthese von Naturstoffen, insbesondere von Alkaloiden, unter physiologischen Bedingungen und ihre Bedeutung für die Frage der Entstehung einiger pflanzlicher Naturstoffe in der Zelle. Angew. Chem. **50**, 779—787, 797—805 (1937).
185. SCHÖPF, C. u. H. BAYERLE: Zur Frage der Biogenese der Isochinolin-Alkaloide. Die Synthese des 1-Methyl-6,7-dioxy-1,2,3,4-tetrahydro-isochinolins unter physiologischen Bedingungen. Liebigs Ann. Chem. **513**, 190—202 (1934).
186. SLOTTA, K. H.: Zur Gewinnung von 3,4,5-Trimethoxybenzaldehyd. J. prakt. Chem. **133**, 129—130 (1932).

187. Slotta, K. H. u. W. Altner: Über β-Phenyl-äthylamine. II. Eine neue Tyramin-Synthese. Ber. dtsch. chem. Ges. **64**, 1510—1520 (1931).

188. Slotta, K. H. u. H. Heller: Über β-Phenyl-äthylamine. I. Mezcalin und mezcalin-ähnliche Substanzen. Ber. dtsch. chem. Ges. **63**, 3029—3044 (1930).

189. Slotta, K. H. u. J. Müller: Über den Abbau des Mezcalins und mezcalinähnlicher Stoffe im Organismus. Hoppe-Seyler's Z. physiol. Chem. **238**, 14—22 (1936).

190. Slotta, K. H. u. G. Szyska: Über β-Phenyl-äthylamine. III. Eine neue Mezcalin-Synthese. J. prakt. Chem. **137**, 339—350 (1933).

191. — — Synthese des Mezcalins. (Eine Berichtigung der gleichlautenden Arbeit von G. Hahn und H. Wassmuth.) Ber. dtsch. chem. Ges. **67**, 1106—1108 (1934).

192. Späth, E.: Über die Anhalonium-Alkaloide. I. Anhalin und Mezcalin. Mh. Chem. **40**, 129—154 (1919).

193. — Über die Anhalonium-Alkaloide. II. Die Konstitution des Pellotins, des Anhalonidins und des Anhalamins. Mh. Chem. **42**, 97—115 (1921).

194. — Über die Anhalonium-Alkaloide. III. Konstitution des Anhalins. Mh. Chem. **42**, 263—266 (1921).

195. — Über die Anhalonium-Alkaloide. V. Die Synthese des Anhalonidins und des Pellotins. Mh. Chem. **43**, 477—484 (1922).

196. — Über das Carnegin. Ber. dtsch. chem. Ges. **62**, 1021—1024 (1929).

197. Späth, E. u. F. Becke: Eine neue Synthese des Pellotins (XI. Mitteil. über Kakteen-Alkaloide). Ber. dtsch. chem. Ges. **67**, 266—268 (1934).

198. — — Die Konstitution des Anhalamins (XII. Mitteil. über Kakteen-Alkaloide). Ber. dtsch. chem. Ges. **67**, 2100—2102 (1934).

199. — — Über ein neues Kakteen-Alkaloid, das Anhalinin, und zur Konstitution des Anhalonins (XIII. Mitteil. über Kakteen-Alkaloide). Ber. dtsch. chem. Ges. **68**, 501—505 (1935).

200. — — Über das Anhalidin (XIV. Mitteil. über Kakteen-Alkaloide). Ber. dtsch. chem. Ges. **68**, 944—945 (1935).

201. — — Über die Trennung der Anhaloniumbasen (Kakteen-Alkaloide XV.). Mh. Chem. **66**, 327—336 (1935).

202. Späth, E. u. F. Boschan: Über Kakteenalkaloide. X. Die Konstitution des Pellotins und des Anhalonidins. Mh. Chem. **63**, 141—153 (1933).

203. Späth, E. u. J. Bruck: Über ein neues Alkaloid aus den Mezcal-Buttons (XVIII. Mitteil. über Kakteen-Alkaloide). Ber. dtsch. chem. Ges. **70**, 2446 bis 2450 (1937).

204. — — N-Acetyl-mezcalin als Inhaltstoff der Mezcalin-Buttons (XIX. Mitteil. über Kakteen-Alkaloide). Ber. dtsch. chem. Ges. **71**, 1275—1276 (1938).

205. — — Über das O-Methyl-*d*-anhalonidin (XX. Mitteil. über Kakteen-Alkaloide). Ber. dtsch. chem. Ges. **72**, 334—338 (1939).

206. Späth, E. u. F. Dengel: Synthese des Salsolidins. Ber. dtsch. chem. Ges. **71**, 113—119 (1938).

207. Späth, E. u. J. Gangl: Über die Anhalonium-Alkaloide. VI. Anhalonin und Lophophorin. Mh. Chem. **44**, 103—113 (1923).

208. Späth, E. u. P. L. Julian: Neue Corydalis-Alkaloide: *d*-Tetrahydrocoptisin, *d*-Canadin und Hydrohydrastinin. Ber. dtsch. chem. Ges. **64**, 1131—1137 (1931).

209. Späth, E. u. F. Kesztler: Synthese des Anhalonins und des Lophophorins. (XVI. Mitteil. über Kakteen-Alkaloide). Ber. dtsch. chem. Ges. **68**, 1663—1667 (1935).

210. — — Über die optische Aktivität des Pellotins (XVII. Mitteil. über Kakteen-Alkaloide). Ber. dtsch. chem. Ges. **69**, 755—757 (1936).

211. SPÄTH, E. u. F. KUFFNER: Die Identität des Pectenins mit dem Carnegin. Ber. dtsch. chem. Ges. **62**, 2242—2243 (1929).

212. SPÄTH, E., A. OREKHOV u. F. KUFFNER: Konstitution und Synthese des Salsolins. Ber. dtsch. chem. Ges. **67**, 1214—1217 (1934).

213. SPÄTH, E. u. J. PASSL: Über die Konstitution von Pellotin und Anhalonidin. Ber. dtsch. chem. Ges. **65**, 1778—1785 (1932).

214. SPÄTH, E. u. H. RÖDER: Über die Anhalonium-Alkaloide. IV. Die Synthese des Anhalamins. Mh. Chem. **43**, 93—111 (1922).

215. SPÄTH, E. u. P. SOBEL: Neue Synthesen des Hordenins. Mh. Chem. **41**, 77—90 (1920).

216. STEINER-BERNIER, M.: Contribution à l'étude du Peyote (*Cchinocactus Williamsii* Lem.). Thèse Doct. Pharm. Paris. 1936.

217. SULTAN, F. W.: Cactine, Active Principle of *Cereus grandiflorus*. New York Med. J. 681—683 (1891).

218. TANAKA, Y., T. MIDZUNO and T. OKAMI: Preparation of Hydrastinine from Narcotine. J. pharmac. Soc. Japan **50**, 559—562 (1930).

219. TOMASO, C.: Gli alcaloidi del peyote. La Chimica **10**, 408—416 (1934); Chim. et Ind. **34**, 138 (1935).

220. TOPCHIEV, K.: Hydrastinine and its Preparation from Narcotine. J. appl. Chem. (USSR) **6**, 529—535 (1933).

221. TORQUATI, T.: Ricerche sulla formazione dell'ordenina durante la germogliazione dei semi d'orzo. Arch. Farmacol. sperim. Sci. affini **10**, 62—71 (1911).

222. ULLMANN, A.: Über Tyramin (*p*-Oxyphenyläthylamin) als wirksamen Bestandteil der Droge *Semina cardui Mariae* (Stechdistelkörner). Biochem. Z. **128**, 402—406 (1922).

223. VOSWINCKEL, H.: Über eine neue Synthese des Hordenins. Ber. dtsch. chem. Ges. **45**, 1004—1006 (1912).

224. WASER, E.: Untersuchungen in der Phenylalanin-Reihe. VI. Decarboxylierung des Tyrosins und des Leucins. Helv. chim. Acta **8**, 758—773 (1925).

225. WASER, E. u. H. SOMMER: Untersuchungen in der Phenylalanin-Reihe. II. Synthese des 3,4-Dioxyphenyl-äthylamins. Helv. chim. Acta **6**, 54—61 (1923).

226. WASTL, N.: Salsoline, a New Drug in the Treatment of Hypertension. Hahnemannian Month. **81**, 243 (1946) [Chem. Abstr. **40**, 5147 (1946)].

227. WEIR, M.: Notes upon the Effects of *Anhalonium Lewinii* (Mescal Buttons). Brit. Med. J. II, 1625 (1896) (cited from ROUHIER).

228. WHITE, E. P.: Alkaloids of the Leguminosae. IX. Isolation of β-phenylethylamine from Acacia Species. New Zealand J. Sci. Technol. **25** B, 139 (1943/44).

229. WINTERSTEIN, E. u. G. TRIER: Die Alkaloide. Berlin: Gebr. Borntraeger. 1910.

230. WOODWARD, R. B.: Biogenesis of the Strychnos Alkaloids. Nature (London) **162**, 155—156 (1948).

231. YURASHEVSKII, N. K.: Alkaloids of *Arthrophytum leptocladum* M. POP. J. Gen. Chem. (USSR) **9**, 595—597 (1939).

232. — Alkaloids of *Arthrophytum leptocladum* M. POP (family *Chenopodiaceae*). II. J. Gen. Chem. (USSR) **11**, 157—162 (1941).

(Received, October 16, 1949.)

Plant Proteins.

By **James Bonner**, Pasadena, California.

With 4 Figures.

Contents.

I. Introduction.

The plant proteins may be divided into at least two large groups on the basis of mode of occurrence and biochemical function. These groups are respectively the reserve proteins of seeds and the functional proteins of the vegetative organs of the plant.

The seed proteins, which may be accumulated to relatively high concentration, appear to be predominantly reserve materials and to serve as a source of nutritive material for the germinating seed. Specific enzymatic or other biochemical functions have with rare exceptions not been as yet associated with the proteins of the seed. Nevertheless, perhaps because of their ready availability, the seed proteins have been extensively investigated over a long period and form at present one of the relatively well-characterized protein groups.

The proteins of the vegetative portions of the plant, those of the growing or functioning leaf, stem, root, etc. have been studied much less extensively than have the seed proteins. It is now clear, however, that the functional proteins are of great biochemical interest since they comprise the enzymatic systems responsible for the wide variety of metabolic processes of which the higher plant is capable. The detailed study of plant growth and plant metabolism will no doubt entail a still more comprehensive study of the individual proteins of the vegetative portions of the plant than has as yet been possible. The study and characterization of the proteins of the vegetative portions of higher plants has undoubtedly lagged in part because of the difficulties inherent in the extraction and preparation of native protein from such tissue. Suitable methods for large scale preparation of plant proteins from leaves and other vegetative organs are, however, now available and within the past few years considerable advances have been made in our understanding of the kinds and roles of the various component proteins of these tissues.

This review will summarize the present information concerning the plant proteins, and, it is to be hoped, may serve to stimulate further investigations of these important materials.

II. Proteins of Cereal Seeds.

The proteins of the cereal seeds and of wheat in particular have been the subject of intensive investigation for many years because of the importance of the wheat protein, gluten, in formation of a cohesive dough in the process of bread making. The proteins of the wheat seed have been divided by OSBORNE (*35*) and others into three principal components, viz. (*a*) the gluten proper (insoluble in water or in dilute neutral salt solutions), which forms the bulk of the seed protein, (*b*) an albumin, viz. leucosin, and (*c*) a globulin. Gluten in turn consists of two major components, glutenin, a glutelin (soluble in dilute alkali) and gliadin, a prolamine. The amounts of these components in typical samples of whole wheat grains is given in Table 1.

Table 1. Component Proteins of Whole Wheat Grains [OSBORNE (*35*)].

Component	% of dry weight	
	in spring wheat	in winter wheat
Glutenin	4,68	4,17
Gliadin	3,96	3,90
Globulin	0,62	0,63
Leucosin	0,39	0,36

Glutenin and gliadin are localized in the endosperm of the grain and make up the reserve protein proper of the seed. Both leucosin and the wheat globulin are on the other hand localized in part in the embryo and in part in the endosperm. It is entirely possible that these two components may make up the cellular proteins of the embryo and of the storage tissue rather than reserve protein. Careful study of the enzymatic properties of these non-gluten proteins should reveal whether or not they do have actual metabolic functions.

Despite the fact that glutelin and gliadin have long been considered as individual protein species, there is now evidence that both of these proteins are in fact complex and consist of several components. Thus gliadin consists of at least two fractions which may be distinguished by their electrophoretic mobilities (*39*). Gliadin as a whole appears, however, to have a molecular weight of approximately 34000 (*24*). Glutenin may be separated into a series of fractions by fractional precipitation from acid solution with neutral salts (*29, 30*). The most soluble fractions have lower apparent molecular weights than the less soluble fractions, the lowest being of the order of 39000 (*29*). In the less soluble fractions the protein molecules are apparently associated into loose aggregates, and accurate molecular weights have not yet been obtained. Both gliadin and glutenin occur in solution as elongated molecules, the axial ratio of glutenin (*3*) approximating 47:1 while the axial ratio of the gliadin fraction has been variously estimated as between 11:1 (*34*) and 26:1 (*3*).

The extensive work on amino acid analysis of the seed proteins of wheat which has been carried out in recent years has confirmed the fact that gluten is particularly rich in glutamic acid, which may occur in proportions as high as 46% of the total in gliadin (Table 2). Smaller proportions of glutamic acid appear to be present in the glutenin fraction.

Table 2. Amino Acid Composition of Gliadin, Glutenin and of Zein [BLOCK and BOLLING (*4*)].

Amino acid	% of weight of protein		
	wheat gliadin	wheat glutenin	maize zein
Arginine	3,2	4,7	1,6
Lysine	0,6	1,9	0,0
Histidine	2,1	1,8	0,8
Tyrosine	3,1	5,1	5,9
Tryptophane	0,9	1,8	0,2
Phenylalanine	2,5	2,0	6,6
Cystine	2,3	1,7	1,0
Methionine	2,3	—	2,5
Serine	0,1	0,7	—
Threonine	3,0	—	2,5
Leucine and isoleucine	6,0	6,0	3,0
Valine	3,0	1,0	3,0
Glutamic acid	46,0	27,2	35,6
Aspartic acid	1,4	2,1	3,4
Glycine	1,0	1,0	0,0
Alanine	2,5	4,4	9,9
Proline	13,2	4,4	9,0
Hydroxyproline	—	—	0,0
Ammonia	5,1	4,0	—
% of total accounted for	98,3	69,8	85,0

A summary of the amino acid components of these proteins together with that of the zein of maize appears in Table 2.

Seeds of cereals other than wheat contain in general a similar complement of proteins. Thus, seeds of barley, rice and rye contain a small proportion of protein soluble in neutral salt solutions and similar to the albumin-globulin fraction of wheat. Glutelin and, particularly, prolamines are known from sorghum, barley, rice, rye and maize. The prolamine of maize, zein, is, like gliadin, not a homogeneous chemical compound but is made up of a series of components which may be distinguished on the basis of their electrophoretic mobilities (*37*). That the prolamines of various species differ from one another in amino acid composition may be seen by comparison of the values for zein with those for gliadin in Table 2.

III. Proteins of Seeds of Dicotyledonous Plants.

The proteins of seeds other than those of the grasses present an entirely different situation, as has been known for many years. In the seeds of dicotyledons (as opposed to the monocotyledonous cereals) the dominant reserve protein is frequently a globulin or globulin complex which is contained in the cotyledons or other reserve organs of the seed. This protein may be extracted from the ground seed with dilute neutral salt solutions and may be precipitated from such solutions by 0,5–1,0 saturated ammonium sulfate, by dialysis against distilled water, or by other means.

In several cases, careful precipitation results in separation of the globulin in the crystalline state, and such crystalline seed globulins have

Table 3. Amino Acid Composition of Various Seed Globulins [SMITH and GREEN (*45*)].

Amino acid	Content of amino acid in % weight of protein					
	Edestin	Pumpkin	Squash	Watermelon	Cucumber	Tobacco
Arginine	16,7	16,2	16,2	17,9	15,8	16,1
Histidine	2,5	2,2	2,2	2,2	2,3	2,2
Lysine	2,3	2,8	3,0	2,9	2,9	1,6
Threonine	3,1	2,6	2,8	2,9	3,6	4,2
Leucine	7,4	8,0	8,0	7,5	9,1	10,5
Isoleucine	6,2	5,1	5,5	5,7	5,5	6,3
Valine	6,6	6,5	6,5	6,4	7,0	6,7
Tyrosine	4,3	4,4	4,4	4,6	4,6	4,1
Tryptophane	1,2	1,7	1,7	1,9	1,9	1,5
Phenylalanine	5,4	7,2	6,8	7,7	6,5	5,7
Methionine	2,2	2,3	2,3	2,8	2,5	2,2
Cystine	1,3	1,1	1,1	1,1	1,1	1,1

provided classical examples of crystalline proteins as in the case of edestin (from hemp seed). Crystalline globulins may also be prepared from the readily available seeds of tobacco, pumpkins, squash, watermelon and cucumber (*51*). The seed globulins have in common molecular weights (*49*) in the range of 200000–400000 and a generally somewhat similar amino acid composition, as is shown in Table 3. The seed globulins differ from the cereal prolamines in a lower content of glutamic acid and in relatively high contents of arginine, leucine, isoleucine, and valine.

Critical work on the homogeneity of the crystalline seed globulins remains to be carried out although edestin appears to be essentially homogeneous in the ultracentrifuge (*2*). Detailed work on the proteins of peanut, soybean, cotton seed, and pea has shown on the other hand that in these cases the globulins present each represent a complex mixture of two to many components. Thus, the peanut contains two principal protein fractions, arachin and conarachin, which differ in isoelectric point and may hence be prepared individually. These two globulins differ also in amino acid composition, arachin being lower than conarachin in both tryptophane and methionine (*10*). Neither of these two globulins represents, however, a homogeneous chemical compound. Thus electrophoretic analysis has shown that both contain approximately 80% of a common constituent, globulin *A*. Arachin contains an additional 20% of a globulin *B* while conarachin contains 20% of a mixture of further minor components (*20, 21*). Ultracentrifugation of peanut globulin by Johnson (*23*) has confirmed the above conclusions as to the heterogeneous character of this protein. Its heaviest component was found to have a molecular weight of 396000. The lower molecular weight fractions are in part closely related to this heavy fraction since the latter was shown to dissociate in dilute solutions to yield a fraction of half the original molecular weight. [Cf. also (*22*).]

The proteins of the pea cotyledon like those of peanut consist of a series of globulins in addition to other minor components. At least three electrophoretically distinguishable components are contained in the globulin fraction (*28, 53*). The soybean likewise contains not one but a series of globulins (*51*).

In the extraction of the seeds of dicotyledonous plants it has been repeatedly observed that only a limited portion, 50–90%, of the total protein can be removed in truly soluble form. The nature of the residual, non-readily extractable protein has long remained obscure. It seems, however, quite possible that this insoluble seed protein may represent material present in the cell in particulate form, perhaps as mitochondria or other particulate inclusions of the cytoplasm. This particulate protein may well be of physiological significance in the seed in connection with the metabolic transformations which occur during germination, for example, in fatty acid oxidation or in other oxidative processes. It is of course now well established that in animal tissues such oxidative processes are in a large measure associated with the particulate matter of the cytoplasm.

Only a limited amount of information is available concerning the enzymatic properties of the seed globulins. Surprisingly enough, the crystalline globulins of curcurbit seeds are reported to have a definite, although low, activity in the decarboxylation of oxaloacetate (*50*). Whether this is actually due to the globulin itself or to an associated minor constituent has not, however, been critically established. It should not be forgotten, in any case, that crystalline urease was obtained by SUMNER as the first crystalline enzyme (*48*) and that this urease is a component of the seed protein of the jack bean. In general, however, it has not been demonstrated that the major seed globulins possess specific enzymatic functions, and it remains possible that they serve primarily as reserve proteins which are made available to the seedling on germination. The methods now available for the study of protein mixtures should make possible investigations of the behavior of the individual seed proteins not only during germination but also during protein synthesis and accumulation in the developing seed.

IV. The Leaf Proteins.

Preparation.

The study of the proteins of green leaves is of special interest since these organs contain much of the enzymatic mechanism for the wide variety of organic syntheses of which the plant is capable. Nevertheless, the mere preparation of native protein from leaves has proved to be much more difficult than preparation of such protein from seed material. The leaf is in general highly hydrated and the protein concentration of the fresh leaf as a whole is correspondingly low. Thus, the protein of the spinach leaf ordinarily makes up less than 5% of the total fresh weight. This is to be compared with the seed in which protein may make up 40% or more of the weight. It has nevertheless been possible in recent years to prepare the native proteins of green leaves and to study not only the biochemistry of the extracted materials but also to characterize the preparations by physical methods.

The proteins of the green leaf may be devided, on the basis of location within the cell, into three major categories,. namely the chloroplastic protein of the green chloroplasts, the nuclear protein of the nuclei, and the cytoplasmic protein contained in the cytoplasm proper. The location of these three major types of protein in a typical plant cell is shown in Figure 1.

Surrounding the entire cell is the cell wall, composed principally of polysaccharides and polyuronides. Within the cell wall is a thin layer of cytoplasm bounded on the outside by the plasma membrane and on the inside by the inner membrane or tonoplast. Imbedded or dispersed in the cytoplasm are the saucer-shaped chloroplasts, typically 3–10 microns in diameter and 1–2 microns in thickness. Of the order of 10–100 such chloroplasts are ordinarily contained in the spongy parenchyma or palisade cells of the leaf. Also imbedded in the cytoplasm is the nucleus, a single nucleus per cell. Nucleus, chloroplasts, and cytoplasm together make up the protoplasm of the cell. The central portion of the cell within the tonoplast is occupied by the vacuole, which contains primarily water together with low-molecular solutes such as inorganic salts, organic acids, tannins, etc. Proteins if present in the vacuole occur there only in small amounts.

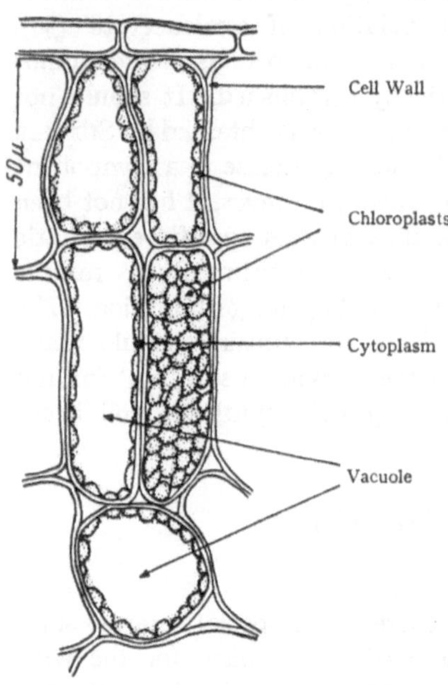

Fig. 1. Diagrammatic representation of a plant cell. After WILDMAN and BONNER (*54*). Cell at right with numerous chloroplasts is an intact cell. All other cells have been cut through and illustrate the peripheral distribution of leaf protoplasm enclosed by the cell wall and surrounding a large vacuole. [From: Arch. Biochemistry *14*, 381 (1947).]

In order to release the proteins from the leaf it is then necessary to grind the tissue in such a way as to rupture the cellulose cell walls, so that the protoplasmic contents may be extracted. In such grinding and extraction it is, however, inevitable that the protoplasmic contents will become mixed and diluted with those of the vacuole. Thus at the very first stage of extraction of the proteins of the leaf (or of any other highly vacuolated tissue) a very considerable dilution of the protoplasmic constituents is brought about. In many cases the vacuole may also contribute materials or conditions (such as excessively low p_H or high concentrations of tannins) which may denature the protoplasmic proteins, and the possibility of such effects must be recognized and guarded against.

The earliest comprehensive studies on leaf proteins are those of CHIBNALL (*11*) whose extraction methods include a preliminary destruction of the semipermeability of the tissue by immersion in ether. The leaves were then pressed to express the vacuolar contents, and the residue ground and extracted with water. The green aqueous extract was then separated from the residue of unground tissue and cell wall debris by a coarse filtration. The extract thus obtained consisted mainly of chloroplasts (or chloroplast fragments) suspended in a clear yellow solution

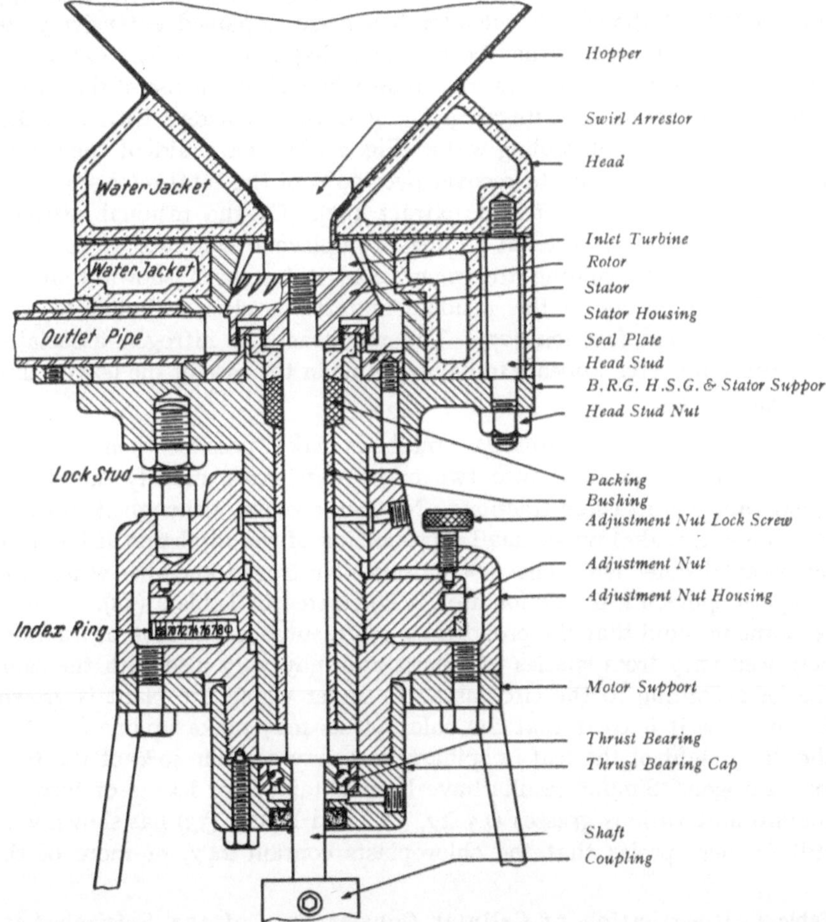

Fig. 2. The colloid mill or high speed centrifugal grinding device used by WILDMAN and BONNER (54) for extraction of protein from leaves and other plant tissues. The plant material is introduced through the hopper, ground between rotor and stator, and pumped out through the outlet pipe. The motor is mounted below the coupling.

composed of the cytoplasmic constituents. To the particulate fraction, removed by filtration or by centrifugation, CHIBNALL applied the name "chloroplastic protein". The proteins of the clear yellow filtrate, or "cytoplasmic proteins", may be freed from associated non-protein material by dialysis or by precipitation either at the isoelectric point, by neutral salts or by protein denaturants.

The general methods of CHIBNALL (*11*) have been used by later workers, although in general the ether immersion has been avoided, more efficient grinding methods have been employed, and the importance of carrying out the entire procedure at temperatures below 5° C. has been increasingly appreciated. Although the blendor has been employed extensively for grinding of leaf tissue it has been shown that more efficient extraction and better control of temperature can be achieved by the use of the colloid mill, a high speed centrifugal grinding device provided with a jacket for the circulation of cooling water (Fig. 2). With the aid of the colloid mill it has ben possible to recover over 90% of the total nitrogen of the leaf in the whole protoplasm extract (*54*). Of the residual nitrogen a portion at least is present as the protoplasmic contents of cells not ruptured by the grinding treatment, and it has been shown that the analytical properties of this residual material is closely similar to that of the extracted fraction (*27*). The unextractable nitrogen associated with cell walls, etc., appears to be negligible in the case of the leaves thus far examined.

The application of CHIBNALL's methods makes possible then a resolution of the leaf proteins into two major categories, namely the chloroplastic and cytoplasmic fractions. Nuclear protein, although it must be also present, makes up so small a proportion of the total that it has thus far escaped detection. The distribution of nitrogen and dry weights in a typical spinach leaf fractionation is illustrated in Table 4 (*54*). It must be borne in mind that the proportions represented in the several fractions may well vary from species to species and may vary also with the same species according to the circumstances under which the plant is grown. In any case it is clear that the chloroplasts may make up one-fourth of the dry weight of the leaf in spinach and may contain 40% of the total leaf nitrogen. Similar results have been obtained for leaves of tobacco, tomato and various grasses (*15–17, 59*), and NEISH (*33*) has shown with still further species that the chloroplasts contain 24% or more of the

Table 4. Distribution of Cellular Constituents of the Spinach Leaf [WILDMAN and BONNER (*54*)].

Fraction of	% of dry weight	% of fresh weight	% of total nitrogen
Cell walls	23,8	2,5	—
Protoplasm	76,2	8,2	100*
1. chloroplasts	26,8	2,7	37,9
2. cytoplasmic proteins	15,8	1,9	39,3
3. non-protein constituents	33,6	3,6	22,8

* The amount of extractable nitrogen is arbitrarily taken as 100%.

leaf dry weight. The cytoplasmic proteins make up on the order of 16% of the leaf dry weight and account, in the example of Table 4, for roughly 40% of the total leaf nitrogen.

Properties of Chloroplastic Protein.

The chloroplast is not a homogeneous body but possesses structure in the sense that it is made up of discrete saucer-shaped bodies, the grana (*14*, *36*). The grana vary in size as between different species but range in diameter from perhaps 0,2 microns to as large as 2 microns in some instances. The grana of the spinach chloroplast are on the average roughly 0,6 microns in diameter by 0,1 microns in thickness. The grana contain the green chloroplastic protein and appear to be suspended in a colorless matrix, the stroma. The entire chloroplast, containing roughly 10 to 100 grana together with the stroma, is further surrounded by a semi-permeable membrane so that the intact chloroplast exhibits osmotic phenomena, shrinking in hypertonic solution and swelling, often times with rupture, in hypotonic solution. The chloroplast itself is frequently found to be readily ruptured by the grinding needed to extract the tissue proteins. Thus with spinach leaves, the chloroplastic protein is usually obtained as a suspension of grana or as irregular flocs of grana rather than as intact chloroplasts. With many species intact chloroplasts can be obtained only by observing special precautions including grinding of the tissue for only a brief period and in an isotonic solution (*15*). With other species, however, such as tobacco, tea, oat and tomato, the chloroplasts are more sturdy and may be readily obtained as preparations of the intact bodies (*14*, *15*, *25*). Intact chloroplasts may in turn be fragmented to their constituent grana by immersion in hypotonic solution or by further grinding.

Nothing is known concerning the composition of stroma and attempts to separate this fraction of the chloroplast have thus far failed (*54*). It seems probable, however, that stroma makes up at most but a small proportion of the total mass of the chloroplast. The composition of the chloroplastic material as a whole (chloroplasts or grana) has been sub-

Table 5. Gross Composition of Chloroplasts or of Chloroplastic Fragments of Leaves of Various. Species.

Species	% protein	% total lipid	% ash	% chlorophyll	Reference
Spinach	40	25	17	—	(*11*)
Spinach	48	37	8	7,6–8,3	(*32*)
Spinach	42–54	26–32	—	6	(*9*)
Spinach	54	34	7	—	(*12*)
Sweetpea	33–50	18–30	—	4–6	(*9*)
Clover	50	22	—	—	(*33*)

jected to repeated study and a summary of the gross chemical composition of a number of species is given in Table 5.

Between one-third and one-half of the total mass of the chloroplastic material is made up of protein, the exact amount varying with species and with the investigator. Roughly one-fourth to one-third of the total mass is made up of a fraction soluble in ether or ether-alcohol mixtures and has been referred to as "total lipid". This fraction includes not only lipids proper but also the chlorophyll and carotenoids of the chloroplast. Chlorophyll itself makes up some 8% of the dry weight while carotenoids, generally less than one-fifth of the chlorophyll in amount, may be expected to make up less than 2% of the dry weight of the chloroplastic material. The balance of the total lipid fraction of the chloroplast is said to include fats, fatty acids, and possibly sterols (*11*). Inorganic materials, including particularly calcium, magnesium and phosphorous, are present in the grana together with smaller amounts of heavy metals, particularly iron and copper.

In the chloroplastic material of the grana the varied components, proteins, chlorophyll, carotenoids, fatty materials, and inorganic components are bound together in a tightly organized structure which may be dispersed in true solution only by rather drastic treatment. Prolonged grinding of grana or of whole leaves results in rupture of a portion of the grana which may then be obtained as a stable green suspension. This material still consists of relatively large particles since they may be detected in the dark-field microscope and may be centrifuged out by moderate centrifugal fields as has been observed by STOLL and WIEDEMANN (*46*). Such suspensions are also inhomogeneous in the ultracentrifuge and are therefore apparently polydisperse [SMITH (*40–42*)]. The slightly turbid green solutions of fragmented grana may be further dispersed by the action of bile salts, digitonin or of sodium dodecyl sulfate (*43, 44, 46*).

The properties of the solution obtained depend on the detergent used. Thus, a sodium cholate treated preparation of ground *Aspidistra* chloroplasts was found by STOLL, WIEDEMANN and RÜEGGER (*47*) to be electrophoretically homogeneous, monodisperse, and to have an apparent molecular weight of 5×10^6. This material contained protein, and pigments, and lipoids in the same proportions as did the original chloroplasts. SMITH and PICKELS (*43, 44*) found on the contrary that digitonin as well as bile salt treatment cleaves chlorophyll from the protein, and that the pigment-free protein may then be sedimented with an apparent molecular weight of only 265000. Sodium dodecyl sulfate cleaves the protein into small fragments although the pigments remain attached. In summary, it is not clear that any mechanical treatment thus far applied has satisfactorily resolved the granum into molecularly dispersed material representative of the original granum in composition. The use of detergents can apparently result in the breakdown of grana fragments into still smaller units. The size of the units obtained as well as their chemical nature (pigment-containing or pigment-free) appears to depend on the detergent used.

It is clear that the granum has associated with it enzymatic properties which suggest the presence of several different types of enzymes. Thus the chloroplasts of some species contain catalase (*33*), an iron-porphyrin enzyme; while a copper enzyme, polyphenoloxidase, is present in the chloroplasts of tea (*25*) and chard (*1*). Grana preparations also contain the enzyme system for the HILL reaction, the photolytic decomposition of water with attendant reduction of ferric iron or of various quinones (*18*, *19*). Starch phosphorylase is also present in the chloroplasts according to YIN (*59*). These enzymatic activities are firmly associated with the grana structure, and the enzymes have not as yet been extracted from the granum and obtained in true solution.

In summary, methods for the preparation of chloroplasts and chloroplastic protein are available and the prepared material may exhibit varied kinds of enzymatic activity. Complete chemical characterization of chloroplast composition and constituents remains as an important unfinished task, while the resolution of the chloroplasts into soluble individual protein constituents presents also an important problem.

The Cytoplasmic Proteins.

The proteins of the cytoplasm of the leaf cell differ from those of the chloroplasts in being freely soluble and may hence be separated and studied by the usual methods of protein chemistry. The cytoplasmic protein fraction is conveniently obtained from ground leaf brei by a preliminary filtration or centrifugation to remove cell wall fragments. The chloroplasts and grana may next be removed by centrifugation at 20000 × gravity. The resultant brownish solution contains not only the cytoplasmic protein but, of course, also the other low-molecular constituents of the cell, sugars, organic acids, amino acids, etc. These may be removed by dialysis or the protein may be precipitated from the solution. The cytoplasmic proteins of several species are denatured and precipitated at $p_H < 4$, and this method may be used if the native protein is not required. Certain of the cytoplasmic proteins may also be precipitated from the whole cytoplasm by ammonium sulfate as will be discussed below. In any case the yield of total cytoplasmic protein obtained has been found to amount to approximately 8 grams of protein per kilo of fresh leaves of spinach, tobacco and others (*54*, *58*).

The amino acid composition of whole cytoplasmic protein of several leaves has been subjected to analysis by various workers including

Table 6. Comparative Amino Acid Composition of Whole Cytoplasmic Protein of Leaves of Various Species [CHIBNALL (*11*)]. (Figures are given as % of total protein N).

Amino acid	Zea mays	Ricinus communis	Phaseolus multiflorus	Dactylis glomerata	Spinacea cytoplasmic	Oleracea chloroplastic
Amide	5,4	5,1	5,4	5,0	5,6	5,1
Arginine	14,4	12,9	14,9	15,5	14,1	13,9
Histidine	2,1	2,2	2,6	2,3	2,2	3,3
Lysine	6,1	6,5	6,1	6,0	6,2	4,7
Tyrosine	2,3	2,6	2,5	2,3	2,7	2,6
Tryptophane	1,6	1,7	1,6	1,8	1,7	1,7
Cystine	1,1	1,5	1,1	1,4	1,4	1,2
Methionine	1,3	1,4	1,1	1,2	1,3	1,3
Aspartic acid	—	5,6	5,2	4,9	5,5	5,8
Glutamic acid	—	6,7	6,7	7,8	6,5	6,5

CHIBNALL (*11*). The results of these analyses, of which a portion are given in Table 6, suggest that the amino acid composition of the cytoplasmic protein of leaves of different species may be remarkably similar. In all cases the protein is characterized by a high content of arginine (about 14–15%), by relatively low content of tryptophane (1,6–1,8%), and by a moderate content of lysine (about 6%). It might be noted also that CHIBNALL (*11*) has shown that the amino acid compositions of the cytoplasmic and chloroplastic proteins of a given species are likewise very similar, at least in the case of spinach.

The cytoplasmic protein of the leaf represents not a homogeneous protein but rather a mixture which may be resolved by electrophoresis and separated by fractional precipitation with salt. The latter method as applied to spinach (*54*) or to tobacco (*58*) leaves results in a sharp separation into two fractions. The more insoluble fraction is precipitated by 0,32–0,38 saturated ammonium sulfate provided that the precipitation is carried out at 0° C. (or lower), and provided further that the initial concen-

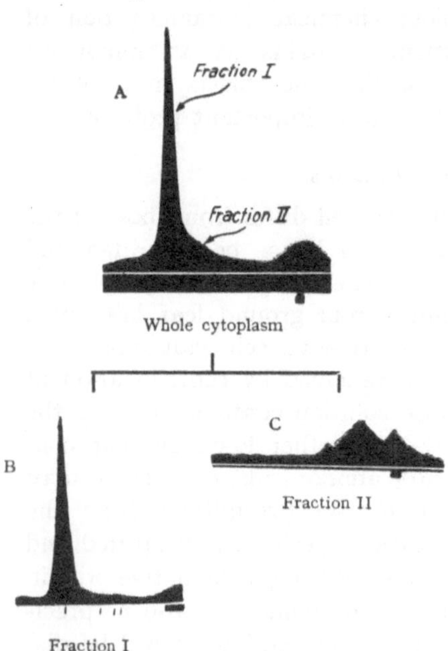

Fig. 3. Electrophoretic scanning patterns of whole cytoplasmic proteins of spinach leaves and of the two fractions obtained from the whole protein precipitated with 0,38 saturated ammonium sulfate. Ascending boundaries only. The blocks at right indicate starting boundaries. After WILDMANN and BONNER (*54*). [From: Arch. Biochemistry *14*, 381 (1947).]

tration of protein is 1% or higher. This fraction, which may be purified by re-precipitation, is electrophoretically homogeneous (*54*); a reproduction of typical scanning patterns obtained from the re-precipitated material is given in Figure 3. This protein, which has been termed merely "Fraction I" protein (*54*) makes up 75–80% of the total cytoplasmic protein in spinach but may amount to as much as 90% in other cases, e.g. in tobacco leaves (*58*). The 10–20% of cytoplasmic protein not precipitated by 0,38 saturated ammonium sulfate, "Fraction II", has been recovered by dialysis of the supernatant solution from the precipitation of "Fraction I". "Fraction II" contains a considerable number of constituents, at least five or six major components being recognizable in the electrophoretic scanning pattern.

The mixture of proteins contained in "Fraction II", (the more soluble of the two principal fractions of spinach cytoplasmic protein), contains numerous enzymes and the activities of isocitric, malic, glutamic and alcohol dehydrogenases, fumarase, aconitase, peroxidase, catalase, polyphenoloxidase and others have been recognized in this material (*8*, *54*). "Fraction I" on the other hand has as yet only one recognized enzymatic activity; it functions as a phosphatase and is capable of the hydrolysis of ATP and of other energy-rich phosphate bonds. It is also active, although less so, toward glycerophosphate and other ester phosphates. Adenosine-triphosphatase activity appears to be closely associated with "Fraction I" protein, and attempts to separate enzyme activity from this protein by fractional precipitation have thus far proved unsuccessful.

"Fraction I" protein of spinach and tobacco leaves contains bound purine derivatives, pentose, and phosphate, these moieties occurring in the ratio of 1:1:2 (*56*, *57*). The whole complex may be split from the protein by one minute of heating in *N*-HCl at 100°, although further heating is necessary to release the phosphate as inorganic phosphate. The two phosphate residues may be differentiated on the basis that one is relatively labile to acid hydrolysis and is completely removed by sixteen minutes heating in *N*-HCl at 100°, while the other is relatively stable to acid hydrolysis as are ester phosphates (*57*). The purine component appears to represent mainly adenine. In many respects, although not in all, "Fraction I" appears to represent then a nucleoprotein. Complete elucidation of the nature of the nucleotide component of "Fraction I" has not, however, been achieved.

A further interesting fact concerning "Fraction I" protein is that in spinach leaves at least, this protein contains associated with it an *auxin*, viz. indole-acetic acid (*54*). The indole-acetic acid is firmly bound, is not removable by dialysis and can be freed only by digestion of the protein by proteolytic enzymes or by alkaline hydrolysis. The amounts of indole-acetic acid recovered by these methods are quite small, being of the order of 0,1 microgram per milligram of protein (*54*). Bound indole-acetic acid does not appear to occur in otner fractions of the cytoplasmic protein, although certain animal proteins do yield indole-acetic acid on prolonged hydrolysis (*38*). While it is possible that the action of indole-acetic acid in promoting plant growth responses may involve a basic effect on the metabolism of energy-rich phosphate (*5, 6, 7*), and even though "Fraction I" contains indole-acetic acid and also contains an adenosine-triphosphatase, it has not yet been possible to directly link indole-acetic acid to the enzymatic activity or function of "Fraction I".

The component cytoplasmic proteins of leaves of other species resemble qualitatively the proteins of spinach and tobacco in so far as they have been investigated. Thus, the electrophoretic scanning patterns of the soluble proteins of pea and bean leaves (*13*) show one principal component together with more slowly moving minor components. Leaves of cucumber and Chinese cabbage also exhibit scanning patterns of a closely similar nature (*55*). It would appear therefore that the presence of a single major or bulk component is a property common to a number of leaf cytoplasms.

Cytoplasmic Proteins of Leaves in Relation to Virus Formation.

A special case of aberrant protein metabolism of leaves is constituted by virus formation in leaves infected with a virus disease. A remarkably large proportion (25% or more) of the total protein of tobacco leaves infected with tobacco mosaic virus can be recovered as crystalline virus protein a few weeks after inoculation. This massive synthesis of virus protein is accomplished without an increase in the total protein of the leaf (*31*). Production of the virus protein must therefore be conducted at the expense of the normal proteins of the leaf.

WILDMAN *et al.* (*58*) have addressed themselves to the problem of describing the behavior of the constituent normal leaf proteins during the synthesis of virus in infected leaves. The methods used in this investigation were similar to those described above for separation of the chloroplastic and cytoplasmic protein fractions of normal healthy leaves.

The general technique consisted in infection of the apical leaf of tobacco plants (variety Havana or Turkish). The mature lower leaves were then removed periodically at times from 0 to 20 days after infection, the leaf samples ground etc., and the individual protein components subjected to determination and electrophoretic examination. It was found that within three days after infection a new protein component became apparent in the electrophoretic scanning pattern of the cytoplasmic protein, as can be seen in Figure 4. This new component was separated from the other proteins by high speed centrifugation or separated directly in the electrophoretic cell and proved to be infectious. It is hence apparently the tobacco mosaic protein itself. With increasing time after inoculation, the virus component increases rapidly in amount for approximately 16 days, after which time it attains a relatively stable concentration at 35–40% of the total cytoplasmic proteins. The scanning diagrams of Figure 4 show, furthermore, that associated with this rapid increase in virus protein is a corresponding decrease in the major component of the cytoplasmic protein, viz. "Fraction I". The minor constituents appear on the contrary to be but little affected. This is true also of the chloroplastic protein of the leaves, as is shown in Table 7.

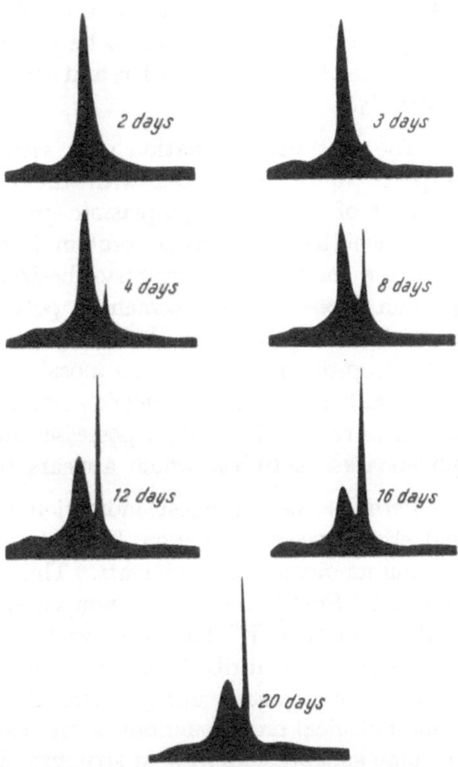

Fig. 4. Course of appearance of tobacco mosaic virus in the cytoplasm of the leaves of Turkish tobacco plants, as followed by electrophoretic scanning pattern of whole cytoplasmic protein of leaves at various times after infection. 1,0% solutions of protein. Arrow represents starting boundary. Time of migration 120 minutes. Descending boundaries only [WILDMAN, CHEO and BONNER (*58*)]. [From: J. biol. Chemistry *180*, 985 (1949).]

Electrophoretic detection of virus in leaves has also been carried out by FRAMPTON and TAKAHASHI (*13*). The scanning patterns obtained by these workers are, however, complicated by the presence of large

Table 7. Effect of Tobacco Mosaic Virus Infection on Protein Composition of Leaves of Havana 38 Tobacco [WILDMAN, CHEO and BONNER (*58*)].

State of leaves	Days after virus infection	Dry weight of leaves %	Dry weight of leaves extracted as cell free juice %	mg. per gram dry weight of leaves		Virus in % of total cytoplasmic proteins
				chloroplasmic protein	cytoplasmic protein	
Normal	5	11,0	64,8	146	87	
Virus infected ...	5	10,8	64,4	144	80	0
Normal	12	10,9	65,7	145	81	
Virus infected ...	12	11,4	63,7	136	84	20
Normal	17	11,2	64,1	161	70	
Virus infected ...	17	12,4	64,4	146	78	40

amounts of a relatively non-mobile component. This non-mobile component has been shown to be related, although in an unknown way, to the use of phosphate buffer, and can be avoided by the use of cacodylate buffer (*59*).

The fact that formation of a virus—tobacco mosaic in this case—appears to be associated with the disappearance of a corresponding amount of a normal cytoplasmic protein suggests at once the question as to whether the normal protein is partially or wholly degraded and then the component parts resynthesized to virus, or whether the normal protein molecules are somehow polymerized to virus with but little alteration. It has been shown above that the normal "Fraction I" is a nucleoprotein as is tobacco mosaic virus, although direct comparisons of the nature of the two nucleic acids are not yet available. On the other hand, „Fraction I" protein possesses an associated enzyme activity, e. g., phosphatase activity, which appears to be absent from the virus.

Perhaps the strongest indication of a close relationship in overall pattern between „Fraction I" protein and tobacco mosaic virus rests on immunological experiments. Thus a specific antibody prepared to purified "Fraction I" (from non-virus infected plants) reacts not only with "Fraction I" but also with crystalline tobacco mosaic virus. Conversely, the antibody to crystalline tobacco mosaic reacts not only with its homologous antigen but also with "Fraction I" (*59*). These immunological cross-reactions of the two proteins are strongly indicative of some general similarity in structure and surface configuration between them.

V. Proteins of Tissues Other than Leaves.

The cytoplasmic proteins of tissues other than leaves have been investigated by modern methods only in a relatively few cases. Thus electrophoretic examination of the proteins of the potato tuber reveals one principal component and additional minor components as in leaves (*13*). The same general methods used in the leaf work have, in addition, been applied to the coleoptiles of etiolated oat seedlings (*56*). In this case also, the protein of the tissue contains associated with it phosphorous which is in part labile, in part stable and resistant to hydrolysis as in the leaf. Although no published work has as yet described the separation and study of seedling proteins, still it is clear that these proteins must differ from reserve proteins of the seed in composition and physical properties. BRIGGS and SIMINOVITCH (*9a*) have prepared the cytoplasmic proteins of tree bark and have shown by electrophoretic examination that it consists of five distinguishable components. Development of cold-resistance is associated with increase in three of these components.

Extension of the methods so successfully applied to leaves for detailed study of the proteins of other tissues will be an important future task of the plant chemist.

VI. Conclusion.

The application of electrophoretic and ultracentrifuge techniques to the study of some seed proteins has shown that these materials are clearly heterogeneous and that even such well-defined proteins as glutenin, gliadin, or the seed globulins may in fact be made up of several component protein species. Whether this fact has biochemical significance is still obscure. A measure of insight into the biochemistry of the plant has however been afforded by the investigations of recent years on the leaf proteins. The quantitative methods now available permit the ready preparation of both the particulate chloroplastic and soluble cytoplasmic proteins of leaves. The cytoplasmic proteins have been further separated into fractions differing in physical and enzymatic properties. The study and separation of these various components of the plant cell has had and is finding its application in such problems of plant biochemistry as the study of photosynthesis, of respiration and plant metabolism, and finally, in the study of virus reproduction.

References.

1. ARNON, D. L.: Copper Enzymes in Isolated Chloroplasts. Polyphenoloxidase in *Beta vulgaris*. Plant Physiol. **24**, 1 (1949).
2. BAILEY, K.: The Denaturation of Edestin by Acid: T. B. Osborne's Edestan. Biochemic. J. **36**, 140 (1942).
3. BARMORE, M. A.: Some Characteristics of Gliadin and Glutenin Indicated by Dispersion and Viscosity. Cereal Chem. **24**, 49 (1947).
4. BLOCK, R. and D. BOLLING: Amino Acid Composition of Proteins and Foods. Springfield, Ill.: C. C. Thomas. 1945.
5. BONNER, J.: Arsenate as a Selective Inhibitor of Growth Substance Action. Plant Physiol. **25**, 181 (1950).
6. — Relations of Respiration and Growth in the *Avena* Coleoptile. Amer. J. Bot. **36**, 429 (1949).
7. — Limiting Factors and Growth Inhibitors in the Growth of the *Avena* Coleoptile. Amer. J. Bot. **36**, 623 (1949).
8. BONNER, J. and S. G. WILDMAN: Enzymatic Mechanisms in the Respiration of Spinach Leaves. Arch. Biochemistry **10**, 497 (1946).
9. BOT, G. M.: The Chemical Composition of Chloroplast Granules (Grana) in Relation to their Structure. Chronica Botanica **7**, 66 (1942).

9a. BRIGGS, D. and D. SIMINOVITCH: The Chemistry of the Living Bark of the Black Locust Tree in Relation to Frost Hardiness. II. Arch. Biochemistry **23**, 18 (1949).

10. BROWN, W. L.: The Threonine, Serine, Cystine and Methionine Content of Peanut Proteins. J. biol. Chemistry **142**, 299 (1942).

11. CHIBNALL, A.: Protein Metabolism in the Plant. Yale Univ. Press. 1939.

12. COMAR, C. L.: Chloroplast Substance of Spinach Leaves. Bot. Gaz. **104**, 122 (1942).

13. FRAMPTON, V. L. and W. TAKAHASHI: Electrophoretic Studies with the Plant Viruses. Phytopathology **36**, 129 (1946).

14. GRANICK, S.: In „Photosynthesis in Plants." Ames: Iowa State College Press. 1949.

15. — Quantitative Isolation of Chloroplasts from Higher Plants. Amer. J. Bot. **25**, 558 (1938).

16. — Chloroplast Nitrogen of some Higher Plants. Amer. J. Bot. **25**, 561 (1938).

17. HANSON, E. A., B. S. BARRIEN and J. G. WOOD: Relations Between Protein-Nitrogen, Protein-Sulphur and Chlorophyll in Leaves of Sudan Grass. Austral. J. exp. Biol. med. Sci. **19**, 231 (1941).

18. HILL, R.: Oxygen Evolved by Isolated Chloroplasts. Nature (London) **139**, 881 (1937).

19. HOLT, A. S. and C. S. FRENCH: In „Photosynthesis in Plants." Ames: Iowa State College Press. 1949.

20. IRVING, G. W., T. D. FONTAINE and R. C. WARNER: Electrophoretic Investigations of Plant Proteins. I. Peanut Meal Extract, Arachin and Conarachin. Arch. Biochemistry **7**, 475 (1945).

21. — — — Electrophoretic Investigations of Plant Proteins. II. Composition of Several Peanut Protein Fractions. Arch. Biochemistry **8**, 239 (1945).

22. JOHNS, C. O. and D. B. JONES: The Proteins of the Peanut, *Arachis hypogaea*. I. The Globulins Arachin and Conarachin. J. biol. Chemistry **28**, 77 (1916).

23. JOHNSON, P.: The Proteins of the Ground-nut *(Arachis hypogaea)*. I. The Isolation and Properties of the Proteins. Trans. Faraday Soc. **42**, 28 (1946).

24. KREJCI, L. and T. SVEDBERG: The Ultracentrifugal Study of Gliadin. J. Amer. chem. Soc. **57**, 946 (1935).

25. LI, L. P. and J. BONNER: Experiments on the Localization and Nature of Tea Oxidase. Biochemic. J. **41**, 105 (1947).

26. LIEBICH, H.: Quantitativ-chemische Untersuchungen über das Eisen in den Chloroplasten und übrigen Zellbestandteilen von *Spinacia oleracea*. Z. Bot. **37**, 129 (1941).

27. LUGG, J. W. H. and R. A. WELLER: Large-scale Extraction of Protein Samples Reasonably Representative of the Whole Proteins in the Leaves of some Plants: the Amide, Tyrosine, Tryptophane, Cystine (plus Cysteine) and Methionine Contents of the Preparations. Biochemic. J. **38**, 408 (1944).

28. MCCALLA, A. G.: Nitrogenous Constituents of Plants. Annu. Rev. Biochem. **18**, 615 (1949).

29. MCCALLA, A. G. and N. GRALÉN: Ultracentrifuge and Diffusion Studies on Gluten. Canad. J. Res., Sect. C **20**, 130 (1942).

30. MCCALLA, A. G. and R. C. ROSE: Fractionation of Gluten Dispersed in Sodium Salicylate Solution. Canad. J. Res., Sect. C **12**, 346 (1935).

31. MARTIN, L. F., A. K. BALLS and H. H. MCKINNEY: The Protein Content of Mosaic Tobacco. Science (New York) **87**, 329 (1938).

32. MENKE, W.: Untersuchung der einzelnen Zellorgane in Spinatblättern auf Grund präparativ-chemischer Methodik. Z. Bot. **32**, 273 (1938).

33. NEISH, A. C.: Studies on Chloroplasts. I. Separation of Chloroplasts, A Study of Factors Affecting Their Flocculation and the Calculation of the Chloroplast Content of Leaf Tissue From Chemical Analysis. Biochemic. J. **33**, 293 (1939).

34. NEURATH, H.: The Apparent Shape of Protein Molecules. J. Amer. chem. Soc. **61**, 1841 (1939).

35. OSBORNE, T. B.: The Vegetable Proteins, 2nd. ed. London: Longmans, Green & Co. 1924.

36. RABINOWITCH, E.: Photosynthesis. I. New York: Interscience Publ. 1945.

37. SCALLET, B. L.: Zein Solutions as Association-Dissociation Systems. J. Amer. chem. Soc. **69**, 1602 (1947).

38. SCHOCKEN, V.: The Genesis of Auxin During the Decomposition of Proteins. Arch. Biochemistry **23**, 198 (1949).

39. SCHWERT, G., F. PUTMAN and D. RIGGS: An Electrophoretic Study of Gliadin. Arch. Biochemistry **4**, 371 (1944).

40. SMITH, E.: Solutions of Chlorophyll-Protein Compounds Extracted from Spinach. Science (New York) **88**, 170 (1938).

41. — The Chlorophyll-Protein Compound in the Green Leaf. J. gen. Physiol. **24**, 565 (1941).

42. — The Action of Sodiumdodecyl Sulfate on the Chlorophyll-Protein Compound of the Spinach Leaf. J. gen. Physiol. **24**, 583 (1941).

43. SMITH, E. and E. PICKELS: Micelle Formation in Aqueous Solution of Digitonin. Proc. nat. Acad. Sci. U. S. A **26**, 272 (1940).

44. — — The Effect of Detergents on the Chlorophyll-Protein Compound of Spinach as Studied in the Ultracentrifuge. J. gen. Physiol. **24**, 753 (1941).

45. SMITH, E. L. and R. D. GREEN: Further Studies on the Amino Acid Composition of Seed Globulins. J. biol. Chemistry **167**, 833 (1947).

46. STOLL, A. u. E. WIEDEMANN: Chlorophyll. Fortschr. Chem. organ. Naturstoffe **1**, 159 (1938).

47. STOLL, A., E. WIEDEMANN u. A. RÜEGGER: Zur Kenntnis des Chloroplastins. Verh. Schweiz. Naturf. Ges. **1941**, 125.

48. SUMNER, J. B.: The Isolation and Crystallization of the Enzyme Urease. J. biol. Chemistry **69**, 435 (1926).

49. SVEDBERG, T. u. I. ERIKSON-QUENSEL: Molekülkonstanten der Eiweißkörper. Tabulae biol. (Den Haag) **5**, 351 (1935).

50. VENNESLAND, B. and R. Z. FELSHER: Oxaloacetic and Pyruvic Carboxylases in Some Dicotyledonous Plants. Arch. Biochemistry **11**, 279 (1946).

51. VICKERY, H. B.: The Proteins of Plants. Physiologic. Rev. **25**, 347 (1945).

52. VICKERY, H. B., E. L. SMITH, R. B. HUBBELL and L. NOLAN: Cucurbit Seed Globulins. I. Amino Acid Composition and Preliminary Tests of Nutritive Value. J. biol. Chemistry **140**, 613 (1941).

53. WETTER, L. and A. G. McCALLA: Canad. J. Res. (in press). Referred to in McCALLA, A. G.: Nitrogenous Constituents of Plants. Annu. Rev. Biochem. **18**, 615 (1949).

54. WILDMAN, S. G. and J. BONNER: The Proteins of Green Leaves. I. Isolation, Enzymatic Properties and Auxin Content of Spinach Cytoplasmic Proteins. Arch. Biochemistry **14**, 381 (1947).

55. — — Electrophoretic Analysis of Normal and Diseased Leaf Cytoplasmic Proteins as a Means of Identifying Plant Virus. Amer. J. Bot. **36**, *930* (1949).

56. WILDMAN, S. G., J. CAMPBELL and J. BONNER: Tissue-bound Acid-labile Phosphorous in Tobacco Leaves, Oat Coleoptiles and Rat Liver. J. biol. Chemistry **180**, 273 (1949).
57. — — — The Proteins of Green Leaves. II. Purine, Pentose, Total Phosphorus and Acid-labile Phosphorus of the Cytoplasmic Proteins of Spinach Leaves. Arch. Biochemistry **24**, 9 (1949).
58. WILDMAN, S. G., C. C. CHEO and J. BONNER: The Proteins of Green Leaves. III. J. biol. Chemistry **180**, 985 (1949).
59. YIN, H. C.: Phosphorylase in Plastids. Nature (London) **162**, 928 (1948).

(Received, January 12, 1950.)

Progrès récents en spectrochimie de fluorescence des produits biologiques.

Par CH. DHÉRÉ, Genève.

Avec 10 figures.

Sommaire.

I. Avant-propos.

Le présent article constitue un complément de la revue intitulée: «La spectrochimie de fluorescence dans l'étude des produits biologiques» que j'ai publiée dans ces *Fortschritte* en 1939. Je l'ai rédigé en me plaçant encore au même point de vue, c'est-à-dire en considérant l'émission par fluorescence en tant que connue dans sa composition spectrale: il s'agit donc toujours d'un exposé de «Spectrofluoroscopie».

D'assez nombreux travaux de fluoroscopie spectrale ont été publiés dans ces dix dernières années. Plusieurs comprennent des déterminations spectrophotométriques. D'intéressantes recherches ont même été faites en 1941/42, par JANSEN (*58*), dans le domaine difficile de la microspectrophotométrie de fluorescence, avec applications en histologie. Un travail postérieur de SJÖSTRAND (*111*) est intéressant au même point de vue. Voir aussi le livre récent de PRINGSHEIM (*93bis*).

Fallait-il s'en tenir rigoureusement à ce cadre ainsi délimité et omettre par conséquent de citer toutes les récentes observations sur la fluorescence où l'examen spectroscopique n'a pas encore été effectué? Je n'ai pas cru devoir le faire dans tous les cas, cela pour diverses raisons: il m'a semblé utile notamment d'attirer parfois l'attention sur ces lacunes.

On constatera que j'ai fait une large place aux réactions de fluorescence qui, particulièrement dans le domaine des stéroïdes, fournissent souvent des résultats bien remarquables. Un certain nombre de ces réactions peuvent du reste servir pour les dosages et sont donc importantes à connaître à ce point de vue.

Enfin un Appendice a été consacré à la question bien distincte de la *topographie spectrale des radiations excitant les fluorescences**, question qui, dans notre bref exposé, a été envisagée à part et en elle-même autant que possible, malgré toutes ses relations avec l'étude des spectres d'émission par fluorescence.

Le lecteur s'étonnera peut-être de trouver, dans cet article, si peu de renseignements sur l'intéressante question des relations qui existent entre la fluorescence et la constitution chimique des corps bien divers dont il est parlé. C'est, il faut le dire, un sujet encore aujourd'hui fort peu connu. Pour s'en convaincre, il n'y a qu'à lire, par exemple, la série des Mémoires de D. BERTRAND (*4*) où seule, il est vrai, la fluorescence visible est généralement prise en considération. Outre de nombreuses remarques critiques, les Mémoires de BERTRAND contiennent d'intéressantes vues originales pour l'interprétation théorique des spectres de fluorescence.

---

* LEWIS (*70*), en 1944, a proposé de réserver le terme de «Spectrofluorescence» pour désigner précisément cette question de la fluorescence le long du spectre excitateur. Nous sommes d'accord avec lui.

Il ne semble pas que le moment soit encore venu de donner un exposé systématique sur ce sujet, la plupart des conclusions formulées jusqu'à présent ne pouvant être considérées que comme provisoires et souvent fort discutables.

II. Hydrocarbones (glucides) et glucosides.

Dans l'ouvrage publié en 1945 par DE MENT (*74*), se trouvent indiqués quelques spectres de fluorescence de glucides*:

Émission par fluorescence (solutions aqueuses)	Xylose	λ 380	à	λ 330 mμ
	Lévulose	380	à	330
	Méthylglucose	400	à	275
	Mannose	380	à	330
	Maltose	380	à	330
	Inosite	500	à	275
	Acide ascorbique	400	à	380
	Dextrine	400	à	275

Il nous a semblé qu'il y avait lieu d'introduire ici les chiffres pour l'acide ascorbique (vitamine C), étant donné que cet acide peut être préparé à partir de divers sucres (xylose, sorbose, etc.).

Les valeurs ci-dessus sont-elles vraiment caractéristiques (au moins d'un groupe)? Nous ne le croyons pas du tout; mais, après quelques hésitations, nous nous sommes tout de même décidé à les citer.

Notons encore que, d'après les observations de FABER et RADNOT (*40*), la fluorescence blanche du glycogène est bien vive et différente de celle des polysaccharides végétaux tels que les divers amidons et la dextrine.

Escorcéine.

De tous les glucosides, c'est l'esculine (glucoside coumarique de l'*Aesculus hippocastanum*) que les physiciens ont surtout étudiée comme corps fluorescent. L'esculétine (esculétol), aglucone de l'esculine, est intéressante au même point de vue. Mais l'escorcéine, produit de sulfonation de l'esculétine, possède une fluorescence encore beaucoup plus

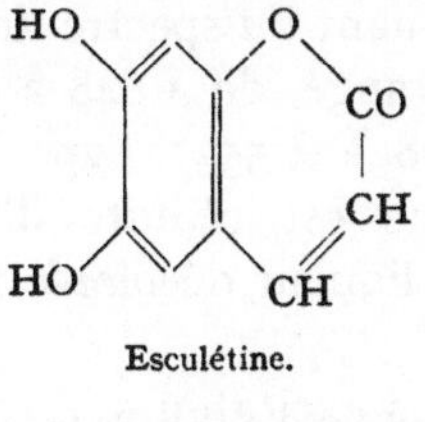

HO O
CO
CH
HO CH

Esculétine.

* Ces déterminations, ainsi que d'autres (aminoacides) qui seront consignées par la suite ont été obtenues, dit l'auteur, par ELLICE MCDONALD, Director of the Biochemical Research Foundation, Newark, Delaware (déterminations spectrographiques, emploi des rayons filtrés de l'arc au cadmium). Ces résultats seraient restés inédits jusqu'en 1945.

forte et beaucoup plus remarquable. DHÉRÉ et BÉRENSTEIN (*28*) ont étudié les propriétés spectrales d'absorption et de fluorescence de l'escorcéine, en opérant sur des solutions acides, alcalines et neutres, et ont été amenés à distinguer trois types de fluorescence:

1° *Solutions acides.* Elles sont orangées et transmettent, sans affaiblissement notable, tous les rayons visibles jusqu'au milieu environ de la plage verte, où commence une assez large bande d'absorption, peu intense, à bords flous. Le spectre de fluorescence est constitué par une bande débutant dans le rouge et finissant dans le vert: Type I (de fluorescence).

Solvant:	Bande d'absorption:	Bande de fluorescence:
Acide sulfurique	λ 541 à λ 474 mμ	λ 649 à λ 558 mμ
Acide acétique	528 à 466	648 à 553

Dans l'acide acétique glacial, l'émission se prolonge jusqu'à λ 547.

2° *Solutions alcalines.* Elles possèdent une très belle couleur pourpre et présentent trois bandes d'absorption. Contrairement à ce que montrent les solutions acides, il y a une remarquable transparence dans la plage la plus réfrangible, à partir de λ 540 environ. Voici les déterminations des bandes d'absorption pour une solution dans la soude *n*/100:

Bande I λ 606 à λ 585; bande II λ 578,5 à λ 567,5; bande III λ 552,5 à λ 541.

La bande I est la plus sombre; la bande II y est presque accolée. La fluorescence est extrêmement forte et d'un rouge orangé; le spectre diffère nettement de celui des solutions acides en ce que la bande brillante est beaucoup plus courte; elle ne va même pas tout à fait jusqu'au jaune proprement dit. Cette bande est comprise entre λ 649 et λ 594: Type II. (On aperçoit de plus dans la plage vert bleu, à partir de λ 548, une faible lueur qu'on pourrait à la rigueur considérer comme une deuxième bande.)

3° *Solutions neutres.* Dans l'eau distillée, on obtient une liqueur orangée dont le spectre d'absorption est, peut-on dire, une combinaison des spectres alcalin et acide. En effet, après les trois bandes des liqueurs alcalines (la 2e et la 3e à peine visibles), on observe une quatrième bande correspondant à la bande unique des solutions acides, quatrième bande étendue de λ 530 à λ 468. Quant au spectre de fluorescence, il comprend deux bandes: l'une dans l'orangé, de λ 648 à λ 591; l'autre (moins brillante) dans le vert, de λ 578 à λ 553: Type III.

Ajoutons que l'escorcéine est réduite d'une façon réversible par l'hydrosulfite de sodium; la liqueur décolorée ne montre plus de fluorescence orangée.

Pour ce qui est des rayons excitateurs, il est intéressant de noter que la fluorescence est excitée par les rayons violets et ultraviolets; et que, de plus, la double raie jaune de la lumière filtrée du sodium ainsi que les deux raies Hg λ 579 et λ 577 mμ excitent fortement la fluorescence des solutions alcalines, mais non celle des solutions acides.

Rutine.

Ce glucoside de quercétine (3,5,7,3',4'-pentaoxyflavone-3-rutinoside), — extrait des feuilles de *Ruta graveolens*, des fleurs de *Sambucus canadensis*, etc., et qui est maintenant utilisé en thérapeutique, — a été étudié, en vue du dosage, par GLAZKO (*47*) dans sa réaction de fluorescence après addition d'acide borique. Les résultats fluorométriques seraient, à plusieurs points de vue, supérieurs aux résultats colorimétriques. La fluorescence est de couleur verte; analysée au spectroscope, elle montre une bande lumineuse comprise entre λ 620 et λ 510 mμ.

III. Stérols et stéroïdes.

On sait depuis longtemps que le *cholestérol* est fluorescent, l'émission étant comprise entre λ 500 et λ 380 mμ, d'après ROFFO (*102*) [entre 536 et 425 mμ, d'après VLÈS (*125*)].

Dans une longue série de publications, ROFFO a examiné la question de savoir s'il existe une dépendance entre l'apparition de certaines lésions cancéreuses cutanées et la formation, en présence d'oxygène, d'un photodérivé du cholestérol (production d'oxycholestérol) remarquablement fluorescent. Bien des résultats anciens de ROFFO, notamment d'ordre physicochimique (que nous avons seuls à considérer ici) ont été fortement critiqués et contestés*; et, au point de vue chimique, on peut dire que la question a été supérieurement et définitivement mise au point en 1941 grâce aux recherches approfondies de WINDAUS (*126*). Mais il convient d'exposer les résultats obtenus quant aux spectres de fluorescence, par ROFFO Jr. (*102*), dans un travail spécial ayant paru à la fin de 1941. L'auteur a opéré sur du cholestérol irradié, pendant 96 heures, avec une lampe de quartz à vapeurs de cadmium. Cet échantillon, en solution benzénique, possédait un spectre de fluorescence à trois bandes, les maximums étant sur λ 446, λ 428 et λ 405 mμ (minimums sur 442,5 et sur 416), bandes que ne présentait pas une solution benzénique, de même concentration, de cholestérol normal (non irradié).

Pour le *1:2-benzopyrène*, très puissamment cancérigène, trois bandes aussi furent déterminées: sur λ 446, λ 426 et λ 414 mμ. ROFFO dit qu'il est très intéressant que chacune de ces deux substances présente trois bandes de fluorescence et qu'elles ont une de ces bandes en commun: celle dont le maximum est sur λ 446 mμ. Il y a dans ce travail une courbe hors texte ainsi que deux reproductions de spectrogrammes de fluorescence.

* Consulter à ce sujet les travaux de BERGMANN (*3*), de KIRBY (*61*) et de PENN (*85*). La communication de PENN (avec reproduction de spectrogrammes de fluorescence) a été critiqué par HIEGER (*53*).

En 1943, KIRBY (*61*) a repris l'étude de la question, mais en considérant spécialement un dérivé du cholestérol obtenu par pyrolyse (chauffage à 270°—300°), dérivé possédant une fluorescence bien prononcée bleu violet. Après purification, la solution dans le *n*-hexane montre de larges bandes de fluorescence sur λ 440, λ 420 et λ 395 mμ. KIRBY fait remarquer que les trois bandes de fluorescence de son produit sont assez voisines de celles appartenant au *20-méthylcholanthrène* (λ 445, λ 425 et λ 400 mμ) en solution dans l'éthanol.

Réaction de SALKOWSKI.

Dans cette réaction classique du cholestérol, l'acide sulfurique montre une forte fluorescence verte bien connue. Mais jusqu'aux recherches de MERKELBACH (*76*) en 1943, on ignorait l'existence d'une fluorescence rouge orangé pour la portion chloroformique. MERKELBACH a fait une étude soignée de ces deux fluorescences, dont il a publié d'excellents spectrogrammes. Il a, de plus, étudié comparativement la fluorescence de l'*acide cholalique.*

Les conditions d'apparition de la fluorescence de l'acide cholalique ont été bien déterminées par BANDOW (*1 bis*).

Pour la fluorescence rouge orangé (couche chloroformique), en utilisant comme rayons excitateurs une plage de radiations vertes, l'émission (modérément intense) s'étendait de λ 670 à λ 580 mμ. Avec LASZT, j'ai repris cette étude (*35*); et, en excitant simplement avec la raie isolée Hg 546 mμ (ou avec les deux raies Hg 577 et 579 mμ), nous avons obtenu des résultats semblables. Sur un point cependant, nous ne sommes pas entièrement d'accord avec MERKELBACH. Celui-ci avait constaté l'existence dans ce spectre d'émission d'une légère discontinuité (non repérée en λ) et avait supposé que l'on avait affaire en réalité à deux bandes de fluorescence (non délimitées séparément par l'auteur). Or, d'après diverses constations que nous avons faites, il ne nous paraît guère douteux que la faible discontinuité dans le spectre lumineux soit simplement le fait de la présence d'une bande d'absorption dans la région correspondante. Nous avons étendu cette étude spectroscopique de la fluorescence dans la Réaction de SALKOWSKI à bien d'autres stérols (*36*): *sitostérol, stigmastérol, ergostérol* et *zymostérol.* Avec le sitostérol, les résultats sont presque identiques à ceux fournis par le cholestérol. Avec les autres stérols, il y a des différences dont le détail, assez complexe, ne peut trouver place ici. Elles sont indiquées dans nos publications.

Enfin un travail de MOIR (*80*) mérite d'être cité à la Bibliographie, car il contient (sans détermination des bandes de fluorescence malheureusement) bien des indications sur les phénomènes de fluorescence apparaissant dans diverses réactions de coloration des stérols; il contient aussi de nombreuses déterminations des bandes d'absorption.

Nous allons maintenant parler de la «Réaction de WINDAUS»; et, à propos de cette réaction, nous dirons quelques mots de la fluorescence de l'ergostérol traité par le sulfate diméthylique ainsi que de la fluorescence du cholestérol et de la vitamine D_2; enfin, dans l'exposé concernant les stéroïdes à fonctions hormonales, on trouvera quelques autres indications sur la fluorescence du cholestérol et de l'oxycholestérol traités par le même réactif.

Réaction de WINDAUS.

Elle est propre aux triols des vitamines D_2 et D_3. Cette magnifique réaction a été décrite par WINDAUS (*127*) en 1942. C'est, faisons-le bien

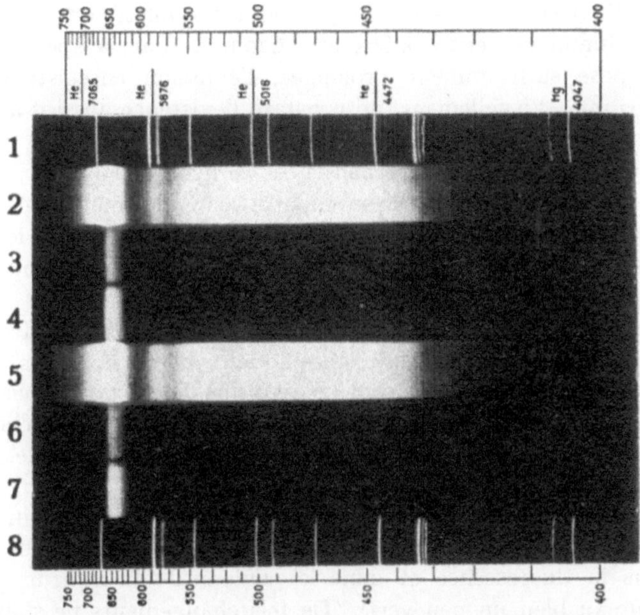

Fig. 1. Réaction de fluorescence de WINDAUS (1942) pour les triols des vitamines D_2 et D_3. Spectrographies de DHÉRÉ et LASZT (*33*), la réaction ayant été effectuée avec du sulfate diméthylique.
Triol D_2: spectres 2 (absorption), 3 et 4 (fluorescence);
Triol D_3: spectres 5 (absorption), 6 et 7 (fluorescence).
Remarque. La raie He 706,5 mμ, bien visible sur le cliché, ne peut être distinguée sur la reproduction. La première raie à gauche est He 667,8 mμ. [C. R. Séances Soc. Biol. Filiales Associées *142*, 17 (1948).]

remarquer, la première connue (dans la série des stéroïdes) des réactions caractérisées par l'apparition d'une fluorescence rouge très intense. D'après WINDAUS, on procède ainsi: dissolution du triol dans 2 cm³ d'un mélange de chloroforme et d'acide acétique glacial avec addition de 3 gouttes d'acide sulfurique concentré. La liqueur se colore en jaune, puis en vert, enfin en bleu foncé; en même temps, apparaît une magnifique et très forte fluorescence rouge. La couleur est bien stable.

Voici maintenant nos propres résultats (*32*, *33*) (Fig. 1) obtenus (avec des échantillons que WINDAUS a bien voulu nous donner) en ajoutant à des cristaux de triol D_2 uniquement du sulfate diméthylique venant d'être correctement purifié. Un début de coloration n'est visible qu'après quelques heures; d'abord jaune, elle acquiert bientôt une nuance verdâtre. Au bout d'un jour ou deux, la réaction est terminée: on a alors affaire à une liqueur franchement bleue, possédant une fluorescence rouge extrêmement forte. Le spectre d'absorption comprend deux bandes: la première, la plus intense, est située entre λ 628 et λ 600 mμ (avec une ombre importante étendue du côté du jaune); la seconde, entre λ 576 et λ 554 mμ. Comme spectre de fluorescence (excitation par des rayons de l'extrême violet et du proche ultraviolet), on observe une bande très lumineuse entre λ 663 et λ 620 mμ. Par la spectrographie, avec une pose suffisamment prolongée, l'émission enregistrée atteint même λ 667 mμ. Visuellement, on constate l'existence aussi d'une faible émission dans le vert, qui impressionne la plaque quand la durée de la pose est assez longue. Cette bande est difficile à délimiter, d'autant plus que le sulfate diméthylique montre dans la même région une fluorescence qui n'est pas négligeable. Résultats semblables avec le triol D_3.

Notons que la double raie jaune du sodium possède un grand pouvoir excitateur pour la fluorescence rouge.

Nous devons enfin remarquer que la réaction exécutée d'après les indications de WINDAUS fournit une liqueur bleue dont les spectres d'absorption et de fluorescence sont bien voisins de ceux que nous venons de décrire. On peut dire que la couleur de cette liqueur est très stable; car, pendant un jour ou deux, aucune altération n'est visible. Par la suite, il apparaît une nouvelle bande d'absorption devenant bientôt plus forte que les deux autres, qui pâlissent, et située plus loin dans le rouge (entre λ 687 et λ 661 mμ). En même temps, il survient des changements dans la fluorescence et dans la couleur de la liqueur (passant du bleu pur au bleu un peu vert). De tels changements ne s'observent qu'à un bien moindre degré et beaucoup plus tardivement (après bien des semaines) quand la réaction de WINDAUS est réalisée avec du sulfate diméthylique.

L'ergostérol (provitamine de la vitamine D_2, appelée calciférol) fournit une remarquable fluorescence verte (avec notable émission dans le rouge) par addition de sulfate diméthylique. Décrivons d'une façon précise cette réaction de l'ergostérol:

En ajoutant à froid, dans un tube à essai, du sulfate de méthyle sensiblement neutre et incolore (et en agitant un peu, de temps en temps), on ne constate directement rien de net pendant les premiers jours; mais, à partir du quatrième ou du cinquième jour, la liqueur exposée au soleil montre une belle fluorescence verte qui augmente notablement d'intensité les jours suivants. En général, pendant 8 jours au moins, la liqueur reste à peu près incolore (très peu d'ergostérol entre d'ailleurs en solution). Par la suite, la liqueur présente une légère coloration d'un rose un peu brunâtre, puis jaune brun; elle possède alors une bande d'absorption à la frontière du vert et du bleu (axe λ 492 mμ environ). Quant au spectre d'émission par fluorescence, il offre dans le vert une bande surtout lumineuse de λ 561 à λ 513 mμ (excitation par des rayons violets et ultraviolets).

Ce résultat est à rapprocher d'une constatation faite, en 1939, par CHRISTIANI et ANGER (*14*), à savoir que l'action combinée de l'acide trichloracétique et du tétracétate de plomb sur l'ergostérol, en solution chloroformique, donne naissance à une vive fluorescence verte, la liqueur ayant tout au plus une très faible coloration rouge. Ces auteurs supposent qu'il peut y avoir formation d'ergostadiène-triol.

Ils n'ont eu que des résultats négatifs avec le cholestérol, ce qui concorde jusqu'à un certain point avec nos propres résultats. En effet, avec du cholestérol commercial (même durée de réaction), il n'y avait pas, directement au soleil, de fluorescence appréciable, sauf en faisant converger les rayons avec une lentille: faible cône verdâtre (présence d'une impureté?). Une moindre solubilité ne paraît pas être la cause de cette différence.

Par rapport aux résultats fournis par l'ergostérol, ceux obtenus avec la *vitamine D_2* manquent tout à fait de netteté. On constate pourtant une faible fluorescence vaguement olivâtre.

Stéroïdes à fonctions hormonales.

Beaucoup de ces stéroïdes ont été examinés en lumière de WOOD; et, comme on l'avait fait antérieurement pour le cholestérol, on a noté les caractères de la fluorescence pour le produit à l'état cristallisé ou après traitement par l'acide sulfurique (en liqueur alcoolique, notamment). Bornons-nous à mentionner, d'une part, les résultats signalés par HARVEY (*51*) ainsi que les nôtres (forte fluorescence bleu violet des cristaux d'équiline, etc.), et, d'autre part, les recherches de BUGYI (*11*, *63*). Il resterait à déterminer les spectres de fluorescence.

Réaction des stéroïdes avec le sulfate diméthylique [Réaction de Mme ROSENHEIM (*103*)].

Il s'agit essentiellement d'une réaction de coloration dans le travail original, déjà ancien (1916), dû à Mme ROSENHEIM. L'auteur avait indiqué cette réaction comme caractéristique de l'oxycholestérol et comparé les résultats obtenus à ceux fournis par le cholestérol. Les spectres d'absorption avaient été soigneusement observés et relevés; mais l'auteur n'avait pas du tout fait une étude spéciale de la fluorescence; elle avait pourtant noté l'apparition, dans certaines conditions, d'une fluorescence bleue avec le cholestérol ainsi qu'une fluorescence verte pour l'oxycholestérol après addition d'un peu de perchlorure de fer.

C'est seulement depuis les recherches de DHÉRÉ et LASZT (*31—36*; commencées en 1946), où cette réaction a été appliquée avec de brillants résultats à beaucoup de stéroïdes (hormones corticales et sexuelles), que cette réaction est devenue vraiment une remarquable «réaction de fluorescence».

Mme ROSENHEIM opérait avec des solutions chloroformiques d'oxycholestérol ou de cholestérol. Nous avons fait aussi de nombreuses déterminations avec des liqueurs chloroformiques. Mais, étant donné que, par irradiation ultraviolette, le chloroforme se décompose facilement au contact de l'air, nous avons préféré traiter directement les stéroïdes étudiés par le seul sulfate diméthylique; c'est ainsi qu'ont été obtenus les résultats que nous allons exposer.

Nous avons constaté (*31*) qu'en ajoutant simplement du sulfate diméthylique à divers stéroïdes, on obtient des liqueurs qui, tout en étant peu colorées, présentent (dans un faisceau convergent de rayons solaires) de belles fluorescences de colorations très variées: *rouge orangé* avec la désoxy-corticostérone, *orangé jaune* avec l'équilénine, *jaune vert* avec la corticostérone, *vert un peu jaune* avec l'équiline et l'oestradiol, *vert relativement franc* avec l'oestrone. (Réactions effectuées dans des conditions identiques.) En lumière de WOOD, on peut observer une fluorescence d'un *bleu violacé* avec la méthyltestostérone, alors que la testostérone fournit une fluorescence *verte*. Plusieurs de ces fluorescences sont fortes; quelques-unes (désoxy-corticostérone, anhydro-oxy-progestérone) d'une intensité prodigieuse. On entrevoit donc les applications possibles de ces réactions aussi bien au dosage qu'à la détection chimique ou histo-chimique de ces stéroïdes, tout spécialement en utilisant, pour l'excitation, des radiations visibles sélectionnées, permettant (comme nous l'avons constaté) d'exclure, dans un mélange, la fluorescence d'un ou de plusieurs stéroïdes.

Ce n'est généralement qu'après une évolution de la réaction plus ou moins longue qu'on parvient à un terme d'une assez grande stabilité. A froid, la plupart de ces réactions évoluent lentement; en chauffant avec précaution, on amène une accélération considérable. Nous avons déterminé aussi les bandes d'absorption, dont on doit tenir compte dans l'interprétation de la structure des spectres de fluorescence.

Hormones surrénales.

Désoxy-corticostérone. Liqueur bleue. Deux bandes d'absorption: très forte entre λ 617 et λ 588 mμ; faible entre λ 565,5 et λ 546,5. Deux bandes de fluorescence: très brillante entre λ 653 et λ 615; notablement moins lumineuse entre λ 550 et λ 515, avec lueur jusqu'au violet. (Tout au début de la réaction, et en lumière de WOOD, bande de fluorescence verte seulement.)

Corticostérone. Liqueur presque incolore, sans absorption notable. Deux bandes de fluorescence, bien brillantes: orangée et verte, avec émission intermédiaire. Tardivement, bande d'absorption dans l'orangé, un peu plus réfrangible que celle de la désoxy-corticostérone; séparation nette alors des deux bandes brillantes. A ce stade ultime, fluorescence orangée de la liqueur (devenue à peine bleue).

Hormones sexuelles.

Oestradiol. Liqueur très légèrement rosée. Bande d'absorption entre λ 536,5 et λ 511 mμ. Bande de fluorescence verte entre λ 569,5 et λ 533 mμ. Cette émission se prolonge faiblement jusque dans l'orangé.

Oestrone. Liqueur un peu bleue à la longue. Bande d'absorption assez faible entre λ 533 et λ 512. Bande de fluorescence intense étendue de λ 565 à λ 530 mμ. Lueurs en deçà et au delà.

Equiline. Liqueur d'abord un peu jaune puis teintée de rose. Fluorescence magnifique, apparaissant, comme celle de l'oestrone, très rapidement. Bande d'absorption entre λ 533 et λ 513 mμ. Au début, autre bande d'absorption entre λ 477 et λ 457 mμ (bande semblable pour l'oestrone). Forte bande de fluorescence étendue de λ 566 à λ 531 (avec prolongement jusqu'à λ 485 mμ au début). Emission notable jusque dans l'orangé.

Equilénine (à noyau naphtalénique). Liqueur acquérant peu à peu une nuance framboisée. Bande d'absorption intense de λ 554 à λ 518. Forte fluorescence comprise entre λ 658 et λ 553 mμ. Réaction rapide.

La fluorescence rouge orangé de l'équilénine n'est pas excitée par la double raie jaune du sodium qui excite intensivement les fluorescences rouge orangé d'autres hormones (désoxy-corticostérone, etc.).

Progestérone. Réaction lente et complexe. A la fin, liqueur framboisée. Deux bandes d'absorption: de λ 624 à λ 577 et de λ 512 à λ 473. Deux bandes de fluorescence: de λ 653 à λ 614 et de λ 577 à λ 503 mμ.

Anhydro-oxy-progestérone. Grandes analogies spectrales avec la désoxy-corticostérone (hydroxy-2 I-progestérone). (Bande d'absorption aussi sur λ 479 mμ.) Vive fluorescence rouge apparaissant bien plus rapidement.

Testostérone. Fluorescence verte, rapide, en lumière de WOOD; verdâtre et peu intense au soleil. Emission dans presque tout le spectre visible avec maximum correspondant à la plage verte. Pas de bande d'absorption nette.

Méthyltestostérone. Superbe fluorescence presque immédiate, rappelant, au soleil, celle de la fluorescéine. Liqueur jaunâtre, à forte bande d'absorption entre λ 508 et λ 478. Bande de fluorescence principale entre λ 564 et λ 510. (La fluorescence d'un bleu violacé, signalée plus haut, ne s'observe qu'à un stade très avancé.)

Nous avons pu nous convaincre bientôt, en procédant avec du sulfate diméthylique rendu parfaitement neutre, qu'un certain degré d'acidité (souvent très faible) est nécessaire dans tous les cas, et qu'en réalité, on a toujours affaire à des réactions d'halochromie. Nous avons donc proposé (*34*) de désigner par un terme spécial: «halofluorie» l'apparition, — pouvant même être prédominante, — de la fluorescence.

L'acide phosphorique sirupeux permet aussi d'obtenir à partir de la désoxy-corticostérone (ou de l'anhydro-oxy-progestérone) des fluorescences rouges assez semblables à celles obtenues par addition de sulfate de méthyle (*32*). Les résultats sont pourtant bien inférieurs. FINKELSTEIN, avec des collaborateurs (*41*), a soigneusement étudié cette réaction pour certaines hormones oestrogènes. Il y a production de belles fluorescences vertes. Il en est de même, comme nous l'avons observé, pour l'équiline.

Réactions spéciales.

En ce qui concerne les fluorescences rouges de *stéroïdes sexuels*, deux réactions très spéciales ont été publiées:

a) *Réaction de* MIESCHER (1946) pour la *cis*-testostérone (*78*) par addition d'acide sulfurique à une solution dans l'acide acétique glacial en présence d'aldéhyde anisique.

b) *Réaction de* PESEZ (1947), pour cette même *cis*-testostérone (*87*), réalisée en ajoutant un peu de réactif glyoxylique à une solution dans un mélange d'alcool et d'acide sulfurique. La fluorescence rouge-feu qui apparaît dans cette dernière réaction, — que nous avons pu observer, — est fort belle. A noter que la *trans*-testostérone exempte de l'isomère *cis* ne fournit ni coloration ni fluorescence, d'après PESEZ.

Les spectres de fluorescence n'ont pas encore été déterminés pour ces deux réactions. Le spectre de fluorescence a été, au contraire, examiné et indiqué pour la réaction suivante qui n'est pas rouge, mais d'un vert jaune.

c) *Réaction de* KLEINER (*62*) pour les stéroïdes à fonction phénolique (oestrogènes naturels). L'oestrone, l'oestradiol et l'oestriol réagissent avec l'anhydride phtalique, et il se forme une substance qui, en solution chloroformique, est d'un beau rose foncé. La liqueur présente une fluorescence jaune verdâtre, le maximum d'émission étant sur λ 557 mμ. Cette phtaléine n'est pas produite en l'absence de OH phénolique; les résultats ont été, en effet, négatifs avec le cholestérol, la testostérone, l'androstérone, la déhydroandrostérone, la progestérone.

d) *Réaction de* BOSCOTT (*6bis*). Cette réaction n'est qu'une modification de celle étudiée par FINKELSTEIN (*41*), dont il a été parlé plus haut: Au lieu d'employer simplement l'acide phosphorique, l'auteur opère avec une solution du stéroïde dans l'acide acétique. Toute une série de fluorescences ont été constatées en lumière U. V., fluorescences allant de l'orangé au bleu suivant les stéroïdes sexuels utilisés. (Pour le détail des résultats, voir la communication originale.)

IV. Protéines (protides) et aminoacides.

Opérant avec un hydrolysat de caséine, REEDER et NELSON (*101*), en 1940, ont observé une bande de fluorescence étendue de λ 530 à λ 410 mμ. D'après ces auteurs, la fluorescence des hydrolysats protéiques est seulement excitée par des radiations comprises entre λ 360 et λ 340 mμ. Leur travail ne contient pas d'autre détermination numérique de l'émission par fluorescence; mais on y trouve de nombreux résultats assez intéressants portant sur la fluorescence de diverses protéines (ovalbumine, fibrine, gliadine, caséine, zéine, etc.) observée dans des conditions très variées.

VLÈS (*125*), en 1943, a constaté que, pour la gélatine, l'analyse spectrale de la lumière émise sous excitation par la raie Hg λ 265 montre une plage continue allant de λ 538 à λ 466 mμ tout au plus, centrée sur λ 490, c'est-à-dire bleu verdâtre. Cet auteur dit que beaucoup de protides lui ont montré que leur fluorescence est excitée par la plus grande partie de l'ultraviolet du quartz depuis les environs de λ 230 (ou 250) mμ jusqu'au visible, la transition et le départ étant de ce dernier côté difficiles à discerner entre la fluorescence et la diffusion par effet de milieux troubles. Ses recherches ont porté sur la sérumalbumine, la gélatine, l'excelsine, l'édestine et la caséine (Fig. 2).

Dans des conditions expérimentales modifiées, VLÈS a même pu mettre en évidence l'excitation de la fluorescence des protides par des radiations U. V. de plus courtes longueurs d'onde encore, jusqu'au voisinage de λ 220 mμ.

Aminoacides. REEDER et NELSON (*101*) disent: «Nineteen amino acids were examined in ultraviolet light for fluorescence. These amino acids were not noticeably fluorescent in the solid state nor in acid, basic or neutral solution.»

Bien des auteurs ont consigné des constatations assez différentes. Nous ne parlerons ici que des résultats publiés dans ces dix dernières années.

Dans le livre de DE MENT (*74*), on trouve citées les déterminations suivantes:

Émission par fluorescence (solutions aqueuses)	Glycine	de λ 500 à	λ 277 mμ	
	Alanine	370 à	300	(!)
	Acide α-aminobutyrique	500 à	325	
	Lysine (bichlorhydrate)	400 à	300	
	Histidine (bichlorhydrate)	500 à	360	

Vlès (*125*) a beaucoup étudié aussi la fluorescence de nombreux aminoacides: glycine, alanine, leucine, acide glutamique, arginine, cystine, tyrosine, proline, histidine; et il a publié des graphiques fort instructifs représentant l'excitation de la fluorescence par des radiations ultraviolettes très diverses (jusque vers λ 220 mμ). Voir à l'Appendice les Figures 8 et 9 (p. 346—347).

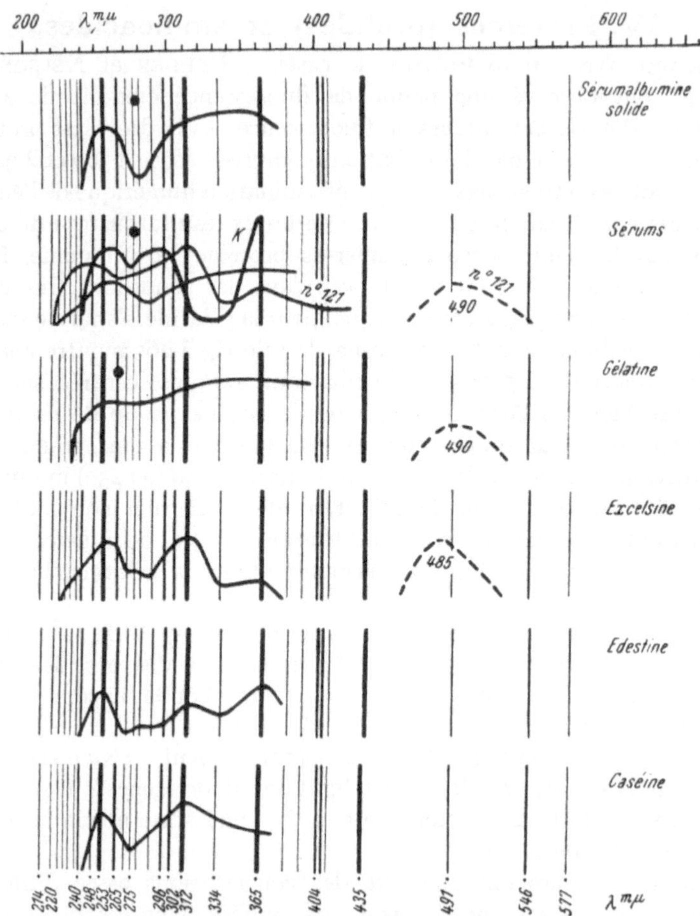

Fig. 2. Graphiques schématisant l'intensité (courbes en traits continus) de la fluorescence de divers protides excitée par les raies ultraviolettes du mercure. Les spectres d'émission par fluorescence en traits discontinus. Le maximum d'absorption dans l'ultraviolet indiqué par le signe. D'après Vlès (*125*). [Arch. Physique biol. Chim.-Physique Corps organisés *16*, 137 (1943).]

D. Bertrand (*4*) a dit, en 1945, que la forte fluorescence du *glycocolle* (glycine) avait été niée par quelques auteurs, mais qu'il l'avait retrouvée avec différents échantillons soigneusement repurifiés. Haitinger (*50*) avait constaté d'ailleurs que la fluorescence du glycocolle en solution alcaline est «leuchtend blau». BERTRAND (*4*) a publié la courbe obtenue, par enregistrement photographique dans le visible, pour le produit à l'état solide. Cette courbe montre deux sommets, les maximums correspondant à λ 519 et à λ 451 mμ. L'alourdissement du radical R de l'acide aminé: $R - CH(NH_2) - COOH$ modifie, d'après BERTRAND, le spectre par déplacement vers les grandes longueurs d'onde; et, dans l'ensemble, l'intensité de la fluorescence décroît avec toutefois une conversion entre la valine et la leucine. Enfin, toujours d'après le même auteur, de façon analogue, si on éloigne les fonctions amine et acide, le spectre change peu pour la position des maximums quand ces fonctions s'éloignent; et, fait surprenant, la cyclisation dans la *proline* ne donne pas un spectre du type d'aminoacide, mais bien d'un type très différent, prouvant ainsi que ce déplacement vers les grandes longueurs d'onde ne doit pas être attribué à un enroulement de la molécule.

Les courbes de fluorescence pour la *phénylalanine*, la *tyrosine* et le *tryptophane* (état solide) figurent aussi dans le travail de BERTRAND (*4*). Ce sont toutes des courbes à deux sommets, les maximums tombant respectivement sur λ 522 et λ 461 mμ (phénylalanine), sur λ 520 et λ 433 (tyrosine), et sur λ 513 et λ 449 (tryptophane). L'auteur estime que la fluorescence visible de la tyrosine est très faible et celle du tryptophane assez forte. Il rattache la fluorescence de la tyrosine à la fonction phénol et celle du tryptophane à l'azote indolique. Comme conclusion, il dit que la règle classique suivant laquelle le groupement carboxyle diminuerait fortement l'intensité de la fluorescence est complètement inexacte pour les acides aminés.

On peut indiquer encore ici les spectres de fluorescence, à l'état solide, de deux importants dérivés de la tyrosine: la tyramine (*p*-oxyphényl-éthylamine) et l'hordénine (diméthyl-*p*-oxyphényl-éthylamine). Pour la tyramine (chlorhydrate), il y a trois maximums: sur λ 729, λ 525 et λ 456; pour l'hordénine (sulfate), deux maximums: sur λ 530 et λ 459 mμ [BERTRAND (*4*)].

Si l'on considère l'ensemble des acquisitions récentes et antérieures dans ce domaine de la fluorescence des protéines et des acides aminés, on a l'impression que beaucoup de résultats sont assez incohérents et plus ou moins incertains et qu'il faudra faire encore bien des recherches approfondies pour répondre définitivement à toutes les questions qui se posent.

Réactions spéciales.

Réaction de HILLMANN (*54*). S'inspirant, comme il le dit, d'un travail de Mme ANDRÉE ROCHE (exécuté sur mon conseil, dans mon laboratoire), HILLMANN a étudié la réaction de fluorescence qui prend naissance quand on ajoute à des protéines de l'*o*-diacétylbenzène (à la place du simple diacétyle utilisé précédemment). Le dosage des protéines, spécialement des peptones, par fluorométrie ainsi exécutée, fournit des résultats excellents, notamment dans l'application à la Réaction d'Abderhalden. Avec une solution de peptone à 1 : 60000, on observe encore une fluorescence bleue caractéristique.

L'auteur a photographié le spectre de fluorescence (spectrogramme reproduit); on voit deux bandes lumineuses, avec maximums d'émission sur λ 520 et λ 425 mμ.

Réaction de Kostir (*68*) (pour le tryptophane). Une solution de tryptophane étant additionnée d'aldéhyde formique puis d'acide sulfurique concentré, on constate d'abord l'apparition d'un anneau à fluorescence jaune. Par agitation, toute la liqueur présente cette fluorescence jaune, qui est permanente. D'après l'auteur, la substance fluorescente pourrait être un dérivé de la 4-carboline.

Réaction de Fearon (tryptochrome). Le tryptochrome, pigment dérivé du tryptophane, possède, d'après Fearon (*40bis*), une intense fluorescence vert orange, stable pendant des mois en milieu acide. La réaction est très simple: il suffit d'ajouter à une solution très diluée de tryptophane dans l'acide acétique à 60—80 p. 100 une goutte de liqueur d'iodate de potassium à 1 p. 100, puis de chauffer jusqu'à ébullition. Le tryptochrome ainsi formé semble être un dérivé de constitution indirubinoïde.

V. Pyrimidines, purines et ptérines.

Pyrimidines et purines.

En 1906, Dhéré (*23*) fit connaître la remarquable bande d'absorption que le noyau pyrimidique possède sur λ 260 mμ environ, bande importante à plusieurs points de vue [cf. Mirsky (*79*)], notamment pour l'étude de la fluorescence des composés pyrimidiques. C'est précisément en reprenant, d'une façon approfondie, l'étude de cette bande d'absorption que Heyroth et Loofbourow eurent l'occasion de constater, en 1931, la fluorescence de l'uracile (2,6-dioxy-pyrimidine). En 1941, Stimson et Reuter (*118*), élèves de Loofbourow, examinèrent la fluorescence de nombreux corps pyrimidiques (uracile, cytosine, quelques aminopyrimidines) et puriques (hypoxanthine, xanthine, guanine et adénine), d'une part, à l'état solide et, d'autre part, en solutions alcalines ou acides, l'acide sulfurique développant plus que l'acide chlorhydrique la fluorescence. La détermination des spectres de fluorescence correspondants devait être entreprise par Loofbourow et Stimson. Mais il ne semble par que ce travail annoncé ait paru. Quoi qu'il en soit, le travail de Stimson et Reuter contient, sous forme de tableau, beaucoup de résultats assez intéressants.

Dans leur travail de 1946, JACOBSON et SIMPSON (*57*) ont dit: «Alloxan, hypoxanthin, adenine and aneurin showed no fluorescence in neutral or alkaline solution. Uric acid and guanine gave weak bands near the mercury line at 4358 Å, in alkaline solution only.»

En parlant maintenant du pigment urique rouge d'HOPKINS, nous aurons une transition toute naturelle pour arriver aux ptérines. Il s'agit peut-être, en effet, d'une sorte de ptérine (xanthoptérine?), comme l'avait suggéré HOPKINS, qui se forme quand on chauffe quelques heures de l'acide urique, dans des tubes scellés contenant de l'eau, à la température de 200°. Le spectre de fluorescence a été déterminé par Miss SIMPSON (*110*) en milieux alcalin (I), neutre (II) et acide (III). Elle a relevé les bandes lumineuses suivantes:

I	λ 620 (faible);	590 et 568 (moyennes);	530 (forte);	465 mμ (moyenne).
II			λ 517 (moyenne);	456 mμ (forte).
III		λ 592 (faible);	520 (moyenne);	459 mμ (forte).

Comparé au spectre de fluorescence de la xanthoptérine, qui sera décrit plus loin, ce spectre du produit HOPKINS présente des analogies, mais aussi quelques différences importantes.

Ptérines.

Dans mon livre (*20*) publié en 1937, j'avais inséré un assez long exposé de la fluorescence des ptérines. Par contre, ne possédant alors aucune donnée sur les spectres de fluorescence des ptérines, j'avais dû renoncer à parler dans mon article (*21*) de 1939 de ces remarquables composés pyrimidiques. Il fallut, en effet, attendre la publication en 1943 d'un travail de HÜTTEL et SPRENGLING (*55*) pour connaître un spectre de fluorescence d'une ptérine convenablement isolée: spectre de l'ichtyoptérine (fluorescyanine de POLONOVSKI) déterminé par Mlle F. PRUCKNER. Ce spectre comprend deux bandes lumineuses: une large bande diffuse, bien visible de λ 475 à λ 450 mμ environ [maximum accusé entre λ 431,6 et λ 432,3 mμ, d'après POLONOVSKI, BUSNEL et PESSON (*89*)], et une très faible bande éloignée, située dans l'extrême rouge, au voisinage de λ 690. Le solvant n'a pas été indiqué. Il s'agirait de solutions de p_H 4 à p_H 8, d'après SIMPSON (*110*). L'étude des variations, dans diverses conditions, de la fluorescence de l'ichtyoptérine excitée par les rayons ultraviolets (compris entre λ 390 et λ 265 mμ) a été faite soigneusement par HÜTTEL et SPRENGLING, puis par POLONOVSKI (*90*).

Signalons que, surtout dans la zone acide, la variation d'intensité suivant le p_H est semblable à ce qu'on observe avec la flavine; mais les réducteurs ne suppriment pas la fluorescence, comme c'est le cas pour la flavine; on constate seulement (solution alcaline) que la fluorescence passe du bleu pur au bleu vert.

L'ichtyoptérine d'HÜTTEL et SPRENGLING (*55*) avait été extraite de la peau de quelques Cyprinidés du Lac de Constance. L'existence de ce pigment (cette ptérine cristallise, en effet, en aiguilles jaunâtres, d'après POLONOVSKI) avait été entrevue d'abord, grâce à sa fluorescence, par HADJIOLOFF. La première contribution importante à la connaissance de cette ptérine, indiquant son individualité, fut apportée par les recherches de FONTAINE (*42*—*43*) bientôt suivies de celles de BUSNEL et de GOURÉVITCH. L'ichtyoptérine (voisine de la 8-désoxy-leucoptérine, d'après POLONOVSKI) fait partie d'un groupe de nombreuses ptérines (naturelles et synthétiques), parmi lesquelles la leucoptérine et la xanthoptérine sont

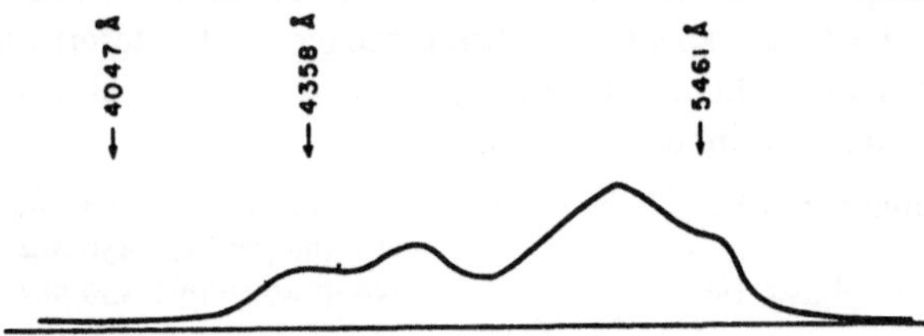

Fig. 3. Courbe photométrique des quatre bandes de fluorescence (repérées par 3 raies du mercure) de la leucoptérine à l'état solide. D'après JACOBSON et SIMPSON (*57*). [Biochemic. J. *40*, 3 et 9 (1946).]

de beaucoup les plus importantes. Les spectres de fluorescence de ces deux dernières ptérines naturelles ont été très soigneusement déterminés, en 1946, par JACOBSON et SIMPSON (*57*); enregistrement photographique avec courbes photométriques des spectres de fluorescence, courbes à grande échelle reproduites dans leur Mémoires. Voici leurs résultats (Tableau 1; Fig. 3).

Dès 1939, JACOBSON (*56*) avait envisagé la possibilité de l'intervention de ptérines dans l'hématopoïèse (formation de globules rouges); et, après examen des spectres d'absorption et de fluorescence de certaines cellules argentaffines se trouvant dans l'estomac et l'intestin, il avait, avec une perspicacité étonnante, admis la présence de ptérine dans l'organisme.

JACOBSON (*56*) disait alors: «The granules of the argentaffine cells have a yellow colour and in ultraviolet light show an orange fluorescence which is composed of two bands with maxima at 610 and 555 mμ.»

Or, on sait maintenant que l'acide folique, doué d'une grande activité antianémique, a précisément pour constituant l'acide ptéroylglutamique. Il convient donc de signaler que les Mémoires de JACOBSON et SIMPSON (*57*) publiés en 1946, contiennent, en plus des déterminations précédemment citées, celles des spectres de fluorescence pour des extraits de tissus et d'organes humains ainsi que pour des extraits de foie d'animaux de boucherie (utilisés en thérapeutique), les fluorescences en question étant (en partie au moins) d'origine ptérique.

Miss SIMPSON (*110*) a écrit: «Examination of a large number of liver extracts suggest that the intensity of fluorescence (évaluée d'après le spectre de fluorescence) might be used as a mesure of the clinical activity of these preparations.»

Tableau 1. Fluorescence de quelques ptérines [JACOBSON et SIMPSON (57)].

Ptérines	*Bandes de fluorescence*	*Intensités*
Leucoptérine $C_6H_5O_3N_5$*		
1° Solution dans NaOH *n/10*	I sur λ 516 mμ	forte
(Fluorescence d'un bleu brillant)	II 462,5	assez forte
2° État solide	I sur λ 620 mμ	faible
(Fluorescence blanc bleuâtre)	II 535	forte
	III 462,5	assez forte (?)
	IV 436	modérée
Xanthoptérine (9-désoxy-leucoptérine) $C_6H_5O_2N_5$		
1° Solution dans NaOH *n/10*	I sur λ 592 mμ	faible
(Fluorescence bleue)	II 516	assez forte
	III 463	forte
2° Solution neutre	I sur λ 620 mμ	faible
(Fluorescence bleue)	II 592	modérée
	III 516	assez forte
	IV 455	forte
3° Solution dans H_2SO_4 (20%) ou HCl *n*	I sur λ 620 mμ	faible
(Fluorescence bleue)	II 592	modérée
	III 523	forte
	IV 460	assez forte
4° État solide	Aucune fluorescence**	

Miss SIMPSON, avec PIRIE (*88*), a en outre étudié une substance à fluorescence bleue existant dans les yeux du *Squalus acanthias* et de l'*Alligator mississipiensis*. Les déterminations des spectres de fluorescence (ainsi que des spectres d'absorption ultraviolets) ont amené ces auteurs à admettre qu'il s'agit de xanthoptérine. Cette ptérine est présente dans la choroïde, et il est intéressant de savoir qu'*in vivo*, elle y préexiste sous une forme non-fluorescente. Toujours d'après PIRIE et SIMPSON, les yeux des Cyprinidés pourraient contenir, comme leur peau, de l'ichtyoptérine. Enfin Miss SIMPSON (*110*) suppose qu'une substance fluorescente qu'elle a extraite des yeux du Vairon (*Phoxinus phoxinus*) est également une ptérine, bien que le spectre de fluorescence présente un type assez aberrant: bande lumineuse de λ 460 à 420 mμ (maximum sur 440) et faible bande sur λ 525 mμ pour la solution neutre.

* Les ailes de *Pieris brassicae* montrent un spectre de fluorescence analogue à celui de la leucoptérine solide avec bande au voisinage de λ 460 mμ relativement plus intense.

** Pourtant les ailes du *Gonepteryx rhamni* (riches en xanthoptérine) montrent un faible spectre de fluorescence analogue à celui de la xanthoptérine en solution bien acide; mais avec bande sur λ 620 plus forte et bandes dans le jaune et le bleu au contraire plus faibles.

Le travail de PIRIE et SIMPSON contient tout l'historique du sujet. Bornons-nous à mentionner les recherches antérieures de GOURÉVITCH (*48*) et celles, très développées, de FONTAINE et BUSNEL (*43* et *44 bis*).

La fluorescence des ptérines contenues dans l'organisme est toujours à l'étude; c'est ainsi qu'en 1946 CASELLA et REGGIANI (*12*) se sont occupés de nouveau de la question du spectre de fluorescence des cellules entérochromaffines chez le Cobaye et ont déterminé spectrographiquement cette fluorescence par une méthode spéciale d'application histologique. Les courbes publiées présentent un maximum d'émission bien marqué au voisinage de λ 558 mμ.

CROWE et WALKER (*16*) ont constaté que, dans le bouillon-toxine diphtérique, coexistent avec la flavine, — qui y a été signalée pour la première fois par DHÉRÉ, MEUNIER et CASTELLI (*37*), — des sortes de ptérines («pterin-like pigments») qu'ils ont pu isoler par l'analyse chromatographique et dont ils ont étudié les spectres de fluorescence par enregistrement photographique (reproduction d'un spectrogramme avec courbe microphotométrique). La courbe reproduite montre deux maximums de fluorescence: I sur λ 531,9 et II sur λ 463,5 mμ (solution aqueuse).

Dans un travail qui vient de paraître (*17*), les recherches de ces auteurs sont étendues à un pigment ptérique («erythropterin-like») formé par le Bacille de la tuberculose humaine. Spectre de fluorescence (sol. chloroformique) étendu de λ 463,5 à λ 358 mμ, avec maximum sur λ 399 mμ.

Remarquons que CROWE et ses collaborateurs ne semblent pas avoir su que GIRAL (*46 bis*), dès 1936, avait examiné la production de ptérine chez quelques espèces de Bactéries.

POLONOVSKI et ses collaborateurs (*91*) ont préparé synthétiquement diverses ptérines artificielles, notamment des ptérines sulfurées dépourvues de fluorescence. Pour une série de ces ptérines douées de fluorescence, l'émission analysée au spectroscope se présentait sous forme d'une large bande débutant avant le jaune et s'étendant jusqu'au violet, spectre de fluorescence peu caractéristique du corps étudié, disent les auteurs.

La question de la fluorescence dans le groupe des ptérines est si intéressante qu'il ne faut pas omettre de relater ce qui a été dit pour l'interpréter. POLONOVSKI, VIEILLEFOSSE et PESSON (*91*) avaient d'abord émis l'hypothèse que la fluorescence serait due à trois doubles liaisons dans le noyau pyrimidique. BERTRAND (*4*) dit, d'une façon plus précise, que ces auteurs ont admis que la fluorescence était liée dans ces ptérines fluorescentes à la structure lactime (*A*) des 2-hydroxyptérines, alors que les 2-thioptérines répondraient à une forme lactame (*B*). Ils en ont déduit que la forme lactime (*C*) devait retrouver la fluorescence caractéristique, c'est ce qu'ils ont vérifié sur différents dérivés éthylés dont ils ont effectué la synthèse.

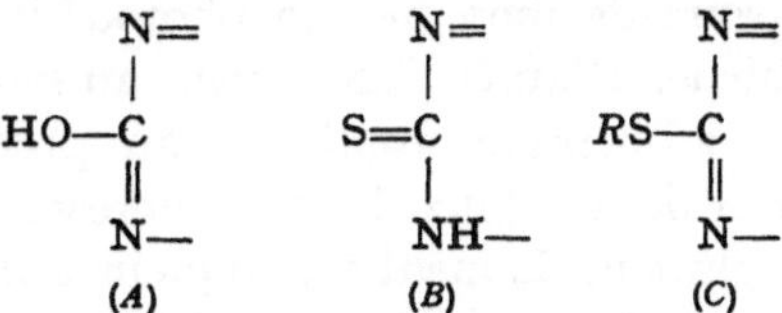

Bertrand ajoute que cela s'explique facilement avec certaines hypothèses qu'il a faites: Dans les 2-hydroxyptérines, seul l'atome d'oxygène situé en 2 est responsable de la fluorescence, à l'exclusion des autres atomes de la molécule et par suite, la forme lactame ne doit pas être fluorescente; d'autre part, le soufre correspond par sa couche électronique externe à l'oxygène et doit par suite présenter les mêmes possibilités de vibrations de fluorescence que ce dernier; la forme (*C*) des thioptérines doit donc correspondre à la forme (*A*) des ptérines.

Et, en terminant cette section consacrée aux ptérines, il faut mentionner la découverte dans l'urine, par Koschara (*66*), d'un corps sulfuré, appelé *urothion*, caractérisé par sa forte fluorescence vert-olive après oxydation permanganique en milieu sulfurique, corps pouvant être rattaché aux ptérines notamment par son spectre d'absorption ultraviolet. Voir aussi l'article de Purrmann (*94*).

VI. Thiamine (aneurine), thiochrome et thiazol.

Sjöstrand (*111*) a publié en 1947 le courbe représentative de la fluorescence de la thiamine (solution) entre λ 670 et λ 420 mμ. Cette courbe présente trois sommets; le plus élevé est très voisin de celui qui appartient au thiochrome et correspond à environ λ 460 mμ. Pour le reste de la courbe jusqu'au rouge, il y a des différences notables entre le spectre de la thiamine et celui du thiochrome.

Dans notre article précédent (*21*), nous avons parlé suffisamment de la belle fluorescence du thiochrome qui a été utilisée par beaucoup d'auteurs pour les dosages fluorométrique (après oxydation) de l'aneurine. Un travail de Dutton et Bailey (*38*) contient une courbe instructive de la fluorescence du thiochrome déterminée spectrophotométriquement.

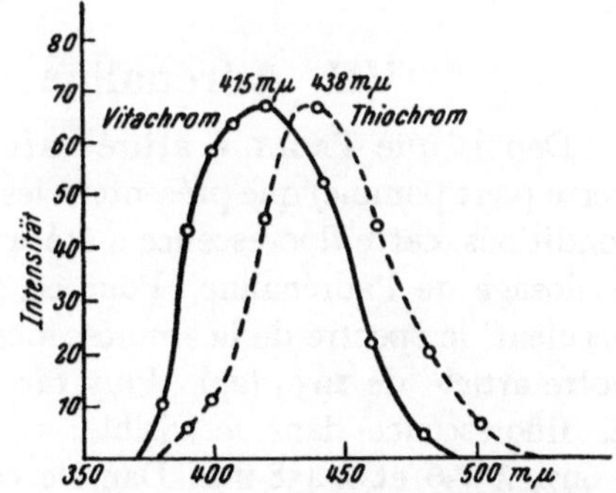

Fig. 4. Spectres de fluorescence (à la même échelle) du vitachrome et du thiochrome en solutions aqueuses de $p_H = 7$. Excitation par la lumière de Wood. D'après Stämpfli (*116*). [Helv. physiol. Acta *1*, 265 (1943).]

STÄMPFLI (*115—116*; Fig. 4) a constaté que le constituant thiazolique de l'aneurine (4-méthyl-5-oxyéthyl-thiazol) fournit, par irradiation avec la lampe de quartz à vapeurs de mercure, un photodérivé possédant une très vive fluorescence bleue. D'après SANZ (*104*), au moyen de l'analyse chromatographique par adsorption appliquée au produit brut de la réaction, on peut préparer des cristaux dont la fluorescence n'est pas bien forte, mais qui, par dissolution, donnent une liqueur à fluorescence bleue extraordinairement intense. KARRER et SANZ (*60*) ont repris cette étude. Ils ont reconnu que le dérivé en question est tout particulièrement facile à préparer par irradiation du 4-méthyl-5-oxyéthyl-2-chlorthiazol. Ils ont indiqué la constitution présumée du principal produit formé dans cette réaction photochimique. Il est, en effet, possible, comme l'a vu SANZ (*104*), en effectuant l'analyse chromatographique, qu'il existe deux autres produits accessoires pouvant être considérés également comme des «vitachromes». Synthèse du vitachrome par KARRER et SANZ en 1944: (*60bis*).

Le spectre de fluorescence du vitachrome (fluorochrome) a été déterminé et figuré par une courbe (comparativement avec le thiochrome) par STÄMPFLI (*116*). Pour le vitachrome, le maximum d'intensité de l'émission est sur λ 415 mμ, alors qu'il est sur λ 438 pour le thiochrome. Le vitachrome, contrairement au thiochrome, possède en milieu acide sa plus forte fluorescence, qui diminue bien par alcalinisation. STÄMPFLI de plus a examiné, notamment en fonction du p_H, comment se comporte, aprés irradiation ultraviolette, la fluorescence de toute une série de dérivés méthylés du thiazol (*117*).

Enfin postérieurement (en 1945), KARRER (*59bis*) a obtenu des fluorescences tout à fait remarquables, de couleurs variées, avec d'autres dérivés du thiazol qu'il avait préparés. La fluorescence de nombreux composés thiazoliques est d'ailleurs connue depuis bien longtemps.

VII. Adrénaline et adrénochrome.

Depuis que PAGET a attiré l'attention, en 1930, sur la fluorescence verte (vert pomme) que présentent les solutions d'adrénaline dans certaines conditions, cette fluorescence a été utilisée par de nombreux auteurs pour le dosage de l'adrénaline. Pour ce qui est de l'adrénaline à l'état pulvérulent, le spectre de la fluorescence ultraviolette a déjà été décrit dans notre article de 1939 (*21*). Plus tard, BERTRAND (*4*) a publié la courbe de fluorescence dans le visible: courbe possédant des maximums sur λ 697, λ 536 et λ 458 mμ. Dans le cas d'une solution aqueuse du chlorhydrate d'adrénaline gauche, il y a deux maximums: sur λ 515 et λ 461 mμ*.

* Ces derniers résultats, encore inédits, m'ont été communiqués par D. BERTRAND.

La question de la fluorescence des produits d'oxydation de l'adrénaline a été examinée à plusieurs reprises par UTEVSKIY et OSSINSKAJA. Ces auteurs ont vu (*123*) qu'en présence de H·CHO, la fluorescence verte, qui autrement apparaît lors de l'oxydation de l'adrénaline en milieu alcalin, ne se montrait pas. Cet empêchement serait dû au blocage de la formation de dihydrodérivés par formaldéhyde.

En 1948, BEAUVILLAIN et SARRADIN (*2 bis*) ont publié de nombreux résultats concernant la fluorescence de l'*adrénochrome* (quinone de l'adrénaline). Cette fluorescence, disent-ils, varie beaucoup suivant le solvant utilisé. La solution dans le dioxane montre une large bande lumineuse dans le bleu, à maximum 465—460 mμ environ. La solution dans le butanol présente une fluorescence verte, disparaissant très rapidement par irradiation ultraviolette. Les courbes de fluorescence spectrale pour ces deux solutions, ainsi que dans le cas de solvants autres (méthanol, acétone, diacétine du glycol, etc.) sont insérées dans ce travail. Les auteurs ont pu isoler, à partir de la solution dans la diacétine de glycol, deux constituants fluorescents ayant respectivement une fluorescence verte et une fluorescence bleue. La solution aqueuse d'adrénochrome est peu fluorescente. Mais, en séchant rapidement une goutte de cette solution sur papier filtre, la tache vire du rouge au jaune et acquiert une magnifique fluorescence jaune vert, qui rappelle celle observée dans la diacétine.

La solution aqueuse du leuco-adrénochrome bisulfitique possède un spectre de fluorescence allant de plus de λ 700 à moins de λ 450 mμ. Il existe trois contreforts à λ 650, λ 495 et λ 470 mμ et un sommet assez étalé entre 562 et 530. Le leuco-adrénochrome ascorbique a un spectre de fluorescence différent, qui ne dépasse guère λ 470 dans le bleu avec un sommet très net à 530 encadré de deux contreforts à 562 et 502 mμ.

VIII. Alcaloïdes.

Beaucoup de travaux ont été faits depuis longtemps sur la fluorescence des alcaloïdes. Il y a encore quelques nouveaux résultats à signaler.

BASTERO BEGUIRISTAIN (*2*) a déterminé les bandes suivantes en 1940 (tous ces alcaloïdes examinés en solution à l'état de sels):

Émission par fluorescence (en mμ)	Morphine	λ 523,2 à λ 413,7	(maximum sur λ	471,5 mμ)
	Atropine	515,4 à 425,0	(„ „	459,2 mμ)
	Aconitine	504,2 à 404,5	(„ „	446,1 mμ)
	Quinine.......	504,2 à 381,5	(„ „	442,7 mμ)
	Strychnine	499,0 à 407,4	(„ „	438,3 mμ)

KRAUL et MEYER (*69*) ont déterminé le spectre de la fluorescence de la *berbérine* excitée par la lumière de WOOD: bande comprise entre 650 et 500 mμ avec maximum sur 560 mμ environ. Les bords de la bande étaient diffus. Il s'agissait soit de chlorhydrate, soit de sulfate de berbérine, soit encore de berbérine isolée à partir de chélidoxanthine (contenant en plus de la chélérythrine) en solutions dans l'éther.

DE MENT a fait en 1944 des recherches étendues sur la fluorescence de la berbérine, recherches qu'il a publiées pour la première fois dans son livre (*74*). De son long exposé, nous ne donnerons ici qu'un résumé. Le chlorhydrate de berbérine à l'état solide montre une forte bande verte de fluorescence entre λ 560 et λ 510, accompagnée d'une bande jaunâtre entre 590 et 560 et d'une bande rouge entre 630 et 590 mμ. En solution dans le *n*-butanol, il y a une émission de λ 560 à λ 480. Quand on dissout le chlorhydrate de berbérine dans l'acide chlorhydrique, il se forme un composé connu sous le nom de chlorure de berbérinium (berberinium chloride) qui, à l'état solide, présente trois bandes de fluorescence: I, dans le vert, de 560 à 510; II, dans le jaune, de 575 à 560; III, dans le rouge, de 620 à 575 mμ. La bande dans le jaune est plus faible que les deux autres. L'auteur a examiné aussi le spectre de fluorescence du composé qu'il désigne sous le nom d'anhydroberbérine-acétone, et il décrit ce spectre d'une façon très complète.

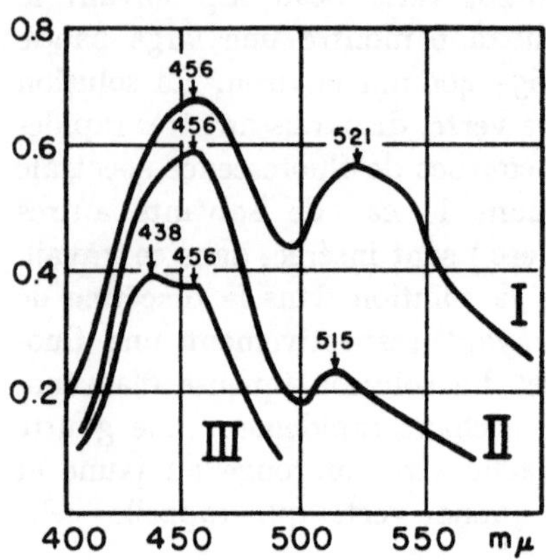

Fig. 5. Courbes de la fluorescence visible du sulfate de quinine: I Etat solide (pose 1 heure); II Solution aqueuse saturée (pose 10 minutes); III Solution aqueuse très diluée (pose 30 minutes). D'après D. BERTRAND (*4*). [Bull. Soc. chim. France (5), 12, 1010 (1945).]

BERTRAND (*4*), en 1945, a déterminé les spectres de la fluorescence visible du sulfate de quinine et de l'yohimbine (principal alcaloïde contenu dans l'écorce de *Pausinystalia yohimbe*). Pour le sulfate de *quinine* solide, il y a deux bandes lumineuses dont les maximums sont respectivement λ 521 et λ 456 (le principal). La solution aqueuse saturée montre aussi un maximum sur 456 mμ (avec maximum secondaire sur 515). Une solution diluée présentait un spectre assez différent: bande lumineuse à sommet ondulé (deux légères élévations sur λ 456 et λ 438) (Fig. 5).

Pour l'*yohimbine* à l'état solide, BERTRAND a enregistré trois bandes d'émission: sur λ 703; λ 564 et λ 467. Il a donné (*4*) une interprétation développée de ce spectre: «La yohimbine possède dans sa molécule quatre atomes pouvant donner des fluorescences visibles, ceux-ci sont groupés par trois: l'azote indolique pouvant donner un spectre type amine aromatique, les atomes d'oxygène de l'alcool secondaire et celui «fluorescent dans le visible» de l'ester méthylique de la fonction acide qui peuvent donner le même type de spectre et enfin l'azote amine tertiaire qui peut être considéré comme appartenant soit au noyau carboline, soit au noyau isoquinoléine. Or, le spectre obtenu correspond ainsi d'ailleurs que le confirme celui du chlorhydrate qui lui est identique, qualitativement et quantitativement, à ce dernier azote considéré comme dérivé de la β-carboline.»

Avec une solution aqueuse de chlorhydrate d'yohimbine, on a enregistré deux bandes de fluorescence: sur λ 513 et sur λ 455 mμ (résultat non encore publié par BERTRAND).

Rappelons que la fluorescence «stark gelbgrün» du chlorhydrate de yohimbine avait été observée par DANCKWORTT et PFAU en 1927 et que MAJIMA et MURAHASI en 1934 ont observé de belles fluorescences bleues avec une série de dérivés de l'yohimbine.

IX. Polyphénols.

Au cours de ses recherches très étendues (huit Mémoires) sur les avitaminoses du groupe de la pellagre, RAOUL a examiné un grand nombre de spectres de fluorescence appartement à des constituants urinaires, quelques-uns notamment dans la série des polyphénols. D'après RAOUL (99), que l'on ait affaire à une urine normale ou pathologique, les spectres de fluorescence sont généralement formés d'une large bande dans le bleu (axe à 450 mμ) et de plusieurs épaulements dans le vert, dont deux particulièrement nets à 560 et à 520 mμ. Les courbes spectrales (tracées d'après des données microphotométriques) accompagnent souvent le texte; mais il est regrettable que les relevés numériques de ces courbes (à petite échelle) soient si rares. On devra se reporter aux Mémoires originaux, fort instructifs, pour plus ample information.

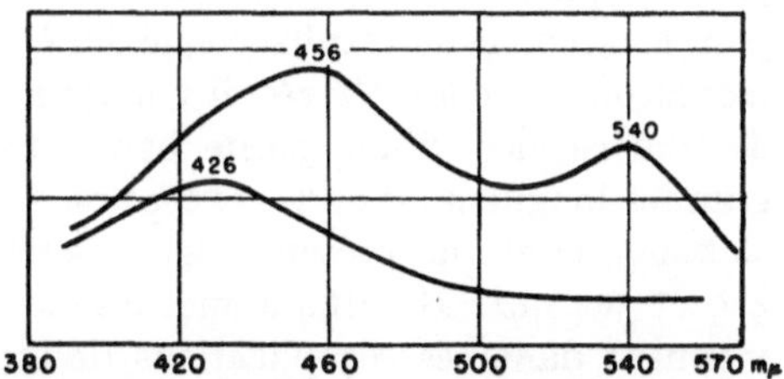

Fig. 6. Spectres de fluorescence du laccol (courbe inférieure) et du diméthyllaccol (courbe supérieure). Détermination microphotometriques de BROOKS (9). [Bull. Soc. chim. France (5), 7, 638 (1940).]

Un résultat très important a été pourtant indiqué: celui obtenu dans l'étude de l'alcaptonurie [RAOUL (98)]. L'alcaptone (acide homogentisique ou dihydroxyphényl-acétique, dérivant de la tyrosine et de la phénylalanine) possède une fluorescence bleue, que l'oxydation augmente beaucoup par suite de la formation d'un composé triphénolique (type oxyhydroquinone 1, 2, 4, celle-ci possédant un pouvoir fluorescent incomparablement plus intense que l'hydroquinone).

Nous allons maintenant considérer à part la fluorescence de deux polyphénols d'origine végétale.

Laccol et moréacol. Ce sont des o-diphénols à longue chaîne latérale éthylénique en C_{16}, isolés du latex d'arbres à laque (*Rhus* sp., *Semeocarpus vernicifera*, etc.). En lumière de WOOD, ils présentent une fluorescence d'un bleu intense, comme l'a constaté BROOKS (*9*), à qui l'on doit une étude très soignée et détaillée de leurs spectres de fluorescence fort remarquables. Pour le laccol $C_6H_3(OH)_2C_{16}H_{29}$, l'émission est comprise entre λ 545 et λ 372 mμ (Fig. 6, p. 335). Le maximum d'intensité offre une structure complexe et comprend quatre petites bandes (433, 426, 419, 413). Il y a aussi deux maximums secondaires.

Pour le moréacol, l'émission est comprise entre λ 537 et λ 380, avec maximum d'intensité complexe (438, 431, 424 mμ) et des maximums secondaires.

D'après BROOKS, la fluorescence de ces polyphénols et de leurs dérivés est due essentiellement à l'influence des doubles liaisons de la chaîne latérale, malgré la présence des OH phénoliques et même malgré l'éthérification de ces derniers.

Son travail contient les reproductions des spectres de fluorescence photographiés ainsi que les courbes microphotométriques correspondantes pour le laccol, le diacétyllaccol, le diméthyllaccol, le moréacol et le diacétylmoréacol. Avec les dérivés, il y a apparition, pour chaque bande continue de fluorescence, d'une petite bande très marquée se déplaçant vers les grandes longueurs d'onde, exception faite pour le diacétylmoréacol, dont la bande continue présente deux cannelures caractérisées par deux axes 419 et 394 mμ. BROOKS admet que les groupes méthyle et acétyle interviennent dans ces complications de structure.

BROOKS (*10*) a observé aussi de curieux changements de structure des spectres de fluorescence après irradiation ultraviolette prolongée.

X. Dérivés de l'indol.

VON DEM BORNE (*6*) a rattaché la fluorescence bleue de l'urine à la présence de monoacétyl-indoxyle et de diacétyl-indoxyle, dérivés apparaissant comme produits accessoires lors de la synthèse de l'indican. Les spectres de fluorescence de ces deux corps ont été photographiés; et, dans son travail de 1938, l'auteur a inséré les courbes microphotométriques qui montrent trois bandes d'émission: I sur λ 554,5, II (plus petite) entre 540 et 520, III sur λ 457 mμ. C'est à l'existence de cette dernière bande bien lumineuse que l'on devrait, selon VON DEM BORNE, attribuer la fluorescence bleue de l'urine. Cette conclusion a été critiquée, en 1943, par RAOUL (*96*), qui fait remarquer que ces deux corps indoliques n'ont pas de vraisemblance biologique et n'ont d'ailleurs pas été extraits de l'urine. Mais les résultats qu'il a obtenus au moyen de la chromatographie ont ramené toutefois RAOUL aux corps indoliques; et il a étudié les spectres de fluorescence des fractions isolées suivantes, qui contribueraient inégalement à conférer à l'urine une notable partie de sa fluorescence bleue: acide β-indol-acétique (peu fluorescent); acide α-oxy-β-indol-acétique (fluorescent); dioxindol (très fluorescent). Les courbes des spectres de fluorescence de ces trois corps ont été publiées par RAOUL.

XI. Acide nicotinique et nicotylamide.

On sait que l'acide nicotinique (si utilisé maintenant en thérapeutique) est l'acide pyridine-3-carbonique. La pyridine ne possède qu'une faible fluorescence visible, comme le dit D. BERTRAND (*4*). D'après lui, le spectre de la fluorescence visible de l'acide nicotinique est du même type que celui donné par la pyridine, mais déplacé vers le rouge et plus intense. Il a déterminé trois bandes de fluorescence avec maximums respectifs sur λ 697, λ 573 et λ 455 mμ.

La fluorescence des sels de l'amide de l'acide N-méthylnicotinique est bien connue. Plusieurs techniques ont été élaborées pour le dosage fluorométrique de l'acide nicotinique et de la nicotylamide. [Voir KODICEK (*64* et *65*).] La nicotylamide est très répandue chez les animaux supérieurs (surtout dans le foie) et chez les végétaux (tout spécialement dans les Levures). Des composés nicotiniques entrent dans la constitution de plusieurs ferments et coferments.

Un intérêt tout spécial s'attache à la nicotylamide en tant que facteur préventif de la pellagre. Dans un travail publié en 1940, NAJJAR et WOOD (*82*) ont étudié la fluorescence de vingt-sept corps pyridiques (dont la pyridine elle-même, l'acide nicotinique et la nicotylamide) en vue de reconnaître quels sont les dérivés nicotiniques excrétés par l'urine après ingestion d'acide nicotinique. Ce travail contient plusieurs spectrogrammes de fluorescence reproduits et les courbes correspondantes. Mais les auteurs n'ont pu répondre qu'imparfaitement à la question posée. Ils remarquent certaines ressemblances avec le spectre de fluorescence d'un nucléotide diphosphopyridique, dont ils publient la photographie.

En 1941, NAJJAR et HOLT (*81 bis*) ont repris cette étude et publié un assez long article sur les corps fluorescents qui apparaissent dans l'urine après transformation de l'acide nicotinique et de la nicotylamide dans l'organisme. Deux constituants sont à considérer séparément, disent-ils, dénommés par eux F_1 et F_2. Le corps F_1 est caractéristique pour l'urine des pellagreux. Sa présence est révélée par une fluorescence bleue directement visible; fluorescence, notons-le, indépendante de la réaction acide, neutre ou alcaline. Au contraire, chez les sujets normaux, on ne constate pas la présence de F_1 après ingestion d'acide nicotinique; leur urine contient pourtant aussi un composé spécial doué de fluorescence, qui a été appelé F_2 (manquant chez les pellagreux); mais cette dernière fluorescence, d'un bleu blanchâtre, n'apparaît qu'après alcalinisation; elle disparaît en milieu acide. Il est d'ailleurs possible que F_2 comprenne deux constituants, d'après des résultats obtenus par ELLINGER et COULSON (*39 bis*) au moyen de l'analyse par adsorption.

RAOUL (*95*) s'est aussi beaucoup occupé de la question, qui a fait postérieurement l'objet d'un très important travail d'ELLINGER (*39*). L'article d'ELLINGER est un exposé surtout critique qui montre toute la complexité du sujet. Il ne peut être convenablement résumé ici. A noter pourtant qu'avec COULSON et PLATT, ELLINGER (*15*) avait constaté, en 1942, que le constituant de désignation F_2, isolé par chromatographie, possède un maximum d'émission par fluorescence sur λ 469 mμ (sol. dans l'isobutanol), celui du thiochrome étant compris entre 460 et 470 mμ. Retenons ici les suggestions relatives à l'étude du spectre de la fluorescence ultraviolette en excitant cette fluorescence avec des raies de plus courtes longueurs d'onde (voisines de λ 260 mμ, par exemple).

A la fin de cette section, il convient de signaler une communication de SINGAL et SYDENSTRICKER (*110 bis*) relative à l'apparition dans l'urine d'un corps spécial à fluorescence bleue lors d'ingestion de *pyridoxine* (adermine, vitamine B_6), substance naturelle (présente dans le jaune d'œuf, la levure, etc.) qui semble posséder précisément une action curative remarquable dans la pellagre. Les relations avec F_1 et F_2 de ce corps fluorescent urinaire (surtout voisin de F_1) ont été envisagées par SINGAL et SYDENSTRICKER, mais ne sont pas complètement élucidées.

Le chlorhydrate de pyridoxine, en solution alcoolique, présenterait une faible fluorescence pourpre d'après HARVEY (*51*).

XII. Uroérythrine (uroroséine).

En 1939, MERKELBACH (*75*) a étudié la fluorescence de ce pigment dont il a publié le spectrogramme de fluorescence. Nous transcrivons ce qu'il a dit:

„Die Fluoreszenzfarbe des Uroerythrin ist rotgelb; dementsprechend zeigt das Fluoreszenzspektrum eine besonders ausgesprochene Intensität im roten Spektralbereich zwischen 633 und 610 mμ. Im übrigen erstreckt sich das Fluoreszenzspektrum bis in den blauen Spektralbereich; eine Abnahme der Fluoreszenz ist im Gelben und im Grünen zu beobachten. Das Fluoreszenzspektrum von Uroerythrin ist übrigens sehr vorsichtig zu bewerten, da Uroerythrin im Licht, und besonders im Ultravioletten, unbeständig ist. Das den ultravioletten Strahlen längere Zeit ausgesetzte Uroerythrin zeigt keine rotgelbe, sondern eine gelbe Fluoreszenz."

XIII. Chlorophylles.

Il n'y a rien de particulièrement important à ajouter sur les spectres de fluorescence des chlorophylles *a* et *b*, sujet amplement traité dans notre précédent article (*21*). Ne manquons pas toutefois d'attirer l'attention sur les recherches bien étendues faites dans ce domaine, en 1943, par ZSCHEILE et HARRIS (*132*).

STRAIN, avec des collaborateurs, a beaucoup étudié la chlorophylle *c* (chlorofucine; chlorophylle γ de TSWETT) en 1942 et 1943 (*119*—*120*). Il déclare être d'accord avec les déterminations de la bande principale de fluorescence qui ont été données à la p. 321 de notre article de 1939. Utilisant un spectrophotomètre à cellule photoélectrique, il a pu déceler l'existence d'une faible bande accessoire sur λ 690 mμ environ*.

On doit à MANNING et STRAIN (*73*) la découverte d'une nouvelle chlorophylle naturelle, qu'ils ont appelée chlorophylle *d*. C'est un pigment vert, contenant du magnésium, qui existe chez un certain nombre d'Algues rouges, en particulier chez *Gigartina agardhii*, d'où on peut l'extraire assez facilement. L'adsorption chromatographique est utilisée pour la purification. Ici, nous n'avons à retenir que ce qui concerne les spectres. Le maximum d'absorption (dans l'éther) sur λ 686 mμ est, par rapport à celui de la chlorophylle *a*, notablement décalé vers l'infrarouge, comme on le voit. Les solutions dans l'éther montrent une fluorescence d'un rouge foncé avec maximum sur λ 693 mμ. Vers λ 750, il apparaît de plus un maximum secondaire assez diffus.

Signalons enfin que tout récemment VAN NORMAN (*83*) a étudié, avec des collaborateurs, les spectres de fluorescence des pigments de deux Algues marines (*Gigartina radula* et *Iridaea* sp.) en opérant d'une part sur les Algues intactes et d'autre part sur des extraits.

Il est établi actuellement que la propriété que possèdent les diverses sortes de chlorophylles d'émettre par fluorescence des radiations comprises dans la plage allant de l'infrarouge au jaune tient à la présence dans leurs molécules de noyaux constitués à partir du complexe magnésien de la porphine (ou de ses dérivés). Or, GRANICK (*48 bis*) a constaté récemment que, dans les cellules de *Chlorella*, il se trouve de la protoporphyrine magnésienne (pigment «pinkish fluorescent», dit-il); et ce fait l'a amené à formuler la conclusion suivante: «The isolation of magnesium protoporphyrin suggests that, after the synthesis of protopor-

* Qu'il me soit permis de rectifier ici un malentendu que je constate à la lecture de ces Mémoires de STRAIN. J'ai toujours pensé, moi aussi, que, — contrairement à ce qu'avait affirmé WILLSTÄTTER, — la chlorophylle *c* pouvait parfaitement préexister *in vivo*. Restait seulement à expliquer pourquoi cette fluorescence, non perceptible en examinant, par exemple, des Diatomées vivantes, devenait intensément visible *en moins d'une minute* (BACHARACH et DHÉRÉ, 1931) après traitement par l'éther. J'ai eu l'occasion de revenir sur la question dans un article (*25*) imprimé au début de 1943 (alors que le travail de STRAIN d'août 1942 m'était encore inconnu). J'ai dit (à la p. 46, note 1): «On peut supposer que l'état de la chlorophylle *c* chez les Phéophycées est comparable à celui de la flavine en combinaison protéique (flavo-protéine), la fluorescence n'apparaissant qu'après décomposition du symplexe.»

phyrin, insertion of magnesium is the next step in the biological synthesis of chlorophyll by *Chlorella.*»

Dans la revue publiée en 1939 (*21*), de nombreuses pages ont été consacrées aux spectres de fluorescence, non seulement des chlorophylles, mais aussi des porphyrines et de ceux de leurs complexes métalliques qui sont fluorescents. Il suffira donc de citer ici un travail de ALBERS, KNORR et FRY (*1*) se rapportant à la question.

XIV. Polyhydroxyanthraquinones et composés similaires.

Hypéricine et pénicilliopsine. Ayant parlé assez longuement de ces pigments dans l'article (*21*) de 1939, nous ne dirons que l'essentiel des

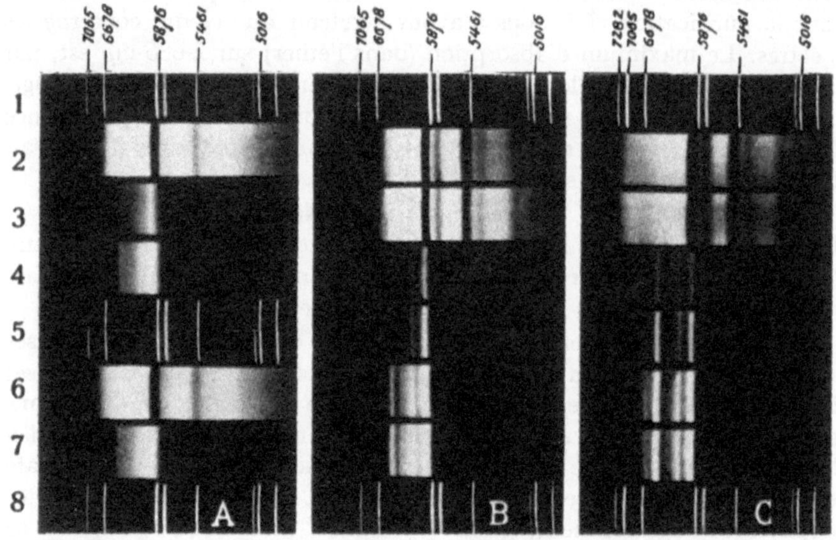

Fig. 7. Hypéricine et Oxypénicilliopsine. Compartiment A: Température ordinaire. Compartiments B et C: — 190° (azote liquéfié). Tous les spectres concernent l'oxypénicilliopsine, sauf les nos. 6 et 7 de A (Hypéricine). Absorption: A, nos. 2 et 6; B, nos. 2 et 3; C, nos. 2 et 3. Tous les autres spectres sont des spectres de fluorescence. DHÉRÉ et CASTELLI (*29*). [C. R. Séances Soc. Biol. Filiales Associées *130*, 593 (1939).]

résultats obtenus par la suite, bien que ces pigments aient été beaucoup étudiés, avec grand succès, depuis cette époque. Les spectrogrammes de DHÉRÉ et CASTELLI (*29*) groupés ci-dessus ont été publiés en 1939, mais seulement après la rédaction de l'article cité. Ils semblent être suffisamment instructifs pour trouver place ici (Fig. 7). (Solutions dans l'éthanol.)

On se bornera à attirer l'attention sur la bande de fluorescence supplémentaire, toute proche de la raie de référence 706,5 mμ (presque à la frontière de l'infrarouge), qui est bien visible dans le compartiment C (spectre No. 7). Solution alcoolique; plaque Agfa total Spektral.

La constitution de l'hypéricine a été élucidée grâce aux travaux de BROCKMANN (*7*) et surtout de MACKINNEY et PACE (*72*). Pour ces derniers, l'hypéricine est «a partially reduced polyhydroxyhelianthrone». OXFORD et RAISTRICK (*84*) ont dit que la pénicilliopsine est probablement un dérivé de l'hélianthrone ou de la naphtodianthrone.

Des recherches sur la fluorescence de l'hypéricine ont été faites par MACKINNEY (*72*) et par SCHEIBE et SCHÖNTAG (*106*). Dans l'ensemble, leurs résultats concordent, d'une façon satisfaisante en somme, avec ceux exposés dans mon article de 1939. Mais on ne saurait omettre de dire que le travail étendu de SCHEIBE (contenant des reproductions de deux spectrogrammes de fluorescence obtenus à 21° et à — 192°) a une importance capitale au point de vue physico-chimique.

On doit aussi attirer l'attention sur la *fagopyrine*, constituant fluorescent (photo-toxique) du Sarrasin, appartenant à la même famille chimique (*7*, *67*).

Lanigérine et strobinine. Comme l'hypéricine et la pénicilliopsine, ces deux pigments appartiennent par leur constitution à la famille des polyhydroxyanthraquinones. C'est ce qu'établissait dès 1936 BLOUNT (*5*), avant que la constitution de l'hypéricine et de l'oxyphénicilliopsine ne fût connue. La lanigérine provient de l'*Eriosoma lanigerum* et la strobinine du *Pineus strobi* qui sont, tous deux, des Pucerons aphidiens. Blount avait déjà noté quelques observations intéressantes concernant la fluorescence de ces pigments; et, grâce à sa générosité, j'ai pu en faire une étude plus complète (*22*).

A l'état solide, la lanigérine ($C_{17}H_{14}O_5$) et la strobinine ($C_{30}H_{24}O_8$), placées dans un faisceau intense de rayons violets et ultraviolets, manifestent une fluorescence rouge. Avec la lanigérine, la fluorescence est d'un rouge orangé et présente un éclat extraordinaire. Le spectre d'émission est du type observé avec les solutions neutres d'oxy-pénicilliopsine, sauf que la subdivision de la bande la plus large et la plus réfrangible n'a pu être constatée avec certitude. Ce qu'on voit donc au spectroscope, ce sont deux bandes lumineuses séparées par un minimum bien distinct. Ces bandes ont respectivement pour axes λ 660 et λ 611,5 mμ. Avec les cristaux de strobinine, la fluorescence est notablement moins brillante, ce qui provient sans doute, dans une certaine mesure, de ce qu'elle est constituée par des rayons de couleur beaucoup plus rouge, impressionnant moins la rétine. L'examen spectroscopique direct n'a permis de déterminer qu'une seule bande lumineuse, assez large, ayant approximativement pour axe λ 630 mμ.

La lanigérine en solution dans l'alcool s'oxyde rapidement au contact de l'air. Ce qui suit se rapporte à une solution déjà oxydée, de couleur jaune orangé. Elle présentait une vive fluorescence d'un vert légèrement nuancé de jaune et d'orangé; tandis que la solution alcoolique de strobinine, de couleur orangée (un peu cramoisie), présentait une fluorescence d'un jaune orangé à peine verdâtre. Ainsi, comme l'avait noté BLOUNT (5), il y a une grande différence de couleur pour la lumière de fluorescence des deux liqueurs; et pourtant l'analyse spectrale montre que les bandes de fluorescence (comme celles d'absorption) sont en nombre égal dans les deux cas et ont à peu près les mêmes axes. Ce qui diffère surtout, c'est l'intensité relative des bandes quand on passe de l'un à l'autre pigment. On se bornera à une description sommaire.

Axes des bandes de la lanigérine ($\lambda\lambda$ en mμ).

Fluorescence: I 634 (ou 635); II 608; III 591; IV 565; V 530; VI 499.
Absorption: I 587; II 563,5; III 521; IV 486.

Pour l'intensité, on constate que les bandes d'absorption I et III sont moins fortes que la bande II; la bande IV est bien forte pour la lanigérine, faible pour la strobinine. Pour la fluorescence, la bande V (verte) est très brillante avec la lanigérine.

Refroidie à — 180°, la solution de lanigérine montre une fluorescence qui est surtout orangée (alors qu'elle est surtout verte à + 18°); on observe alors les bandes suivantes (et quelques autres du côté ultraviolet). Fluorescence: I λ 644,5; II λ 612; III λ 596; IV λ 567; V λ 533; VI λ 503 mμ.

Ce spectre n'a pas paru bien stable. Après refroidissement dans l'air liquide, le spectre de la strobinine n'a pas été autant transposé vers l'infrarouge (l'axe de la première bande était sur λ 639 mμ).

Il est intéressant de signaler que, dans la plage comprenant le rouge, l'orangé et le jaune, les spectres de fluorescence de la lanigérine, de la strobinine et de l'oxypénicilliopsine (ou de l'hypéricine) présentent, en solutions alcooliques, une grande ressemblance; mais, à la température ordinaire, le spectre de l'oxypénicilliopsine est un peu plus près de l'infrarouge. En déterminant l'axe de l'intervalle correspondant au minimum d'émission qui sépare les deux premières bandes de fluorescence, on a: lanigérine solide λ 639; solution d'oxypénicilliopsine λ 626,5; solution de lanigérine λ 623; solution de strobinine λ 621,5 mμ.

Ces constatations m'ont amené à rechercher si les solutions alcooliques d'oxypénicilliopsine cristallisée et d'hypéricine [de MACKINNEY et PACE (72)] ne possédaient pas, elles aussi, quelques bandes de fluorescence dans le vert. Ces bandes ne sont pas perceptibles à la température ordinaire dans les conditions habituelles d'observation; mais on en aperçoit plusieurs après refroidissement à — 180°.

En terminant, on nous permettra de faire remarquer que ces quatre pigments anthraquinoniques sont avec les porphyrines les produits biologiques naturels actuellement connus possédant dans le visible les spectres de fluorescence les plus «nettement structurés» (surtout à très basse température), — spectres qui, tous, ont été d'abord déterminés dans notre laboratoire. Tout à côté, se placent certains pigments chlorophylliens dont nous avons aussi beaucoup contribué à étudier les spectres de fluorescence.

L'examen d'une solution alcoolique de *citro-roséine*, — pigment anthraquinonique (ω-oxy-émodine) produit par le *Penicillium citroroseum* et préparé par POSTERNAK et JACOB (*93*), — m'a montré (*24*) cinq bandes lumineuses émises à la température d'ébullition de l'air liquide.

XV. Lampyrine.

Dans un article datant de 1943, METCALF (77) a fait connaître un curieux et intéressant pigment à fluorescence d'un rouge orangé dont il a reconnu la présence chez de nombreux Lampyrides (Coléoptères).

Ce pigment, qu'il a appelé lampyrine, est insoluble dans l'eau, dans le méthanol et l'éthanol, dans l'éther, le chloroforme, le sulfure de carbone, la pyridine, l'acide acétique. Mais il se dissout dans l'acide chlorhydrique à 5% (et davantage), donnant une liqueur colorée en rose, douée d'une brillante fluorescence orangée. Le spectre de fluorescence s'étend de λ 655 à λ 563 mμ, avec maximums sur 635 et 600 mμ. Après addition d'un excès de potasse, on obtient une liqueur d'un bleu sombre avec fluorescence rouge feu comprise entre 672 et 622 mμ, le maximum d'émission correspondant à λ 650. Par irradiation, la liqueur pâlit peu à peu et tourne au vert jaunâtre. Il existe une forte bande d'absorption dans la plage jaune-vert (maximum sur λ 565 mμ) pour la liqueur chlorhydrique.

METCALF (77) a comparé le spectre de fluorescence de la lampyrine avec tous ceux que j'ai décrits dans la même région. En tenant compte de la solubilité et en considérant que le spectre de fluorescence ne correspond à aucun de ceux actuellement connus, il est arrivé à la conviction que la lampyrine est un nouveau pigment bien à part. Par ses caractères de solubilité, la lampyrine pourrait, dit-il, sembler voisine des ptérines; mais, tandis que les ptérines donnent une réaction de la murexide positive*, la réaction est négative avec la lampyrine.

Le travail contient bien d'autres indications d'ordre chimique et beaucoup de détails zoologiques qu'il n'y a pas lieu de donner ici. Il suffira de dire que les recherches de l'auteur ont surtout porté sur *Photinus marginellus* et que, chez l'animal vivant, la fluorescence apparaît déjà dans certaines régions lors de l'observation avec le microscope à fluorescence. Notons encore que la lampyrine a pu être préparée par METCALF à l'état de cristaux présentant une fluorescence rose saumon. Espérons que, sans trop tarder, la constitution de ce remarquable pigment pourra être élucidée.

---

* L'érythroptérine ne donnerait pas la réaction de la murexide.

XVI. Carotène, phytofluène et vitamine A.

J'ai déjà eu l'occasion (*21*) de parler de la fluorescence du carotène, $C_{40}H_{56}$ (qui, chose curieuse, avait complètement échappé à Tswett). Il ne semble pas que d'autres déterminations spectrales de la fluorescence aient été faites par la suite. Il est intéressant de noter que Straus (*121*), dans ses belles recherches sur les chromatophores, a constaté que les chromatophores contenant du carotène (*Daucus carota*) présentent en lumière ultraviolette une fluorescence «hellgrün» après purification extrêmement poussée.

Phytofluène, ce polyène naturel incolore (ou pâle) $C_{40}H_{64}$ ($\pm$ H_2) doit être placé tout près des carotènes [Zechmeister et Polgár (*129*); Zechmeister et Sandoval (*130—131*)]. Les solutions dans l'hexane ou l'éthanol montrent en lumière diffuse du jour une fluorescence vert bleuâtre. En lumière ultraviolette, le phytofluène dissous ou adsorbé présente une très intense fluorescence d'un gris verdâtre. On observe des phénomènes semblables en examinant le phytofluénol, qui est un dérivé naturel du phytofluène [Zechmeister et Pinckard (*128*)].

Schaerer, avec des collaborateurs, s'est beaucoup occupé de la fluorescence de la *vitamine A*, $C_{20}H_{29}OH$ (*105*). On trouvera, dans mon article de 1939, mes résultats obtenus avec de la vitamine A relativement pure. Schaerer n'a opéré que sur du Vogan. Il reproduit le spectre de fluorescence qu'il a obtenu et remarque: „Leider ist das Luminescenzspektrum des Vitamins A, wie es z. B. das Vogan zeigt, sehr uncharakteristisch. Es findet sich ein kontinuierliches Spektrum zwischen 500 bis 600 mμ . . . Unser Leuchtstoff entspricht mit großer Wahrscheinlichkeit einer Form des Vitamins A. Sicher ist jedoch, daß das Vitamin auch in anderer, nicht lumineszierender Form, im Körper vorkommt.“

Greenberg et Popper (*49, 92*) ont étudié la fluorescence de la vitamine A, au moyen du microscope à fluorescence. La fluorescence verte des foies de Mammifères et de Poissons de mer serait due à la vitamine A_1; tandis que, chez les Poissons d'eau douce, les cellules hépatiques montreraient une fluorescence jaune brun, attribuée à la présence de vitamine A_2.

Au point de vue chimique surtout, les travaux de Sobotka et collaborateurs (*112—114*) sur la fluorescence de la vitamine A sont pleins d'intérêt. L'augmentation très rapide de la fluorescence par irradiation dans le cas des esters (acétate de vitamine A, etc.) doit être signalée.

XVII. Pénicilline et streptomycine.

On ne sait encore que bien peu de chose sur la fluorescence de ces deux corps; mais, étant donné l'importance extraordinaire de ces antibiotiques, on indiquera ce que l'on connaît. Helander (*52*), en 1945, a signalé la fluorescence verte de la pénicilline; puis, dans un travail étendu publié postérieurement (*52 bis*), il a eu l'occasion de donner des indications complémentaires: A l'état pur et sous forme de poudre, la pénicilline présente, dit-il, en lumière U. V., une assez forte fluorescence d'un vert jaunâtre. Après injection chez la SOURIS (sol. au millième), cette fluorescence peut être décelée sur de minces coupes de tissus, la couleur passant au jaune puis au jaune brunâtre par chauffage (5 minutes) à 150° puis à 200°. SCUDI et JELINEK (*107*) ont décrit une méthode fluorométrique pour la détermination microphotométrique de la pénicilline. Mais il ne s'agit pas d'une méthode basée sur la fluorescence directe de cet antibiotique (technique compliquée à voir dans le travail original).

Les observations de FREDIANI (*45*) sur la fluorescence des sels de streptomycine à l'état solide doivent retenir l'attention, bien que l'intervention d'une impureté ne soit pas absolument exclue.

Appendice: La topographie spectrale dans l'excitation des fluorescences (Spectrofluorescence).

Beaucoup de fluorescences visibles sont excitées par des radiations visibles. Beaucoup de fluorescences visibles ou ultraviolettes sont excitées par des radiations ultraviolettes. Nous attirons, de plus, l'attention sur le fait qu'il peut y avoir (on en connaît maintenant de nombreux exemples) émission par fluorescence de rayons infrarouges dans la plage infrarouge voisine du visible (DHÉRÉ et AHARONI, 1930). Enfin, dans cette même plage invisible, ne pourrait-il pas y avoir émission de rayons infrarouges par excitation uniquement au moyen de rayons également infrarouges de moindres longueurs d'onde? J'ai eu l'occasion déjà (*26*) de parler de cette question en 1944. Malheureusement, je ne vois pas d'observation à citer en exemple pour ce dernier cas qui ne semble pas avoir été pris en considération par les auteurs jusqu'à présent.

On ne peut songer à donner ici un exposé détaillé des problèmes de spectrofluorescence. Je n'ai pas été sans en parler dans mon livre de 1937 (*20*). J'y ai décrit notamment l'ingénieux dispositif imaginé pour cette étude par V. HENRI et R. WURMSER (cf. *19*). Dans un important travail de VLÈS (*125*), deux dispositifs d'une application très pratique sont décrits et figurés: l'une des techniques est un repérage à vue, l'autre un repérage photographique (les excitations étant réalisées par les raies du spectre du mercure); voir aussi GERMANN et WOODRIFF (*46*).

Après avoir précisé théoriquement le problème des radiations excitatrices posé à partir de la règle de STOKES, VLÈS examine d'une façon critique les résultats qu'il a obtenus. Maintes remarques de l'auteur, concernant notamment l'apparence antistokes de l'excitation des fluorescences des protides, sont très instructives. Il discute en particulier l'application possible de la loi de Wawilow quand la fluorescence de certains protides est excitée par des radiations de l'ultraviolet éloigné, jusque vers λ 220 mμ. Les observations de Vlès n'ont pas porté que sur des protides et des acides aminés (voir Figures 2, 8 et 9); elles ont

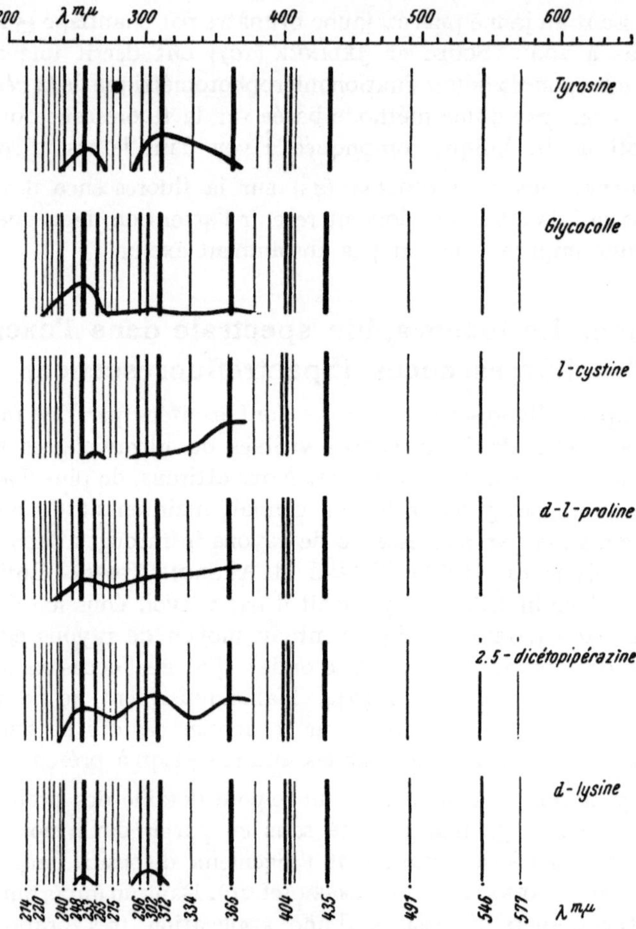

Fig. 8. Courbes représentant, pour quelques acides aminés et la dicétopipérazine, l'intensité de la fluorescence excitée localement par les raies ultraviolettes du mercure. D'après Vlès (*125*). [Arch. Physique biol. Chim.-Physique Corps organisés *16*, 137 (1943).]

été étendues à bien d'autres corps (le verre y compris). Nous reproduisons ici la classification des domaines d'excitation qu'il a proposée d'après l'ensemble de ses observations:

A. Substances fluorescentes ayant *une seule* bande d'excitation (*monoflores*); celle-ci peut être:

a) *progressive et étendue* sur tout l'ultraviolet (verre d'urane, ombelliférone, gélatine); ou bien:

b) *limitée* à une portion déterminée de l'ultraviolet, soit le proche ultraviolet (huile d'arachide, ac. nucléique), soit l'ultraviolet éloigné (verre).

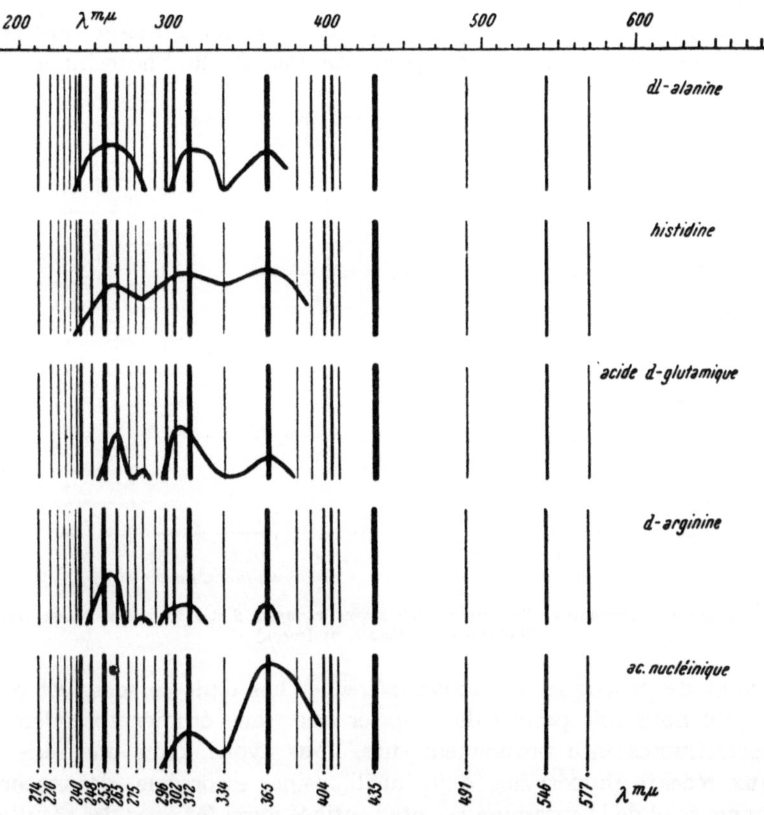

Fig. 9. Courbes représentant, pour quelques acides aminés et l'acide nucléique, l'intensité de la fluorescence excitée localement par les raies ultraviolettes du mercure. D'après VLÈS (*125*). [Arch. Physique biol. Chim. Physique Corps organisés *16*, 137 (1943).]

B. Substances fluorescentes ayant *plusieurs bandes* d'excitation (*polyflores*):

a) les différentes bandes *excitent une même fluorescence*, c'est-à-dire se rapportent à un même chromophore (*homoflores*: quinine, bande rouge de la chlorophylle, tyrosine, sérumalbumine); ou bien:

b) les diverses bandes d'excitation *excitent des domaines différents*, c'est-à-dire s'adressent à des systèmes chromophoriques distincts (*hétéroflores*: naphtalène; chlorophylle a).

Comme le dit VLÈS au début de son article (*125*) (et cela reste vrai même après son travail), nous ne possédons encore que très peu de données sur la topographie spectrale des radiations excitatrices des fluorescences. A signaler en passant l'intérêt que présentent, dans cet ordre d'études, les recherches d'ANDANT exposées dans notre article (*21*) (fluorescence des alcaloïdes).

Cet appendice a surtout pour but de souligner l'intérêt que peut présenter cette connaissance au point de vue de la Photochimie bio-

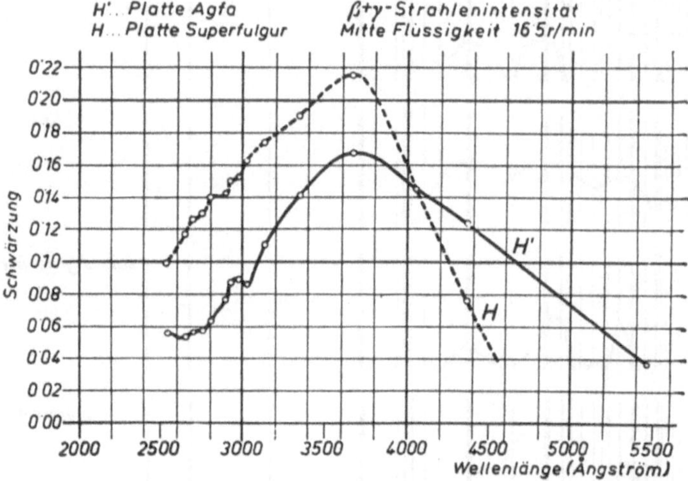

Fig. 10. Spectre de fluorescence de l'eau par excitation avec les rayons β et γ du radium. MAIER (*72 bis*). [Radiologia Austriaca *2*, 91 (1949).]

logique et de provoquer de nouvelles recherches dans ce sens. Et pour finir, qu'il nous soit permis de rappeler que, sans étudier spécialement la spectrofluorescence proprement dite, nous avons, dans une série de travaux récents (hypéricine, oxypénicilliopsine, escorcéine, désoxycorticostérone, triol de la vitamine D_2, etc.), utilisé aussi (et avec des résultats souvent excellents), pour l'excitation pratiquement monochromatique des fluorescences, la double raie jaune du sodium ainsi que les deux raies du mercure λ 579 et λ 577 mμ.

Les lignes précédentes sur la Spectrofluorescence étaient remises à l'imprimeur quand nous est parvenu un important et tout récent travail de Mme HILDA MAIER (*72 bis*), travail qui constitue précisément une extension considérable de la question examinée dans cet Appendice. Ainsi qu'il vient d'être dit, VLÈS a poussé l'étude de l'ultraviolet excitateur jusqu'à λ 220 mμ environ; or, dans ses recherches, Mme MAIER a opéré non seulement avec le rayonnement du radium constitué par les rayons β et γ associés, mais encore avec les seuls rayons γ qui, on le sait, diffèrent des rayons ultraviolets proprement dits par leur longueur d'onde beaucoup plus courte. Pour la première fois, sauf erreur, ce travail apporte, dans ce domaine, des spectres de fluorescence photographiés; les résultats principaux qu'il y a lieu de signaler ici se rapportant à des échantillons de sérum, à l'hématoporphyrine dissoute ainsi qu'à l'eau et l'alcool. Comme le montre la Figure 10, il apparaît pour l'eau une bande de fluorescence assez forte et pratiquement continue, comprise entre λ 230 et λ 500 mμ, avec maximum bien marqué sur λ 366 mμ environ. Même résultat par irradiation avec les rayons γ seulement.

Avec l'hématoporphyrine, il n'y a pas eu d'enregistrement d'une émission correspondant à celles décrites et figurées dans notre précédent article (*21*, pp. 325 à 330). Voici d'ailleurs ce qui dit Mme MAIER: »Die Fluoreszenzspektren von Stoffen, wie dem Hämatoporphyrin (solution dans HCl à 5 p. 100) oder dem in biologischer Hinsicht wichtigen menschlichen Serum, dürften hauptsächlich durch die Fluoreszenz des Wassers bedingt sein. Veränderungen gegenüber dem letzteren sind hauptsächlich durch Absorption der Fluoreszenzstrahlung in der Flüssigkeit bedingt*.»

La technique spectrographique, d'une grande difficulté, est amplement décrite par l'auteur, qui a publié les reproductions de plusieurs de ses spectrogrammes. Il faut ajouter que ce beau Mémoire contient aussi une discussion approfondie d'ordre physique pour les diverses constatations relatées.

Bibliographie.

1. ALBERS, V. M., H. V. KNORR and D. L. FRY: The Absorption and Fluorescence Spectra of *ms*-Tetra-(3',4'-methylene-dioxyphenyl)-porphine of HCl Number Four and of Its $Ag_5C_5H_5N$, Zn and Ni Complex Salts. J. chem. Physics **10**, 700 (1942).

1 bis. BANDOW, FR.: Über die Absorptionsspektren organischer Stoffe in konzentrierter Schwefelsäure. Biochem. Z. **301**, 37 (1939).

2. BASTERO BEGUIRISTAIN, J.: Untersuchung der Alkaloide auf Grund ihrer Fluoreszenz- und Absorptionsspektren im ultravioletten Spektralbereich. Quimi appl. **1**, 73 (1940).

2 bis. BEAUVILLAIN, A. et J. SARRADIN: L'adrénochrome. Etude des produits d'évolution et des leucodérivés. Bull. Soc. Chim. biol. (Paris) **30**, 478 (1948).

3. BERGMANN, W.: Über vermutliche Beziehungen zwischen Cholesterin und cancerogenen Stoffen. Z. Krebsforsch. **48**, 546 (1939).

4. BERTRAND, D.: Fluorescence visible et structure chimique. Bull. Soc. chim. France (5e série) **12**, 1010 (1945). — (Suite de neuf Mémoires.)

* J'avais eu l'occasion, il y a une douzaine d'années, d'étudier sommairement, dans le laboratoire du Professeur FR. DESSAUER (et avec un de ses assistants), la question de l'excitation possible d'une fluorescence rouge ou orangée pour diverses solutions d'hématoporphyrine exposées à l'action des rayons X. Nos constatations visuelles furent négatives. Mais cette étude devrait être reprise.

5. BLOUNT, B. K.: The Chemistry of Insects (Part. II). Examination of the Wooly Aphis and of the White Pine Chermes. J. chem. Soc. (London) **1936**, 1034.

6. BORNE, KR. VON DEM: Die blaue Fluoreszenz von Urin im ultravioletten Licht durch Indoxylacetylverbindungen. Acta med. scand. **97**, 311 (1938).

6*bis*. BOSCOTT, R. J.: New Colour and Fluorescence Reactions in the Steroid and Synthetic Oestrogen Series. Nature (London) **162**, 577 (1948).

7. BROCKMANN, H.: Lichtkrankheiten durch fluoreszierende Pflanzenfarbstoffe. Forsch. u. Fortschr. **19**, 299 (1943).

8. BROCKMANN, H., FR. POHL, K. MAIER u. M. N. HASCHAD: Über das Hypericin, den photodynamisch wirksamen Farbstoff aus *Hypericum perforatum*. Liebigs Ann. Chem. **553**, 1 (1942).

9. BROOKS, G.: Recherche sur les relations entre la fluorescence et la constitution chimique du laccol, du moréacol et de leurs dérivés. Bull. Soc. chim. France (5e série) **7**, 638 (1940).

10. — Sur les modifications des spectres de fluorescence et de la structure moléculaire du laccol, du moréacol et de leurs dérivés irradiés. Bull. Soc. chim. France (5e série) **7**, 643 (1940).

11. BUGYI, B.: Capillary-luminescence Investigation of Various Hungarian Commercial Vitamin and Hormone Preparations. Magy. Gyógyszerésztud. Társaság Értesitöje **14**, 365 (1938).

12. CASELLA, C. e M. REGGIANI: Spettro di florescenza delle cellule enterocromaffini di Cavia. Boll. Soc. ital. Biol. speriment. **22**, 480 (1946).

13. CASTELLI, V.: Recherches sur la fluorescence de quelques pigments d'origine microbienne. Dissert. Fribourg (Suisse) 1941.

14. CHRISTIANI, A. v. u. V. ANGER: Über einen empfindlichen Nachweis von Ergosterin und über eine Unterscheidung von Ergosterin und Ergosterinester. Ber. dtsch. chem. Ges. **72**, 1124 (1939).

15. COULSON, R. A., P. ELLINGER and B. S. PLATT: The Comparative Effect of the Elimination of Fluorescent Pigments in Human Urine of Nicotinamide and Related Compounds. Biochemic. J. **36**, XII (Proceed.) (1942).

16. CROWE, O'L. M. and A. WALKER: Fluorescence and Absorption Spectral Data for Pterin-like Pigments Synthesized by the Diphteria Bacillus and Isolated by Chromatographic Analysis. J. opt. Soc. America **34**, 135 (1944).

17. — — Pterin-like Pigment Derived from the Tubercle Bacillus. Fluorescence and Absorption Spectral Data for Erythropterin-like Pigment Isolated by Ultrachromatographic Analysis. Science (New York) **110**, 166 (1949).

18. DECKER, P.: Über die Flügelpigmente der Schmetterlinge. Nachweis und Bestimmung von Leukopterin. Hoppe-Seyler's Z. physiol. Chem. **274**, 223 (1942).

19. DHÉRÉ, CH.: Nachweis der biologisch wichtigen Körper durch Fluoreszenz und Fluoreszenzspektren. In: ABDERHALDENS Handbuch der biologischen Arbeitsmethoden, Abt. II, Teil 3, S. 3097 (Liefg. 420). Berlin-Wien. 1933.

19*bis*. — L'absorption des rayons ultraviolets par les acides nucléiques au point de vue chimique et au point de vue cytologique. C. R. Séances Soc. Biol. Filiales Associées **101**, 1124 (1929).

20. — La Fluorescence en Biochimie. Paris: Les Presses Universitaires 1937. (Les Problèmes biologiques XXI.)

21. — La spectrochimie de fluorescence dans l'étude des produits biologiques. Fortschr. Chem. organ. Naturstoffe **2**, 301 (1939).

22. — Spectres de fluorescence de deux pigments provenants de Pucerons aphidiens: la lanigérine et la strobinine. C. R. Séances Soc. Biol. Filiales Associées **131**, 672 (1939).

23. Dhéré, Ch.: Sur les bandes d'absorption ultraviolette dites „Bandes de Soret". Arch. Sci. physiques natur. (Genève) (5e période) **23**, 137 (1941).

24. — Sur les propriétés optiques de l'hypéricine: spectres d'absorption et de fluorescence. Comparaison avec l'oxypénicilliopsine et quelques autres pigments. Boissiera (Genève) **7**, 423 (1943).

25. — Michel Tswett. Le créateur de l'analyse chromatographique par adsorption. Sa vie, ses travaux sur les pigments chlorophylliens. Candollea (Genève) **10**, 23 (1943).

26. — La spectrochimie de fluorescence dans l'étude des produits organiques. Schweiz. Chemiker-Ztg. **27**, 164 (1944).

27. — Sur une remarquable réaction colorée de l'ergostérol décrite par Charles Tanret. C. R. hebd. Séances Acad. Sci. **228**, 604 (1949).

28. Dhéré, Ch. et M. Berenstein: Étude des spectres d'absorption et de fluorescence de l'escorcéine. C. R. hebd. Séances Acad. Sci. **223**, 934 (1946).

29. Dhéré, Ch. et V. Castelli: Détermination comparative des spectres d'absorption et de fluorescence de l'oxypénicilliopsine et de l'hypéricine à la température ordinaire et à la température d'ébullition de l'azote liquide. C. R. Séances Soc. Biol. Filiales Associées **131**, 669 (1939).

30. Dhéré, Ch. et A. Gourévitch: Sur la teneur en flavine du bouillon-toxine diphtérique. C. R. Séances Soc. Biol. Filiales Associées **130**, 593 (1939).

31. Dhéré, Ch. et L. Laszt: Réactions de fluorescence de quelques stéroïdes (hormones surrénales et sexuelles) avec le sulfate neutre de méthyle. C. R. hebd. Séances Acad. Sci. **224**, 681 (1947).

32. — — Sur quelques réactions de fluorescence dans la série des stéroïdes. Chimia (Suisse) **1**, 203 (1947).

33. — — Contribution à la spectrochimie d'absorption et de fluorescence dans le groupe des vitamines D. C. R. Séances Soc. Biol. Filiales Associées **142**, 17 (1948).

34. — — Halochromie et halofluorie. C. R. hebd. Séances Acad. Sci. **226**, 809 (1948).

35. — — Étude spectrochimique (absorption et fluorescence) de la réaction de Salkowski: cholestérol. C. R. Séances Soc. Biol. Filiales Associées **142**, 1422 (1948).

36. — — Étude spectrochimique (absorption et fluorescence) de la réaction de Salkowski: sitostérol, stigmastérol, ergostérol, zymostérol. C. R. Séances Soc. Biol. Filiales Associées **143**, 444 (1949).

37. Dhéré, Ch., P. Meunier et V. Castelli: Sur la détermination, par l'analyse spectrale, de la fluorescence du bouillon-toxine et de l'anatoxine diphtériques. C. R. Séances Soc. Biol. Filiales Associées **127**, 564 (1938).

38. Dutton, H. J. and G. F. Bailey: Modification of Cenco Spectrophotometer (Fluorescence Spectra). Ind. Engng. Chem. **15**, 276 (1943).

39. Ellinger, P.: Nicotinamide Methochloride and Its Fluorescent Derivatives. Nature (London) **155**, 319 (1945).

39bis. Ellinger, P. and R. A. Coulson: Urinary Elimination Products Following Ingestion of Nicotinamide. Nature (London) **152**, 383 (1943).

40. Faber, V. u. M. Radnot: Über die Fluoreszenz des Glykogens. Z. allg. Pathol. u. pathol. Anat. **76**, 82 (1940).

40bis. Fearon, W. R.: Tryptochrome: a Pigment Derived from Tryptophan. Nature (London) **162**, 338 (1948).

41. Finkelstein, M., S. Hestrin and W. Koch: Estimation of Steroid Estrogens by Fluorimetry. Proc. Soc. exp. Biol. Med. **64**, 64 (1947).

42. Fontaine, M.: Sur la teneur en flavine des divers organes de l'Anguille. C. R. hebd. Séances Acad. Sci. **204**, 1367 (1937).

43. FONTAINE, M. et R. G. BUSNEL: Flavine et substances à fluorescence bleue dans la peau des Téléostéens. C. R. Séances Soc. Biol. Filiales Associées **138**, 370 (1938).

44. — — Répartition de la flavine et des substances à fluorescence bleue dans la peau et les écailles de quelques Poissons d'eau douce. C. R. hebd. Séances Acad. Sci. **206**, 372 (1938).

44*bis*. — — Répartition des flavines et de quelques autres pigments fluorescents dans la peau et les yeux des Téléostéens. Bull. Inst. Océanographique (Monaco) No. 782, 1 (1939).

45. FREDIANI, H. A.: Fluorescence of Solid Streptomycin Salts. Science (New York) **107**, 344 (1948).

46. GERMANN, F. E. and R. WOODRIFF: Cross-prism Investigation of Fluorescence. Rev. sci. Instruments **15**, 145 (1944).

46*bis*. GIRAL, F.: Sobre los liocromos caracteristicos del grupo de Bacterias Fluorescentes. An. Soc. españ. Fisica Quim. **34**, 667 (1936).

47. GLAZKO, A. J., F. ADAIR, E. PAPAGEORGE and G. T. LEWIS: Fluorophotometric Determination of Rutin and Other Flavones. Science (New York) **105**, 48 (1947).

48. GOUREVITCH, A.: Études sur les substances à fluorescence bleue. C. R. Séances Soc. Biol. Filiales Associées **127**, 214 (1938).

48*bis*. GRANICK, S.: Magnesium Protoporphyrin as a Precursor of Chlorophyll in *Chlorella*. J. biol. Chemistry **175**, 333 (1948).

49. GREENBERG, R. and H. POPPER: Differentiation between Vitamin A_1 and Vitamin A_2 by Means of Fluorescence Microscopy. J. cellular comparat. Physiol. **18**, 269 (1941).

50. HAITINGER, M.: Die Fluoreszenzanalyse in der Mikrochemie. Wien u. Leipzig: Haim & Co. 1937.

51. HARVEY, N. E.: Luminescence. In: O. Glasser, Medical Physics, p. 691. Chicago. 1944.

52. HELANDER, ST.: Detection of Chemotherapeutics in Thin Sections of Tissue by the Aid of Fluorescence Microscopy. Nature (London) **155**, 109 (1945).

52*bis*. — On the Concentrations of Some Sulfanilamide Derivatives in Different Organs and Tissue Structures. A Histopharmacological Study by Means of Fluorescence Microscopy with some Aspects on the Possibilities of Obtaining High Concentrations of Chemo-therapeutics in Certain Tissues. Acta physiol. scand. **10** (Supplem. XXIX), 1 (1945).

53. HIEGER, J.: Fluorescence of Methylcholanthrene. Nature (London) **149**, 300 (1942).

54. HILLMANN, G.: Über die Fluoreszenzreaktion des o-Diacetylbenzols mit Eiweiß und Eiweißabbauprodukten und ihre Anwendung auf die Abderhaldensche Abwehrfermentreaktion. Hoppe-Seyler's Z. physiol. Chem. **277**, 222 (1943).

55. HÜTTEL, R. u. G. SPRENGLING: Über Ichtyopterin, einen blaufluorescierenden Stoff aus Fischhaut. Liebigs Ann. Chem. **554**, 69 (1943).

56. JACOBSON, W.: The Argentaffine Cells and Pernicious Anemia. J. Pathol. Bacteriology **49**, 1 (1939).

57. JACOBSON, W. and D. M. SIMPSON: The Fluorescence Spectra of Pterins and Their Possible Use in the Elucidation of the Antipernicious Anemia Factor. Biochemic. J. **40**, 3 and 9 (1946).

58. JANSEN, M. T.: Microspectrophotometrie van Fluorescentielicht. Dissert. Utrecht 1942.

59. JONES, R. N. and J. R. JANNESON: Fluorescent Constituents of Liver Lipid Extracts. Nature (London) **160**, 677 (1947).

59bis. KARRER, P. u. FR. FOSTER: Über 4,4',5,5'-Tetraphenyl-dithiazolyl-2,2' und 4,4',5,5'-Tetradiphenyl-dithiazolyl-2,2'. Helv. chim. Acta **28**, 315 (1945).

60. KARRER, P. u. M. C. SANZ: Über die Photosynthese eines fluoreszierenden Stoffes der Thiazolreihe (Vitachrom). Helv. chim. Acta **26**, 1778 (1943).

60bis. — — Synthese und Konstitution des Vitachroms. Helv. chim. Acta **27**, 619 (1944).

61. KIRBY, A. H.: Attempts to Induce Stomach Tumors; Effect of Cholesterol Heated to 300° C. Cancer Research **3**, 519 (1943).

62. KLEINER, J. S.: A New Color Reaction for the Phenolic Steroïds (Naturally Occuring Estrogens). J. biol. Chemistry **138**, 783 (1941).

63. KOCSIS, E. A. u. B. BUGYI: Nachweis des Folliculins im filtrierten ultravioletten Licht. Microchim. Acta **2**, 291 (1937).

64. KODICEK, E.: Some Applications of Fluorimetry in Vitamin Analysis. Analyst **72**, 385 (1947).

65. KODICEK, E. et D. K. CHANDHURI: Une méthode fluorométrique de dosage de la nicotinamide. Bull. Soc. Chim. biol. (Paris) **31**, 249 (1949).

66. KOSCHARA, W.: Die Isolierung des Urothions. Hoppe-Seyler's Z. physiol. Chem. **277**, 284 (1943).

67. KOSTIR, J. V.: Fluorescent Buckwheat Pigments. Chem. Listy Vedu Prumysl **40**, 127 (1946).

68. — Fluorescent Reaction of Tryptophan. Nature (London) **160**, 266 (1947).

69. KRAUL, R. u. H. MEYER: Über einen Nachweis von *Chelidonium majus* mit Hilfe der Quarzlampe. Arch: Pharmaz. Ber. dtsch. pharmaz. Ges. **281**, 287 (1943).

70. LEWIS, S. J.: Spectrofluorescence, with Special Reterence to Sugars. J. Soc. chem. Ind. **63**, 157 (1944).

71. LOOFBOUROW, J. R.: Fluorescence: Methods. In: O. Glasser, Medical Physics, p. 446. Chicago. 1944.

72. MACKINNEY, G. and N. PACE: Hypericin, the Photodynamic Pigment from St. John's Wort. J. Amer. chem. Soc. **63**, 2570 (1941).

72bis. MAIER, H.: Spektrographie der durch Radiumstrahlen erregten Fluoreszenzstrahlen und Probleme des Wirkungsmechanismus der durchdringenden Strahlen in der Radiumtherapie. Radiologia Austriaca **2**, 91 (1949).

73. MANNING, W. M. and H. H. STRAIN: Chlorophyll *D*, a Green Pigment of Red Algae. J. biol. Chemistry **151**, 1 (1943).

74. MENT, J. DE: Fluorochemistry. New York: Chemical Publ. Co. 1945.

75. MERKELBACH, O.: Die Fluoreszenz des normalen und pathologischen Urins. Helv. medica Acta **6**, 535 (1939).

76. — Fluoreszenz der Gallensäure und des Cholesterins. Helv. medica Acta **10**, 67 (1943).

77. METCALF, R. L.: The Isolation of a Red-fluorescent Pigment, Lampyrine, from the Lampyridae. Ann. entomol. Soc. America **36**, 37 (1943).

78. MIESCHER, K.: Über Steroide (49. Mitt.). Helv. chim. Acta **29**, 751 (1946).

79. MIRSKY, A. E.: Chromosomes and Nucleoproteins. Adv. Enzymol. **3**, 1 (1943).

80. MOIR, J.: Cholesterol and Phytosterol and the Spectroscopy of the Colour Reactions of the Sterols in General. J. South African chem. Inst. **12**, 16 (1929).

81. MUMM, O.: Ein Beitrag zur Frage nach den Zusammenhängen zwischen Fluoreszenz und chemischer Konstitution. Ber. dtsch. chem. Ges. **72**, 29 (1939).

81bis. NAJJAR, V. A. and L. E. HOLT, Jr.: The Excretion of Specific Fluorescent Substances in the Urine in Pellagra. Science (New York) **93**, 20 (1941).

82. NAJJAR, V. A. and R. W. WOOD: Presence of a Hitherto Unrecognized Nicotinic Acid Derivative in Human Urine. Proc. Soc. exp. Biol. Med. **44**, 386 (1940).

83. NORMAN, R. W. VAN, C. S. FRENCH and F. D. MACDOWALL: The Absorption and Fluorescence Spectra of Two Red Marine Algae. Plant Physiol. **23**, 455 (1948).

84. OXFORD, A. E. and H. RAISTRICK: Penicilliopsin, the Colouring Matter of *Penicilliopsis clavariaeformis.* Biochemic. J. **34**, 790 (1940).

85. PENN, H. S.: Spectra of Lipoid Fractions from Human Noncancerous and Cancerous Tissue. J. chem. Physics **10**, 145 (1942).

86. PENN, H. S. and J. KAPLAN: Fluorescence of Cancerous and Noncancerous Lipoids. Nature (London) **160**, 18 (1947).

87. PESEZ, M.: Nouvelles réactions colorées en série stéroïde. Bull. Soc. chim. France (5e série) **14**, 911 (1947).

88. PIRIE, A. and D. M. SIMPSON: Preparation of a Fluorescent Substance from the Eye of the Dogfisch *(Squalus acanthias).* Biochemic. J. **40**, 14 (1946).

89. POLONOVSKI, M., R. G. BUSNEL et M. PESSON: Propriétés biochimiques des ptérines. Helv. chim. Acta **29**, 1328 (1946).

90. POLONOVSKI, M., S. GUINARD, M. PESSON et R. VIEILLEFOSSE: Étude sur les ptérines de synthèse. — Étude physico-chimique des thioptérines. Bull. Soc. chim. France **12**, 924 (1945).

91. POLONOVSKI, M., R. VIEILLEFOSSE et M. PESSON: Étude sur les ptérines de synthèse. I. Thioptérines. Bull. Soc. chim. France **12**, 78 (1945).

92. POPPER, H.: Distribution of Vitamin A in Tissue as Visualized by Fluorescence Microscopy. Physiologic. Rev. **24**, 205 (1944).

93. POSTERNAK, TH. et J. P. JACOB: Sur le pigment du *Penicillium citroroseum.* Helv. chim. Acta **23**, 237 (1940).

93bis. PRINGSHEIM, P.: Fluorescence and Phosphorescence. New York and London: Interscience Publ. 1949.

94. PURRMANN, R.: Pterine. Fortschr. Chem. organ. Naturstoffe *4*, 64 (1945).

95. RAOUL, Y.: Mesure de l'excrétion de corps à fluorescence bleue au cours du traitement nicotinique. Bull. Soc. Chim. biol. (Paris) **25**, 266 (1943).

96. — Nature de certaines substances à fluorescence bleue de l'urine. Bull. Soc. Chim. biol. (Paris) **25**, 271 (1943).

97. — Répartition et rôle éventuel de certains corps à fluorescence bleue. Bull. Soc. Chim. biol. (Paris) **25**, 279 (1943).

98. — Polyphénols de l'urine à fluorescence bleue. Bull. Soc. Chim. biol. (Paris) **26**, 355 (1944).

99. — Fluorescence „protéique" et fluorescences accessoires de l'urine. Bull. Soc. Chim. biol. (Paris) **26**, 361 (1944).

100. — Parallélisme des tests indoliques et de la fluorescence de l'urine normale. Bull. Soc. Chim. biol. (Paris) **26**, 494 (1944).

101. REEDER, W. and V. E. NELSON: Fluorescence Associated with Proteins. Proc. Soc. exp. Biol. Med. **45**, 792 (1940).

102. ROFFO, A. E., Jr.: Espectro de fluorescencia del colesterol irradiado con Cd y del 1-2 Benzopirene. Bol. Inst. Med. exp. Estud. Tratamiento Cáncer (Buenos Aires) **18**, 949 (1941).

103. ROSENHEIM, M. CHR.: A New Colour Reaction for "Oxycholesterol". Biochemic. J. **10**, 176 (1916).

104. SANZ, M. C.: Das Chromatogramm des Vitachroms und seine Isolierung. Helv. physiol. Acta **1**, 7 (Verhandl.) (1943).

105. SCHAERER, E., J. RECHENBERGER, H. GOCKEL u. K. PATZELT: Beitrag zur Frage des Vitamin-A-Stoffwechsels. Virchow's Arch. pathol. Anatom. Physiol. klin. Med. **305**, 360 (1939).

106. SCHEIBE, G. u. A. SCHÖNTAG: Lichtabsorption und Fluoreszenz des Hypericins. Ber. dtsch. chem. Ges. **75**, 2019 (1942).

107. SCUDI, J. V. and V. C. JELINEK: A Rapid Micromethod for the Fluorometric Determination of Penicillin. J. biol. Chemistry **164**, 195 (1946).

108. SIMON, F. I.: Über Fluoreszenzbestimmungen am Harn und ihre Verwendbarkeit in der Krankenuntersuchung, besonders in der Krebsdiagnostik. Acta med. scand. **100**, 553 (1939).

109. — Das Fluoreszenzverhalten von Tumorextrakten in Zusammenhang mit einer neuen Methode der Krebsdiagnostik. Schweiz. med. Wschr. **74**, 195 (1944).

110. SIMPSON, D. M.: Application of Fluorimetric Analysis to the Study of Pterins. Analyst **72**, 382 (1947).

110bis. SINGAL, S. A. and V. P. SYDENSTRICKER: The Effect of Pyridoxine on the Urinary Excretion of a New Fluorescent Substance. Science (New York) **94**, 545 (1941).

111. SJÖSTRAND, F.: The Fluorescence Microspectrographic Localization of Riboflavin and Thiamin in Tissue Cells. Acta physiol. scand. **12**, 42 (1946).

112. SOBOTKA, H., S. KANN and E. LOEWENSTEIN: The Fluorescence of Vitamin A. J. Amer. chem. Soc. **65**, 1959 (1943).

113. SOBOTKA, H., S. KANN and W. WINTERNITZ: Fluorophotometric Analysis of Vitamin A Esters. J. biol. Chemistry **152**, 635 (1944).

114. SOBOTKA, H., S. KANN, W. WINTERNITZ and E. BRAND: Ultraviolet Absorption of Irradiated Vitamin A. J. Amer. chem. Soc. **66**, 1162 (1944).

115. STÄMPFLI, R.: Über den photochemischen Zerfall von Vitamin B_1, Thiochrom und Thiazol. Dissert., Bern 1942.

116. — Über die photochemische Herstellung von Fluorochromen zur Vitalfärbung. Helv. physiol. Acta **1**, 265 (1943).

117. — Über fluoreszierende Bestrahlungsprodukte des Thiazols. Helv. physiol. Acta **1**, 54 (Verhandl.) (1943).

118. STIMSON, M. and A. M. REUTER: The Fluorescence of Some Purines and Pyrimidines. J. Amer. chem. Soc. **63**, 697 (1941).

119. STRAIN, H. H. and W. M. MANNING: Chlorofucine (Chlorophyll γ), a Green Pigment of Diatoms and Brown Algae. J. biol. Chemistry **144**, 625 (1942).

120. STRAIN, H. H., W. M. MANNING and G. HARDIN: Chlorophyll *C* of Diatoms and Dinoflagellates. J. biol. Chemistry **148**, 655 (1943).

121. STRAUS, W.: Über Chromatophoren. Reinigung und Analyse der Farbstoffträger aus Mohrrüben und Spinatblättern. Helv. chim. Acta **25**, 179 (1941).

122. SUFUE, N.: Quinine Derivatives in the Transplantable Tumor. — I. On a New Fluorescence Method. Jap. J. med. Sci. **13**, Part IV, pp. 1 et 46 (1940). (Nom écrit aussi SOFUE.)

123. UTEVSKIY, A. M.: Products of Oxidation of Adrenaline and the Structure of Sympathins. Advances mod. Biol. (USSR) **18**, 145 (1944).

124. UTEVSKIY, A. M. and W. O. OSSINSKAJA: Biochemistry of Epinephrine. Effect of Formaldehyde on Formation of Red and Fluorescent Products of Oxidation. Eksper. med. **1**, 1 (1940).

125. VLÈS, F.: Recherches sur l'excitation de la fluorescence des protides. Arch. Physique biol. Chim.-Physique Corps organisés **16**, 137 (1943).

126. WINDAUS, A., K. BURSIAN u. U. RIEMANN: Über die Photooxydation des Cholesterins. Hoppe-Seyler's Z. physiol. Chem. **271**, 177 (1941).

127. WINDAUS, A. u. U. RIEMANN: Über die Einwirkung von Bleitetraacetat auf einige Sterinderivate. Hoppe-Seyler's Z. physiol. Chem. **274**, 206 (1942).

128. ZECHMEISTER, L. and J. H. PINCKARD: On Naturally Occurring Colorless Polyenes. Experientia **4**, 474 (1948).
129. ZECHMEISTER, L. and A. POLGÁR: On the Occurrence of a Fluorescing Polyene with a Characteristic Spectrum. Science (New York) **100**, 317 (1944).
130. ZECHMEISTER, L. and A. SANDOVAL: The Occurrence and Estimation of Phytofluene in Plants. Arch. Biochemistry **8**, 425 (1945).
131. — — Phytofluene. J. Amer. chem. Soc. **68**, 197 (1946).
132. ZSCHEILE, F. P. and D. G. HARRIS: Studies on the Fluorescence of Chlorophyll. The Effect of Concentration, Temperature and Solvent. J. physic. Chem. **47**, 623 (1943).

(Reçu le 24 octobre 1949.)

Namenverzeichnis. Index of Names. Index des Auteurs.

Sachverzeichnis. Index of Subjects. Index des Matières.